Part Six
THE ESTABLISHMENT TURNS A DEAF EAR

Part Seven
THE BANDAZHEVSKY AFFAIR.
AN INNOCENT MAN UNJUSTLY TREATED

Documents illustrating WHO's dereliction of duty

PROLOGUE

Who in this world is able to answer the terrible obstinacy
of the crime if not the obstinacy of the witness?
ALBERT CAMUS, Actuelles II

"The future belongs to those with the longest memory".
FRIEDRICH NIETZSCHE

The authors of this book are the victims of the Chernobyl disaster that took place on 26 April 1986, the peasant communities whose voices I recorded in the north of the Ukraine and in the forests in the south of Belarus. They are the millions of people who eat caesium-137 every day in their food. They are the young mothers, contaminated themselves, who unknowingly are a source of poison for the new confident life forming inside them. They are the children who, even if they appear healthy when they are born, are condemned to become ill as they grow up, because they eat radionuclides morning, noon and night... They are the "liquidators", the unsung heroes of Europe, who suffer all the unknown illnesses of the atom. Hundreds of thousands of people are ill, tens of thousands have died young and continue to die in unimaginable suffering... And they are the doctors and scientists, the few who have not submitted to the nuclear lobby, who, fortified by their knowledge, are engaged in a battle for the truth. As an Italian journalist, a Russian speaker, educated in France, it was by chance that I began to pass on these forbidden truths, as part of a fragile human chain made up, in the East, of activists in a country trapped in radioactive contamination and in the West, by activists who support them against scientific lies. It is with the help and the contribution of these men, women and children that I, wishing only to bring clarity and truth, am able to present the documents and testimonies that are in my possession thanks to them. Because lies and secrecy surround the worst technological catastrophe in History and threaten the future of humanity.

This book also recounts the struggle of two Belarusian scientists who risked their careers, their health, their own personal safety and that of their families to come to the aid of the population that had been contaminated. The physicist,

2

Vassili Nesterenko, and the doctor and anatomopathologist, Yury Bandazhevsky, forced into the role of dissidents because the International Atomic Energy Agency in Vienna prohibits the recognition of the harmful effects of low dose ionising radiation on health, were persecuted for their opposition to the official dogma. The first was a member of the Academy of Sciences in Belarus, but with his career in ruins, he continued to fight independently to protect children from radiation in the contaminated villages[1]. The second was condemned to eight years in prison by a military tribunal, after he revealed the pathogenic effects on vital organs of low doses of radioactive caesium incorporated through food. Amnesty International recognised him as a prisoner of conscience.

The international organisations still refuse to verify the validity of the research findings of these two scientists and will not even consider the radioprotection measures that they recommend that might, at least, save the lives of the 500,000 children living in the contaminated villages in Belarus.

---

[1] Fate did not decree that the scientist V.B. Nesterenko should live to see the book appear in English. He died on 25th August 2008. *"One can only marvel at the unbelievable determination, energy, talent for organisation, and indomitability of this great man who created Belrad under conditions of unprecedented tyranny and repression from those in power who fear the truth about the terrible consequences of the Chernobyl disaster. Today, we can say that Vassili Nesterenko has entered the ranks of the great humanists of our time: Schweitzer, Ghandi, Sakharov..."* Words spoken at the funeral oration by A.V. Yablokov

PART ONE
IGNORANCE

Chapter One
MESSAGE IN A BOTTLE

On the morning of Thursday 29th November 1990, four and a half years after the accident, we had a meeting with Alla Tipiakova at the school in Poliske, 68 kilometres west of the nuclear power station at Chernobyl. We had come to film a conversation with the pupils in her class. When we arrived, she was waiting for us outside in the drizzling rain. She got into our bus and made it quite clear that the step she was taking would probably end her career as a teacher. Then she asked us to film what she had to say.

A. Tipiakova.—I've been a primary school teacher for thirty five years. I know everything about the lives of the children in my class. There are 22 pupils in the class. Half the class is always absent. One little girl is in hospital permanently. Over the last two years her diabetes has got so bad that she needs two injections of insulin every day. She has been given every available treatment, but there is no real hope. I have a boy who, over the last year, has had such violent asthma attacks that he almost suffocates and spits blood. All of the children, one hundred per cent, have got abnormal blood chemistry. We have seen their clinical records. Nearly all the children, about 80% of them, have diseases of the thyroid. Hyperplasia of the first or second degree. All of them have dizzy spells. All of them experience nausea. Fainting is very common. Nosebleeds have become an everyday occurrence. Yesterday, when I spoke to them to prepare them for the meeting with you, they had tears in their eyes and told me that they faint in the street. They all know where their heart is. Children who know where their heart is! Because they experience stabbing pains in the heart. One little girl told me: "I feel as if my head is blowing up like a balloon. Then it tightens up, everything creaks and I feel as if I'm going to die". Another girl told me that her blood pressure is either too high or it's too low and she has to lie down. There is a little boy, Seriozha, who has started to stammer. He is already so sickly, so weak. He is one of twins and one of them already thinks of himself as being older. The children wanted to tell you all about it. Because they know…

Q.—Why can't we meet them?

—That's what I wanted to explain… I had told the children, very calmly, that some people from another country were coming: "A television company. You can talk to them about your family, your situation, your thoughts, your troubles, and about your health. You can confide in them". The children agreed. They were very pleased about it, especially as it was something new, something they had never done before. This morning I went to talk to the head teacher about it. I told her that someone was coming to see us, and I asked permission to meet you. She categorically refused: "How could you invite foreigners in to hear about the children's health without my authorisation?" I replied, "They aren't foreigners. They're people who have come to help us!"

—But that's exactly what you were doing, weren't you, asking for permission?

—Yes. I told her about it straight away. Before I'd even taken my coat off, I went into the staffroom and gaily announced: "Today, we're going to have a little show…". "What! We're working and you're having fun?" After a telephone call to someone, I don't know who it was, she called me into the staffroom a second time. And then she told me categorically no. So there had been a "no" from the regional office of the education authority. They told me there was to be no direct contact between you and the children. That upset me beyond all measure. I know my children, they weren't going to complain about anything… And even if they had, why shouldn't they? We can't keep on with this slave-like mentality, is time to break free. We told the children it was cancelled and they went home very disappointed.

But I want to talk to you about something else. For a long time, we've been living here hoping that we were going to be evacuated, hoping to escape this circle of death. And at the beginning we were promised the earth. Pretty soon we noticed that nearly all the local managers had got themselves housed elsewhere while we were being lulled to sleep with fairy tales. We were told everything was normal, everything was fine and that there was no reason in theory not to live here. The adults will manage somehow. We have our roots here, this is our land… we will die here and be buried here… But we have to save the children. And now the children know, because they hear their families talking about it, that it is very unlikely that Poliske will be evacuated. They don't trust us any more because they've been told lies. Today one of my children said to me: "If I had to write an essay today about a happy childhood based on the last four years, I couldn't do it". There is no happy childhood round here. They have forgotten how to smile, they have forgotten joy. They aren't the same, they have become aggressive, disobedient. They don't do their homework like they should. All of this is so hard for them that even we don't understand our own children any more.

We really need help to save our children, get them out of here. We need to appeal to international opinion. We are deeply grateful for the gifts that they send us. Our children have been to France, Italy, Germany, Cuba, Bulgaria. The United States also has sent us food, vitamins for the children. But these are crumbs to keep us quiet. The reality is that if we stay here, we are condemned to death, we are hostages and we will all die here, our children with us.

Thank you. Send our good wishes to everyone who is helping us. We know that most people on earth are generous and understanding. In spite of borders, political ambitions and obstacles, they've opened up their hearts to us. But these are emotional gestures, heart felt.... But it's as if the policies, that might actually resolve our problems, are being blocked somewhere and at the political level no-one's interested in us.

I want to tell you a little bit about my grandson. He is one year and nine months old. He knows all about health centres. He has chronic amygdalitis, chronic rhinitis, chronic pharyngitis, all sorts of bacterial infections. He's so young and he already has all these illnesses. When you ask him "Where does it hurt?" he opens his mouth: "My throat hurts". He's less than two years old. Sometimes he doesn't speak for a couple of days and then on the third day, he speaks in a baritone voice, and the next day he makes little cries like a mouse. His parents are so worried about him. We are doing everything we can to save his life. And what about families with lots of children? In my class, I have families of five, six children. Three or four, is the norm. What about families where the father and the mother are terrified for each one of them? Their lives are condemned...I want to tell the whole world. I want to tell anyone who has a heart. We no longer believe our children can be saved if they stay here. You have to help us. Maybe there are centres in Europe that could look after them properly. They could invite the mothers and their little children. Here, no-one will have them. This is the most vulnerable group in society, young children and babies. They need medicines and special care.

The truth, only the truth, everything I've just explained to you, that any of these children could tell you. Tell other people what you see and what you hear, and shoot. Film it. Let this truth resound round the whole world! I don't want to say bad things. I belong here in this village. I grew up here. I teach the children here. I don't want to paint everything black. I know people are doing a lot. But the way it's being done is crazy, it's impulsive and chaotic. It's not solving the real problem.

This is definitely the end of my teaching career. It doesn't matter.

In 1998, I was sent to the Chernobyl territories again by Swiss television and I went to find Alla Tipiakova. I found only her daughter and family who were living in Kiev. I left them a cassette recording of her voice, her face. Alla Tipiakova did not lose her job, but her life. Some time after she delivered this desperate message, she died of cancer. But she left us with this lamentation, this painful whisper, which is still ringing in our ears.

Barely five days after the explosion of reactor number 4 at Chernobyl, the physicist Bella Belbéoch wrote an article that appeared on 1st May 1986 in the journal 'Ecology'. "Over the next few days we can expect an international conspiracy on the part of experts minimising the number of victims that will be caused by this catastrophe. The pursuit of civil and military programmes will impose on all countries a tacit complicity that will override ideological or economic conflicts". [1]

In the autumn of 1990, when I made my first visit to the contaminated areas of Ukraine and Belarus to make a documentary about the accident at Chernobyl, I knew very little about nuclear physics, radioactivity or the interests of the nuclear industry. I did not have the expertise of a physicist like Bella Belbéoch. But one of the privileges of our profession is that we learn a lot in the course of our inquiries. In fact the very ingenuousness of a fresh approach can sometimes shed light on questions that have become too familiar to the experts.

Our duty is to inform. Before I present the testimonies and documents recorded for the five documentaries I made between 1990 and 2002, and the material uncovered over a period of five years when I was passing on information on a daily basis between independent scientists and doctors from the East and West, I will present the crucial pieces of evidence that might have helped Alla Tipiakova understand why the children in her class had been abandoned by the rich countries of the West. Neither destiny nor providence can be blamed for the fate suffered by these children. The responsibility is entirely and exclusively human.

---

[1] Quoted in *Tchernobyl, une catastrophe*, Bella and Roger Belbéoch, Editions Allia, Paris 1993; Tokyo 1994.

Chapter II

MEDICINE AND THE NUCLEAR ESTABLISHMENT

Everything to do with the disaster at Chernobyl, its causes and its consequences, must be made public. We must have the absolute truth.

ANDREI SAKHAROV, May 1989

I'm going to speak frankly, because life is short.

ANDREI SINIAVSKY, *A Voice from the Choir*

Humankind made the leap into the nuclear age some decades ago, but still does not understand the multiple consequences on human health of chronic exposure to low level radiation, or about the toxic effects of artificial radioactive substances that have been disseminated in large quantities into the environment by both civil and military nuclear activities.

The chronic incorporation into the organism of radionuclides, by millions of people (caesium-137 only disappears after three centuries) condemned to feed themselves with radioactive food, is a completely new phenomenon resulting from the Chernobyl disaster, of which humanity has no experience. The same can be said of Gulf War Syndrome in Iraq and of the health problems experienced by the Serbian and Albanian populations in Yugoslavia where tonnes of uranium-238 (described as "depleted" but having a half-life of four and a half billion years) were dropped. If research on these matters has been undertaken in secret by the scientific establishment, it has so far been reluctant to share the benefits of its findings with the planet's populations.

Why has the World Health Organisation (WHO) done nothing in the territories around Chernobyl? Why has it handed over its radioprotection work to the promoters of nuclear power plants? Public opinion is largely unaware of the conflict of interest between the two specialist United Nations agencies directly responsible for managing the consequences of the Chernobyl catastrophe on the health of the contaminated populations.

An agreement signed between WHO and the International Atomic Energy Agency (IAEA) in 1959 prevents WHO from acting independently in the nuclear domain without the IAEA's consent. The IAEA, made up of physicists, not doctors, whose main objective is the promotion of the nuclear industry around the world, is the only specialist United Nations agency that reports directly to the Security Council. It imposes its *diktat* on WHO, whose objective is *"the attainment of all peoples to the highest possible level of health"*.

Today, the two agencies still only recognise as a consequence of the accident at Chernobyl the deaths of 44 firemen in the first hours following the disaster, (two from trauma and one from heart failure), 203 cases of acute radiation exposure and 200 easily avoidable thyroid tumours. The UN predicts a total of 4000 deaths that *could be verified as resulting from exposure to radioactivity from the accident.*[2] In contrast, the Office of the United Nations for Humanitarian Affairs shares Kofi Annan's view that estimates 9 million as the total number of people affected in the long term by radioactivity and confirms that the tragedy of Chernobyl has only just begun. Kofi Annan, not being a scientist, was sharply reprimanded by the director of the Scientific Committee of the United Nations Scientific Committee on the Effects of Atomic Radiation (UNSCEAR), whose task it is to evaluate radiation doses world wide, their effect and the risk they carry.

These facts and contradictions erupted at the international conference in Geneva in 1995, and then at the conference in Kiev in June 2001, that we filmed.[3] Here we witnessed the fury of the nuclear agencies and of their accomplices from the former Soviet Union on hearing the revelations of researchers and doctors in the field about the radiological causes of the health catastrophe in the contaminated

---

[2] Joint WHO/ IAEA/ UNDP communiqué, 5th September 2005. The previous day, the official death toll from the accident was still being put at 32 by the three agencies.

[3] At this conference the chief medical officer for the Russian Federation stated that nearly 30% of the liquidators listed in their medical-dosimetric register were ill and about 10% had already died. This official register brings together information on the 184,175 liquidators. Estimates of the total number of liquidators, who were summoned from all parts of the Soviet Union to construct the sarcophagus and decontaminate the area, vary between 600,000 and 800,000 young men in good health (the associations defending the rights of the liquidators put forward the figure of a million). Their average age was 33 in 1986. Information about the disaster at Chernobyl was a state secret during the first four years (the last years of the USSR) and the doses of radiation that they received were systematically minimised, so official figures today can only be wrong, by default. The survivors of this army were dispersed over the 11 time zones of the former Soviet Union, many are untraceable statistically and, thanks to the disinformation from both the Kremlin and the UN nuclear agencies, they do not know why they are ill or what is causing them to die so young. The official numbers registered by the Russian Federation allow us to estimate that the total number of liquidators who are ill is between 200,000 and 300,000 and that some 60,000 to 80,000 have already died.

territories. The data and recommendations for radioprotection presented by independent scientists to protect hundreds of thousands of contaminated children were cast aside with disdain and arrogance. They refused to discuss it.

A deliberate scientific crime has been going on for twenty-eight years at the heart of Europe, sanctioned at the highest level, against the background of disinformation and general indifference of the technologically advanced Western civilisation. In order to preserve consensus about the nuclear industry, the nuclear lobby and the medical establishment are knowingly condemning millions of human guinea pigs to experience new pathologies in their bodies in the vast laboratory of the contaminated territories of Chernobyl. The dictator Lukashenko, safeguarding his own position, is simply the local administrator of policies that emanate from the Permanent Members of the Security Council of the United Nations. (United States, France, Great Britain, China and Russia) and are "legitimised" by the experts at the agency in Vienna. Europe is an accomplice, investing millions of euros in the programme CORE (Cooperation for the Rehabilitation of Conditions of Life in the areas of Belarus contaminated by Chernobyl), set up in 2003. This "humanitarian" programme, whose covert aim, according to one of its administrators, is "to occupy the territory", has formulated no scientific projects concerning the health of the contaminated inhabitants. It shares the same objective, as we shall see later, as that expressed by the United Nations agencies responsible for the atom, in agreement with the Soviet Union, from the first days following the accident. "The 'consensus' report of the United Nations on Chernobyl published in February 2002 broadly reflects the proposals put forward by the nuclear lobby: the economy of the contaminated territories needs to be revived, the local people given help to 'develop' them and even to recolonise them. These optimistic proposals are not based on any measures of the levels of radioactivity in people or in locally-grown food produce. 'Chernobyl is over.' 'It is the economic crisis and stress that are responsible for the problems encountered by the inhabitants.' They need to be reassured and to resume normal life again". (Solange Fernex in *The Ecologist*, February 2003.)

The influence of the military-industrial nuclear mafia over the health of the population started, as we will see, in the 1950s. If we compare two documents published by WHO, one in 1956 and the other in 1958, we can see quite clearly the change in direction taken by this specialised United Nations institution before its final submission to the nuclear lobby in 1959. The first document is a seriously worded warning against the choice of developing nuclear power, presented by a group of distinguished genetic scientists, one of whom was H.J. Muller, who received the Nobel Prize in 1946. The second document is a report of a study group analysing "questions of mental health (for the population) posed by the use of atomic energy for peaceful purposes". Representing France in this study group was Dr M. Tubiana, a cancer specialist at Villejuif.

First document 1956: "...genetic heritage is the most precious property for human beings. It determines the lives of our progeny, healthy and harmonious development of future generations. As experts, we affirm that the health of future generations is threatened by increasing development of the atomic industry and sources of radiation...We also believe that new mutations that occur in humans are harmful to them and their offspring". (WHO, *Genetic effects of radiation in humans. WHO Study Group Report*, Geneva 1957).

Second document from 1958: "However, from the mental health point of view, the most satisfactory solution for the future peaceful use of atomic energy would be to see a new generation of people who would have learnt to accommodate ignorance and uncertainty and who, to quote Joseph Addison, an English poet of the 18th century would 'sit astride the hurricane and direct the storm'". (*Technical Report* No 151, page 59, WHO, Geneva 1958.)

How apt a programme for Chernobyl, on which Professor Michel Fernex, a keen observer of WHO's abandonment of the nuclear issue, comments in measured terms. "This justification for keeping people in ignorance illustrates the contempt in which the world's people are held, that goes against the letter and the spirit of the Constitution of the World Health Organisation".[4]

---

[4] "La catastrophe de Tchernobyl et la santé" (The Chernobyl disaster and health) in *Chroniques sur la Biélorussie contemporaine*, L'Harmattan, 2001.

Chapter III

THE ATOMIC TRAP

1. THE ACCIDENT

I do not know Grigori Medvedev, Chief Engineer at the Chernobyl nuclear power station from its inception, on a personal level, but I read his book[5] while at the site of the tragedy itself during my first visit to Chernobyl. I was sent there in 1990, to make a documentary about the disaster and try to help people to understand the situation four years on. Uncover the facts, understand myself and help others to understand why the subject was mired in deception, and the victims involved in protracted misery. Medvedev, who wrote the book, and Volodymyr Tykhyy, a young Ukrainian physicist who gave me a copy of the Russian edition, provided the key. They helped me to get closer to what had happened, and what was still going on, in the villages, among the liquidators who were ill, in the health centres where children with leukaemia and heart problems were receiving care. I read and reread *"The Chernobyl notebook"* in between pre-shooting and shooting. I spent my nights reading and then, because I spoke Russian, I was able to decline the kind invitations made to journalists by the Soviet television companies of Ukraine and Belarus, and instead went to meet people and find out about the real situation first-hand. It was autumn 1990 and together, *perestroika* and Chernobyl were driving the USSR towards implosion. Every day I saw new people, new places and situations; I had first hand contact with people who were trapped and looking for a way out.

Everything about the first hours of the accident has been recounted in Medvedev's book. It is described in masterly detail, minute by minute, from both a technical and a human point of view: the dramatic struggle to control the fire and its immediate effects in the days following the explosion, the slow grinding cogs of the Soviet bureaucracy, the errors made by the scientists and designers of the reactor, the censorship, the lies and the pressure exerted by the hierarchy. But

---

[5] *"Cahier de Tchernobyl"* (The Chernobyl notebook), published in France with a preface by Andrei Sakharov under the title *" La vérité sur Tchernobyl "* Albin Michel 1990.

curiously, Medvedev makes no mention of one aspect of the disaster at Chernobyl: that is the possibility of a nuclear explosion that Soviet nuclear scientists were afraid might happen, during the ten days of the fire. Unlike the physicist Vassili Nesterenko who I interviewed for the first time in 1998, during my second visit to Chernobyl.

From 1977 to 1987, Nesterenko was the director of the Institute of Atomic Energy at the Academy of Science in Belarus, of which he is a member. Holder of over 300 scientific patents in nuclear physics and in radiation safety, he had access in the Soviet era, to towns that were forbidden to others for military reasons. He invented and built a mini reactor that could be transported by helicopter and used to set off intercontinental missiles, to counterbalance the mobility of American submarines. He flew over the burning power station at Chernobyl and came up with the idea of injecting liquid nitrogen under the foundations of the reactor in order to put out the fire. This idea was put into action a few days later by a team of miners, in a hellhole of fire and radiation. Badly irradiated himself, it is incredible that he was still alive. He was the only surviving member of the four passengers of the helicopter. From the first moments following the accident, he told the Soviet authorities that they needed to protect the population. No-one listened to him. Accused of sowing panic, removed from his post as director, he was subjected to pressure from the KGB and survived two attempts on his life.

Nesterenko *knows* what radioactivity is and he adores children. He saw them evacuated in their thousands, snatched from their mothers' arms and thrown into train carriages at Gomel railway station. Watching the scene brought back his own childhood in rural Ukraine, under German occupation. He told me how the Wehrmacht, withdrawing from his village because they were under attack from the Red Army, gathered up the women and children in front of the tanks, hoping to protect themselves with this human shield. He saw the same terrified children, exhausted from hours of travelling, arriving at his Institute near Minsk. It was not the Nazis that were pushing them through.

He began to measure the levels of radioactivity emanating from their bodies. When he saw the dosimeter needle blocked at the far end of the scale, he knew he would never again use his scientific skills in the service of a technology that could produce such a disaster. It was at that precise moment that he decided he would never work in that scientific area again unless it was to help the victims.

In a recent interview, Nesterenko confirmed that the Soviets had feared that if the fire had not been put out on the 8th May 1986,[6] there might have been an atomic explosion of such power that it would have rendered Europe uninhabitable. French scientists have cast doubt on this hypothesis, claiming that it was impossible.

---

[6] Interview with Galia Ackerman in *"Les silences de Tchernobyl"*, Autrement, 2004.

I asked Nesterenko to give a scientific explanation for the Soviet physicists' fears. He replied on the 17th January 2005.

2. THE THREAT OF A NUCLEAR EXPLOSION

Dear Colleagues,

[...] I am therefore going to try to reconstruct the events chronologically, using my archives (notes from 1986).

I am very familiar with the construction of the RBMK reactor in which thousands of tonnes of graphite are used as a neutron moderator. We know that when the reactor is functioning normally, all the graphite is contained within a steel cylinder. Slowing down the neutrons in the graphite provides 6 to 7% of the total power of the reactor. In order to maintain the working temperature of the graphite at between 500 and 600 degrees C, the graphite cylinder is filled with an inert gas: a mixture of nitrogen and helium. The heat transfer liquid (water) circulates in the interior of the graphite assemblage.

We know that the accident occurred when errors were made by the staff in the course of an experiment that was dangerous from a nuclear point of view: the experiment was to investigate how, in the case of an emergency shut down of the reactor, it might be possible to use residual calorific emissions to produce extra electricity.

The absorbent rods used in this reactor were shortened and were missing the graphite ends that should fill the channel at the moment when the rods were taken out of the reactor core; as a consequence, when the rods were removed from the active zone, the channel filled up with water (caloporter fluid).

The protocol for the experiment had been submitted to the Minister, the Chief Engineer of the Chernobyl power station, the academic Nikolai Dollezhal, and the scientist responsible for the power station, Anatoli Alexandrov. Despite not having received a positive reply in writing, the management at the power station decided to go ahead with the planned experiment on the 25th April 1986.

The RBMK reactor is particular in having relatively low fuel enrichment (1.8% in uranium-235) and very high positive coefficients of temperature, particularly at lower power levels.

In summer 1986, after the accident, the minister for mechanical construction, E. Slavski, showed me the whole experiment program. According to this program, they needed to lower reactor power to 800MW and then, at this level of power, after the security rods had been lowered, observe the running of the turbo generator through inertia to determine the quantity of electricity produced.

When the experiment began, the reactor power fell to between 60 and 80 MW, and, in accordance with the laws of physics, the reactor fell into an "iodine hole". In this situation, they should have shut down the reactor, waited two or three days until the short lived iodine isotopes had disintegrated and the power had returned to normal.

According to statements made by those who took part in the experiment, the personnel at the power station took the compensatory rods out from the heart of the reactor, and switched on the complementary circulation pumps in order to pump water into the reactor. The radiolysis of the steam in the channel produced an explosive mixture of hydrogen and oxygen which caused the first thermal explosion at the heart of the reactor.

The result was a deviation in neutron flux, and water that had filled the channels when the absorbent rods were removed started to boil. In two to five seconds, the power of the reactor increased a hundredfold. Ceramic fuel elements (uranium dioxide) with low calorific conductivity deteriorated rapidly under enormous thermal pressure.

It is known that water decomposes most effectively in fuel bursts. There then followed a second detonation of this explosive mixture that tore through the hermetic covering of the graphite and blew off the upper concrete slab (weighing about 1200 tonnes; even today it is still leaning at an angle of 60°). The graphite reservoir was now open to the air. When it burns in air, graphite can reach temperatures of between 3600 and 3800 degrees C. At these temperatures, the zirconium covering the fuel elements and the reinforced tubes in the graphite played a catalytic role, like a spark plug, contributing to the subsequent development of the accident.

The 1700 active channels of the reactor contained 192 tonnes of uranium (1.8% enriched uranium-235). In addition, the maintenance channels contained used cartridges that had been discharged from the reactor.

With graphite burning at such a high temperature, the channels containing the combustible material began to melt (like electrodes in a voltaic arc) and the melted combustible material began to flow downwards and infiltrate the areas where electric cables were housed.

The reactor rests entirely on a concrete slab about 1 metre thick. Underneath the reactor, were solid concrete chambers built to store radioactive waste.

As the personnel continued to pour water into the reactor with the circulation pumps, the water filtered down, of course, into these reinforced concrete underground chambers. A huge risk became apparent. If the material, in the process of fusion, pierced the concrete slab under the reactor, and reached these concrete chambers, it could create conditions liable to cause an atomic explosion. On the 28th and 29th April 1986, the joint members of the department of Physics of Reactors at the Institute of Atomic Energy at the Academy of Science in Belarus, of which I was the director, made calculations that showed that 1300 to 1400 kg of a mixture of uranium, graphite and water constituted a critical mass capable of causing an atomic explosion of 3 to 5 megatons (50 to 80 times more powerful than Hiroshima). An explosion of this magnitude could cause massive radiation lesions in inhabitants within a radius of 300 to 320 km, including the town of Minsk, and contaminating all of Europe to the extent that normal life would be impossible.

I wrote a report on the results of these calculations on May 3rd 1986 at a meeting with N. Sliunkov, first secretary of the Central Committee. What follows is the assessment of the situation that I presented at this meeting: the probability of an atomic explosion was not high because at the moment of the thermal explosion the heart of the reactor had been broken up and dispersed not simply inside the reactor but all around the reactor. They asked me why I could not give a 100% guarantee that no atomic explosion would happen at Chernobyl. I replied that in order to do that I would have to know the state of the concrete slab underneath the reactor. If the slab was unbroken, had no cracks or splits and if no holes appeared later, it would be possible to confirm that there would be no atomic explosion.

But I can tell you one thing with absolute certainty: thousands of railway carriages were brought to the area around Minsk, Gomel and Mogilev and the other towns within a 300 to 350 km radius of Chernobyl to evacuate the population should the need arise.

It was thought that the explosion could occur around the 8th or the 9th May 1986. That is why all possible steps were taken to extinguish the burning graphite before this date. Tens of thousands of miners were brought in to Chernobyl, as a matter of urgency, from Moscow and Donbass, to dig a tunnel under the reactor, install a coil of tubing, and cool down the concrete foundation slab under the reactor to prevent cracks forming. The miners worked in an inferno (high temperatures and high levels of radiation) to save the concrete slab. It's impossible to truly appreciate the work these men did to prevent a potential nuclear explosion. They worked with no regard for their own personal safety. Most of these young men are ill, a number of them have already died in their thirties.

I compared Nesterenko's text with the descriptions in Medvedev's book where he talks about liquid nitrogen being used to "cool down the concrete slab" rather than putting out the fire, and he adds: "I was afraid that the core would pierce the concrete slab on which it rested and that it would collapse into the condensation pool, thus provoking a terrible *thermal* explosion". [Author's italics.] Medvedev is not talking about an atomic explosion. I put the question again to Nesterenko, who replied in more detail on the 6th February.

I can remember quite clearly being called to KGB headquarters in Minsk on the 1st May 1986, where I had a telephone conversation by "high frequency link" (they had a high frequency signal) with Legasov, who was at Chernobyl. He told me that they were thinking of putting liquid nitrogen into the collapsed reactor. They thought that, as it evaporated, the liquid nitrogen would displace the air (oxygen) and that the graphite would stop burning. But amongst the scientists there was a rumour circulating that when liquid nitrogen is introduced into the active zone (the core of the reactor) it might cause an explosion. People were saying that this had happened before in England. I told Legasov that at our Institute at Sosny, in the research reactor IRT-M, there is a horizontal channel to test materials and studies had been made of magnetic and mechanical characteristics using different samples of metals at low temperatures. In order to do this, we had introduced liquid nitrogen into the channel. No explosion had taken place. I was asked to verify this information, with other people who had participated in the experiment. I also called (it was 2 o'clock in the morning) B. Boiko (who was the director at that time of the Institute of the Physics of Solid Bodies and Semi-conductors at the Academy of Science in Belarus), who confirmed that there had been no explosion. They then let me know that they were sending a helicopter to take me to the site so that I could assess the situation there.

After the accident, while the reactor was still burning (the fire lasted ten days) hundreds of scientists came to the site. They were all studying different things and proposing different schemes to control the situation. You know that 6000 tonnes of lead were poured onto the ruins of the reactor, and this lead evaporated and covered the surrounding area. It is one of the causes of the high level of lead found in children's blood in the Bragin, Khoiniki and Narovlia districts. One scheme involved miners digging a channel under the foundation of the reactor. During our helicopter flight over the reactor, we tried to locate a route for the lorries carrying liquid nitrogen to reach the ruins of the reactor, displace the air and extinguish the burning graphite.

We know that after the explosion in the reactor, the personnel at the power station continued to pump water into the reactor using all eight pumps at the same time. It is clear that the whole sub-foundation under the concrete base of the reactor was flooded with water. All the scientists feared that if the lava consisting of the materials from the active zone pierced the concrete base, and if the combustible material and the graphite flowed into the concrete chambers underneath the reactor, this critical mass could lead to an atomic explosion. The primary objective was to stop the graphite burning. One of the solutions was to displace the air (containing oxygen). That was why they wanted to bring in liquid nitrogen. Many thousands of tonnes of sand and dolomite were dropped on to the reactor from helicopters. The graphite stopped burning after ten days. But afterwards, there were further outbreaks of fire from this melted mass, between the 15th and 17th May, and radioactive emissions which we detected at a distance of 70 to 100 km from Chernobyl, in the Narovlia region in Belarus.

According to witness statements from all the chemists (Legasov, Gidaspov...) there were two thermal explosions on the 26th April. The fuel elements that exploded and formed a mass of tiny particles created the ideal conditions for the radiolysis (decomposition) of the water vapour into oxygen and hydrogen. When this mixture reaches a certain concentration (7–9%) it explodes.

Under these conditions, ie if between 1,300 and 1,400 kg of water, graphite and uranium had come together, an atomic explosion could have taken place. My opinion is that we came close to a nuclear explosion at Chernobyl. If it had taken place, Europe would have been uninhabitable.

There is a false and dangerous idea currently in the West: with the reactors at Chernobyl shut down, there is no longer any risk of an atomic explosion. But as long as there is combustible nuclear material in the interior of the damaged reactor, it presents a danger not only for Ukraine, Belarus and Russia, but for all the people of Europe.

The European people should, in my opinion, be eternally grateful to the hundreds of thousands of liquidators who, at the cost of their own lives, saved Europe from the worst possible nuclear catastrophe.

V. NESTERENKO,
Corresponding Member of the Academy of Science of Belarus,
Professor, Doctor of Technical Science, and liquidator at the accident
that occurred at the nuclear power station at Chernobyl in 1986.

For thirty six hours, the inhabitants of Pripyat, 2 km from the power station, knew nothing of the volcano spreading its immense invisible evil over objects and beings. The Soviet media remained silent. The spring was exceptionally warm that year. Walking in the town parks, mothers took off their children's shirts.

3. PRIPYAT: NUCLEAR TRAP

"I often think about Pripyat" writes Medvedev. "Life was great in this 'nuclear' town. I watched it being built with my own eyes. When I left to work in Moscow, three residential areas were already being lived in. It was a clean pleasant little town. The quality of life was very good. Newcomers never ceased to marvel at it. Many military personnel wanted to move there for their retirement. Occasionally, after endless problems and endless requests, they won the right to come and live in this paradise, that combined the beauty of nature with every convenience of city life". [7]

Before the accident, Pripyat had a population of 56,000. Today it is deserted, overgrown by vegetation, and will remain forever a ghost town. The classical music emanating from the loud speakers on a wet foggy morning in November 1990 added a touch of surreal madness to this shipwreck from the Soviet era. Yulia Lukashenko stood in front of the house she had abandoned, now surrounded by brambles, and recounted the day of the accident. The day after the accident the people of this town were uprooted from their homes for ever. She had asked and had received permission to accompany us to Pripyat in order that she might "reconcile herself". She was visiting her home for the first time in the four years that had elapsed. I watched from a distance as she held a piece of paper against the outside wall and wrote in capitals "I'M SORRY!" She was crying.

Q.—Why are you saying sorry? You are not to blame for what has happened to your house.

Y. Lukashenko—I'm a human being. Like all other human beings, I'm responsible for what happens to our planet, the Earth. I am, in a way, to blame for this house. Do you understand? I'm not judging myself as an individual, but in general, as part of humanity. And I feel guilty…

OK. I feel better now. Look over there. I can see the toys my children threw out the window when they chased us out of here. That makes me feel better. I am at home again… *(She turns to face the front of the house)* Look here, these little red and yellow cubes, and next to them a little yellow wheel. Their toys. There, in the hole, that's our pen and pencil. I feel good. *(She smiles apologetically)*

—When did you understand what had happened at the moment of the accident? In the first few hours?

—No. Two weeks later. We only knew that it had been a serious accident two weeks later. But we didn't know how serious. The morning of Saturday, 26th April,

---

[7] *Op. cit.*

we knew something had happened at the power station. Lessons were interrupted, which was very unusual, discipline was very strict at the school—the teachers had been called in to the office and they were told that during recharging at the power station, there had been a mechanical failure and a fire had started. "The fire is under control (I'm quoting them) and there is no danger. But as a precaution, the children need to be kept at school, they should not go out, close all the windows, wash the floors, (which we did) and explain it to the children. Don't let them out of your sight".

Blinded by panic, the people who were responsible for the accident could not accept reality, and for the first thirty-six hours persuaded themselves that the reactor had not been damaged. The decision to evacuate the population was suspended, spontaneous initiatives to protect the children were rejected so as not to alarm people, the town of Pripyat was surrounded by a cordon of armed police, and people were forbidden to leave. It was only on 27th April at two o'clock in the afternoon that the inhabitants were evacuated.

During this time, vaporised nuclear fuel was being discharged into the atmosphere and being deposited as a fine dust of radioactive particles which was being inhaled by the people of Pripyat. Water tankers were already washing the town. It was lovely weather. Everyone thought it was in preparation for the 1st May celebrations, a bit early perhaps. But the answer was very different. Samples had been taken from the roads, from the air, and from the dust in the gutters and revealed that 50% of the radioactivity was Iodine-131. On the roads, the level reached 50,000 microsieverts per hour. The average level of natural radioactivity considered "normal" is 0.11 μSv/h. This is the result of radiation from the cosmos and from the earth's rocks, and is compatible with life. That morning at Pripyat, radioactivity from Iodine-131 was 450,000 times higher than natural background levels. And radioactive iodine, if it is not immediately blocked by a dose of stable iodine, attacks the thyroid gland especially in children.

Y. Lukashenko—During the second or third lesson, I can't remember now, it was around 10–11 o'clock, they started to distribute tablets, for us and for the children. From that point on, we knew it was really serious.

We went back to the house. We had a powerful telescope. We went out on to the roof of our house and spent a lot of that day watching what was going on at the power station. It's about 2 kilometres away as the crow flies. The glow of a fire. We couldn't see any flames, but you could see the fire glowing from the town. We invited the neighbours up there. The telescope passed from hand to hand. We could see the children playing down below in the sand, in their underwear. It was a really hot day. The Mums were out with their pushchairs, and the little ones had no clothes on.

For the last twenty years there has been a deliberate policy of disinformation that has led the world to believe that the disaster at Chernobyl was not, in the final analysis, a very serious accident. Ten days after the accident, the Vice-President of the USSR Council, M. Shcherbina, gave assurances at a press conference convened in Moscow that levels of radioactivity in the immediate vicinity of the reactor were about 150 μSv/h, when in reality it was 150 *million* μSv/h. The resulting policy decisions had cruel consequences for the people in the Western part of Russia, in Belarus and in northern Ukraine.

Today, the town of Pripyat has been abandoned, but the radioactivity expelled during the ten days following the reactor explosion was dispersed by wind and rain over an immense area, and even 200 kilometres from the town, people are living with levels of radioactivity higher than those we measured in the fairground at Pripyat, which were themselves 36 times higher than normal background radiation.

4. THE CAPTIVES OF POLISKE

In the five years following the accident, WHO never came to Poliske, the town in which Alla Tipiakova was born, where she taught and where she died. WHO did nothing to protect her children. Yet the constitution of this United Nation's agency stipulates that it must satisfy the following obligations:

- to act as the directing and co-ordinating authority on international health work;
- to furnish appropriate technical assistance and, in emergencies, necessary aid upon the request or acceptance of Governments;
- to provide information, counsel and assistance in the field of health;
- to assist in developing an informed public opinion among all peoples on matters of health.

Today, Poliske has been wiped off the map, abandoned, invaded by brambles and overrun with wild animals. But in 1990, when we were filming, there were still 13,000 people living there. Radioactive dust had been deposited there in irregular patches measuring anything between 5 to 40, 50 and even 300 curies per square kilometre. Like Pripyat, this town should have been evacuated immediately, as Nesterenko had asked. But that would have spelled the end of an agricultural economy and a thriving small scale industry. In particular it would have led to the demise of the local managerial class. The reaction of the authorities was speedy and energetic: between 1986 and 1990, more than 40 million roubles (at that time 0.60 roubles was the equivalent of 1 dollar) was invested in the area to encourage the population to stay. New houses were built, new schools, a swimming pool, a water supply was established, gas and heating in all the flats, salaries doubled…

Eight years later Professor Nesterenko gave me his opinion.

V. Nesterenko—I consider it a crime. The level of radiation permitted for professionals is ten times higher than the level permitted for members of the public and that is not because we are in better health. We were selected because we had a better resistance to radiation and we had been trained in radio-protection. Ordinary members of the public do not know how to protect themselves. Just to give you an example, during my visits to Chernobyl, I happened to see a man on a tractor, happily working his radioactive soil, rolling a cigarette and smoking it without washing his hands.

Today in Minsk there are 31,000 people who were evacuated from the contaminated areas. They are on a medical register and undergo an annual examination. Among this group, cancer of the digestive system and of the lungs is six times higher than the other inhabitants of the town and thyroid cancer is 33.6 times higher. Why? Because they weren't evacuated until 1992–1993. Up until then they were living in the contaminated zone, working their fields and filling their lungs with radionuclides. Today they are paying the price.

Over a number of years, the Soviet authorities, sanctioned by the complicit UN agencies, transformed Poliske into a radioactive trap. In 1990 agricultural production did not stop; quite the reverse. Despite very high levels of radioactivity in the fields, the village executive committee took pride in maintaining production at the same level as before the disaster. A board of honour was erected in the main square of the village honouring and celebrating the year's achievements. Accompanied by Rostislav Zatkhei, a doctor and a militant in the Ukrainian Green Party, we visited Vladimirovka, a nearby village that had been evacuated. We found a copy of an old newspaper *The Flag of Communism,* lying around in the rubble of an abandoned *isba* (wood cabin). It was dated 18th June 1986. Zatkhei had this to say about it.

R. Zatkhei—Reading this article you can see that the fate of the people was the last thing the authorities were concerned about. The first secretary of the Party, Primachenko, said that all efforts should be concentrated on hay making, on preparing tools for the harvest, and on getting the shops ready to stock all the produce. And all this in 1986, when everything round here was crackling with unusually high radioactivity. The produce was harvested, of course, and in all likelihood, was mixed with uncontaminated produce from clean areas. This mixture is still being distributed all over the Soviet Union. The low dose radiation in these products is exposing all of us to a creeping genocide, a slow death and increasing illness for future generations.

Poliske is quite a large settlement, the district centre, and in order that it remain the administrative centre, it was decided at the time, not to evacuate it, unlike all the villages in the surrounding area. They were little villages, whereas Poliske was preserved as an administrative centre. And despite official approval of the policy of evacuation, the same attitudes remain. Even today, if people from Poliske want to leave and move to a cleaner area, they no longer receive any legal protection. They have no status. They have nothing. Whereas the people who stay receive a great many benefits. They have twice the normal salary, free food, special clothing, they can buy consumer items, when most people have nothing: furniture, television sets, etc. They have a lot of privileges, for instance, they can buy a car without going on a waiting list.

Q.—Why did you come here in 1987, into this contaminated zone?

—I studied medicine at Lvov in the West of Ukraine, which was not contaminated. In 1987, after university, we were assigned posts by the state. Initially I was sent to the Odessa province. But I felt, probably instinctively, that the situation here was serious, because there was no information in the press about any problems, so I asked to come to this area. I was assigned to work here, in Poliske. I wanted to come to the Kiev region, where the tragedy had taken place. It's hard to explain. It was an instinct. I thought I could be useful. A human impulse—to rise to the challenge. But to begin with, when I arrived, it was really hard to get involved because we didn't know anything. What levels of radioactivity were there? All the sources of information, and the medical establishment repeated "Everything is normal, everything is OK, there's no problem". And I couldn't argue back, we had no data, no measuring instruments. But when we managed to find a dosimeter and made our first measurements, and they were very high, unbelievable and when various changes in the children's behaviour and in their health began to appear, then we sounded the alert. It was at that point that we were able to step in and to press more and more strongly for action to be taken.

—What were your political views before you began working here? Were you a communist?

—More anti-communist, because my parents had been arrested. My father and my mother died very young in the camps and that really scarred me. My sister and I were adopted by some people who had taken part in the big political struggles in the years 1939–1940, in Western Ukraine.

—What difficulties have you encountered in your struggle today? Do people make it hard for you?

—When my wife was still living here with our child, she often got threatening phone calls. They told her I would be attacked on my way home. They broke the door down, cut off the telephone, things like that. But that's not the most significant thing. You can deal with that, we'd already experienced it at Lvov. The biggest difficulty, in my opinion, is the fact that people are too slow to realise the gravity of the situation.

—What illnesses are you coming across?

—A general deterioration in health, in children as well as adults. Serious nose bleeds that are very difficult to stop. One in two children has clouding of the crystalline lens in the eye. This leads directly to cataracts and blindness in children. Then there's leukaemia. When we met in August, I talked to you about a little girl who is no longer with us. She died, Irina Soubota. Diagnosis: leukaemia. Many different changes appeared in the blood. Mono-cellular formations, very large cells: new components in the blood and a variety of pathological changes that then lead to changes in the whole organism. An increase in the size of the liver and other similar symptoms.

—Are you able to get results for all these tests despite the censorship?

—I've got to say that at the local level, there are doctors who have had the courage to tell the truth right from the start. They did everything possible to tell people the truth. But it was very difficult at the time because of the pressure on them from the Party machine. But at a local level doctors have done their job well.

I started to have health problems after my wife and I visited the heavily contaminated areas. Using a dosimeter we looked for heavily contaminated areas. We visited evacuated villages. I spent a lot of time in these places because we were also studying architecture; we wanted to get a picture of the whole area. I probably accumulated a large quantity of radionuclides during this time. So I get tired, I get haemorrhages: three days ago I had a tooth out and it bled for three days non-stop. I had to take medicines to stop the bleeding. So there we are. My symptoms aren't acute. It's there inside me, the poison, just waiting to get me.

At the end of this interview Rostislav Zatkhei took us to Tarassy, a small village close by that has not been evacuated.

R. Zatkhei—Immediately after the accident, they decided to build a new school for the village, a nursery school, a bit further away, public baths, a laundry and other buildings. The same thing happened in many of the other villages in the district and in Poliske itself. It was ridiculous. An enormous new school for 190 children when there were no more than about fifty children of school age in this village. The nursery school was built for 50 children when there are only 12 children of nursery age in the area. When we asked someone to explain the decision, the local authorities said that we should be "looking to the future", getting a new perspective on things. I suppose they were looking to the future when all that building work went on in Poliske, a second school completely rebuilt, although the decision had already been taken to evacuate the town: there will be no more children here; no one will ever live here again. And yet, the work carries on. They are planning to rebuild the stadium, with three indoor swimming pools. They carry on endlessly investing money simply in order to persuade people that there is no danger, and that you can live here. "Look. We're investing money, we are competent at what we do, we're building all this for you, the people of the town, to make life better for you!" It would have been better to build all this in an uncontaminated zone, where people will move to, put all that money into a place where children can go to nursery school and school safely, can swim in the swimming pool, and make use of these stadiums.

There is a travel agent that organises trips to the Chernobyl nuclear power plant for well fed Westerners in need of a thrill, while Europe is investing 2 million euros in a programme that recommends "radiological quality" in areas that need "rehabilitation", ignoring completely the health catastrophe that is hitting the local population. The thesis of this European programme, which refuses to administer a pectin-based prophylactic for children, is that if people obey the experts, they can live a *good* life in these territories. [8]

5. THE NURSERY SCHOOL IN POLISKE

In the documentary *Nous de Chernobyl (We of Chernobyl)* that we made in 1990, there is a scene showing a group of rather unhappy children, slowly crossing the concrete yard of the nursery towards the exit. The voice-over describes how "for four years, the children here walked across an invisible nuclear furnace, as radioactive as the ghost town of Pripyat. It was only in 1989, after the first political elections in the USSR, that the local councillors reacted to the parents' protests. But instead of closing the school, they sent in decontamination teams that had to recover the yard three times with a thick layer of concrete, without succeeding in eliminating the radiation. Even today the level of radioactivity is 17 times higher than natural background levels".

Here are the testimonies of three teachers at the school:
—They wore masks while they worked. On the roof and in the yard, they wore masks, while the children were playing there.
—They re-did the roof ten times, replaced the fencing, they covered the ground several times with asphalt, but in any case it never went away and we are still here.
—There are places here that are so radioactive nobody should be allowed to enter. Here, for example, behind the shed.....where the children play. Over there, near the oak tree, the level of radioactivity is terrifying. And the children play there too…Where can they go?
—In summer, it's full of dust, sand. The children breathe it all in. They got us to take up all the carpets in the school because they are contaminated. The children

---

[8] CORE Programme, Part Four, Chapter III.

have to sit on the bare floor. We took all the fluffy toys away because they absorb radioactivity. But how can we live like this? The children need to run around, to play. We're treating them like prisoners. We're not living any more, we are prisoners.

—They often have nose bleeds. They all get headaches. Pains in their armpits. Their eyesight is deteriorating. They complain about being tired all the time, of having no energy. They are often sick. They have countless problems.

Chapter IV

THREE ENCOUNTERS

1. SVETLANA SAVRASOVA

Before leaving for Chernobyl in November 1990, I asked Irina Ilovaiskaya-Alberti, who was director at the time of *Pensée Russe* (*Russian Thought*), a newspaper set up during the first wave of Russian emigration to Paris, and first published in 1947, to recommend someone who could be useful in my investigation. She suggested that I meet Svetlana Savrasova, a young journalist from Belarus that she liked and respected. We arranged a meeting over the telephone and arrived at Svetlana's apartment in Minsk, where she had brought together a group of people wanting to share in her venture: an association of 25 young people that had only just come together, at her instigation, to accompany children when they travelled abroad for their decontamination.

Q.—Did you all know each other before you formed this association together?

S. Savrasova.—No, none of us knew each other. My husband was the only one I knew.

Q.—How long have you been meeting together?

S. Savrasova—For four months, since July 1990.

Q.—How did it come about?

S. Savrasova—We all came here through a chance event.

Q.—Came where?

Tolik—Here, at Sveta's house. It was Sveta who began the action. She is a journalist. One of her articles, which was turned down by her own newspaper, was published in Poland. The Polish readers reacted very warmly.

S. Savrasova—When my editor asked me to write an article about the children of Chernobyl, I wanted to avoid turning it into something sensationalist about the new monsters that could appear on Earth. I went to a school in Khoiniki, met the children and asked them to write me a letter as if they were writing to an old friend. We chatted for quite a long time and they trusted me. I found these letters very moving. In my article I emphasised the seriousness with which the children had set about the task. You could see from their letters how deeply they had thought about the subject. While they were writing, I noticed one girl wiping away a tear, another who was searching for her handkerchief, a boy in the front row whose chin was trembling. "They are reliving the evacuation", explained the head teacher, as she also took her handkerchief from her bag. "It was a week after the accident. The coaches were waiting in the square. They separated children from their parents. They were about 12. Crying, screaming... They had to drag the mothers from the coaches by force. They were banging on the windows, holding on to the wheels..."

One boy of 16 wrote a letter that I can't forget: "I am 16. I live in the Poliske region. This is my country. I don't want to leave here for any other place in the world. I know that because of the 3 roentgens that I received on 30th May I will soon die. But I have discovered the wonderful world of literature. I want to stay in this world, right to the end of my days. It's my Mum's death that hurt me the most. She died a year after. I felt hurt by the doctors who looked after her and did not seem to care... On 30th May, the whole school went out into the fields to harvest the potatoes. Among the potatoes we found families of rats with dead babies. We wondered what could have happened. Why had the baby rats died? Overhead the clouds were passing". At the end, he wrote, "I forgive all the adults responsible for this tragedy. I forgive the technicians at the power station, because like me, they are dying of leukaemia". For me this letter is like a Lord's Prayer, something I can't turn my back on. You can ignore a newspaper article, but not this boy.

Q.—So, that is what started it all off?

S. Savrasova—Yes. I think the Polish people reacted, not to my work as a journalist, but to these letters written by adolescents. Because it is impossible to express pain the way a child can. No adult finds such simple words. It is pure living emotion.

Q.—Is that when they began to invite children to come and stay with them?

S. Savrasova—One day I came back to the house, opened the letter box and found seven letters. Polish stamps, written in Polish, this was all new to me. I didn't understand anything. I didn't even know the article had been published. I got Larissa, who works as an interpreter here, to translate the letters. I found out they needed people urgently to accompany groups of children. These were among the very first. And we came, Gianna and I. We had to leave that same day. It was a terribly difficult journey. I remember coming back in tears. But you couldn't refuse... Something had happened inside me. It's impossible to describe it in words.

Seriozha—Tolik rang me up to tell me about this initiative. We were studying at the same institute. Vera was with us too. I came round to Svetlana's house... and we had to leave the next day. We had to leave straight away. For me, it was seeing the children, their faces, how they behaved, when I heard them express themselves, when they told us what they thought, that I began to understand what the Chernobyl catastrophe really was. I understood it better. It was probably the direct contact with these children. I saw what the real consequences of Chernobyl were. This is a wonderful project also because it really helps us to understand ourselves. It's the sort of thing that we really don't meet in our day to day lives. It's completely different from the normal every day grind, with meetings that no-one really needs. The "system" is entirely absent, ideology has no role any more. It's a different world, more natural. And you feel better.

S. Savrasova—I know loads of people who never think about Chernobyl. Just like me actually, at the beginning. I never thought about it in the first three years following the accident. I only went into the zone three years later. I was completely uninvolved. Hearing about it upset me. I never read anything about it, not plays, books, newspaper articles. It was only later, when my own child was directly affected that I had to open my eyes. At the beginning the desire to think that everything was alright was so strong that despite the allergies and the other problems she had, I didn't want to make the connection. The shock we had all experienced made us lie even to ourselves. Today I have no more illusions, and that makes me unhappy sometimes. We'd been told so many times that we were the best, we were the happiest. That helped us to live. It may not have been true but it made life easier.

However, immediately after Chernobyl the order was given to persuade pregnant women to have abortions. It wasn't said openly but I know because I went to the clinic for consultation because I was pregnant. It was like a mowing machine—everyone passed through it. Their arguments were unbelievable. "If you want to give birth to a two headed baby, go ahead". The women were given this message right from the start. They had to just accept it. I had a friend who had just gone through a seven month sterility treatment. She got pregnant and they made her have an abortion. She still hasn't got any children.

First we got Chernobyl, then the men in white coats. There are things that go on in this country that can't be explained by normal logic. The decision was probably made "to be on the safe side" by some civil servant at the Ministry of Health. To avoid the thing they feared, that millions of monsters would be born. They had no idea how things were going to turn out. The authorities take decisions without taking any scientific data into account, I think Chernobyl has shown this very well.

For the first year after the accident, it was officially forbidden to leave the area so as not to create panic. You have to remember that in the USSR we had interior passports and if you wanted to go somewhere else, you had to have a travel permit, authorisation. People who left areas that were too contaminated could not obtain the legal right to settle anywhere else, they were fined, a child would not be accepted in school or nursery, a mother could not join the library, a grandmother could not buy her medicines because she was not registered at the dispensary. They were pariahs. Today, you are allowed to leave, but wherever you go, attitudes are the same, you'll be shunned like the plague, your child will have to sit away from the others in the refectory or in the crèche, he'll have to sleep separately from the others. I've known several families from the area who left and came back after two years, because they were rejected everywhere they went, hounded out. They heard all these ludicrous stories about how people from the contaminated zone glow at night. Ordinary people are naïve, it's how they react to things. When the children go to Russia to the pioneer camps, they come back with better blood results, but they all came back in an anxious state because of the way they are marginalised.

Q.—Did you feel that people wanted to go and live somewhere else?

S. Savrasova—Yes, but now we're seeing different reactions in people. In the first year, everyone would have left. Now everyone talks about international aid. When someone has lived all their life in one house and has accumulated belongings and only receives 500 or 300 roubles compensation for the family, when at Bhopal, in India, they know each victim gets 25,000 dollars, they start to fight, not for a new house in the uncontaminated zone any more but for social justice and that has a negative aspect. People would rather their child receive a certain amount of radiation if it gets them a little compensation money. And then on top of all that, there is a big psychological problem. People start developing a kind of begging mentality. It can lead to situations where someone expects a subsidy while their child is dying.

Q.—In what spirit do you organise children's travel abroad?

S. Savrasova—Going abroad really does help their health. For me as a woman that's the important thing. A month in Poland, for instance, prevents them falling ill for three months, once they return to the contaminated zone. And I want the people who look after the children in Europe to realise that Chernobyl is not our problem, yesterday's problem, but their problem, tomorrow. Radioactivity doesn't stay in one place. It takes no notice of borders. Today it affects our children, tomorrow it could be yours.

I was walking through the village of Khoiniki. A woman came running over to me. "Did you take some of the milk?" I didn't understand. "If you took some, throw it away!" According to the existing norms, each child should receive half a litre of clean milk every day. The milk is delivered once or twice a week. Sometimes the milk that was thought to be clean is replaced by radioactive milk. And the women were running from one yard to the next warning people not to give it to the children. Imagine the state of mind of those mothers who have just given some milk to their child! The farms in the zone are still in production. The cows look quite healthy. When someone asked a minister why these farms were still working, he said "In our socialist system, there should not be any unemployment. After all we can't pay people to do nothing!" So the authorities would prefer us to produce radioactive milk and meat. If clean milk, meat and other food products can be brought in to the people living in these dangerous areas, why not wood from forests near Moscow or Leningrad? I can well imagine a woman in these peasant villages stocking her fire with these radioactive logs. The wood burns, a bit of ash falls into the soup. Instead of bacon flavoured *bortch* she makes…I don't know…strontium flavoured *bortch*. With the radioactivity inside these stoves, it's like having a reactor in the house. The stoves have been transformed into household reactors. The woman empties the ash into a bucket and then spreads it all over her vegetable plot. That's what everyone does, her mother and grandmother before her. No-one will ever have told her that she should bury the radioactive waste. And then the hens come along and forage among all this radioactive waste and then they lay radioactive eggs. This is how the contamination perpetuates in the zone, the process never stops. And they're doing it all with their own hands, these poor wretches living in these villages like guinea pigs.

Q.—Do people understand the whole truth today?

S. Savrasova—I don't think so. Where I live, I know people don't understand it. And anyway, what is the truth? All of us have our own truth. The children are growing up and every one of them has their own version of the truth. A girl of 14 wrote to me: "I go out in the garden and I see my sister and her friend splashing around in the slush. Two little girls playing around in radioactive mud. I'm so frightened for them! We've already had a life, but they've got their life in front of them! How is it possible that they are there playing around in this radiation?" This is a girl of 14 who writes that she has "already had a life". This is one more "truth" about Chernobyl. There are many destinies and many truths. Some want information, others don't. You can give every possible explanation to one person, but if they're only interested in their vegetable plot, it's a waste of time. A different person will insist on finding out more and eventually will understand something. Take me for example. I don't really know the full extent of the disaster. When I was studying at the Institute of Radio Engineering the government was putting out an enormous amount of propaganda about the dangers of nuclear weapons. Now most days the disinformation makes me feel as if I am falling down a well and life drifts by, leaving us suspended and anxious. I'm not sure that even with the information I get, with my training and through the contacts I have with scientists and with the children, that I understand the true scale of the tragedy. I don't think so.

This year, I was able to send 1500 children on holiday to Poland. We wanted to take blood samples from all the children, but many of them refused because they really couldn't face it again. But when it came to filling out their medical records, it turned out that there was nothing to fill in. The hospital gave them a certificate to say they didn't need any medical treatment, they were doing fine, they were in good health. When I said that all the children should be examined in the hospital and the results written into their records, I was told that there weren't enough doctors to do this, that half of them had left and the other half were so overworked that they couldn't take that on as well. But doctors arrive from Moscow all the time; they take samples of blood from these children and do not report the results of their analyses. They do loads of tests on people but no-one tells us anything. I make enormous efforts to resist the thought that the whole area is an immense human laboratory. It's a terrifying thought. Our own National Front of Belarus[9] has stated that there is an agreement at international level that includes organisations for the development of the nuclear industry. An enormous experiment could be going on here right now.

Svetlana Savrasova was the first person we filmed in the Chernobyl territories. Naturally I did not believe this hypothesis to be real, a hypothesis that she herself tried hard not to believe. I interpreted her remark as a "rumour", a kind of collective psychosis among a population traumatised by the accident and deprived of reliable information by a totalitarian state. I had already come across the phenomenon in Armenia in 1988, when the country was devastated by an earthquake. Glasnost had awakened in the citizens the desire for national sovereignty and the possibility of escape from under the Soviet yoke. The earthquake of Leninakan was perceived by many Armenians as a deliberate attack from Moscow to sabotage the move towards independence. The mainstream Soviet media had no credibility with the people, and some quite extraordinary stories began to circulate, including the idea that the Soviets had deliberately set off underground explosions. On the other hand, Svetlana's hypothesis about an enormous laboratory experiment, however unbelievable, did remind me of an unpleasant conversation I had had the week before in a restaurant on the top floor of the hotel in Minsk. A German guest, obviously recognising a fellow Westerner had installed himself at my table and we began a kind of "foreigners abroad" conversation. He wasn't actually a journalist. He described himself as an economic advisor for the area. We ended up having a heated argument. He seemed to have taken it upon himself to lecture me about the real consequences of the Chernobyl accident: the effects of radiation on health were exaggerated, lies told by a poverty stricken population to elicit sympathy and maybe even funds. The stress caused by the economic situation, alcoholism and the cover up by the communist authorities had not helped their state of health, he acknowledged, but it had nothing to do with Chernobyl. I was struck by the contempt and categorical assurance of this well-dressed Westerner, well fed with uncontaminated food, who felt able to deny what had not yet been studied. "Coloniser, occupier, an official on mission", these were the words that came to mind. I would meet many others like him in the next few years. This set me thinking. An official working for whom and for what? Had he been sent in a professional capacity, like an agent in a war situation? Unlikely. More plausible, a well paid official, with a certain approach to reality, whose ideology corresponded to his

---

[9] Independence Party

superiors. I was reminded of Hannah Arendt's concept regarding Eichmann of "the banality of evil". The notion of some nuclear experimental *lager* (concentration camp) still seemed far fetched, but it chimed with the categorical arrogance of my Western dinner guest. The interview with Yury Shcherbak in Kiev that I conducted soon after finally cleared the matter up for me. He too surprised me with a quote, that I had not heard before, from a French professor and I was reminded of Hannah Arendt again.

2. YURY SHCHERBAK

Yury Shcherbak, doctor, writer, leader of the Green movement in Ukraine in 1990, and elected deputy in the first "democratic" Supreme Soviet of the USSR, was one of the first to unmask the lies and fraudulent information concerning the accident at Chernobyl. Twenty-five years have gone by since this interview. It remains a valuable historical document revealing not only the hopes raised during this brief period of democratisation but also the convergence, that began in the first weeks after the accident, between Western experts and the Kremlin, that Bella Belbéoch had so clearly predicted.

Y. Shcherbak—1986 was a huge shock to the whole of our society, and it was only in 1987 that we began to realise what the explosion at Chernobyl had done to our people and to our country. It was at this point that opposition began to form. At the beginning the opposition wasn't specifically political but, let's say, ecological. We made a series of key demands, a series of principles that could be reduced to this: they needed to tell the whole truth about the extent of the accident and of its potential consequences; they needed to tell people the truth about the degree of contamination of the environment and of food products; they needed to tell the truth about the possible health effects both now, and in the future, over a long period, twenty to thirty years from now; and they needed to ask for help from the international community. And there was another series of demands made by the opposition. The system ignored them totally. Because they had no intention of telling the truth to their citizens, and they were not going to transform the Chernobyl accident into a subject of international discussion. They organised a good number of conferences, but they never gave any concrete figures. Western specialists came here, but our people told them lies. The Ministry of Health lied cynically to the people whose lives it was supposed to protect. This situation lasted for years. But the problem of Chernobyl grew; it was no longer possible to contain it. The situation in the country was changing. The number of people who were willing to speak freely was increasing. Political organisations were multiplying. Then finally, after three or four years, the State and the Communist Party began to realise they were losing ground. Of course, Chernobyl was being used by the opposition forces in the fight against the Communist Party. So, because they were the ruling party, the Communists understood that this was a very serious issue. They had lost their seats in many regions in the elections to the Congress of People's Deputies in the USSR. In the Ukrainian Parliament a third of the deputies elected to serve belong to the opposition parties. So the Communists also began to meet the demands of society, and revealed a series of facts and figures, and called for international aid.

Today, on the eve of the fifth anniversary of the disaster, we know much more about its consequences than in the earlier years when there was total secrecy. Government policy has changed in that regard. Furthermore, the government has acknowledged that it was a global accident of enormous proportions, the most serious accident of the twentieth century. They have also acknowledged something that they previously denied—that they cannot deal with the problem by themselves and have now appealed to the international community for help. The appeal was made at the UN forum by Ukraine, Belarus, members of the UN and the Soviet Union. So today, the problem of Chernobyl is out in the open. It has been presented to the international community, which had already begun to forget about it, in a completely different light. In the early years, especially in 1986, Soviet propaganda had claimed that we had dealt successfully with all the consequences of the accident, that everything was fine, there was no tragedy. This policy was maintained for three and a half years. People in the West were even surprised to see the issue back on the table. Why have you started to talk about Chernobyl again today? Why ask for help suddenly after three, four years? You didn't need help before? We had to explain that there had been a change in government policy. Thanks to the pressure from the political opposition.

Q.—Is the whole truth being told now?

—No. We still do not know who it was in the first three days that gave the order not to talk about the accident to the international community. Exactly who was it? Was it Gorbachev? Ryzhkov? The Politburo? The Council of Ministers? Who? We don't know. We don't know the circumstances that prevailed during those days.

—Maybe it wasn't anyone?

—You're right...

—Maybe it was just the system.

—It's possible that it was just the system. It's anonymous, it functions all by itself, it's a self-regulating system. It's perfectly possible. We set up a commission within the Supreme Soviet of the USSR to investigate all the circumstances of the accident. A similar commission was created in the Supreme Soviet of Ukraine. Ukrainian MPs have access to documents in Ukraine, but we will have access to even more crucial materials. I think some things will become clearer although some key witnesses, who knew everything, are no longer with us. For instance Shcherbina, Legasov...

—Who is Professor Ilyin?

—Professor Leonid Ilyin is the former vice-president of the Academy of Medical Sciences, director of the Institute of Biophysics at the Academy of Medicine, a very secretive organisation, and he chaired the USSR's Committee on Radiological Protection. So he is a scientist, a radiologist, and a medical doctor. Where is he from? He belongs to the military-industrial complex that produces atomic weapons. These doctors are sworn to secrecy in the same way as the engineers, physicists and chemists working in this area. They study the effects of nuclear explosions. They study the diseases of people working in these enterprises. All of them are covered by official secrets legislation. And Dr. Leonid Ilyin is a very brilliant representative of the military industrial complex, working on the medical side. He is a nuclear scientist in a white coat. Over the issue of Chernobyl, he proved to be extremely reactionary. There is evidence—entirely convincing—I saw the secret documents myself—that Leonid Ilyin and Professor Israel advised the Government of Ukraine and the Ukrainian Politburo against evacuating children during the first days of May.

—These are the official advisers?

—They are scientists. Professor Leonid Ilyin is a qualified scientist, an expert in his field. When we were investigating the responsibility of individual politicians, we asked who had advised them. In all states there are always advisors. A politician is not required to understand radiology. We know just how important this advice can be. Look at, say, Lysenko, the famous academic who advised Stalin. We know who Khrushchev turned to for advice and what resulted from it. So, these are the men who advise Gorbachev and Ryzhkov. So we need to look very carefully at the role they play. They play a very important role. And Ilyin's role today is so obvious and clear to us that our Green movement has declared him *persona non grata* in Kiev and in Ukraine. We won't tolerate his visits and we organise demonstrations against him.

—Does he still have influence?

—Certainly. He was awarded the title Hero of Socialist Work for the Chernobyl affair. It's Leonid Ilyin and his staff who set the norms for external radiation that people can, according to them, safely receive over a lifetime. The 35 rem[10] allowed during decontamination work, is the norm they set. It was they who developed this theory of 35 rem over a lifetime, a theory which has provoked passionate feelings and unheard of tensions in the contaminated areas where residents were not evacuated. It is a political theory. Ilyin has acknowledged that when he set these levels his calculations were not based on medical evidence but on political and economic considerations.

---

[10] Former unit of measurement for absorbed radiation dose by a living body (100 rem = 1 sievert). See Glossary.

—Isn't this limit of 35 rem the level recognized internationally for workers in nuclear power plants who have no direct contact with sources of radiation?

—Yes, exactly. Recently this dose, introduced by Professor Ilyin following the accident had been adopted in Great Britain but there, they reduced the annual dose from 1 to 0.5 rem per year. In other words, it's alright for children from Polessie to receive the same annual dose as workers in English nuclear power stations. With this enormous difference, the workers in England eat food that is absolutely clean and uncontaminated. They are subjected only to a short exposure to external gamma rays. The really serious problem around Chernobyl today is internal radiation. The two are not comparable. The Academy of Sciences of Ukraine, the Minister of Health in Ukraine, the Minister and the Academy of Sciences in Belarus are categorically opposed to this theory. And there are also many radiologists in Moscow who know this theory does not stand up to criticism.

—What is his theory and what is your opinion of it?

—According to Ilyin's theory, a man who receives 50, or even 100 rem will not suffer any consequences...

—Is that true?

—No. Because there is no strict rule like this in medicine. In medicine and biology, there is a very wide range of different individual responses. There is no single response. This is one of the reasons why it isn't true. It is also not true because we now have data on the action of low doses of radioactivity. At Sellafield (Windscale) in England, for example, at the nuclear reprocessing plant, there are data on the harmful effects of even one rem!

—What is this theory of 35 rem? To begin with, they ignore any previously absorbed doses and claim that over an average seventy year life span, people can receive 0.5 rem every year. They will accumulate the 35 rem dose over a lifetime. Thus, someone living in the contaminated territories, consuming radioactive products can continue living there and will not need to be evacuated as long as they receive no more external radiation. Beyond 35 rem, they need to be evacuated. This reasoning is totally abstract. It completely ignores another aspect. If the children who live in these peasant communities can't go into the forest, walk in the grass, swim in the river, drink milk from the cow, if they live in a sterile world so as not to exceed this threshold, if the land they live on cannot even be cultivated, because the level of radioactivity of the soil is too high, then all these arguments about 35 rem are completely absurd, it is not a normal life. And anyway we can't simply put aside the fact that there are regions in Ukraine where people received anything up to 10 rem in the first year, maybe 5 rem the second. They could have already accumulated an enormous dose in the two or three year period following the accident. It's a very different situation when a young man in good health goes to work in a nuclear power station—special medical measures are in place. Applying this reasoning about

35 rem to a child who has received this dose of radioactivity, it's unacceptable, it's simply criminal.

—What is the attitude of the International Agency of Atomic Energy? Do you share their view?

—We have a very negative opinion of the IAEA. We said this to Hans Blix, the director, and to Mr. Gonzales. They met the members of our Green movement here and we expressed all our discontent. Of course, they explained that the agency is there to serve the governments that fund it, including the USSR, so they can't interfere in the internal affairs of different countries. In other words, they are perfectly aware of the dreadful things that go on in these different countries, but can not say anything. Even so, we did ask them why they believed the lies told by those in power. Why did they never conduct in depth studies of the nuclear power stations that were still in use? Why have they still not made any assessment of the situation following the accident at Chernobyl? They sent specialists to collect soil samples and to assess the health of the population. We suggested they call in independent experts, people that we can trust, who have no connection with any nuclear mafia. They agreed to this in theory but so far I do not know of anyone we recommended that has been called to serve in their commission. There is currently in Ukraine and Belarus great mistrust, I would even say hostility, towards them.

—Do you think they are under pressure from the nuclear *lobby*?

—Naturally! It is a nuclear lobby in the truest sense of the term. It is an international nuclear lobby. That's why we believe it is absolutely essential to undertake independent assessments with independent experts.

—We can assume, for example, that they work closely with Leonid Ilyin's group?

—Of course! It's a direct link. They are colleagues... I'll tell you something else. Representatives from the nuclear industry, from the United States and France, have visited us here and many of them share Ilyin's ideas. It's understandable. They received their training in their own military-industrial nuclear complex. They think the same way. I asked a French professor what he considered the most important aspect of the accident, and he said. "It's very interesting! I could never have set up this sort of experiment in my laboratory, but now I can observe it". You can imagine the cynicism and behaviour of these people!

—So the people who live here in the contaminated territories are just guinea pigs?

—These are poor people, brought low by fear. They don't care what happens to them, it's their children they're worried about. These people don't know what to believe—because they have always been lied to. They are helpless. They don't even know what a neutron is, or a white blood cell. They are lacking the most basic information. You can't blame them for that. They are peasants living off the land. Now they feel threatened by everything around them. A threat that you can't see, hear or touch. And they are confused.

Chernobyl has shown that it is possible to destroy the world and life on earth in a totally peaceful manner. Dropping bombs or making nuclear weapons—it is generals and politicians who make that decision. It's their choice, their decision. But in this case, no one intended any harm. And yet if we look at the consequences and the scale of the disaster, it's proved to be a very successful way of wiping out humanity, destroying the genetic heritage of a nation, of many nations. It's a bomb launched into the future. It's very serious. The world has not really grasped it yet. As long as it doesn't actually affect anybody personally, it's very hard to understand. It's very hard to understand that nature can become hostile to man. Nature has become our enemy. The grass, trees ... The forest that has always protected man has become a threat, a danger. Water can also present a danger. For people born into this natural environment and who feel so close to it, it is very difficult to see it from this angle. Everything has suddenly changed. It's very serious.

We have to bear in mind constantly the enormous concentration of power in a nuclear reactor, which can combine with unforeseen circumstances. For example, Europe and the world do not know that two people, two firefighters who were on the roof, and to whom I talked later, rescued the whole gigantic structure, which makes up the Chernobyl nuclear power plant, from the fire. Because on the roof of the engine room where the turbines are located, the bitumen caught fire. And they extinguished the fire. Two people. If it had carried on burning, the turbines would have exploded. There was a lot of oil and hydrogen. It's hard to imagine what the consequences would have been for the whole world. In other words, a series of unforeseeable events can create a catastrophic event. Humanity should take a much more serious attitude vis-a-vis the accident at Chernobyl. It's a problem for the whole of mankind, but in particular for the highly industrialised areas. Because it is possible to imagine a sequence of events over which we lose control. If by chance damage occurs in an energy unit, for example, and is transmitted to a chemical unit, then the situation becomes uncontrollable. What happened here was the first real manifestation of a force beyond our control capable of destroying life on the planet.

The sarcophagus is a cause for alarm because of the way it was constructed. It was built under very difficult conditions, with the constant threat of high levels of radioactivity. It was impossible really to build a completely secure structure. For the moment it's holding. But there are a lot of cracks, a lot of movement. And inside there are 35 tonnes of radioactive dust, highly radioactive. If the building collapses, it will release a new cloud of radioactive dust with very serious consequences and there will be widespread panic. So there are problems with the sarcophagus. And they have to be resolved. How? Either we dismantle the whole thing—but this is something of which we have no experience—or build a huge sarcophagus over the existing sarcophagus, which would cost between 1.5 and 2 billion dollars. A sarcophagus that could last two hundred years.

If we lived in a normal European country, I would be optimistic. But the situation in the Soviet Union at the moment is so unstable that it is hard to imagine a future without bloodshed, violence, or totalitarianism. Democracy ... we are realising gradually that society needs to mature before it will become democratic. We have not reached that level of maturity. Our democrats lack culture. I had a conversation with someone today. They fight amongst themselves just like the previous politicians from the totalitarian era. Democracy is a self imposed inner discipline. We haven't got it. In the heat of the political struggle, the struggle for power, all these problems of survival have been forgotten. The old reactionary forces will fight for power by whatever means. At the moment they're trying to get rid of Gorbachev. In Moscow, the situation is very worrying...

That was November 1990. Today, the situation is a lot worse. Here is a report from the Novosti Agency in Moscow dated April 29, 2005:

"The experts are sounding the alarm. The sarcophagus covering the damaged reactor at the Chernobyl nuclear power plant is in disrepair, the walls are cracked, and the concrete roof has subsided, according to the daily newspaper *Troud*.

Rain and snow, falling onto the solidified nuclear fuel could provoke a nuclear chain reaction, say scientists. The walls of the sarcophagus are permanently hot, glowing can clearly be seen inside. If the roof collapses, tonnes of radioactive dust will be ejected into the atmosphere up to an altitude of 2 km, with fallout in Ukraine, Russia and neighbouring Belarus.

As early as April 2004 the Ukrainian nuclear scientist Valentin Kupny, the former deputy director at the Chernobyl power station, said that the reactor covering could collapse at any moment. The academician Dmitry Grodzinski who has been the general supervisor of the plant for sixteen years, says that the nuclear fuel inside the damaged reactor is getting hotter, and instruments have detected an increase in neutron currents and leakage of radioactive dust; there are 170 tonnes of nuclear fuel under the sarcophagus which is fissured with more than 1 square kilometre of holes and cracks. The structure was built very hurriedly and the concrete was laid without steel reinforcement. According to Dmitry Grodzinski, there are more than 800 deposits of radioactive waste in the area. There are hundreds of thousands of cubic meters of radioactive material. The material dates from the accident and was supposed to remain there only for five or six years. Now americium is escaping. It is more dangerous than strontium. The Pripyat and Dnieper River whose waters are used to irrigate the surrounding agricultural land contain strontium.

There are increasing numbers of mutations in the area. Piglets are being born, blind or dicephalous (two headed). We see monsters hatching, not chickens. More and more children with Down's syndrome are being born. Cases of childhood thyroid cancer are a thousand times higher than before the accident. Levels of radiation are now insignificant but the genetic instability is continuing—said Academician Grodzinski.

Today plans have been finalised for a second sarcophagus to cover the first. It will enclose the reactor for another hundred years. But it cannot be constructed just yet because the levels of radioactivity are too high and there is not enough money to fund the project, which will cost an estimated 750 million dollars".

3. ANATOLI VOLKOV

I came across Anatoli Volkov in the newspaper *Izvestia,* which took advantage of Gorbachev's policy to publish information that was almost subversive in the opinion of communists, painfully enduring *glasnost.* The article recounted the passionate struggle against official disinformation of a solitary man in the contaminated forests and fields of Belarus. He came up against the hostility of the local authorities, but gained the support of the local people and of the military that had been drafted in to "decontaminate" the territories.

We had little time to meet Anatoli Volkov, when I finally got hold of him on the telephone late one evening in Pinsk, 450 kilometres from where we were staying. The next day we were leaving Belarus to continue our work in the Ukraine. Travelling overnight, he met up with us at dawn on Monday 19 November at the station at Krasnopolie, a little town near the Russian border, about 250 kilometres north of Chernobyl. It was rumoured that when the radioactive cloud had passed over Bryansk, travelling in a north easterly direction, chemicals had been fired at it from helicopters (in a process called cloud seeding) to cause it to rain and, by so doing, avoid radioactive fallout over Moscow. Volkov knew this highly radioactive area. Right up to the beginning of 1990 the Moscow radio-protection service, directed by the eminent specialist Ilyin, had maintained an almost complete news black-out on the extent and the exceptional intensity of the fallout in the Bryansk region and on the way the exposed population had been forced to remain in situ.

Anatoli Volkov—The meeting had been fixed for the following morning. I got up early, at 7 o'clock, and within two hours I had taken measurements of the whole village. I went into the flood meadows, then here near the school, then to the petrol station which had the highest level of radioactivity. I transferred the information quickly onto a map. My map was ready. I arrived at the meeting well armed to demonstrate that the reality was very different from what they were being told.

Q—They didn't know what you were doing?

—Nobody knew. We went into the office of the president of the area executive committee. He was a good man, ready to deal with all of this. The person in charge of the decontamination brings me up to date. "We've already done all that, we've sent the report to the centre. Everything's OK here". Without saying a word, I unroll my map, read out the figures and go on the offensive. "What do you think you're doing?!" He says "But it's far away from the houses!" "The people here don't know where you've dug up the soil and where you've spread sand. They live here, they move around!" The truth was there in front of them: "You think you can solve the problem by lying about it. You're deceiving people. You have no right to do that!" I've sworn an oath to the people of Belarus and I will never betray them. "You can do what you like to me but I'm going to tell them the truth". The tractors belong to the people, the earth belongs to the people, the people paid for my education. How can I lie to them? I could never bring myself to do that. I told them how it was. That's all.

Then, one day, a secret service official contacted me out of the blue, a soldier from the army. He said: "Anatoli Grigoritch, is there anything you need? We have given orders that no-one is to cause you any trouble". Because people were already trying to intimidate me with phone calls, but I carried on with my work and these people gave me a lot of help with my measuring. No-one bothered me after that. I made radiological maps of the region.

—Was it local officials that put obstacles in your way, or did it come from higher up, from people like the scientist Leonid Ilyin?

—No. I don't think it came from Ilyin. It was just the political climate here in the Mogilev region. "Nothing serious has happened. There's no problem here, we mustn't alarm people, we're quite a distance from the reactor, etc". When I first arrived, Kostiukevich, the vice-president of the executive committee, kept a close watch on me. So I decided that there was only one thing to do and that was to invite him to the guest house in Cherikov where I was staying. It was already midnight when I unfolded the maps to show him. "Think about what you are doing! Why are you stopping me doing this? Just have a look at this". He listened very attentively and then he left. The next day there was another phone call. I was invited to the plenary meeting of the local party committee. The meeting was in progress. The secretary Leonov—a very interesting man—introduced me: "You need to listen to this man". They listened. The facts speak for themselves. All of them—after all, not everybody is an enemy of the people—were very worried.

—They were acting in this way out of ignorance?

—Simply, they did not understand the real situation. The meteorological service and all these people from Moscow like Ilyin kept telling them that there was nothing seriously wrong, everything was normal, that the local people would not be contaminated etc.

—How is it that you knew this area so well?

—I knew this area, the Polessie plain, very well because for a number of years I studied, for economic reasons, the effects of industry on ecology and sanitation in the Pripyat river basin, where the accident happened. I knew that it was a very vulnerable region, because there are a lot of marshes, a lot of underground water and surface water and great hydrometric mobility throughout the region. In addition this basin is directly linked to the Southern drainage basin, where 40 million people live, right down to the Black Sea, along the Dnieper where there are a series of dams at hydro-electric power stations. And naturally, because the water is so abundant, any contamination will move through the landscape very easily. Today the radioactive silt is being transported into the Dnieper by its tributaries, the Pripyat river, the Sozh, the Nesvich, the Iput', the Bessiad', the Braginka, the Kolpita and the Pokot'. The Kiev reservoir is a "time bomb". The water is clean but the silt 'glows'; 60 million tonnes of silt.

I studied this area for twenty five years. I knew about everything except the radioactive contamination. True, I had the data about radioactive contamination from global fallout before the accident but I was only interested in them up to a certain point. I was mainly interested in chemical pollution and its impact on the water supply in this area. When the accident happened, and it happened right in the middle of this area, nobody invited me here. However I knew that no-one knew this basin and these migration channels better than me. For example, the flood plains form a perfect route for migration: all migration will pass through them. The laws of geochemistry are immutable. They last for centuries. These laws will govern the migration of radioactive isotopes.

Of course, when I arrived, there was even a certain amount of incomprehension about my work. What was the point of all these survey reference stakes? Why bother to take measurements when the really urgent thing was the evacuation of residents and saving the animals? The evacuation of the inhabitants was very moving. They didn't understand what was happening to them. They all had this lost look. It was terrible. First, they loaded up the children, then the old people, and then the animals. They took them away…. At the same time, a helicopter brought me the survey stakes, threw them down, they were put in place, and I traced out my map. No-one was interested in the map, but I knew that it would come in use, that we would have a basis from which to work, that we would know where the fallout had been in the first few days, in the next few days, ten days after, and that it would make our calculations easier. I knew how important it all was.

—What do you think of the attitude of the USSR scientists?

—I have to say that many of the scientists showed an interest, in their own particular field—especially biologists, ecologists—we shouldn't underestimate it.

—But did they have any idea what had happened?

—Not at all. They had no experience of this problem. For that matter, nor had I. But out in the field, when I started to study every square metre of land, I realised the seriousness of the situation. I began to realise that not even the USSR Academy of Sciences had maps like the ones I had made. I showed them mine….

—Did they welcome you straight away as a colleague?

—Not at all. They completely ignored me. There were times in fact when they banned me from the area. Because I had started to go public with my information.

This area is very radioactive. I've measured all of it. Here, near the abandoned school where we are standing at the moment, it measures 200 curies. That's more or less the level around the Chernobyl power station. Over there they've got plutonium and other heavy substances derived from uranium, like strontium etc. Here, we've got caesium-137. The same caesium that they have around the power station. We discovered paradoxically that areas far away like this one, 200 or 300 kilometres from the power station are contaminated. That means there could be radioactive areas that we know nothing about. We've already lost four years. To protect the children, they removed the surface soil three times but children need space... they can't stay on a piece of ground the size of a pocket handkerchief. They move, they run, they go to the woods to get blackberries, they play ball, they go everywhere. You can't clean up a patch of land and say everything's alright. The whole place is dangerous. The school will stay empty for ever: 200 curies!

After I had measured the levels of radioactivity emanating from children's bodies with a spectrometer, I decided to tell the truth to everyone, to every doctor. The Polessie region is very rich genetically, with people of Polish, Belarusian and Ukrainian origin. The children are really beautiful. And there's no problem of alcoholism here. People have still got work here. We are in the West. When I saw all these children arriving in their thousands to have their levels of radioactivity measured...Their Mums had dressed them up beautifully, the little girls were so pretty with their dark eyes and ribbons in their hair... They took their turn on the spectrometer and I would read the verdict on the screen, "radioactive", "radioactive"... For me it was a real tragedy. I looked into their eyes and I could not bear the thought that we were lying to them, that we were saying nothing. It was a tragedy in the making and no-one was taking responsibility. I was very upset by it. It was at that moment, looking into their sad eyes—because those children knew they hadn't come for a party, that I made a pledge that I would never betray them and I would make the truth known whatever the price. And I said to the doctor "Make a thorough examination, see if there are any who are not contaminated, but above all, try to get these children away from the contaminated zone, and save their genetic heritage". Because we could lose everything. A day could come when there is no-one left, just monsters, instead of these beautiful people that have always flourished here.

A peasant woman who lives in the countryside here came to talk to me. She said: I get no pleasure any more from raising pigs. I take meat and sausages to my children in the town, but as soon as I have gone, they throw them out. My daughter-in-law throws it out straight away, my daughter waits till I've gone. My meat is no use to anyone. I never see my grandchildren here in the village. They don't bring them here. They're frightened. If I can't hear children's voices, then it means that there's no future here". This is what she said, this peasant woman, who can't read or write. She never hears children's voices now. "Before, the place was full of children, running here and there, in the gardens, to school, there was always the sound of their chatter. Their voices disappeared. Life stopped". Those were her words.

One day an old woman called me into her *khata (cottage)*.[11] She complained of constant headaches. In her stove which she stocked with wood from the forest the level of radioactivity was 300 milliroentgens per hour. All the forest in these contaminated areas is radioactive. The wood shouldn't be used for anything, constructing houses, making furniture, or even for firewood. The peat 'glows' as well. What can they heat their houses with?

I've been to the 'dead zone' several times (the zone within a radius of 30 kilometres of the power station where entry is forbidden) which has been transformed into a nuclear dump. Everything's been piled in on top of everything else—agricultural machinery, contaminated clothing, furniture. The abandoned houses "glow" like candles. When fires break out in the peat, it makes the problem worse. The smoke carries the radioactivity to other places further away. How do you fight against this sort of evil?

—There seem to be two attitudes. There's yours: all the people must be evacuated, no-one can live with even 1 curie, as you say: and then there's Ilyin's theory, which says the opposite, that not only can you live here, you can live a good life here.

—I'd like to see Ilyin live here himself. I know him very well. He's part of the mafia that should have been sacked a long time ago. They have really damaged our country, these Chazovs, Ilyins and the others. The country's medical profession has shirked its responsibilities. Only the military helped me because they saw the absurdity of the decontamination. I had this heavy heart. I kept saying "Be careful. Don't deceive the people, because they haven't got any power, they don't know anything".

—But while they were misinforming the public, they were also making huge efforts and spending enormous amounts of money on decontamination measures. What did you think of these measures? What did they do? What was the outcome?

—I want to emphasise again, that more than anyone else, it was the military that helped me in my work. Without reservation. The generals, the colonels, anyone who could, helped me without any discussion. I want to express my gratitude to them for that. They were in charge of the clean up operation, but they knew perfectly well how useless it was. They could see it achieved nothing. Right from day one I told all the assembled officers that it was useless. I said "Send a regiment, do your work, and afterwards I'll measure the levels. I'll show you that even if you worked here for a thousand years you would never achieve anything". It's a huge area, it's impossible to remove all the earth, impossible to carry it away. Burying

---

[11] Ukrainian peasant house. In Russian, *isba*.

it all somewhere else is just as dangerous because you are just concentrating the radiation in one place. It was only after two years that the military thanked me for showing them the uselessness of this work that had cost millions: 28 million roubles (the equivalent of dollars) were spent every year just in the area around the nuclear power station.

This useless work cost them their lives. Hundreds of thousands of contaminated "liquidators" are ill, tens of thousands have died and continue to die. In 1986, their average age was 33.

—What useless things did they do?
—Washing the roads. Then they would take up a layer of soil near the houses and cover it with sand. But a little further on, it was just like before.... They washed the roofs and the houses, the radioactivity flowed into the streams, then into the river. It accumulated in the silt. It became concentrated in other places. They would take down a roof, bury it somewhere, and the concentration of radioactivity would be in a new place. In short, someone had to show how absurd this work was and that's what I did. People understood and now they're not doing it any more.
—They buried the pines from the "red pine forest".
—I know the story of the "red pine forest" very well. Look, I've even got it marked here on my map. These were pines that were very close to the power station that turned red through the effects of the radioactive fire. I recommended doing nothing and letting nature do its work. But they brought in powerful machines from America, felled it all and chain sawed it. They dug trenches and buried the trees. This wood prevented humidity from reaching the surface layers and so there was a problem of sand erosion. There is sand all around here. This area is the delta of the river Pripyat, and it is a very large alluvial plain, there are even some sand dunes. Nothing grows any more. They tried to reforest the area, but the trees grow poorly because the humidity can't rise up through the buried wood. It wasn't the right thing to do.
—Is there a risk of contamination of underground water?
—The contamination could end up in underground water. There are large geological faults near the power station and it could end up in the underground water. We should never have interfered in the natural process. We should have left the pine forest as it was. Nature would have sorted itself out. This area should be reforested as soon as possible to avoid any further wind erosion. In April and in October, we have very powerful storms. Cars have to have their headlights on because of the dust. So of course the radioactivity is dispersed over long distances and contaminates other areas.
—Are the ditches where the radioactive waste was dumped dangerous?

—It wasn't properly thought through. In countries that have nuclear power stations, nuclear waste dumps have stone foundations, or a concrete or lead base, etc. Here, we simply selected large natural ravines and threw all the waste in there. From there, the radioactivity can filter through. In the Polessie region the soil is very light and the water table is very close to the surface so contamination is possible. So yes, the ditches present a danger too.

We absolutely have to tell the whole truth. Our newspapers always reported that 50 million curies had been emitted. The same figure as the IAEA. But it was around a billion curies! A billion curies! Nothing like this has ever happened before anywhere in the world. Even during the most serious accident in Great Britain, only 20,000 curies were emitted. There's no comparison.

—It's because the USSR didn't want to acknowledge the full extent of the disaster to the rest of the world?

—Of course. The figures speak for themselves.

At that time neither I nor Anatoli Volkov knew that in August 1986, at a conference held in Vienna behind closed doors[12], the West had forced the Soviets to divide their estimates of the health consequences of the disaster by 10.

Q.—They sacrificed their inhabitants.

A. Volkov—That's what I believe. They wanted to allay the so-called "radiophobia" by building new houses, farms, take people's mind's off the subject in whatever manner they could. But I had made all these measurements. I could see. Eventually, I couldn't hold out any longer and I published the map. And when the local people saw the map they said: "Get everyone here now, meteorologists, all the best brains, all the people responsible and bring Volkov. The people will decide who is right". I came but no-one else did. I showed them my map and I said: "Take the map—I'm giving it you for free. It cost me three months work, I'm giving it to you". Later, my data was recognised. They closed the *sovkhoze* (Soviet state run farm). They began evacuating people. So, step by step... The situation is completely different now. I can say this in all confidence because I am a Member of Parliament. I was elected. People voted for me because I behaved honestly. At Chernobyl, no-one knew me, and I only learnt... that I had been elected deputy over the phone. When I became deputy I made the same pledge—that I would not go back on my word. I would only speak the truth.

—Do you think the scientists who come from abroad are objective, or do they represent the interests of the nuclear industry?

---

[12] See Chapter VI, p. 65

—I've worked with the IAEA. They are highly qualified specialists but they represent the agency. They cannot deviate from the agency dogma. There's no doubt they are high level experts, but this is a completely new situation for them. Just imagine a reactor stuffed to the brim with 192 tonnes of radioactive fuel burning for ten days! And then tonnes of lead, gravel and sand is thrown in on top. And then all that evaporates and is ejected into the atmosphere again. We artificially increased the radioactive emissions. And no-one says a word. But I've seen it all, on the surface of the soil near the reactor, in the power station, and throughout the whole area. More than 300 artificial radioactive elements were expelled from the reactor. They did their dirty work and many have now disappeared. All that's left are the long lived isotopes. Science is mute today in the face of this complexity.

—Foreign scientists too?

—The scientists are saying nothing. They've never encountered anything like this. They measure the levels someone has accumulated so far, how much external radioactivity there is, ask them how they feel and say "You can live here".

—In your opinion, is science capable of dealing with the consequences of an accident on this scale, independently of any interest or pressure?

—The pressures are enormous at present. I'll give you an example. I met Hans Blix the director of the IAEA, in an official capacity at the Supreme Soviet. I asked him, "Tell me, are you aware, as director of the IAEA, that 17 million people, Russians, Ukrainians and Belarusians, including 5 million children are living in the contaminated areas? In Belarus alone, we're talking about 2 million people, including 500,000 children". He told me that until 1989 he knew nothing about it. How is it possible that the director of the IAEA did not know about it immediately? He told me: "I only knew about the technical aspects of the accident". They always pretend they know nothing about it.

You're asking me if it is possible to deal with the problem scientifically. I think the answer is yes. If all the scientific capability of the world, all the honest scientists came together, and if we created a study centre here. It would require an enormous amount of work, in depth research. At present, we have scarcely begun to look at the problem and the little we have achieved is frankly pretty mediocre. Whereas what is needed here is scientific research in all fields, research on humans, in agriculture, detailed knowledge of the radioactivity in each area, and to draw conclusions from it for the whole world. No-one will give us the necessary finance. We need the technical equipment, we need to buy the best in the world. We need money. Here in the Soviet Union, no-one will finance us.

—And no-one abroad either?

—No-one else will; it was Hans Blix who told us that. Because we are one of the great powers with a large army, and money goes to developing nations. Today, no-one will give us anything. But we mustn't lose hope.

—Do you think Chernobyl ushered in a new era for humanity?

—Humanity has not understood. We haven't understood that this could be the end of everything. If the fire in reactor 4 had not been put out and if reactors 3, 2 and 1 had caught fire, Europe would have been uninhabitable. The whole of Europe would have been covered with radioactive material, and there would have been nothing we could do. That's where nuclear energy can lead.

We have lost all these territories and we won't be able to return there for many years. The people here have lost all their "joie de vivre". Before, you know... Belarus was a really beautiful country... there was the hay making, people extracted the juice from silver birches to make a drink, it was a great life. The Belarusians are a very interesting people. Now they're no longer happy. They're afraid of everything. The soil is radioactive, the potatoes, the wild strawberries as well....they're frightened of everything. Perhaps there is a general consciousness, as you say, beginning to emerge. If this poison covered the whole of the world, what sense would life have?

—Are the foreign scientists that come here aware of this anguish?

—It's as if they've come from another planet to observe us. We feel like hostages here. All these experts draw conclusions, but not ones that are any use to us. We need concrete evaluations. Their conclusion is simple—there's nothing so terrible here, we can carry on living here. As a deputy, I put in a request to the Supreme Soviet, asking that the IAEA experts' conclusions be discussed in detail by the Supreme Soviet. Because the IAEA exerts a great influence. If it presents its findings at the United Nations, it will be very difficult for us to prove anything to the contrary. They don't say how things are. These experts have not done any research. They are defending the interests of the nuclear industry.

—Can modern science meet the challenge that has been presented to humanity by the disaster at Chernobyl?

—I don't think so. We're not ready yet. And yet only science can provide the answers. It's here that people need to work, to listen, to see the people's suffering. Then at least, the scientists would understand. No country can face this problem alone. We need to mobilise everyone to come and work here, immediately, so that the international community as a whole realises what has happened here.

—Who would supply the money?

—I think there are honest people in the world. They will give what they can. I do believe that. That's why they need to be informed. But if no-one hears or sees anything, no-one will ever know. We need to use this accident to help the whole world. If we treat Chernobyl too lightly, it will be the end for all of us.

—Don't you think that the scientific world should admit that it is unable to give unequivocal clear answers that can be understood by all, to the challenge represented by Chernobyl, and undertake research to meet this challenge?

—We need to study the whole issue in depth. Starting with humanity itself. It should be a lesson in what not to do. A lesson about the risks of nuclear arms and of civil nuclear energy for people and for the ecology of the planet. You say quite rightly that the environment that has been created here is entirely new. Man can't adapt to it. We struggle as best we can. We can't go and live on the moon. We are condemned to living here. What can we do? We have to find a solution. Maybe the people of Belarus should go and live in the other "clean" republics, abandon this area. Everyone living in areas contaminated by 30, 15, 10, even 5 or 1 curie should leave. The area should be reforested and then abandoned: but having said that, the area needs to be looked after in ideal conditions, with its rivers and forests, like a botanical garden, and research needs to be undertaken here. We can't just create an enormous nuclear waste dump open to the skies in the centre of Europe. If we polluted it, if our actions have caused the land to become ill, it's our job to nurse it back to health. I mean invest everything in it.

I'll say it again: there is no-one in the world that really understands the situation. No-one has access to all the information about the problem, in all its complexity. On the other hand, there are research centres across the world with enormous wealth...

As I listened to this desperate appeal by Anatoli Volkov, I was unaware that two other solitary and extraordinarily competent men were launching themselves in the same struggle: the physicist Vassili Nesterenko, an expert in radioprotection and Yury Bandazhevsky, the youngest doctor of pathology in the USSR, whose work would create a nightmare for the nuclear industry and would lead to his imprisonment. The two men had not yet met each other. They were destined to come together some years later and join others from the West, to oppose the cover up and the deceit surrounding the disaster at Chernobyl.

Chapter V

THE STRATEGY OF IGNORANCE

"The environment that has been created here is something completely new. Human beings cannot adapt to it. We struggle as best we can. We can't go to the moon; we have to live here. What else can we do? We have to find a way out. We need to study the environment and do research which is vital for the international community…" This was as much as Anatoli Volkov was calling for at that time. But back then in November 1990 neither of us was aware that in the course of a meeting held in Vienna between 25th and 29th August 1986—or in other words, only four months after the explosion at the Chernobyl nuclear power station—the West had already settled on the "reassuring" figure of 4000 as the number of deaths that were likely to result from the accident,[13] rather than the 30–40,000 initially put forward in the report presented by the Soviets. The Soviets had come under pressure to divide their predicted death toll from the accident by ten: 4000 extra deaths from cancer were "acceptable" for the Western experts. It meant they could dispense with any serious scientific research into the health consequences of Chernobyl.

Shocked by the disaster, the representatives of the "Evil Empire" had apparently been sincere in their analysis of the situation, but the figures they were putting forward would have sounded the death knell for the nuclear industry. The strategy of ignorance was set in motion. Even today, the minimisation of this "ordinary accident, just like any other" is cloaked in a semblance of scientific credibility. Apparently there are no ill effects on health: "[…] from the point of view of health generally, positive perspectives should be adopted as regards the future health of the majority of people (around Chernobyl)".[14] This explains the complete absence of any initiatives coming from WHO in the five years following the accident. Their *experts*, sent periodically to the USSR, "validated", in their pseudo-scientific reports,

---

[13] A figure that was repeated in the joint WHO/ IAEA/ UNDP communiqué, issued on 5th September 2005 to mark the 19th anniversary of the disaster.

[14] "Exposures and effects of the *Chernobyl accident*" (Annex J), 49th Session of UNSCEAR, Vienna 2–11 May 2000. This statement was repeated at the Kiev conference in June 2001 by Norman Gentner from UNSCEAR; see the film *Nuclear Controversies*, W. Tchertkoff, Feldat Film, (*https:// www.youtube.com/watch?v=MZR_Fvp3RrQ*).

the minimising of the consequences of the disaster and the reduction in the number of evacuations of inhabitants from the contaminated territories.

Professor Legasov, the man who had flown over the burning reactor with Nesterenko, was the Soviet representative at the meeting which took place behind closed doors at Vienna. The minimisation of the figures was imposed against his will. He committed suicide in April 1988, on the second anniversary of the disaster at Chernobyl.

In August 1986, the Soviet scientists gave up their independent judgement and submitted to the West's political thesis. Based in ignorance, a dogma was established from which the nuclear agencies constructed an apparently scientific version of the truth and imposed it from their position of authority.

Based on what logic?

1. A SCIENTIFIC ARTIFICE

The argument that is put forward to deny any link between the new diseases observed among the inhabitants of the contaminated areas of Chernobyl and the chronic low dose internal radiation that they receive every day through their food, is that this causal relationship is *impossible*: it does not fit with the correlation between levels of radiation and the pathologies found in the Japanese survivors of the atomic bombs, who were exposed to *very high external doses of radiation* at the moment of the explosion. Compared to the doses at Hiroshima, the doses at Chernobyl are *insufficient* to cause these diseases. This has become a dogma, an accepted truth, an axiom, a postulate, a matter of principle that is not open to discussion.

But to explain Chernobyl today, using Hiroshima as a model, is an evasion of the truth, a subterfuge that has no scientific basis. It is a mistake to compare these two disasters. The two events and the mechanisms by which they cause us harm are different.

There was no atomic explosion at Chernobyl. There were two thermal explosions within seconds of each other and there was a fire that burned for ten days. Background radiation around the power station today is low. However, enormous quantities of artificial radioactive elements were ejected by the thermal explosions and were dispersed over great distances by wind and rain. Some of them will disappear only over the next few centuries: caesium-137 and strontium-90 in three hundred years; plutonium-239 in two hundred and forty one thousand years. These elements contaminate the environment, plants, animals and human beings. They have destroyed the health and life of hundreds of thousands of young *liquidators* who ingested or inhaled radioactive particles while working around the power station, and they will contaminate future generations, the descendants of the inhabitants, who are becoming more and more ill as a result of their consumption of radionuclides over the last twenty-eight years. The nature of this event is not comparable to what occurred in the two Japanese towns.

In order to understand how it differs from Hiroshima we need to know something of the parameters governing atomic explosions. For example, there are a number of variations depending on altitude:

- underground explosion (no material ejected into the atmosphere);
- surface explosion (the most polluting, giving maximum fallout);
- low altitude explosion, at a height from the ground, equal to the radius of the fireball, resulting in local fallout;
- medium altitude explosion (at Hiroshima—600 meters) at a height greater than the radius of the fireball. Little local fallout[15];
- high altitude explosion, well above the radius of the fireball. Global fallout only.

It should also be understood that during an atomic explosion the temperature can reach 100 million degrees Centigrade, in other words much hotter than the sun, whose surface temperature is 6000 degrees. After this type of nuclear explosion at an average altitude (Hiroshima), most of the radioactive elements are sucked up very rapidly in an updraft—at twice the speed of sound—into the stratosphere, and are then dispersed from the mushroom cloud over the rest of the globe with minimal vertical fallout at the blast site, compared to Chernobyl[16].

---

[15] However, in Hiroshima a large number of deaths and serious illnesses were recorded in people who arrived in the city from other areas and who were not directly exposed to radiation during the explosion. One possible explanation is the effect of contamination by residual radioactivity from neutron induction, created by the explosion of the bombs: all the rubble, the ash, even the stones become temporarily radioactive. In addition, M. Horio and T. Kikushi reported in "Radioactive dust from nuclear explosions" (in *Bulletin of the Chemical Research Institute Japan*, July 1955) that fission products had fallen in the western part of Hiroshima on the day that the bomb exploded and that at Nagasaki, a radioactive dust obviously containing certain fission products covered the whole area of Nishiyama. (quoted in *Atomic Park*, Jean-Philippe Desbordes, Actes Sud, 2006). The lies and the cover up began in 1945. In addition to the ban on publishing any personal testimony, the U.S. occupation forces, put in place the ABCC (Atomic Bomb Casualty Commission), an organisation which collected information, in the utmost secrecy, about the effects of radiation on humans, in particular, survivors of irradiation (*hibakusha*), without offering them any medical treatment. After the Americans left in 1952, the organisation was renamed RERF (Radiation Effect Research Foundation) and the work was continued by Japanese scientists. (cf. *Hiroshima, 50 ans*, Autrement, 1995, p 80).

[16] Description by Maurice E. André, former officer of the NBCR (nuclear, biological, chemical, radiological) working exclusively for the Belgian Air Force.

2. PROXIMITY EFFECTS

What is the principle or the basic mechanism of radioactivity? Radiation loses energy as it passes through a material. It is this energy, absorbed into the material, which is measured to evaluate the level of radiation, or *absorbed dose*. It is measured in rad (radiation absorbed dose) or gray (1Gy = 100 rad). The Hiroshima model, created by physicists studying the effects of *uniform acute external high level doses of radiation on the whole body*, is mathematical, reductionist and simplistic. It applies exclusively to the concept of dose as applied to a mass of 1 kilogram or more. It cannot be applied to biological mechanisms (which have not been studied) caused by low internal exposure doses, or isotopes, or hot particles, incorporated at the microscopic (histological) level in the cell. This model denies what it has not studied because it is incapable of apprehending it within its own parameters. At present the experts refuse to consider any new models: they block the way to the *validation* of alternative research and they continue to deny the evidence of a health catastrophe in the areas contaminated by Chernobyl. The difference between the two "phenomena"—Hiroshima and Chernobyl—is that in the case of internal low dose contamination (Chernobyl), it is not the value of the "dose", but the direct proximity effect of the material concentration of radioactive elements in the body tissues that must be taken into account. It is not the job of a physicist to describe processes at this level but of a pathologist. An honest physicist would simply acknowledge the need for this research. This is how Maurice E. Andre explained the difference between the two approaches in an article published in 1976, ten years before Chernobyl:

> The technical aspect, which will be developed below, reveals that a tiny speck of plutonium, a micron in diameter (a millionth of a metre), causes death by simply lodging in the lung. This speck will deliver 100,000 rad per year to the area of the lung surrounding it, a tiny area, limited by a sphere with a diameter of about a tenth of a millimetre, with the speck at its centre.
>
> But first I need to explain the trick used by pro-nuclear scientists to mislead scientists from other fields and the general public. Before examining that calculation, I will illustrate its falseness by means of an example where the false logic is more obvious. Here is the example: it is possible to assert that a bullet is not dangerous. All that is needed is to make an abstraction of the point of impact (which obviously absorbs all the kinetic energy of the projectile) and to hypothesise that all the kinetic energy of the bullet is absorbed by a larger area—for example the entire surface of the body—in which case, it can be shown that the flesh will not be ruptured. In this example, it is very easy to spot the false logic which fails to take into account the real situation in which a bullet hits a very precise area rather than the whole body or a whole organ. The bullet ruptures the flesh because it concentrates all its energy in a small area: given an equal amount of energy, the smaller the area, the more certain the rupture.

In the case under review, it is seriously misleading to the public to employ a calculation in which the energy released by a speck of plutonium in a given time is supposedly spread throughout the lung whereas in reality it is concentrated very precisely in a tiny area of the lung and is therefore extremely dangerous and can cause death.

For the benefit of non-scientists, a speck of Pu-239, a micron in diameter, lodged in the lung will damage the tiny sphere surrounding the speck because it is being bombarded approximately every minute (1414 times every thousand minutes, to be exact) by alpha-emitting radionuclides travelling at about 20,000 kilometres per second.

Under continuous attack the body is unable to repair the affected area, however small. It is as if one were to ask some builders to construct a house around someone with a machine gun who is firing a bullet once a minute, without any warning, in any direction.

In this analogy, the builders are the biological material that is drained towards the destroyed area to make repairs, and the house under construction is the area of the lung that needs to be repaired. The radioactive speck is the machine gunner firing at regular intervals continuously for a number of years (a particle of plutonium has a half-life of twenty-four thousand years, a very long time in comparison to a human life, and will continue to emit radioactivity at the same rate up to that point). This type of continuous intense bombardment may be on a very small scale but this makes no difference to the fact that it will, come what may, lead to cancer of the lung.

It is on this basis that we can say that local and repeated radiation is damaging and leads to the development of necrosis. No matter how small, if an area has been subjected to intense ionisation for a sufficient time, cancer will proliferate throughout the whole body. It is in fact the body's reaction to the exhaustion it experiences from trying to repair one very specific site that has been destroyed innumerable times.

The phenomenon we are considering is, in reality, completely different from the case of total radiation (brief and intense, resulting from the explosion of an atomic bomb or from accidental exposure to a significant amount of radioactive material through inadequate protection) when all regenerative ability is exhausted (100% mortality) which occurs if the human body receives more than 600 roentgens.

Yet even today we still hear *experts* quoting figures that relate specifically to the brief intense total radiation (experienced at Hiroshima, and based on old military statistics) when talking about the dangers of nuclear power stations. The radioactive contamination of food and of the atmosphere by the nuclear industry, as well as the global fallout that has resulted from *peaceful* (or otherwise) nuclear tests, has a completely different effect on the body.

We are being exposed to intense localised radiation, chronic and internal, sometimes damaging a specific organ (localised effect) where certain radioactive elements concentrate (for example iodine in the thyroid gland) and sometimes damaging the whole body and originating in the pulmonary area (through the inhalation of dust from the atmosphere into the lungs or through a wound). The familiar tables of statistics, varying from 25 to 600 roentgens, cannot be applied in the case of this type of chronic radiation. The threat to the public today from the nuclear industry comes from this insidious contamination, constant, permanent, cumulative, chronic and generalised. The wind can carry radioactive dust 1000 kilometres in twenty four hours, and no-one can be sure they will not breathe in "their" speck of plutonium or some other radioactive gift. This attack, and it really is an attack—a serious act of aggression, is, as I have said, extremely insidious because it is painless and consequently undetectable. It is therefore an attack on every one of us, wherever we are, and whatever we are doing. The nuclear industry's war against the human race began several years ago and urgent measures are needed to rid ourselves of it for ever, or we are condemning ourselves to death. Some people will find what I am saying hard to take, but we have to be realistic. We will act because we will be forced to. Can the truth really be swept under the carpet just because those in charge of these huge capital sums have no thought for future generations or even, for that matter, for their own real interests in the here and now?[17]

3. ORIGINS IN THE MILITARY

The ignorance that is described in this chapter was not deliberately conceived as such in the beginning. It resulted from the outcome of the Second World War and the balance of terror between the two Cold War superpowers. In July 1945, the US acquired nuclear weapons (the Manhattan Project), spurred on by the fear that Nazi Germany would develop them first. Three bombs were conceived: the first was used as a test to confirm all the principles of the atom, and the other two were unleashed on Japan. The USSR began their own identical nuclear programme to match the military might of Washington, and on 29th August 1949 to the great surprise of the Americans, revealed their nuclear strength, with an atmospheric test[18]. And so the arms race was set in motion, based largely on reciprocal feelings of extreme threat. It is no coincidence then, that five years after the first nuclear massacre in history, the Americans began to reconstruct and study the doses received by survivors from Hiroshima and Nagasaki at the moment of the explosion. To ensure their supremacy, they needed to perfect their arms and understand the effect on the enemy and the risks posed by nuclear testing. But the Americans only did half of the scientific work needed to assess the effects of an atomic explosion on life. They focused exclusively on the military point of view: virtually all their research is limited to the consequences of a war scenario. They did not concern themselves with what happened in the hours and days following an explosion, nor with the fate of people far away who were subjected to the supposedly "low level" global fallout. Rosalie Bertell notes:

> They wanted to know how many people would be killed rapidly and how many would be disabled from fighting. These were the main thoughts of the people who did the research, and these were the main calculations that they made. They were

---

[17] Maurice E. André, *Études et Expansion*—Trimestrial, N°276, May-June 1978.

[18] See Jean-Marie Collin, " *L'atome militaire* ".

not concerned with miscarriages, nor foetal deaths, neonatal deaths; they were not concerned by sick people or children, nor by many of the long term effects which were not tabulated. They were very selective in their research, and in the damage they admitted.[19]

This is how the "Hiroshima model" was established and limited to the effects of an instantaneous nuclear flash. For sixty years it has been the only official model in existence to recognise pathologies caused by radiation. It only takes into account the flash of gamma rays, followed a fraction of a second later by intense heat and a violent wind, which in themselves have nothing to do with radioactivity. At the same time, we remain in the dark about the effects of inhaling and ingesting the tonnes of radioactive material dispersed around the planet, initially by the two bombs dropped on Japan, then by the 500 atmospheric nuclear weapons tests, by Chernobyl (where more than 400 artificial radioactive elements were expelled during the fire at the power station) and finally by the so called "depleted" uranium weapons (with a half-life of four and a half billion years) used in modern warfare.

> I tried to indicate that the number of radiation victims is now conservatively estimated at 32 million people. We are talking about workers, we are talking about the Japanese population, we are talking about the victims of above-ground nuclear weapons testing and the various accidents and incidents that have occurred. Chief among them is the Chernobyl accident, which was a huge disaster.[20]

The bomb at Hiroshima and the fire at Chernobyl are not comparable. One does not explain the other. The first case involves the very high external exposure to the flash of immaterial gamma rays, and the second involves the proximity effects of microscopic particles incorporated into living tissue. But the second case does not exist for "science". It is not part of the approved science. It is not published in *validated* reports. It falls outside the parameters of the "Hiroshima dogma" and thus the worldwide nuclear consortium that unites the Pentagon, the Security Council, the IAEA, UNSCEAR, WHO, CEA, AREVA COGEMA, CEPN and SIEMENS, can claim

---

[19] Rosalie Bertell—*Permanent People's Tribunal on Chernobyl*, ECODIF, 1996. Epidemiologist and researcher specialising in radiation, Rosalie Bertell is Coordinator of the International Medical Commission on Chernobyl (IMC—Chernobyl) Toronto.

"The Permanent People's Tribunal (successor to the Russell Tribunal) is part of a legal tradition intended to give a voice to those who have never been allowed to speak. For 50 years, victims of radiation have been forced to remain silent: those from Hiroshima, from Bikini, just like those from Mururoa, the soldiers forced to watch the nuclear tests, and then dying a slow death, like "human guinea pigs", made to absorb plutonium so that its effect on the body could be studied..." Excerpt from the Preface of the Proceedings of the session on "Chernobyl" (Ed. ECODIF, 1996). This session was held in Vienna in April 1996, when the International Atomic Energy Agency ended its negationist conference, "One decade after Chernobyl", on the other bank of the Danube.

[20] Rosalie Bertell, op. cit.

that at Chernobyl *"there is no correlation between the pathologies observed and radioactivity"* (Gentner—UNSCEAR, Conference at Kiev 2001). For the experts who uphold the "accepted scientific view", only the very high doses of gamma radiation during the flash are pathogenic. According to them, at Chernobyl, *a priori*, there can be no correlation because the levels are too low.

Today the Cold War cannot be used as a justification. With the evidence of the health catastrophe at Chernobyl, ignorance has become a criminal strategy and a lie by omission on the part of the civil and military nuclear lobby that has subjugated the UN agencies responsible for the atom and the health, for its own ends. As for the nuclear states, whose ministers are not necessarily *experts*, they find, within the UN agencies, the scientific backing they need to legitimise their suicidal political decisions.

Chapter VI

INSTITUTIONS OF IGNORANCE AT CHERNOBYL

The ICRP has its origins in the International Committee on X-Ray and Radium Protection, set up in 1928 to develop regulations for radiologists and technicians to ensure the safety of both doctors and patients during examinations. It was only in 1950 that this committee became the International Commission on Radiological Protection (ICRP) that we know today.

The current mission of the ICRP is defined as follows on its website: "The recommendations of the ICRP are not mandatory but provide a benchmark at an international level, because of their scientific value, and their cautious and realistic approach to the problems of radioprotection. As a result, the ICRP's recommendations have provided guidance for the establishment of regulations adopted by large international organisations, such as the United Nations (UN), the World Health Organisation (WHO) and the International Atomic Energy Agency (IAEA), as well as to the European Community and many other countries. The authority of the ICRP explains the international character of radioprotection measures, which, because they are based on the same recommendations, are similar in Europe, in the United States, in Russia, in China, etc [...] The methodological approach was initially based mainly on health data, but today needs to consider also technical, economic and social issues: it has become multi dimensional". (ICRP Publication 60)

But we must not forget that the origins of the ICRP are not medical but military and American. Set up in 1950, little is known about the history of this "non-governmental organisation, though recognised by the United Nations" [sic], behind the façade of independence that it presents. Its history also begins with the bomb. It dates from the era of American research at Hiroshima and Nagasaki.

This organisation was born out of the secrecy of the atomic weapons and national security, therefore secrecy is embedded in the formation of this organisation. ICRP consists of a main committee which elaborates all the definitions and takes all the decisions. From 1954 to 1991 the members of this committee were 13 men, since then one woman (Professor A.K Guskova) joined[21]. They are self appointed and self perpetuated. They are the ones who make the recommendations for radiation protection standards, which have then to be adopted by every nation. They are the source of the regulations applied by the IAEA, indeed applied in a very cruel way after the Chernobyl accident and elsewhere.

It is very important to look in detail at the documents of the ICRP. I was very taken aback that in their 1990 report, they actually talked about "transient effects of radiation", effects that they didn't consider serious enough to receive compensation or to be recognised. But these are exactly the kinds of problems that people experience which they are trying to present to the world's opinion. The existence of these problems is continually denied by the IAEA.

Meanwhile the ICRP who recognised them as "transient effects", has stayed very quiet in the background, because their professional credibility is on the line: they can't honestly say that these effects do not occur and are not related to radiation. The engineers and physicists at the IAEA are the ones who are mandated to talk about health effects.[…] the IAEA, which acts like a police force irrationally enforcing mechanisms which were already in place.[…] Chernobyl is presented as a scientific problem, but this is not so. It is fundamentally about repression, about political decisions that are causing the appalling problems that we are seeing now. […]

Since 1951, the myth that you can't study low level radiation has prevailed. 1951 represents a very significant date: that of the opening of the above ground nuclear weapons test site in Nevada.[…] At that time there was a concerted effort to declare that low level radiation wasn't harmful and that there was no way to prove that any effects were connected with them[22].

1. AMERICAN ORIGINS OF THE ICRP

Below is an abridged version of a text by ECRR (European Committee on Radiation Risk) that details the origins of the ICRP in the American military[23].

In 1946 the US government, having tested the bomb and used it on Japan, clearly recognized the sensitive nature of nuclear science. It outlawed the private ownership of nuclear materials and set up the Atomic Energy Commission (AEC) to administer the area. At the same time, it established the National Council on Radiation Protection (NCRP), immediate forerunner of the ICRP.

There is now ample evidence that the NCRP was under pressure from the AEC to fix exposure limits which would not cause blocks to research and development.

---

[21] See Part Six, Chapter II, page 417, the Kiev Conference, filmed in June 2001

[22] R. Bertell, *op. cit.*, statement to the Permanent People's Tribunal, during its hearing on the Chernobyl disaster, in 1996.

[23] ECRR, Ed. Green Audit, 2003. ISBN:1 897761 24 4, p. 29 and following pages.
http://www.euradcom.org/2011/ecrr2010.pdf

The NCRP was formed by reviving the US Advisory Committee on X-ray and Radium Protection that the medical profession had originally established to provide itself with advice on radiation protection. Now that there was a new source of risk involving the military, government, and private companies with research contracts, it was clearly necessary to rapidly set up a body with sufficient credibility to claim to be the ultimate authority on radiation risk. There was a pressing need to revise the existing limits for exposure to X-rays and extend these to the new risks from external gamma rays which resulted from weapons development research and nuclear bomb test exposures. There was also a need to develop exposure limits to internal radiation from the host of novel radioisotopes which were being discovered, produced and handled by workers, and discharged into the environment.

The NCRP had eight sub-committees looking at various aspects of nuclear risk, but the two most important ones were Committee One, on *external* radiation limits chaired by G. Failla, and Committee Two, on *internal* radiation risks chaired by Karl Z. Morgan. Despite the fact that the NCRP had set an acceptable limit for external exposure in 1947 (at a level 20 times higher than that accepted for workers today), it was not until 1953 that the full report from the NCRP was published. The reason for this delay was that Morgan's Committee Two was finding it very difficult to agree on values and methods which could be easily applied to determining the doses and risks from the many radioisotopes which could become sources of *internal irradiation to organs and cells within the body*. Part of this difficulty had to do with lack of knowledge at the time of the concentrations and affinities of the radioisotopes for the various organs and their constituent cells. (This was subject of nine years of research by Professor Bandazhevsky in the contaminated the territories of Chernobyl[24]; *author's note*.) The NCRP became tired of waiting for a resolution of these problems and in 1951, its executive committee summarily ended Committee Two's deliberations and insisted that its report on internal emitters be prepared for publication, possibly on the basis that some guidance on risk was necessary. (At this moment, the Nevada atmospheric test site was opened, *Author's note*.)

This was the time when the radiation-risk black box was sealed up. Its internal workings had been constructed under pressure for the rapid development of some convenient methodology for defining exposure. The model cannot deal with small volumes and inhomogeneities of dose[25] and for this reason, is unsafe to apply to internal irradiation.

But the problem today is that this is the black box for radiation risk which represents the model used by the ICRP. It developed out of the NCRP. The Chair of the NCRP, Lauriston Taylor, was instrumental in setting up an international version of the NCRP, perhaps to divert attention from the clear evidence that the NCRP was associated with the development of nuclear technology in the US and also perhaps to suggest that there was some independent international agreement over the risk factors for radiation. The new body was named the International Commission on Radiological Protection.

---

[24] See Part Three, chapters I, II and following, p. 164.

[25] Yury Bandazhevsky carried out research at a cellular level (*small volumes*) and has shown that the concentration of radionuclides differs from one organ to another (*inhomogeneities of dose*).

So the circle was complete. It is this sealed *black box*—the Hiroshima dogma— in which remain imprisoned the independent researchers and victims of the cover up. The ICRP, which in 1928, acted exclusively and uncompromisingly on behalf of doctors and patients, with human health as its absolute imperative, was subverted during the fifties by the political and economic constraints of war and the nuclear industry. *It became multi dimensional.* Its independence was no longer credible.

Taylor was a member of the ICRP committee and the NCRP Chairman at the same time. The NCRP committees One and Two were duplicated on the ICRP with the identical chairmen, Failla and Morgan. The interpenetration of personnel between these two bodies was a precedent to a similar movement of personnel between the risk agencies of the present day. The present Chair of the ICRP is also the Director of the UK National Radiological Protection Board (NRPB). The two organizations have other personnel in common and there are also overlaps between them and UNSCEAR and the BEIR VII (Biological Effects of Ionising Radiation). This has not prevented the NRPB from telling the UK's regulator, the Environment Agency, that UNSCEAR and ICRP are *"constituted entirely separately"*, a statement which the Environment Agency accepted. Thus credibility for statements on risk is spuriously acquired by organisations citing other organisations, but it can be seen as a consequence of the fact that they all have their origins in the same development and same model: the NCRP/ICRP post-war process. This *black box* has never been properly opened or examined. A full and reliable history[26] of the development of radiation risk standards is to be found in Caulfield 1989. Taylor himself has described these developments in some detail (Taylor, 1971) and in an interview on the development of radiation risk in the post-war period, Morgan, who left both the NCRP and ICRP said of these organizations and their satellites, "I feel like a father who is ashamed of his children".[27]

Since 1955, the ICRP has received its scientific expertise from the United Nations Scientific Committee on the Effects of Atomic Radiation (UNSCEAR), a body set up by the UN General Assembly to evaluate radiation *dose*, its effects and the risks it poses worldwide. This committee, which brings together eminent scientists from 21 countries, does not itself establish norms or make recommendations, is the source of information about radiation, providing the basis on which the ICRP and the corresponding national commissions establish norms and recommendations, in particular those who have developed a nuclear industry.

These two organisations constitute the scientific and technical arms of WHO's policy in the nuclear field. But WHO itself is not independent, being subject to veto from the IAEA, in any initiative which might damage the agency that promotes the

---

[26] Taylor, 1971.

[27] ECRR, *op. cit.*, extract abridged.

development of nuclear power stations. The agreement between the WHO and the International Atomic Energy Agency, adopted on 28 May 1959, states in Article 1 that "Whenever either organization proposes to initiate a programme or activity on a subject in which the other organization has or may have a substantial interest, the first party shall consult the other with a view to adjusting the matter by mutual agreement", and in Article 3, that "The International Atomic Energy Agency and the World Health Organization recognize that they may find it necessary to apply certain limitations for the safeguarding of confidential information furnished to them".

2. THE AGENTS OF SCIENTIFIC IGNORANCE AT CHERNOBYL

WHO ARE THEY?

> The driving force behind the whole diabolical sequence of events that transformed an accident into a catastrophe, and that catastrophe into a never-ending tragedy can be located in this pact between the WHO and the IAEA. A tragedy directed by a small number of individuals who, day by day, decide the fate of millions of people[28].

Eminent specialists, scientists and experts, UN officials who deal with the nuclear industry and its consequences, are co-opted, as we have seen, from national nuclear authorities and public institutions responsible for radioprotection, whose own directors have generally come directly from those very authorities. The power vested in these individuals gives them authority beyond question (other than by their peers) by virtue of their scientific competence as experts from these international organisations. Professor Leonid Ilyin, for example, pillar of the nuclear establishment in Moscow, who headed the Soviet delegation in 1986 and 1987 at the various conferences of the IAEA and UNSCEAR on Chernobyl, was himself a member of the ICRP.

> It is worth noting at this point that WHO extended the official immunity offered to UN officials to its expert advisers. The enjoyment of this "immunity from legal process of any kind [...] in respect of words spoken or written and acts committed by them in the performance of their duties" is reinforced in a special annexe detailing the privileges granted to experts within the organisation, which stipulates that *this immunity from legal process shall continue to be accorded notwithstanding that the persons concerned are no longer employed on missions for the United Nations* (author's italics). It is understandable that international officials or experts mandated by international bodies need special

---

28 Yves Lenoir, *"Tchernobyl, l'optimisation d'une tragédie"*, Bulle Bleue, Amis de la Terre, CEDI, Environnement sans frontière, 1996.

protection during the performance of their duties. On the other hand, the fact that legal action can never be taken against them, leaves the organisation (WHO) with no meaningful control *a posteriori*—a situation which goes against any principle of law or democracy. The three experts, Pellerin, Waight and Beninson, who were sent in official capacity to the areas contaminated by radioactive fall-out following the an accident at Chernobyl, need never lose any sleep even though they deliberately and powerfully put at risk the lives of hundreds of thousands of people[29].

WHAT DID THEY DO AND HOW?

They did not go and listen to Alla Tipiakova's children nor to the despairing doctors treating them. They did not follow Anatoli Volkov's advice: "It's here that people need to work, to listen, to see the people's suffering... "

They went to the aid of the Soviet nuclear establishment. Without studying the real situation in the "huge laboratory" that so delighted the French professor, and without conducting any research, they explained Chernobyl in terms of Hiroshima. From afar, on the basis of abstract mathematical models, they extrapolated from the effects of the atom bomb on the bodies of Japanese survivors.

It took five years, between 1986 and 1991 to bring the Soviets round to their point of view. Their collaboration was needed to make the lie acceptable. It was no easy task and was achieved in stages through the period of *perestroika* and the dissolution of the USSR, up to the publication of the "International Chernobyl Project", a group of international experts directed by the IAEA, whose final report, presented in Vienna in May 1991, would claim that the radiation has no effect on the health of the population.

3. THE POLITICAL CONTEXT WITHIN THE SOVIET UNION

At this time the Soviet Union was going through the most turbulent and arguably the most dangerous period of its entire history: the Western powers feared that this great nuclear power would implode. Thanks to the initial opening up under Gorbachev, a freedom of expression, unknown since 1917, existed alongside the secret and brutal activities of an all-powerful KGB. While articles of the penal code severely punishing "anti-Soviet propaganda" were still in force, "informal" associations and improvised newspapers opposed to the regime were tolerated. The campaign for the first "democratic" election of the first Soviet of deputies of the people of the USSR, set in motion by Gorbachev, lasted four months, from December 1988 to 26 March 1989. It was accompanied by an incredible verbosity and a profusion of publications, discussions, associations, fears and hopes. Publication of Pasternak's *Doctor Zhivago* was finally authorised; yet photocopying was still against the law

---

[29] Yves Lenoir, *op. cit.* See p. 71 of this book.

for the general public, and there were no telephone directories. Neither served any purpose in a police state where freedom of thought and private initiative had been abolished. After seventy years of repression, bureaucracy and official deceit, collective passivity and apathy had paralysed society. While the West was going through its third if not fourth technological revolution, the Soviet Union was on the brink of economic and social collapse: 35 million Soviets "worked" full time queuing in front of empty shops in the most resource rich country in the world, while the State mafia plundered what was left of the economy at every level—in the villages, towns, cities and regions—by means of an alternative black market, through systems of privilege and favour, and through rackets and murder. Faced with this crisis of self destruction, the majority of the ruling Communist party members, following Gorbachev's lead, finally agreed unwillingly, to recognize that the true, the primary wealth of the country resided in its people, the individual human being, trampled underfoot for three generations. The more forward thinking members were fearful that it might be too late, and all of them wanted to remain in power. We know that in August 1991, the system collapsed, almost without a shot being fired, the masses recovering themselves from the rubble, free to survive in misery while the state's financial treasures disappeared through the back door to the Cote d'Azur, Switzerland, the Bahamas and other *off shore* havens.

Once the diarchy of the Cold War had disappeared and the Soviet and post-Soviet Ministries of Health were subdued, the nuclear lobby in the West emerged victorious, with sole responsibility, the master and promoter of the crime of non-assistance to people in danger for the twenty eight years since Chernobyl.

For the record, and to set the historical facts against scientific artifice, I will make use here of the chronology of the subjugation of science to the Western nuclear powers, recounted bitterly and in meticulous detail by Bella Belbéoch in "Western Responsibility in the Health Consequences of Chernobyl disaster in Belarus, Ukraine and Russia". [30]

4. STAGES OF SUBMISSION TO THE LIES

Much has been written and in great detail about the lies told by the Soviet government[31]. This is not new, given that the country's leaders have always lied to their people, starting with the original lie in 1918, when Lenin promised "soviets, peace and land", continuing with the massacre of sailors at Kronstadt in 1921—

---

[30] B.Belbéoch, *"Radioprotection et Droit nucléaire. Entre les contraintes economiques et ecologiques, politiques et ethiques"* sous la direction d'Ivo Rens et Joel Jakubee, SEBES Ed.Georg, 1998, p. 247–261

[31] Alla Yaroshinskaya—*Tchernobyl, vérité interdite* Arte-Éditions de l'Aube, 1993.

under orders from Lenin and Trotsky, when the Soviets were finally stripped of their power—and up to the Chernobyl disaster.

> "Recently, Savchenko, former Minister of Health of Belarus, replying to the Supreme Soviet Commission investigating the veil of silence in the years following Chernobyl, reported that they had been summoned by Prime Minister Ryzhkov, who told them, *"It's not secret, it's top secret, all doses and all information about the Chernobyl tragedy"*. That's the whole approach. That is why no dose measurements were made, or if they were, they had to be minimised without fail. The lies and the silence were ordered from above". [32]

It is much more interesting now to focus our attention on the lies told by the free and democratic West which condemns, at Chernobyl alone, millions of disinformed peasants and their descendants to suffering and to a terrible death from the effects of radiation.

Four months after the catastrophe, an international conference is organised by the IAEA in Vienna (25–29 August 1986) to analyse the consequences of the Chernobyl accident. The key figures are D. Beninson, President of the ICRP and Argentina's Director of Atomic Energy, and M. Rosen, Director of Nuclear Safety at the IAEA.

Working group meetings take place behind closed doors and no journalists are admitted. Discussions are tough. There is too great a discrepancy between the Soviet predictions and the figures that the West find acceptable. V. Legasov presents a voluminous document consisting of a general report and 7 annexes.

The stumbling block for the West is Annex 7, consisting of 70 pages, devoted entirely to *"Medical and Biological Problems"*. In the absence of any measurements of individual dose or medical data, the Soviets have predicted, using abstract mathematical calculations, an extra 40,000 cancer deaths among the 75 million inhabitants of European Russia.

Equally abstractly, and equally in the absence of any measurements of individual dose or medical data, the West decides this figure is too high—a trial of strength which had absolutely nothing to do with science, since Chernobyl was an absolutely unique event in the history of science and anyway, only four months have elapsed since the disaster, during which time no real research nor any systematic measurements have been undertaken.

At the press conference the next day, August 26th, Dan Beninson, chairman of the study group on the health consequences of Chernobyl says that the Soviet figures are *"extremely overestimated"*. Morris Rosen, Director of the Safety Division at the IAEA, puts the upper limit at 25,000 deaths, still without any objective data.

---

[32] The liquidator Anatoli Borovsky in *Nous de Tchernobyl*, TSI 1990.

And the dispute is not over. Two days later, the limit has been reduced to 10,000 and for Beninson 5,100 at most, even though he still has no scientific information at his disposal to make this claim. But Beninson is president of the ICRP and his opinion carries weight: according to him, Soviet figures are too high because the levels of internal contamination by radioactive caesium have been "overestimated".

Since this international conference, Annexe 7 has disappeared from circulation. Very few people ever saw it. Since then, neither Soviet nor Western experts ever referred to it again. It is as if it never existed.

Over the following months a number of official reports are churned out and relayed by the press. In a European Community document (COM, 607, October 1986) experts claim that the internal radiation dose has been overestimated by a factor of 10, and the director of the IAEA who made a five-day visit to the USSR in January 1987 (five!) announces that "the first post-accident assessments of health effects were too pessimistic and should be decreased by a factor of 5 to 7".

On 8th October 1986 *Le Monde* writes: "The accident at Chernobyl has increased *the audience and credibility of the IAEA*, as M. Gerard Errera, the governor for France says [...] In the days following the accident at the Ukrainian nuclear power station, the IAEA,—through its Director-General—showed a clear determination to manage the situation [...] There need be no doubt that from now on, even greater emphasis will be placed on nuclear safety". Thus, WHO is sidelined.

The juggling of the figures, which had started in Vienna, and had no scientific basis, now intensifies. In order to be credible and arrive at the same conclusion, the downward revision needs to come from the Soviets themselves.

In May 1987, during a conference organised by WHO in Copenhagen (13–14 May 1987), A. Moiseyev asserts, in abstract terms again, that "the positive trend in the radiological situation" allows a reduction in the collective external dose by a factor of 1.45 and in the collective internal dose by a factor of between 7 and 10.5 in relation to initial estimates. We know, that at the same time, a significant quantity of milk from farms in Belarus, containing well over the accepted limit of caesium-137, was withdrawn from local consumption and mixed with clean milk from areas further away. We also know that this "democratisation" of dose, by increasing the number of individuals contaminated by low level radiation, does not change the final outcome, statistically.

In September 1987 at the IAEA headquarters in Vienna, L. Ilyin and O. Pavlovsky present yet another dishonest report on the radiological consequences of the accident entitled "Data analysis confirms the effectiveness of large scale interventions to limit the effects of the accident". It states that 5.4 million people including 1.7 children received prophylactic iodine to protect their thyroid against radioactive iodine. This is completely false. One of the reasons why Professor

Nesterenko[33] was removed from his post as director of the Institute for Atomic Energy at the Academy of Science in Belarus was precisely because he had called for a massive programme of iodine prophylaxis, which in order to be effective, should have been dispensed in the hours immediately following the accident. His recommendation was rejected; he was accused of sowing panic, and was removed from his post.

According to Ilyin and Pavlovsky's report, there has been no increase in morbidity in children, no difference in the health of children in contaminated and control regions and, for the first time, the syndrome of "radiophobia" is invoked to explain the overall increase in morbidity among inhabitants of contaminated areas. One wonders what sort of "radiophobia" is experienced by a foetus that has miscarried, a newborn baby, or a child.

Finally, it is apparent that the statistics in the report have been crudely manipulated: the effective dose is calculated on the basis of a seventy year life span for the *whole* population of the USSR (278 million inhabitants), whereas Annexe 7 in 1986, considers only the contaminated population of the European part of the USSR, i.e., 75 million people, and this results in a collective dose which is 18 times lower.

In April 1988, at the conference in Sydney, Ilyin re-evaluates slightly the dose that he had previously divided by 18 (report by Ilyin and Pavlovsky). But at its General Assembly in 1988, UNSCEAR decides to "average" Ilyin's two doses and reduce the dose provided by Legasov in "Appendix 7" from 1986, by dividing it by 9. Two years after the Vienna conference, as Bella Belbéoch commented "Beninson must have been pleased. Ilyin and Pavlovsky were both signatories to Annexe 7. So this is a really criticism of themselves". Already the victim of persecution in Moscow, Legasov committed suicide, four days after Ilyin's announcement at the Sydney conference.

In September 1988, in order to avoid large scale and costly evacuation programmes, the USSR Council of Ministers rules on the criteria for "safe residence" and "safe life" and adopts Ilyin's theory of a permissible limit of "35 rem over seventy years".

The rem is a unit of measurement of absorbed dose of radiation which expresses the biological damaging effects of radiation on the body. In order to evaluate the biological impact of radioactivity, the absorbed dose is assigned a coefficient which describes the *destructive efficacy* of radiation on living tissue. Alpha particles (α), which are very efficient, have a coefficient 20 times higher than beta particles (β)—or gamma rays (γ)[34]. Over millions of years, biological mechanisms have

---

[33] See Part Two, Chapter I. p. 82 and following pages.

[34] See Glossary, p. 599 "Ionising radiation α, β, γ".

adapted to constant low level background radiation, composed of rays from the sun and the cosmos and traces of radioactive isotopes that were present in rocks when the earth was formed, and at very low levels, in the potassium consumed by living organisms. Far more important and destructive are the effects of artificial, radioactive emissions produced by the nuclear industry and dispersed across the planet in enormous quantities by military weapons testing, nuclear power stations and the Chernobyl accident. During the first acute phase when the reactor was on fire and during the first weeks when the "sarcophagus" was being constructed, levels of radiation were still very high. Today, the radioactivity is not in the form of immaterial background radiation but consists of radioactive particles released into the environment, and then ingested through food or inhaled. Once inside the body, they begin their destructive action on the cells.

With the scientist Ilyin's seal of approval (he is a medical doctor!), the threshold of 35 rem, which was only allowed for workers in the nuclear industry who in principle were protected, became the acceptable level of radiation for everyone: children, pregnant women, the old, and the infirm. In reality, the permissible level for the general public recommended by international radiation protection authorities is five times lower. Given the soviet system of "internal passports" which restricts people to residence in the same area for the whole of their lives, the principle of 35 rem lifetime dose, established a veritable radioactive gulag, in which 9 million people, according to Kofi Annan, were condemned to struggle every day with levels of radioactivity up to 5 times higher than natural background radiation, with terrible "proximity effects" on organ tissues, through the incorporation of radionuclides and "hot particles".

This measure was presented as if it conformed to international recommendations and was contested by a group of Belarusian scientists from the Academy of Sciences. Yury Shcherbak recounted how the combination of *glasnost* and the silence and lies about the Chernobyl disaster was transforming society. Instead of 35 rem, these scientists demanded a 7 rem over 70 years, which corresponded in fact to the limit that was internationally recommended by the ICRP since 1985 (the Paris declaration): an annual dose of 0.1 rem.

March-July 1989: political tension in the USSR. The disaster is recognised and denied.

Disagreement between the Moscow experts and the Belarusian scientists—in other words between the central powers and the academicians supported by public opinion—emerged during debates in the Belarus Soviet in March 1989, right in the middle of the electoral campaign for the first democratically elected Soviet of deputies of the people of the USSR held on 26th March 1989. A special session was devoted to the "35 rem in a seventy year life-span", in June at the Academy of Sciences of Belarus in Minsk.

Meanwhile the totalitarian system is reaching crisis point both at the top and at the periphery of the empire. The nationalist aspirations of non-Russian Republics are awakening, fuelled by the citizens' desire to break free and claim their rights. But unrest is also being fomented and manipulated by forces hostile to the idea of *perestroika* in order to justify repression of the growing democracy. In the interval that separates the election of the USSR parliament on 26th March and the session of Congress fixed for the 26th May, there is a massacre, on 9th April 1989 on Boulevard Roustaveli, of Georgian demonstrators at prayer in front of the government palace at Tbilisi, in Georgia. The massacre is prepared and decided on in Moscow, by the Conservative Ligachev, the Minister of Defence, Yazov and by the head of the KGB, Chebrikov, while Gorbachev is visiting Great Britain.

But this Soviet-style Tiananmen Square has the opposite effect of the events in Peking. The institutional reforms introduced by Gorbachev allow people to express themselves in a legitimate forum, a few weeks later at the first Congress of Deputies. During twelve day of televised debates, the repression in Tbilisi and the lies and crimes of the State are denounced—morally and politically—before a television audience of 300 million people. At last, the accident at Chernobyl begins to appear in the eyes of the world for what it really is: a catastrophe, the dimensions of which can no longer be kept secret on grounds of national defence. Nevertheless, its effects on the environment and on the health of the population will be denied authoritatively by UN Agencies, a few months later.

In May 1989, during a debate at the Supreme Soviet of the USSR, deputies accuse the political office of the Communist Party (the Politburo of the Communist Party of the Soviet Union) and the government of the country of covering up the real situation after the accident at the Chernobyl power station, and of failing to implement adequate protective measures at the time of the accident and thereafter. Those in power defend themselves saying they did not known anything about the contamination in certain regions and that the scientists had not put forward any proposals to protect the inhabitants. It is at that moment, during a conversation on the telephone with Andrei Sakharov, that Nesterenko tells him that in his archives, he has copies of letters and reports, dating from the first few days after the accident and covering a whole year (a thousand letters and notes), that he had sent to the authorities in which he reported radiation levels in all the contaminated areas, called for the evacuation of all inhabitants within a 100 kilometre radius around the nuclear power station, and for the mass distribution of stable iodine to children to protect their thyroid gland. Naturally, Sakharov suggests that he publish these documents. Nesterenko reminds him that all material concerning Chernobyl is a state secret and that publishing them without authorisation would be suicide, but on the other hand, the deputies of the Supreme Soviet of the USSR could get them declassified. "Declassify information about Chernobyl and I will publish them" says

Nesterenko. Sakharov organises the declassification. A selection are then published in Nos 5,6 and 7 of the newspaper *Rodnik (The spring)* in 1990.

June 1989.To shore up the breach opened by the Belarusian scientists who are demanding a lifetime dose of 7 rem instead of Ilyin's 35, the same three members of WHO, will be sent to the Minsk Academy of Sciences in June, to attend a special session devoted to "35 rem over seventy years". P. Waight is director of WHO's working group on Radioprotection; Dan Beninson, president of the ICRP; Professor Pellerin was head of SCPRI—*Service Central de Protection contre les Rayonnements Ionisants*, attached to the French Ministry of Health. These heavyweights, with the institutions behind them, tip the balance and regally announce their conclusion. On the 28th May 1998, nine years later, Professor Nesterenko recounts these events to me:

> The experts from the ICRP arrived in Belarus in June 1989. A meeting took place that day at the Academy of Sciences. There were many contributions. I intervened to describe the situation and put forward my arguments. One member of the ICRP retorted: "Should the 3 million inhabitants of Madhya Pradesh in India (where the disaster at Bhopal took place) have been evacuated? I replied that the Indian Government had not asked for my opinion and I had no lessons to teach them but that it was my duty to advise my government: we must evacuate the inhabitants of all areas where an uncontaminated food supply cannot be guaranteed. Because today, 80-90% of the radioactivity comes through the food supply. I was told: "In any case, you haven't got the money to evacuate all those people, so you will have to put up with 70 to 100 rem rather than the recommended 35". They kept saying things like that. The only explanation I could find for this kind of behaviour was that these people were only interested in obtaining experimental data on the health impact of low dose radiation. It is the scientific data that they are interested in, not the health of the people. It is cynical, but that's how it is. They treat us like guinea pigs. I consider it a crime[35].

The report of the three Western experts was published in *Sovietskaya Bieloroussia* on 11th July 1989 under the title "The experts' point of view". The report disparaged those scientists who opposed the life-time dose limit supported by Soviet authorities: "Scientists who are not well versed in the health effects of radiation have attributed all observed biological and medical disorders to radiation exposure. These disorders cannot be caused by ionising radiation" (WHO, 1989). They added that "If they had been asked to fix a limit on the lifetime cumulative dose they would have chosen dose-limits two or three times higher than 35 rem". The legal immunity enjoyed by the "experts" from WHO and from ICRP allowed

---

[35] *Le piège atomique*, Film TSI, May 1999.

them to say anything with impunity. These "experts" invoked psychological factors and *stress* to explain these biological disturbances.

Bella Belbéoch comments "It is frightening the way these *experts* deny to the Chernobyl disaster the sad privilege of being a new phenomenon in medical *experience* and close the door on the only possible approach to understanding this new phenomenon; first, record all biological and medical information[36]". This is what Anatoli Volkov was asking for and what Vassili Nesterenko and Yury Bandazhevsky, among others, would undertake, in a systematic and global fashion, refusing to submit to the official dogma.

At the end of July 1989, during the session of the Supreme Soviet of the Socialist Republic of Belarus, the health and political authorities make use of the report by Waight, Beninson and Pellerin, that appeared in June, to override the objections of academicians from their own country, who were considered ignorant about radiological questions by the professors named above.

But the rebellion still worries the powers that be, both in Moscow and in Vienna. It needs to be discredited by the Soviets themselves.

On 14th September 1989, a group of 92 "scientists working in medicine and radioprotection in relation to the situation created by the Chernobyl accident" write a joint letter to Mikhail Gorbachev. Among the signatories was Ilyin of course, but also S. Yarmonenko—at the top of the list, A. Gouskava, M. Savkin and the Ukrainian, V. Bebechko. Twelve years later, in June 2001, we would film them at the international conference in Kiev, aggressive, and still obstinately anchored in negationism[37].

The "92" defend their decision to impose a dose-life of 35 rem: "At every stage of its development, this limit has been established through systematic consultation using the rigorous expertise of different competent international organisations such as the IAEA, WHO, UNSCEAR which have examined it from all angles and given their approval".

Models for the calculation of dose limits and for radioprotection criteria used by the international nuclear lobby are all based on a single, fallacious scientific argument, as set out below:

> The justification for setting a limit of 35 rem as a life-time dose is reinforced by the findings of the long term monitoring of the health of the population exposed to very high levels of radiation—inhabitants of Hiroshima and Nagasaki, who were victims of the atomic bombs [...] It has been shown that increases in solid tumours in Japan

---

[36] Bella Belbéoch *op. cit.*

[37] Cf. Part Six, Chapter II, p. 417

only occurred with doses of radiation (an instantaneous massive exposure) above 100 rem, and in leukaemia and myeloma, above 50 rem. No increased level of genetic alteration was observed following irradiation.

Some of the arguments used by the signatories to counter a life-time dose of 7 or 10 rem are very important, [as Bella Belbéoch notes.] The authors invoke the deep psychological stress and damage to health that would be caused by "the displacement of hundreds of thousands of people (up to 1 million) [...] If the life-time dose of 7–10 rem were to be accepted as a criterion for being re-housed, this would affect the inhabitants of several large towns and regional centres". Further on, doubts were expressed about the quality of medical care that could be guaranteed "in a plan to evacuate a million people".

One thing is clear. If a life-time dose limit of 7–10 rem had been adopted in September 1989, hundreds of thousands of people, possibly a million, would have had to be evacuated. If we compare these estimates with the official figures of inhabitants living in areas subject to radiological control, we see that it includes all of them (contamination exceeding 5 curies/km²)[38].

The journalist N. Matukovsky (*Izvestia* 26th March 1990) interviewed E. Petryaev, Doctor in chemical sciences at the faculty of Radiochemistry at the V. I. Lenin State University of Belarus, who told him:

It is impossible to set a *threshold* over a lifetime for the following reason: to establish dose limits they only take into account caesium-137. But the affected areas were also contaminated with strontium, plutonium and a great *bouquet* of transuranian elements which, in the form of micro-dispersed aerosols, can penetrate the human body through respiration. The worst are the *hot particles* measuring a micron or more. In the Southern areas, in the Gomel region, there are between one and ten per square centimetre of soil. And all of these are in addition to the 35 rem!

Quite honestly, the *scientific* controversy isn't about curies or rem, it is about the numbers of inhabitants and which areas need to be evacuated. To be even more precise, it's about money, about how many billions of roubles they are prepared to spend on the evacuation—10, 15 or 20 billion? What hypocrisy! And we preach that human life has no price...

The USSR Congress of Deputies sets up a commission to analyse the causes of the accident at the Chernobyl nuclear power station and to assess the responsibility of the government for failing to implement effective protective measures in the post-accident period. A standing group of experts, rebels against Moscow's authoritarian and non-scientific stance, including around 200 specialists from the three Republics affected by the disaster, representing different scientific and technical domains, work at the heart of this commission. In the group are about twenty deputies,

---

[38] Bella Belbéoch *op. cit.*

including Yury Shcherbak, Alla Yaroshinskaya, and the writer Ales Adamovich, who along with Andrei Sakharov will support the creation of Vassili Nesterenko's independent institute Belrad. Nesterenko will direct the work of the commission in Belarus and then, from 1991, the independent Committee that brings together 200 specialists from Belarus, Ukraine and Russia that took over from the Commission of the Supreme Soviet, on the dissolution of the USSR.

In October 1989, the Soviet nucleocrats ask their colleagues in the West for help once again. To regain legitimacy in the eyes of their people who are protesting through their elected representatives, the Kremlin asks the IAEA to put together a team of international experts to evaluate the consequences of the accident and the effectiveness of the counter measures undertaken by the authorities. Since the discussion that took place behind closed doors in August 1986 in Vienna, where Legasov's Annexe 7 had been thrown into the bin, the central soviet power has rejoined the ranks. It has silenced the dissident scientists and is now asking for something in return. The IAEA sets up an international consultative committee, the International Chernobyl Project, headed by Doctor Itzuro Shigematsu, the representative of RERF—Radiation Effects Research Foundation[39]—an organisation financed by the American and Japanese governments to study the consequences of radiation on the victims of the atomic bombs in Japan. In accordance with the plan approved by the International Consultative Committee, between March 1990 and January 1991, around fifty missions visit the USSR, including, apart from the Soviets, 200 experts from 25 countries representing among others the IAEA, UNSCEAR, WHO, FAO and the Commission of the European Community. Their final report will be presented at the IAEA Conference in Vienna, 21st-24th May 1991. The message from the scientists is that *radiation from Chernobyl has had no ill effects on the health of the population.*

On 21st May 1991, the IAEA unveils a 57 page summary of its report, presenting the scientists' basic conclusions (IAEA 1991a). The publication of the report itself is delayed; neither the press nor independent experts have access to it before October 1991 (IAEA 1991b). In the meantime, the summary of the report and its accompanying press release inspire the following headlines in the newspapers. "UN Committee blames psychological stress not radiation for the problems around Chernobyl" (*Washington Post* 22nd May 1991), "Consequences of Chernobyl are psychological not physical" (Associated Press, 21st May 1991). The newspaper publishes this quote: "Of course, the inhabitants of the contaminated areas think they are ill", declares Lynn R. Anspaugh, who was responsible for the medical section of this study. "All that comes, not from the radiation, but from the fear

---

[39] See note 15, p. 53.

that it inspires…" For its part, Associated Press does not tell us anywhere that L. R. Anspaugh works for the US Department of Energy, to which the former Atomic Energy Commission is now attached. He is one of the principal authors of the "zero risk model" for the evaluation of the consequences of accidents at nuclear energy installations.[40]

---

[40] John W. Gofman—*Chernobyl accident, Radiation Consequences for this and Future Generations,* 1993.

PART TWO
KNOWLEDGE

Chapter I

VASSILI NESTERENKO:
A PHYSICIST LOYAL TO HIS PEOPLE

Galia Ackerman, the translator of Svetlana Alexievich's book, "*Voices from Chernobyl: The oral history of a nuclear disaster*"[41], remarks on the similarity between Vassili Nesterenko and Andrei Sakharov, the inventor of the Soviet H-bomb and winner of the Nobel peace prize. "Like Sakharov, Vassili Nesterenko worked for the Soviet Army in the nuclear domain ... But the time came when the regime that he had served so faithfully appeared to him (as it had to Sakharov) in a different light: after Chernobyl. At that point, he did something quite unheard of in what was still a totalitarian state... he decided, without any approval from his superiors, to stop the work he was engaged in at the Belarus Institute of Nuclear Energy ... Instead, he called on his staff to study the consequences of Chernobyl and to develop aid programmes to help the victims".[42] Dismissed from his post, he lives a precarious life financially, unlike other academics in Belarus, and fights tirelessly on behalf of the nearly 500,000 Belarusian children still living in the contaminated territories.

1. NESTERENKO'S CHOICE

We met Vassili Nesterenko in May 1998, on our second mission to the Chernobyl area for Swiss Television in Lugano. Our curiosity was aroused by the sudden volte-face made by the Corresponding Member of the Academy of Sciences of Belarus, who renounced his faith in the development of the atom, renounced the privileges of a brilliant career and, putting his own life in peril, entered into conflict with the Kremlin. What had happened to him? Where was he from? Who was he? In extreme situations, the moment of decisive action presents us with a choice. Either the moment defines us, or we ignore our conscience and dodge the issue. He was the only witness to

---

[41] Svetlana Alexievich was awarded the Nobel prize for Literature in 2015 for her "polyphonic writings, a monument to suffering and courage in our time".

[42] *Les Silences de Tchernobyl*. Autrement, 2004.

this turning point. Nesterenko talked to us about his decision with neither affectation nor reticence, as a man, loyal to his own people, to whom he felt indebted. It almost seemed as if he had not needed any special courage to make this break. In fact as we shall see later, he practised courage every day for over twenty years, in his fight against the nuclear lobby and against the physical injuries to his body that resulted from the radioactivity. He was one of the 800,000 Chernobyl liquidators (some say 1 million) who exposed themselves to radiation to protect us, the people of Europe.

V. Nesterenko.—There's an old Russian saying that "The bringer of bad news will always have his throat cut". But I have no choice; a sense of duty is ingrained in the Russian *intelligentsia*. Even if it's no longer fashionable today. When I was dismissed from my post as director in 1987, I carried on working in radioprotection in a laboratory at the Institute of the Academy. But I faced continual harassment, with people saying that it was not my job, that the appropriate ministries were dealing with it, that I should devote myself to science. I disagreed because I owe a lot to the people—whether Belarusian, Russian or Ukrainian, because it was they who paid for my education and for my standard of living, they subsidised us, the brains, the elite of the nation. I grew up in Ukraine, I studied for twelve years in Moscow and I've lived here in Belarus for 35 years. At the age of 33–34, I defended my doctoral thesis and was made a professor. In 1972, I was elected to membership of the National Academy of Sciences of Belarus, was given an apartment and a magnificent standard of living. Whatever the people could give, they gave. Since this tragedy at Chernobyl, people are suffering; it's obvious that our duty is to use our skills and expertise to assess the situation properly and help them to protect themselves against the radiation.

That was what I wanted to devote myself to but at the Academy I was prevented from doing so. One day, Andrei Sakharov telephoned me to ask me to publish the data I had collected from the Mogilev region because the authorities were refusing to accept the levels of contamination. "Send us your data; they must be published, because Professors Israel and Ilyin are telling Moscow that they knew nothing about the contamination in the Mogilev region". "What? I sent them our maps as soon as they were produced. And I keep copies of all of my classified 'top secret' correspondence with the government on the subject!" I had four volumes of correspondence with the political authorities, each about 250 pages long. I asked Sakharov who was a deputy at the Supreme Soviet at the time, to get information about Chernobyl declassified. In May 1989, the Supreme Soviet declassified a number of state secrets and I presented my documents to the newspaper *Rodnik* for publication. In summer 1989, I was summoned to the Central Committee and told to withdraw them from publication; otherwise it could bring about the fall of the government. I told them that people needed to know the truth and I refused to withdraw them. Soon afterwards, I started getting threatening phone calls: "Get out of Belarus or you'll get what's coming to you!" I didn't take it all that seriously. This was during the euphoria of *perestroika* and I thought no-one could stop me saying anything. A little bit later, I got a phone call from the Procurator-General of Belarus, Nikolai Ignatovich, who we had known for a long time—the laboratory had nominated him as a candidate at the elections to the Supreme Soviet. He said: "Did you give your data to the newspaper?" "Yes". "You go to work at the institute every day in your car, don't you? Don't drive for a few months". "Hold on a minute, Nikolai Ivanovich[43], I know you're a politician now, but that's bit far fetched, isn't it?" "It's my duty to warn you", and he hung up. That was on the 25th August.

On the 8th September, I was driving, as usual, to the Institute (in the new town of Sosny) taking the shortest route—through the Drazhnia quarter on the outskirts of Minsk—the road had three lanes in one direction and three going in the other. I was in the right hand lane, the other lanes were empty. At the crossroads the lights were red, so I braked, and out of habit, I glanced into the rear view mirror. I saw an ambulance driving up behind me in the right hand lane, going very fast. It didn't have its flashing lights on and it didn't look as if it was going to stop or move into the other lane. I took off the handbrake, put the car into gear and accelerated. I just had time to move forward a few metres into the crossroads when the ambulance hit me at high speed. The impact pushed me about thirty metres beyond the crossroads. If I had not moved a little bit further forward, I would have been hit by the traffic coming the other way which had already started moving. People would have said "the professor wasn't paying attention, he braked too soon and it cost him his life". I suffered whiplash and still feel the effects of it.

A few days later, the telephone rang again: "Maybe you'll shut up now". I hung up.

Ilsa Nesterenko , Vassili's wife, intervenes at this point.

I. Nesterenko.—I feared for his life. I don't know how many threats we received! They telephoned the house; I would pick up the phone. They made threats: "Anything might happen…Stop interfering!" I can't remember the details now. I just remember the phone ringing and their voices. I'd put the phone down before they finished, I'd throw the receiver down and I'd be shaking inside. And of course, anything could happen. An accident for example. It's not a time we like to remember. One day, much later, when everything had calmed down a bit and we felt a bit easier, Vassili told me, "I was in such a state I felt like driving the car into a wall and to hell with all of it. It was only the thought of you, left to bring up Aliosha all by yourself that stopped me". At this time, what struck me most was him returning to the house,

---

[43] Nesterenko is referring to the Procurator-General mentioned five lines above.

pushed to the limit, and, at a loss, saying: "They might as well be deaf. I might as well talk to the wall. They don't react to anything I say. You'd think they'd be curious at least, ask questions, express some objections, tell me I'm wrong... Nothing. They are deaf. Like this wall".

V. Nesterenko.—The same thing happened recently. Last year, in 1997, I was preparing a speech to Parliament protesting against the construction of new nuclear power stations in Belarus when the telephone rang. "Don't forget 1989". Again, I didn't take much notice. My speech to Parliament was on the 15th. On the 16th, I had to go to a garage about 20 kilometres from here. I had an old 1984 Peugeot. As I started the engine I heard a strange noise and the steering wheel started to shake. I thought it was coming from the right wheel. I tested the bolts, they were nice and tight. I drove off again and as I got onto the motorway, suddenly the left wheel came off and rolled in front of me.

I used to think physics was the most advanced of all the sciences, because so much had been achieved for humanity in this domain. But... that was before Chernobyl. Before I understood that this technology was not compatible with humanity's level of morality at the present time. What should the government do in these situations? They need to tell it how it is. The following proposition should have been put to the United Nations: "The accident that happened to us, unfortunately, won't be the last; there will be other accidents in other countries. We need to set up an insurance fund. Help us today and in return, we will help other countries, should the need arise".

Those who failed to take this decision at the time have a moral duty towards the victims and should have to answer before a court why they did not take necessary precautions to protect the population. Even today, they still maintain their silence so as not to lose face and to avoid the accusation that they have stayed silent for all these years. But then again, if the State told the truth today, people would expect them to take necessary measures, and they cannot afford to do that, which explains their ostrich-like behaviour: "You can live in this area, it's not dangerous". I can understand why older government officials allow themselves to lie because when the serious consequences happen they will no longer be in power, they won't be alive. But there are many young people in the hierarchy and I don't understand them. Sometimes I get the impression that the information I send, my letters and bulletins, to government, never reach the high-level politicians, that it gets blocked by these older civil servants.

The international community hasn't done anything either and I find that completely incomprehensible. I am absolutely convinced that the United Nations should be directing at least half of the IAEA's activities towards radioprotection. Today that organisation spends all its time on nuclear arms non-proliferation, and it does not do that very effectively either. The Indian nuclear weapons tests proved how useless it was. As for radioprotection, the IAEA does nothing. In 1989, the UN launched a financial appeal for Chernobyl. They expected to get 700 or 800 million dollars in aid for Belarus. That would have been an appreciable sum. But Japan gave 10 million and that was it. The famous conclusions of the international panel of experts that claims there have been no serious health consequences, came at exactly the right time to show that no aid was necessary. Countries with nuclear power have no interest in recognising the health effects of the catastrophe. Here, the victims have not been compensated, but if one of those countries had the misfortune to have a similar accident, they would have to pay millions in compensation to the victims. They have to avoid a legal precedent. In the West, the population density around nuclear power stations is five times higher than in the USSR. According to German experts, if an accident like Chernobyl happened in the Hanover region, more than 8 million people would be affected and the economic damage to Germany would be enormous. These are not my figures, these are estimates made by German experts, but I don't think the public is kept informed about these issues. In my opinion, the international nuclear lobby does not want to recognise the true scale of the disaster here, because if they did, nuclear energy could no longer rightfully be pursued.

2. A VOICE IN THE DESERT

V. Nesterenko.—I came up against two dramatic situations that changed my view of things radically: the general level of irresponsibility and the contamination of children.

I was in Moscow… the accident happened on Saturday 26th April. It had been hushed up. Even I didn't know about it and I was the director of a nuclear energy institute. On Sunday, I took the plane—I did this frequently, every week I went to Moscow on assignment—and I had a little Geiger counter, a dosimeter that I always carried with me. Suddenly I saw something: what the hell was that? I thought it must be broken, and I put it in my pocket and forgot about it. I landed at Moscow, arrived at the Kremlin, went to see Petrovich (who worked for the military industrial commission) and said "I need to see the head of department, he has to see me. I need to speak to him, to sort something out". "I haven't got time, Chernobyl is on fire". It was Monday 28th. I said to him, "Seriously, Yury Petrovich, it's Monday, remember; yesterday was our day off". "No. Really. The reactor is on fire". When I realised, I understood the seriousness of the event, because I knew about the accident at Three Mile Island in the United States. There had been the accident at Cheliabinsk that I knew well professionally, with the problems of radioprotection that it had posed. I also knew the reactor at Chernobyl, I knew that it did not have a protective concrete covering, that everything was exposed to the open air. I grabbed the telephone and called the president of the Academy in Minsk. They did not know what had happened, but our Geiger counters had been registering high levels of radioactivity and they were worried that it could have been caused by our Institute. Secrecy had already been imposed on the subject of Chernobyl. I said to him: "Don't bother looking for a fault there; it's in the south, our neighbours in the south". That's all I could say to avoid jamming the line with the word *Chernobyl*. "There's been an accident, you need to inform the authorities, and evacuate the inhabitants in the south". He replied "You know our Sliunkov. I won't be able to convince him. Call him yourself". He was the first secretary of the Central Committee of the Communist Party of the Soviet Union in Belarus; in fact he was president at the time, the number one. OK. I telephoned his office. I called once, twice, three times. His secretary answered: "He's busy! He's busy! He can't talk to you". "Tell him that it's urgent!" I was shouting. Finally someone put me on the line. I said to him: Nikolai Nikitovitch, there's been a bad accident. In the south, on the border; there is an emergency situation, you need to immediately relocate residents; you must distribute preventive iodine to the inhabitants within a 300–400 km radius of the nuclear power station". "What are you getting so upset about? I know all about it. I've been told all about it, there's been a fire, and it's been put out". "That's not true, the graphite will keep burning for months; if it isn't put out, it will burn for a long time". "Alright, come back tomorrow, we'll have a look". Our conversation took place at about ten, ten thirty in the morning, Monday 28th. I called the Institute straight away: "Inform the schools immediately, and our colleagues, so that they can pass it on to their families. You need to close all the windows, wash with water, don't let the children go outside, don't send them to school". In short, simple protective measures. And distribute iodine as a preventive measure. Everyone who worked at the Institute always carried an assortment of iodine tablets. Everyone needed to take them straight away.

I took the plane back that evening. My chauffeur came to the airport. I made a quick detour to the house to leave some tablets for Aliosha.

The nuclear power station is 320 kms from Minsk. I took measuring equipment and I left that night. First I went to Bragin, then returned to Narovlia via Mozyr. I examined the situation. At Khoiniki for example, there was about 15,000–18,000 microroentgens per hour (μR/h). At Bragin the level reached 30,000–35,000 μR/h. And at Narovlia as well, the level was 25,000–28,000 μR/h. That is thousands of times higher than background radiation. According to all our criteria, it meant we needed to evacuate the population. All of this happened on the night of the 28th. I picked up some food somewhere, some eggs, took a sample of soil with a bit of grass and took it all back for analysis at the institute. At dawn on the 29th I was already back. And at 8 o'clock in the morning, I was at the Central Committee. I wanted to see the people who made the real decisions. But there was no-one there, no-one... So then I went to the institute, which is in Sosny, about 20 kms from Minsk, to get the spectrometric results of the samples. They revealed that there were high levels of contamination by radionuclides in the soil and in the food products. We needed to take protective measures urgently.

I walked across town holding the dosimeter high up and then low down: there was a big difference. The radioactive pollution was falling on Minsk. It was hot. Outside people were selling *pirojky*, meat, ice creams, everything. I went back to the Central Committee. Kusmin, the secretary at the Central Committee for Science, tried three times to get Sliunkov to see me. But he wouldn't. So I made my mind up: I went into the antechamber thinking: "This time I'm going in. I'll force my way into the office". They might shout at me, they might chase me out but I had no choice. It was about five o'clock. I'd been waiting since eight in the morning! I saw Nil Gilevich, a well known poet in Belarus. He came out, I was walking round in circles, he recognised me. "Hello, hello, how are you? I've had a really good talk with Nikolai Nikitich about developing culture in Belarus: an hour and a half". "There won't be anyone left to appreciate culture if we don't evacuate the population straightaway". "What are you talking about? Sliunkov says everything's under control there, everything's fine". I went in and I started to tell him about this and then about that. He said: "What are you talking about? You're wrong. It's all under control…" I made him listen: "It's not true, you've been misinformed!" and I saw a flicker of fear in his eyes. He put in a call to the Prime Minister. According to the law here, emergency situations, evacuations, etc can only be declared by the president for Civil Protection, in other words, the Prime Minister of the Republic. He picked up the phone and said: "Mikhail Vassilich, I've got Nesterenko here and he's telling me different things from what you've been told. Listen to him". I could hear them in the receiver, one voice in particular, saying: "Why is he spreading panic like this? His people are all over town, sowing panic. Make him stop. Order him to stop". "At least see him". "He must get his dosimetrists out of town!" "See him". I went there, to the Council of Ministers. I arrived at about 6.30 in the evening. Remember it was already the 29th April. A big room, on the third floor, I think. They were sitting there, the Prime Minister, the Deputy Minister, the Chief Officer for Civil Protection, the Chief Medical Officer and also the Mayor of the town. The technicians from my radioprotection services had evaluated the situation, gone all over town measuring the level of radioactivity in people's thyroid, which is an ideal indicator. You put the dosimeter against the thyroid and you can see how much iodine it has already incorporated. They had convinced the Chief Medical Officer to prepare 700 kg of iodine solution.

We had asked that the Mayor be invited because it was he who could give the order to distribute iodine in the town…But when I saw that they were afraid, I said: "Make it simple: add a solution of iodine to the drinking water supply where chlorine is added. Do the same thing in the dairies. The whole population will be protected automatically". This is what we wrote to them, from the Academy. They were all sitting there, and it was the Prime Minister who needed to make the decision. I came in and he said to me in an irritable tone: "What is it you want to tell us?" I replied that they needed to evacuate a radius of 100 km at least, distribute preventive iodine, forbid the open air sale of food products, close the uncovered area of the market, cancel the May-Day parade, and... in short, I told them all the things they needed to do, and how I had dealt with the situation at the institute. On the top of the table I saw they had a map which they had marked with arrows, like a military operation, showing the spread of radioactivity. I told them that I had been there, what levels of radioactivity we had discovered, that in this area it was 28,000 $\mu R/h$, in this other area, 18,000, elsewhere nearly 30,000... On 29th April, near the building of the Central Committee of the Party and the Council of Ministers we measured a level of 800 $\mu R/h$.

Meanwhile the Minister of Health, Savchenko, telephoned Moscow to speak to Leonid Ilyin, who was radioprotection advisor to the government, and asked him to comment on my proposals. Ilyin's response: "You don't need to do anything" Turning to me, Savchenko said "The doctor says it's not serious, and you're sowing panic". At that moment, the President said in a steely tone: "Comrade Nesterenko, you do what you think right at your Institute! But we'll cope here without you!" I told him, "Mikhail Aleksevich, I have already given the necessary orders at the Institute. It's part of my professional duties. I don't need your authority there. They have already taken iodine as a preventive measure. "We'll think about it" he said. They carried on with their discussion blocking me out. I left the room and I said to myself, "No, I'm not going to leave it at that!" I went back to the institute and wrote a memo to the government. I coded it, because we had already been told that all information about the radioactive contamination was secret.[44] And I told the head of the main department at the institute: "Tomorrow morning, 30th April, Sliunkov should find this report on his table!" I wrote on the 30th April, then on 3rd May, then on 7th... wherever we travelled, the information we were receiving got worse, I was seeing a catastrophe and I continued to send them reports.

It was difficult for anyone with a sane mind to imagine this degree of irresponsibility... When scientists do their work, they behave in a responsible way. But there was one thing we hadn't taken into account. It's like under socialism, when we thought we had got rid of material motives: "People should be morally responsible" I thought everyone would act responsibly in the development of these new technologies. Could you have imagined what they did at Chernobyl, when they

---

[44] As Mikhail Gorbachev testified: "The accident had no precedent. It produced ten times more fallout than Hiroshima and contaminated millions of victims. It was a tragedy and I would have liked to have more knowledge of the situation, more time to react, and better warning. *It is difficult to imagine, but at the beginning, no one had any idea, not even the scientists, of the scale of the disaster.* They thought they could control the situation. And we, in the government, were unclear about what we should do ... [...] No one is safe". Taken from an interview in the monthly *Jonas*, 24 August 2001. The disinformation continues. Politicians are prisoners of their ignorance locked in secrecy. (Author's note)

blocked the security rods which were there for precisely that eventuality, to stop the process and avert disaster in the case of an accident like Chernobyl? We have a saying in this country: "There's no defence against an idiot". I repeat, I don't think humanity has the maturity at present to deal with nuclear technology. I visited the areas around Chernobyl, and I saw so many children, we measured their levels, I remember…

While we were taking measurements, it was tragic, seeing the local people who had been directly in contact with radioactive sources. All my life I've been taught that you must only approach sources of radioactivity wearing special clothing, that you must protect people, that you need biological protection. People were walking along and getting irradiated. The fact is that we do not have sensory organs that can detect radiation. We saw that the people were completely defenceless. I got into very great conflicts with the authorities at this time.

When the documents from Chernobyl were declassified by the Supreme Soviet, V. Nesterenko published a selection of his reports to the government of Belarus and the USSR, chosen from about 1000 pages of copies that he saved from his archives, about the urgent measures that needed to be taken. Here is the first of these reports, published with a brief introduction, in the newspaper *Rodnik*, in July 1989.

For nothing is secret that will not be revealed.

CHERNOBYL
THE DOCUMENTS TELL THE STORY

The disinformation surrounding Chernobyl and above all the eternally reiterated, and now familiar, fable about "the silence of Belarusian scientists" at the time of the tragic accident[…] have compelled me to publish these documents and prepare this material for the newspaper *Rodnik*.

It is not hard to guess, I assume, who and for what reason needs to shift the blame for their own faults onto innocent people. We are quite familiar with these games. But in this case the stakes are much too high. We are talking about the future of our nation, its very existence. This is precisely why I cannot allow myself to calmly observe the way today's high level politicians distribute the winning cards in this game, how suddenly the trump card finds itself in the hands of their faithful ideological adepts, the guardians of power with their monopoly over the truth.

Telling lies about Chernobyl today makes the risk of another Chernobyl more likely. I therefore consider it my duty to make these documents public. I hope that everyone will then have the opportunity to find answers for themselves, without suggestions from others, to our eternal questions: "Whose fault was it?" and above all "What is to be done?"

V. Nesterenko

30th April 1986
To N.N. Sliunkov

Following a request from the Director of general epidemiological health of the Socialist Soviet Republic of Byelorussia, the Institute of Nuclear Energy at the Academy of Sciences of the Socialist Soviet Republic of Byelorussia undertook a spectrometric analysis of samples from the outside environment (soil, water, milk), given to us by the centre for Epidemiological Health at Minsk and by those territories close to the accident at the nuclear power station of Chernobyl in the Gomel region. The spectrometric and radiometric analyses of the soil in the villages and districts of Khoiniki, Narovlia and Bragin show that the inhabitants of these districts were exposed to increased levels of radioactivity from the moment of the accident up to the 30th April 1986. Our evaluations of the predicted radiation dose in these territories based on the results of soil sample analysis indicate that there could be considerable health consequences for the inhabitants.

In consideration of the estimated radiation dose of the inhabitants and the isotopes found in the soil samples, it appeared necessary to study and to put measures in place as soon as possible in order to:

1) distribute preventive iodine to the inhabitants of those regions of the Republic that have been exposed to radiation;

2) take health and hygiene measures to decontaminate skin, clothing, local amenities and housing;

3) limit the provision and sale of untreated food produce;

4) protect uncovered sources of drinking water;

5) organise health education for the population of the Republic;

6) conduct clean up operations in the towns of the Republic.

V.N.

Report of 7th May 1986

To N.N. Sliunkov

(...) In our opinion it is necessary to:

1) extend the evacuation zone to 50–70 km from the site of the accident [...];

2) set up a team of radiobiologists to work with colleagues from the Institute of Biophysics of the USSR Ministry of Health to evaluate the radiation dose received by the population;

3) set up a commission made up of scientific representatives, specialists from agro-industry, from the Ministry of Health, Food Industry, Trade and Services in order to take all necessary measures to evaluate the radioactive contamination of cereals, vegetables and other local products of vegetable or animal origin and make recommendations on soil use in the future;

4) ask the Ministry of Health of the USSR to establish radiation monitoring levels for the post-accident period.

V.N.

In the 1950s. Students harvesting potatoes. V. Nesterenko, first on left.

The young Nesterenko

The young Ilsa

Vassili and Ilsa in Moscow.

Vassili Nesterenko
and A. Alexandrov

V. Nesterenko, director
of the Belarus Institute
of Nuclear Energy.

Vassili Nesterenko travels around the territories with a Human Radiation Spectrometer.

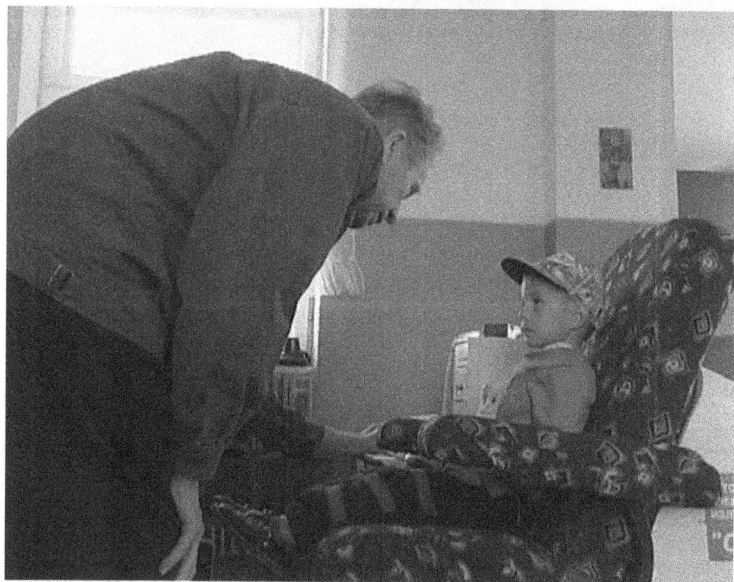

Taking measurements
of the children
in Olmany

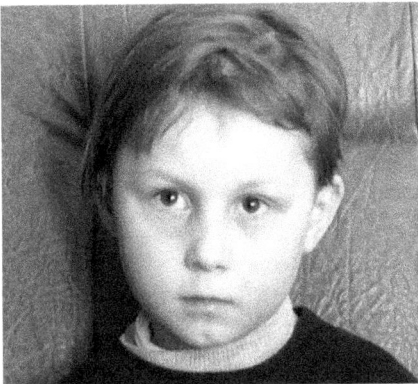

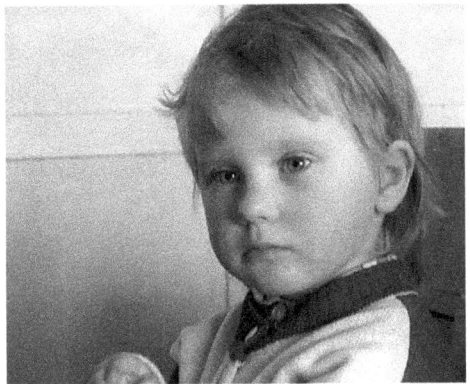

Ilsa and Vassili Nesterenko

3. HELL

V. Nesterenko.—During the night of 31st April and 1st May, I was called to the centre of communications at the Committee for State Security which is equipped with special telephone links, and at the other end of the phone was Valery Legasov. He was already at Chernobyl. I confirmed that using liquid nitrogen to put out the fire would not cause an explosion. "Perfect, a helicopter will come to fetch you. We will develop a plan together to put out the fire". The helicopter landed near my *datcha* and I left. If, as believers say, hell exists, I've been there. We were suspended at an altitude of about 300 metres. In the cabin it measured about 100 R/h, an enormously high level. It was dawn, when the morning sun creates very clear contrasts. Visibility was good. I had the impression that about half of the graphite content of the reactor had escaped. Through the thick smoke we could only see the concrete walls, shining a little in the sun. The smoke was monstrous, a dark red colour rising up at least 100 metres. The helicopter circled round Block 4, we couldn't circle directly above, of course—that would have meant certain death. The smoke fanned out so we were affected anyway also. At that time the cabins were not protected. I even tried to lean out to get a better view, and got my face burned by caesium. At that level, you can feel the radiation physically. When the radioactive particles touch the skin, it feels like an intense sun burn. We were all extremely tense, perhaps because of these unusual circumstances. When I tried to lean out, Legasov grabbed me by the neck: "What are you doing? Think of your son! He's only 12!" Our families were friends. I said to him: "If something happens to us, the state will help him, won't it?" And he grabbed me by the neck again and said "You're joking!" Now, I see how the liquidators are being treated....Most of the people who were there with me are dead. Legasov has gone; Professor Karasev is no longer with us. They're all dead.

We stayed above the reactor for about 15 minutes. Then we got ourselves decontaminated. They have to put a tube down into the trachea and wash the lungs, because obviously we had inhaled particles. The burns lasted three years. After that, the effects of radiation manifest themselves differently. In the stomach there are fermenting agents, digestive enzymes that break up different foods. Unfortunately some of those enzymes were destroyed. This was the case for most of the liquidators who received high doses of radiation. I know there are clinics in Germany and in Switzerland where you can have them restored. It takes about three months. But at the moment I can't afford it. In spite of my various titles—Emeritus Professor, Doctor of Science and Technology, member of the Academy—and my allowance from the government for outstanding merit, I only get 85 dollars a month. As you can see, that sort of money wouldn't even pay for a day in this type of clinic.

After flying over the reactor, we decided, with Legasov, to throw away the dosimeters that we had used because we had not kept to the recommended limits and the people who had let us stay out so long would have been punished. I think we received about 100 rem. All I know is that when I set out I weighed 84 kg and two months later I weighed no more than 61 kg.

The reactor burned for ten days. I was in no doubt about the seriousness and the scale of the accident. As the person in charge of radiological situations at the Academy of Sciences, I immediately assigned more than 1,200 of my colleagues at the Institute to evaluate the contamination and to establish maps of the radioactivity. In the months following my first letter to Sliunkov, the 1000 pages of reports to the government were of a similar nature: "We measured the radioactivity in such and such a place; here are the results; we propose the following measures…"

It had almost no effect. Only one of Nesterenko's recommendations was implemented: they washed the streets before the May-Day parade. While transcribing the above, a thought came to me spontaneously. Imagine if Nesterenko, who has battled in this human desert between East and West for twenty years, had simply never existed. Would these 2 million Belarusian peasants, including 500,000 children, even exist for us, for the world, to remind us about what has happened, what is to come, for historical memory? Without this witness, who, with his high level of expertise, dared to defy the nuclear powers, these children, these women, these men would have been ignored, annulled, "disappeared" one by one in silence. Annihilated by the lobby of deceit, that is preparing more destruction of normal life on Earth, a destruction that is statistically probable. "We're all going to die, so what's new?" This is the implicit calculation of indifferent officials towards the human guinea pigs. They arrive for a few days, are paid a great deal of money, then leave, having delivered their scientific verdict to passive, complicit governments. It would be up to us alone to transform this demented dream into a nightmare for the nuclear lobby. But, in the words of Doctor Zatkhei, the activist from Poliske[45], "People are too slow to realise the seriousness of the situation".

V. Nesterenko.—What really changed my life was the shock of seeing children contaminated. We had been insisting since the beginning of May on their evacuation. But it was only after the 10th May that they started the evacuation. I am thinking again of one of those night-time meetings in the room at the Central Committee office, where I and the president of the Academy were insisting that measures be taken, that the children be evacuated, and no-one supported us. Soon after that, I went to Khoiniki to see what was going on there. On my return, the president of the Academy had brought together all the institute directors to hear my report. As I was recounting what was happening there, I actually broke down. I started to cry because I felt that my words were falling on deaf ears, nobody would do anything.

---

[45] See page 24.

But I think my anguish must finally have got through because around the 10th May, things started to move.

On one of my missions, I found myself at Gomel, when they were evacuating the children. What I saw shook me so decisively that it changed the course of my life. At that point, it was no longer possible to leave either Kiev or Gomel. People were quite literally hurling themselves onto whatever transport was available, to get out of town. Parents and children had been taken to the station where convoys of trains were waiting. I do not know how they got them there; they probably had to use force. Children were being snatched from their mothers, thrown into the train, sent off to unknown destinations in the Urals, in Bashkiria, or Udmurtia. They loaded tens of thousands of children into these trains, and then they left. It brought back my childhood memories of the war: the Germans were retreating and they used terrified women and children to protect their troops... they put them... in front of the tanks, (*he represses a sob*) so that our planes did not bomb them. They were crying and screaming....In the end, the Germans were surrounded but our forces bombed nevertheless, and a huge number of people were killed. It took a week to bury them. At Gomel I relived this scene: children were being grabbed, loaded onto trains, and sent off, amidst sobbing, screaming, and hysterical cries for help. That was just my first experience...

Later, there was another. A never ending line of coaches was passing by on the ring road in the town near the *datcha* where we lived. I got in my car and drove to the nearby sanatorium, where the coaches had stopped. The children were getting out, exhausted. You can imagine, after a journey of six or eight hours, without anything to eat or drink, dressed any old how, completely done in, worn out. I watched them coming towards me. When I measured their levels, I found they had all been irradiated...

The third time that I got upset was when I had been called in by the president of the Academy, to examine children sent to a pioneer camp at Rakov, not far from the institute. Each child was carrying a toy and wearing clothes that we had to monitor with the dosimeter for radioactivity: everything was contaminated; it all had to be thrown out. We bought them new clothes. We had to take all of it away and throw it in the container for radioactive waste. At that moment, I thought that if this technology caused so much suffering to hundreds of thousands of people, it had no right to exist.

It is not possible to employ scientists of a sufficiently high quality at every nuclear power station to ensure that all regulations are followed. At Chernobyl, if they had not blocked the security systems, this accident would never have happened. But it did happen. There is nowhere in the world where man's moral conscience is equal to the challenge posed by such dangerous technology.

From 1989 to 1993, I presided over the joint, independent expert committee of Russia, Belarus and Ukraine, which brought together 200 scientists, doctors, geneticists, agricultural specialists, physicists and radioprotection specialists, a group which was to distance itself from the international committee of experts. The latter had claimed in its conclusions to the International Chernobyl Project, that the accident had had no significant health consequences. Hans Blix, the director of the IAEA, had stated that humanity could tolerate an accident like Chernobyl every year. What blasphemy given that in Belarus alone, 2 million people, including 500,000 children, out of a total population of 10 million, live in areas contaminated by Chernobyl and the government still cannot guarantee their safety. So long as we are unable to deal with the consequences of Chernobyl, we have no right to continue to develop atomic energy.

It appears that the countries that have nuclear power have learnt nothing from the accident at Chernobyl. In reality, we cannot guarantee the safety of people living within 200–300 km of a nuclear power station. Take the example of Olmany in the Brest region, more than 200 km from the power station. The level of contamination in food products there is so high that it constitutes a real danger to the health of the inhabitants. The children there are seriously ill. I've been to England, France, Germany and the United States… I've visited nuclear power stations there and they all say that everything is fine and that our problems are due to design faults in our nuclear power stations. Their protective measures are aimed only at the population living within a 20–30 km radius of the power stations. It reminds me of Professor Alexandrov, the scientific director at Chernobyl, who said one day: "This power station is so safe, you could put it in Red Square in Moscow". Thank goodness they didn't… Whatever safety measures you have, there is no guarantee against human error. Someone could always block the automatic security system, neutralise it as they did at Chernobyl, and the catastrophe would happen again in spite of everything. You can understand the position of governments who have invested a lot in the development of nuclear energy and want to see some return on their investment. There are also enormous financial interests at stake in the construction of nuclear power stations. It's a lobby. You can't expect objectivity from the IAEA because it owes its existence to the contributions made by countries that have nuclear power stations. What I can't understand is the position of doctors, whether at the WHO or at the ICRP (International Commission on Radiological Protection).

4. A HEALTH CATASTROPHE

THYROID CANCER, BREAST CANCER, DIABETES, NUCLEAR AIDS

V. Nesterenko.—I have worked for forty years in the nuclear industry and since 1990 I have worked full time with doctors in radioprotection. Today, the whole world recognises it: ten to fifteen years before Chernobyl, we would come across only 2 or 3 cases of thyroid cancer among children in Belarus, but today, in 1998, we have already operated on 920 children. In other words, there are more children suffering from thyroid cancer in Belarus today than anywhere else in the world. This phenomenon appeared very early on. Every year we come across 80 to 100 new cases of childhood thyroid cancer in Belarus and the numbers are not dropping. In twelve years, the children have grown up, and in that group of adults, the number of cases of the illness is also increasing. Today in 1998—I know these figures— more than 3000 adults have been operated on for thyroid cancer. In second place is breast cancer in women. This phenomenon had already been observed following the accident at Cheliabinsk, and now we are seeing it here, in Belarus. Childhood diabetes is becoming much more common, particularly in the Gomel region. It is very important to emphasise that radioactivity affects the whole immune system. And it isn't only the short-lived isotopes that appeared in the first few days; above all it has to do with the fact that for all these years, the inhabitants have been eating food contaminated with long-lived radionuclides. The organism becomes incapable of fighting off infection. It's a kind of nuclear AIDS.

EYES
For example, let's take the village of Svetilovichi, about 80 km from Gomel. It is about 180 km from the Chernobyl power station. The children in the village have, on average, an accumulation of 150–200 Bq per kilogram of body weight. But some have more than 2000 Bq/kg. These children have been examined by doctors: 23% of the children in the village have a cataract or impaired vision caused by radiation. An examination of the eyesight of 661 children in the Vetka region revealed that 48% of them had a cataract and need specialist treatment.

HEART
When radioactive caesium enters the organism in food products, it acts as a substitute for potassium and lodges in the cardiac muscle and other areas. The electrical conductivity of the myocardium is altered. In the same village of Svetilovichi, more than 84% of the children suffer heart arrhythmia. Unfortunately, they are not only candidates for heart attack but will have heart problems from now on.

BLOOD PRESSURE
Almost half of the children suffer major alterations in their cardiovascular system caused by the constant accumulation of radionuclides in their bodies. In general their blood pressure is too high. Where else would you come across children of 13–15 years old with blood pressure of 18–19 cm/Hg? That is what we are seeing, unfortunately.

DIGESTIVE SYSTEM, DIGESTIVE ENZYMES
Doctors examined children's gastrointestinal tract using gastroscopy to see what was going on in the oesophagus, stomach, and duodenum. They found that 80%

of the children had gastritis or even a stomach ulcer. They found that the mucus lining of the stomach, even in children 12–15 years old, was atrophying rapidly and looked like the stomach of a 70 year old. Radiation has a wasting effect on the body. These children are already suffering today and will become very ill in later life.

—You have this problem too, don't you?

—Yes, it is the case with most of the liquidators. Studies like those conducted by Professor Burlakova in Moscow, for example, have shown significant increases in cancers of the digestive system as well as a loss of digestive enzymes. It is a characteristic of the liquidators. I lack all these digestive enzymes as well, to such an extent that I no longer know what an apple is, or a cucumber or a tomato....I can count on the fingers of one hand the items I'm allowed to eat. I've lived like that since 1986, I've got used to it.

LEUKAEMIA

Those people who were at the scene of the accident and received very high exposures to radiation began to develop leukaemia from that moment. That was to be expected. But we must expect other cases of leukaemia for another reason. It wasn't only caesium that was dispersed across the area. There was a lot of strontium in the fallout. When strontium enters the body—it is the chemical equivalent of calcium—it concentrates in the bone and penetrates the bone marrow, the seat of our hematopoietic system. Unfortunately there are already cases of leukaemia in Belarus. Now is the time that it will begin to develop.

FERTILITY, THE PLACENTA

Another fact that has been verified by doctors. I listened to the gynaecologists' report at the annual conference of Professor Bandazhevsky's Institute of Medicine at Gomel in December 1997. We know that the placenta of a pregnant woman protects the child from heavy metals: it captures them. It also captures caesium, of course, but nature had not foreseen that this captured metal would emit gamma rays. This is how the foetus is irradiated, the pregnancy retarded, that problems arise during birth, and what is particularly serious, the child will experience problems throughout its development. This is what doctors have verified in experiments at Gomel. The doctors examined about 250 young girls from the town of Gomel, aged between 18 and 23 years old. The permanent accumulation of radioactive elements in the body through the ingestion of contaminated food causes damage to their genital organs; 20% of these young women will never be able to have children.

BRAIN

Over a period of about ten years, Professor Kondrashenko monitored the psychological development of children living in the areas contaminated by Chernobyl. This is what he established. 42% of the children in this area had delayed psychological development of between 3 and 4 years, leading to mental retardation; 6000 children were born mentally retarded. It is their brains that have been affected. The newspaper *Izvestia* reported recently that 600 people have already presented themselves at the psychiatric hospital in Kiev: "Help us. We can't read any more. We can't count money any more". Later I talked about this with the doctors at the psychiatric institute, and they explained the phenomenon by the following mechanism: when the radionuclides enter the brain tissues, the immune system reacts to these tissues as foreign bodies and it starts to attack them. In fact they simply devour them. Professor Kondrashenko believes that if nothing is done, if clean food is not provided and the children are not protected from this radiation, there will be no longer be anyone capable of completing higher education in three or four generations in Belarus. This is what he has published.

Chapter II
THE LIQUIDATORS SACRIFICED

The liquidators have been sacrificed and betrayed, at least four times by those they saved: 1. They did not receive adequate information or protection. 2. The price they have paid, and continue to pay, in terms of their lives and their health, has not been recognised. 3. They receive no medical care. 4. They have been forgotten by the world and cast onto the scrap heap of History.

The graphite and uranium scattered all over the roof of the nuclear power station after the accident was giving off up to 20,000 roentgens per hour. Held in the hand, a piece of graphite would give off the equivalent of a lifetime's accumulated dose from natural background radiation, in just a second and a half. According to the most recent estimates, between 800,000 and a million young men, known as the "liquidators", were mobilised from all over the USSR to deal with the exploded reactor, to put out the fire that burned for ten days, to enclose it in an improvised "sarcophagus", in conditions of terrifying radioactivity, where the remains of 200 tons of melted nuclear fuel had solidified like lava through a surreal labyrinthine ruin of concrete and twisted steel, and to "clean" the contaminated territories—that is to say the fields, the roads, the houses in the villages—all the areas on which the radioactivity had rained.

They battled with the radionuclides with their bare hands, using spades and water cannon. The barrel of the Danaides, an absurdity, a saturnine moment from ancient mythology. These victims of torture, exposed to enormous levels of radioactivity, given no official recognition, are ill. Tens of thousands have already died, and more are in the process of dying.

People often wonder why such an enormous number of people, a veritable army, were employed to deal with an accident at a power station. Surely this sort of accident, which is statistically predictable, is an inherent risk factor in the functioning of the industry? The absence of any protective measures that could deal with the scale and the duration of the radiological event that had taken place at Chernobyl demanded military logic: unable to offer effective protection against such high levels of radioactivity, the Soviet authorities chose to spread exposure over as large a number of people as possible, over a time period that, depending on the risk, might be counted in minutes, or even seconds. Many received doses of radiation that exceeded the scale measured by the Geiger counters. This calculation was a mental abstraction, without any forethought or preparation, to which has to be added improvisation on the part of panic stricken officials, with a disdain for human life and the practically unlimited freedom of totalitarian power to mobilise the masses to serve their cause. A series of propitious circumstances for those in power. The "conscripts", denied protection and information, were used to close, with their bodies, the breach, that had opened up in the presumptions of modern technology. It was, historically, the last luxury the Soviet system could afford to pay itself, and us, for our protection. We have to ask ourselves how the "free world" will manage things in the event of a major accident in the West, that the IAEA itself estimates as (statistically) probable.[46]

Twenty years have passed. The IAEA, UNSCEAR and the WHO have, incomprehensibly, excluded the liquidators from their statistics, and still claim that only 32 people have died, maybe 40, as a result of the accident at Chernobyl. There has been no monitoring, no epidemiological research among this cohort of 800,000 men exposed to enormous levels of radioactivity. Covered by the silence of the scientific establishment, the nuclear states, and the "international community", look the other way, and are simply waiting for the liquidators to disappear without making waves. You cannot even say they have been "forgotten", because to forget someone you have to recognise that he existed in the first place. No, these anonymous individuals, dispersed over the eleven time zones of the former Soviet Union, are excluded from the human community. They do not exist, these people who saved us and who were simply asking "to be treated like human beings".

Here is the testimony of six liquidators that we interviewed during our first meeting, in 1990.

1990

1. MOBILISATION

Piotr Shashkov.—It was 5th June. I was working the third night shift with my team. At 2 in the morning, they came to get me: "Shashkov, the service chief wants to see you in his office". "What's it about?" "I don't know". I got there and found

---

[46] The IAEA conference entitled "A decade after Chernobyl" held between 8th and 12th April 1996 in Vienna, discussed among other things "…the measures to be taken in a future accident, which is an inevitability, with the explicit aim of reducing costs for the industry responsible". Michel Fernex, who was present at the conference, reported this in "La catastrophe de Tchernobyl et la santé", (The Chernobyl Catastrophe and Health. See page 564) "Mechanisms of Injury", "Radiation Dosimetry", "Epidemiological Approaches" and "Future Accidents" were the four themes of the conference entitled "Biological Effects of Radiation Injury" organised in 1996 in Minsk under the auspices of the European Commission and the USA Department of Energy.

a colonel, a major and a captain. They said to me "Comrade Major Adjutant, you have been called up for service". "Straight away?" "Yes, straight away. Get changed. The coach is waiting for you". I knew there had been an accident. I asked "Where am I going?" Answer: "Liquidation work" "What's the point of changing? I'll go in my work clothes. They'll be thrown out afterwards anyway". I left in my work clothes. I couldn't refuse. The summons I had received was marked red.....I was under military orders. After the accident, the Ministry of Defence declared a "state of war" and general mobilisation. The summons was underlined in red. If I had refused, I would have faced a military tribunal.

M. Boikov.—In our country, all the men are reservists and can be called up at any moment. The system made great use of this, to use them as an army to undertake agricultural work, construction work, repair work...It's free labour. Instead of paying people, you conscript them and send them anywhere there is a need for man power.

Defence of the country is the sacred duty of every Soviet citizen. It is written into the constitution. Refusing to comply is the equivalent of desertion. You have no choice.

Alexander Grudino.—To begin with, I had no idea I was being taken to the nuclear power station. We were all brought together; I thought we were going into the fields to harvest the wheat. When we arrived at Bragin, I understood everything. We weren't coming back. We're talking about the Soviet Union. You receive the call, you go and fight. Like in Afghanistan, or Grozny... Dying for your country. I wasn't at all frightened but I was aware of what I was doing. We were going to war, really. Only it was a war without bombs or shells... We had no idea what sort of war it was, in fact.

2. THE WORK

P. Shashkov.—We had to get onto a coach and were taken to the base. There we were given bulldozers. Because I was familiar with all military engines—I had been a tank driver—they gave us armoured infantry tanks adapted specially for the occasion: a digger at the front and a tip truck behind on caterpillar tracks. I had to dig up a layer of earth 20 cm deep, make a big pile and load it onto a lorry that took it away immediately. Then my place was taken by another lad. It took about forty minutes in all. That was a lot, the level of radioactivity was very high. But they said, "The armoured sides will protect you".

I also worked on the roof of the power station for four days. The first day I was told to cut through the side of a concrete slab to allow water through. With a pick axe and a sledge hammer, it took about three minutes. The second day, seven of us lifted up a concrete slab and dropped it down. The third day we had to dismantle a ventilation pipe. And on the fourth day it was a piece of graphite. There was no shovel. I had to pick it up in my hands and throw it down. I knew pretty well what dose I'd got, but they marked it lower. I protested. They told me: "Get out of here and don't say a word". I also worked in the clothing depot, where the radiation level was very high. The special overalls that I gave to the "comrades" were all contaminated.

For all our good work on the roof of the power station, we were entitled to a diploma that the colonel gave us when we got back. While we were there, we were told to "run like a dog then flee like a hare" because you absolutely could not stay there for longer than three minutes. In three minutes, the dosimeter registered 30 roentgens. But they never marked down more than 5–7 R in our records. The first day, my dosimeter registered 34 and they marked it 9. The second day, about 30 and again they marked it 5. The third day, it registered nearly 40 and they marked it down as 2! What they wrote in the records was about ten times less than what we received. I pointed it out to the colonel: "We know better than you what's got to be done. Get out!" I myself was an adjutant-major. I said to him: "You've got no right! Write down what I received!" "Get out of here and don't let me see you again!" Full stop. That's it.

Victor Kulikovsky.—Me too. I tried to find out what was really going on. I felt the effects of the radiation straight away: nausea, dizziness, feeling ill, sudden weakness. And I really wanted to know how many rem I had absorbed. But at the radiometry office, they took the dosimeter from me as soon as possible, and took it away behind the door, then came and told me: "Your average dose is 11.92 rem". I said: "That's not true. I'm going to see your superior". "Go ahead". I found him; he sat there in his armchair, sneering: "Ha Ha! You should be happy with what they've written down. If you make a fuss, they'll mark down even less".

A. Grudino.—I worked at the power station twice, once on 25th August and once on 30th August. Before they built the sarcophagus, they had to clear up all the graphite and uranium rubble that had been thrown onto the roof by the explosion. They had already tried to do it using robots but the radiation prevented them from working: the electronic instruments melted inside, the robots blocked and would not move. So they decided to send men in. We went in wearing only our soldier's uniforms, gas masks to protect our faces and motorbike goggles. Then I worked on the reactor itself. I had to clean up the roof of the machine room. We had to throw the graphite and uranium off the roof, sometimes using our hands, sometimes using shovels. The first time I stayed there for about 40 seconds. Within this time lapse, you had to run up, grab a shovel if there was one to hand, and if there was no shovel, you had to pick up the piece of graphite with your hands. Then throw it into the reactor. Then get out as quickly as possible. A siren would sound to tell you when your time was up. If I spotted a piece on the ground, I would try to take it as well, even if the siren was going, then I would rush back down. The second time I went up, the radioactivity measured between 800–900 roentgens an hour.

We made ourselves a sort of armour made out of sheets of lead to try to protect ourselves a little from the radiation.

Q.—Are you saying that four months after the accident, you still hadn't been given any real protective clothing?

A. Grudino.—No, we had nothing other than what we put together ourselves. When we came off the roof, we took our shoes off and the next person put them on, if they were the right size, when it was his turn to go up.

Victor Kulikovsky.—I worked directly on Block 4. We were installing lighting. The work carried on night and day. Our job was to ensure there was lighting at night. We were pulling up cables from electricity panels; 5–10 people got hold of the cable and then ran with it as fast as possible and threw it down. The other team took over. We had literally two minutes to do it. There was a ladder there. And from this ladder, directly against the wall of the reactor, from the first few days, was where we climbed to install the lighting.

Q.—And you didn't stay longer than two minutes?

V. Kulikovsky.—No…that was the time indicated by the dosimetrists: "There is this level of radiation, so you have this amount of time". That's it. But how can you do your work in two minutes: run there, do it, run back? It isn't realistic. We stayed for five minutes, seven minutes, sometimes as long as ten. Some of the dosimetrists had to pull us away. They loaded us onto the bus and drove us away. They really had to chase us off. Today, the place where we were working is underneath the sarcophagus.

One day, after work, I went to the depot to get a change of clothes. The dosimetrist came over and measured the radiation. "Normal" he said. "How can it be normal if I climbed up the wall?" "Normal. For you, it's indicating normal". Sometimes it took five or six of us to push our way through to the manager's office in this dosimetry service. It was only after we threatened to complain to HQ that they gave us papers certifying that our clothes had been contaminated and that we needed to exchange them for clean ones.

When we went to the canteen in town, in Chernobyl, we had to take everything off apart from underpants. "Everything off! Your clothes are radioactive!" That lasted till they set up a place for us to change our clothes. We ate our meals dressed only in our underpants. Then we put on the same togs and went to work for a couple of days again at the power station. "Your clothes are clean…"

Anatoli Saragovets.—My first job was to measure the levels of radioactivity in the villages in preparation for digging up the top layer of radioactive soil. They gave me a Geiger counter. Everywhere I measured, it blocked. The radiation was too high. I gave the counter back so I didn't have to see this horror. "Take it. Give me something else". They gave me a big shovel and I did a bit of work. Then they put me in a lorry, a water container that was being used to wash out any radioactive waste in the ditches and on the roads in the villages. One day, next to a road that went through areas that were all closed off, I saw a huge pond, full of contaminated water. The water was dark green, really frightening to see. The firemen beckoned me forward and they pumped this same water into my lorry. I asked them: "Where do you want me to put this water if it's contaminated?" "Take it to the village and don't argue". And that was the water the firemen were using to wash the houses. Afterwards they measured the levels of radioactivity at zero point something: "It's liveable…" But the next day, the control units had come by in the night, they sent me back to the same village: the levels of radioactivity were too high there….We'd done a bad job.

Q.—Do you think the work you were doing was absurd?

A. Saragovets.—Completely absurd. You dig up contaminated earth from around a house. Then you wash the house and the contaminated dust drains away with the water into the gutter. Then you dig a channel to drain the bad water away but all around the channel everything is contaminated. The people who lived there saw perfectly well how useless it was. When I asked my superiors "What's the point of this idiocy?" they said: "Don't discuss it. Just do what you're told".

Anatoli Borovsky.—I was battalion chief. We were decontaminating the villages. We were digging up earth with shovels and loading it onto lorries. Then it was taken away and put into ditches and buried. Obviously the dust was flying all around us and we were breathing it in. Our detachment was working in a region between Khoiniki and Bragin. We asked to be provided with mechanical excavators, made in Minsk, little bulldozers on tractors. They would have been very useful. An official from the Central Committee came and promised us everything under the sun. He kept looking at his watch as if he was very busy and had to leave any minute. We never got anything. We carried on working with shovels. But now I would like to tell you about a scandalous fact, the way the State betrayed us…

3. STATE SECRETS

A. Borovsky.—Savchenko, the former Health Minister, in a recent reply to the Supreme Soviet Commission that is investigating the cloak of silence that has surrounded the events at Chernobyl for so many years, has stated that they were called in by Ryzhkov, the Prime Minister of the USSR, who told them that all information about Chernobyl was top secret. That was the reason why the calculation of dose was never done, and when measurements had been taken, they had to be lowered. The order to lie came from high up.

Q.—Can you explain the logic behind this behaviour?

A. Borovsky.—During the seventy three years of communism, our blessed State has always undertaken extensive, grandiose schemes. The first was world revolution. As a result, all of us, I'm not sure why, felt we owed something to the State. We were eternally in its debt. Never the other way round. And that continues through inertia up to the present day. The academic Ilyin's famous thesis, according to which it is normal to accumulate 35 rem over a lifetime has its origins in this attitude. This is where it all comes from.

V. Kulikovsky.—Just as people are all different, so communists are all different. For example, some want to evacuate the contaminated territories, some are against. And again, at Chernobyl in 1986, some were in favour of sending men onto the roof of the reactor, some were not. I remember this colonel shouting "I will get the roof built with partisans!" meaning all of us, who had been called up for the decontamination. "I will build it!" And that's what happened; he built a cover for the reactor with our lives.

A. Borovsky.—This year, before the sowing season, the Supreme Soviet adopted a resolution that forbade agricultural activity in any areas contaminated at levels above 15 curies per square kilometre. Three days later, the Central Committee of Belarus and the Council of Ministers, 99% of whom are communists, met together and took the decision to plant seed. An inhumane decision; anti-human in fact. And the land was planted.

M. Boikov.—We set up a non-governmental independent association, the Pripyat Union, which brings liquidators together. We get help from business. But the Communist Party has never given us any help.

4. HEALTH

A. Borovsky.—My state of health.....at the moment, I have neurovegetative dystonia, cardiac neurasthenia, that all of us have, everyone who was at Chernobyl. I suffer very badly with my stomach. I have a kidney problem whereas before I had no problems with my kidneys. I have become very irritable. I feel permanently exhausted, which they have defined as "asthenic syndrome". It was diagnosed recently when I was hospitalised at the radiology clinic.

M. Boikov.—In 1989, I worked as a specialist technician repairing the washing machines that were used to wash the contaminated clothing. These machines were used non-stop and had accumulated enormous levels of radioactivity. This was at Bragin. I lived in this contaminated area for six months until it was evacuated in 1991. I have heart problems too, neurovegetative dystonia. I have a lot of headaches and I feel exhausted.

A. Grudino.—I am registered as having category 2 invalidity. I have so many illnesses that I can't enumerate all of them. At 35, I am like an old man of 70. Diseases of the blood, stomach ulcer, shrinking of the blood vessels in the brain, permanent fatigue, I feel sleepy all the time. No energy. Dizziness. I can't walk far. After about a kilometre, my hands and feet are swollen and I have to find somewhere to sit down. I have become a little old man of 70.

A. Saragovets.—In November, I lost feeling in my left hand, then in my left arm, then the whole left side, then my legs became paralysed. No-one knew what it was.

They refused to recognise any link with the decontamination work. I continued to work because I have to feed my family. I drove the trolley bus with one hand and one foot; I was driving people and I didn't say anything. At rush hour there could be 150 to 200 people in my trolley, sometimes more. In the morning I left early to get to the bus stop. When I got off, I walked behind everyone else so that no-one would see me walking. My left leg would do nothing I told it to; I had no feeling in it; it was like a foreign body. It dragged on the ground.... I did the best I could, I made a joke of it, right up to the day I lost consciousness and they brought me back to the house. Now I can't walk at all. I have dizzy spells, but that's nothing... it's my legs really. They don't work at all. I have to hold on to the wall. And I have just turned thirty.

Q.—Are you getting any treatment?

A. Saragovets.—(laughing) From whom? We were in hospital in April. As soon as we started our hunger strike in May[47], they made us leave the hospital. Nobody is interested in us now. Not the doctors, nobody. We are society's rejects.

V. Kulikovsky.—In 1986, I worked for two months at the reactor. We don't really know exactly how much radioactivity we received. In my documents, as I said, they marked down 11.92 rem. I was in hospital at the Institute of Radiology in Minsk. I was in such a bad state that they didn't even want to take me. I got worse when I went to the pulmonary hospital at Minsk. They couldn't do anything there so I was sent back to the radiology institute. The blood test showed I had received 100 roentgens.

I have about twenty different illnesses; when the neuropathologist at the Borovliany hospital saw my records, he said: "What's all this, you must be an old man to have been diagnosed with all this!" I'm 35.

The medical profession has rejected us. "We can't help you". When we finally got them to recognise the link between our illnesses and the contamination, the doctor at the clinic said: "What are you doing here? They've recognised that radiation is the cause of your state of health. We can't cure you. Go home".

P. Shashkov.—In any case, they don't have the medicines we need.

V. Kulikovsky.—The professor told us quite openly. "These medicines can only be bought with foreign currency, from abroad. They're not for people like you".

A. Grudino.—(*He holds the papers in his hands, looks at them and says*) It's like looking at a piece of waste paper. They gave us these diplomas just after we'd been on the roof, for all our good work.

In the documentary "*The Sacrifice*", we included a clip from a Soviet news programme, in which you see a colonel giving out diplomas to the liquidators,

---

47 Hunger strike to obtain recognition of the correlation between their illnesses and radioactivity, in order to get free health care and subsidies set aside for liquidators.

who have just come down off the ruins of the roof at block 4 and are looking a little intimidated: "And to this brave young man too, I am awarding you this certificate and I wish you good health, well-being…and may you continue in your work with enthusiasm".

A. Grudino.—At the time, I was proud to receive this certificate, like everyone else. I thought that this document would give my children the right to live in a clean area, that I would have priority if I wanted a telephone installed, or wanted housing. We were promised everything but that's all they were… promises.

A. Borovsky.—As an officer, I needed to keep up the morale of my soldiers. My men understood perfectly well the importance of the task, to save the population and they didn't ask for anything in return, but they didn't think they'd be forgotten. Today, we're of no use, we're just a burden. We upset people because we are asking simply to be treated like human beings. In my opinion, we have paid all our debts to the mother country. Now, it's her turn to do something for us. But she wishes we weren't here. When we're gone, she'll be relieved.

A year after this first meeting, Lieutenant Colonel Borovsky died.

Chapter III
CITIZENS UNBOWED

Corresponding member of the Academy of Sciences in Belarus, the liquidator Vassili Nesterenko took us to the Institute of Nuclear Energy, situated in "Sosny" the little scientific community, built in the 1960's, eight kilometres from Minsk, where Nesterenko worked for twenty six years (1963 to 1989) including 10 years as director (1977 to 1987). Functioning at a reduced level, the institute now had the same air of abandonment that you find in many places in the former Soviet Union that were once of strategic importance. In the deserted forecourt at the edge of the forest, Nesterenko continued his account of the dramatic events he experienced here in his last few months after the disaster.

1. PERSECUTION

V. Nesterenko.—After the accident at Chernobyl, when we had already stopped work on the mobile mini-reactor, there were hundreds of cars parked here. They arrived from everywhere in the Southern part of the USSR to have food products tested. Our laboratory worked night and day. We had a special radioprotection service. It was the first to report that there was radioactivity in Minsk.

We tried to work on the 'Pamir' reactor again, but already our main focus was to use most of our staff, more than 1000 institute employees, to draw up the first maps of the contamination. By the end of May we had completed the map of Gomel, showing contamination by caesium and iodine, and by the end of June, of Mogilev. We put forward our proposals. There were instruments in this building and we explained to the government how to use them to help the inhabitants. As we were explaining the situation in Gomel, Sliunkov[48] suddenly turned to me and said: "What equipment are we going to be able to send there?" I told him: "This is to protect the people here. For other places, you'll have to ask Moscow". Equipment was requested and we sent it off. As regards Mogilev, I wrote a letter at the end of June. It caused great irritation. They had scarcely made the decision for the Gomel

---

[48] First Secretary of the Central Committee of the Communist Party, Belarus.

region and evacuated some of the inhabitants and here I was bringing up another problem. I told them they needed to monitor 50 villages and that the Ministry of Health needed to determine the degree of danger for inhabitants if they continued to live there. They would need to implement the distribution of preventive iodine, and all the rest, of course. I wrote all this in my letters to the government. So they decided to get rid of me.

As it was a military Institute—subject to defence secrets—I had a second in command, a Lieutenant General Budakovsky. He had worked in Prague, in central Asia, in short, I think he was probably a KGB collaborator. He was called in to the Central Committee and was told: "Your director is causing us enormous problems with his letters about Chernobyl, we need him removed. There must be things going on at the institute that are not satisfactory. Write to us about it. We'll send a commission from the Party to undertake an inspection". He wrote in August. On 2nd September, a commission from the Central Committee arrived. The commission was restricted to 75 people in all. They brought together all the staff and asked them to make accusations: "If you have any complaints about the director, write to us". There were about 200 to 250 people in the large assembly hall; there were officers, and the intimidation began: "No-one leaves the room until they have presented their deposition".

I continued to send letters to the Belarus government, then I wrote directly to Gorbachev and to the Prime Minister of the USSR, but my letters were returned. I started to have problems with my health, because after flying over the burning reactor, my stomach ulcer reopened and I finished up in hospital. They tried to treat me there while the commission continued its methodical work at institute. I was given a lot of support by the President of the Academy, Nicolai Borissevich. He was an academician, not only in Belarus but in the USSR and now works in Russia. He defended me as much as he could, but I saw that it was going to turn out badly. So then, because I knew V. Legasov—we had worked together—I said to him:" Vladimir Alexeich, they won't let you work in Moscow, come to Minsk. This institute should dedicate itself entirely to Chernobyl and I will work for you in whatever post". I knew quite well they were going to sack me. He agreed, he got on well with the president. But a week after our conversation—they had started to persecute him in Moscow—he committed suicide.

In March, Borissevich was removed from his post as president. He said to me: "You'll be next". I was seriously ill in hospital; I had lost a lot of weight. It was my heart, of course, and my stomach. I was called in… An official from the Central Committee came in and said: "Your absence at the institute is causing problems; you need to go to Moscow, sort out some financial questions. You need to take an exeat and fly to Moscow". In our legal system, there is this rule. As long as I was in hospital, they could not sack me or make any administrative changes. I came out of hospital, I went to the meeting—the academician Platonov was there, he had been made president of the Academy. Up till then, he used to tell me: "The people of Belarus will always remember what you did. You warned us of the danger; it's just a pity that you wrote to Moscow. You should never wash your dirty linen in public". I told him that I had written to the Central Committee here and had never had any response. That's why I had written to them. So I went to the meeting and they removed me from my post "for failure to complete experimental work on the Pamir nuclear reactor on time". I didn't finish on time, because we were conducting very detailed tests, but they wanted to hurry things along. Secondly, for having created, as they expressed it, a bad atmosphere among the staff. Because some of the staff had begun to write to various authorities, including Moscow, when they saw how unjustly I was being criticised. For example, Devoino and others wrote to Moscow to defend me. There weren't many, most of them betrayed me actually…This is how they saw the situation: my institute was the most dynamic in the Academy and they thought that whoever took over as director would keep all the funding and the accommodation. But these things were not there for my pleasure. Behind it all was the project, and it was the project that got all the money so that we could pay the staff good salaries, build them houses, a school, a nursery.

—We in the West always thought that the people of the USSR were still afraid to make independent choices against the wishes of the Party and that you paid dearly for it.

—I would say that many terrible things happened at that time, including my hair turning white, because after all, they accused me of having caused panic, and the Party punished me. That affected me very badly. A kind of negative force field developed around me: I was a wicked person. A faulty personality, an alarmist, etc. But in reality people's reactions were not so clear-cut. There was one Party member called Kusmin; he was secretary of the Central Committee. He said: "Why do you take it so seriously? These punishments handed out to you, during this so-called *perestroika*, the Academy's blame, you should think of them as medals: you should think of it as "the Red Medal for Work", the regional committee of the Party blames you; think of it as the Order of Friendship of the people. The Central Committee blames you, think of it as the medal of the Order of Lenin. So you have lots of decorations, don't let them get you down". *Perestroika* was challenging all the values that we had held before. He made me see the idiocy of this cult of decorations and ritual recognition. I often remember Brezhnev, who loved to display all his medals on his chest. I understood the vanity behind all these honours and rewards. He said: "Don't worry, do your work as if nothing had happened! All these reprimands are as meaningless as an array of medals pinned to your chest. Really, you should think of them as marks of distinction. Your work was so good that they had to punish you". That was how he saw it. He was a very interesting man. He's still alive; he fought in the war. Whereas I fretted about it. I believed in the Party. I belonged to that era. I had become an outcast in society where before

I had received all the honours. Because it's very important for a scientist that people say that your work is useful. That was the general opinion. Then, suddenly, what I did was of no interest to anyone. That was what really upset me. Later I realised also that the government could not give 100% help to the people—the economic costs were calculated later. That's the reason that the government needed to prove that the negative effects of the radiation were insignificant. That hasn't only just started. It started under Gorbachev.

Q.—Were you disappointed in the Party?

Nesterenko.—Yes, when there was the putsch, to get rid of Gorbachev, in 1991, I realised that it just a power struggle, they didn't care about the people. When I realised that, I sent my card back to the Party. That was when we went to Germany, myself, Ilsa and Aliosha, I had already left the Party. When I was there I faced a crisis of conscience: I was asked to stay. But I had already set up Belrad, it was already ongoing. I told them that I and my staff were working on a number of projects, and that I couldn't do that to them, I had to go back. And I went back.

Q.—Before, you believed in the Party?

Nesterenko.—Yes, absolutely. Our information about the West was limited and we were convinced that our system was more just. Take my case. In my family, my father was an electrician and my mother was illiterate. I think that in another system, I would never have received this further education and become part of the elite. You can criticise it as much as you like but if you had a brain, in this system you would receive further education and achieve your potential. That's already disappearing now. I myself never had any problems. By the seventh grade I already knew that I was good at maths and physics. I wanted to study at the Faculty of Missile Technology. They didn't let me, because when I was 7, I lived in an area that had been occupied by the Nazis. I regretted it very much, but today I'm really glad that I chose nuclear technology and not rockets.

2. INDEPENDENCE

Nesterenko.—During this period when I was being persecuted and obstacles were being put in the way of my radioprotection work, the writer Ales Adamovich suggested that I set up an independent institute. Material about Chernobyl was no longer a state secret. Nobody believed the official information any more. The Prime Minister of Belarus at the time was someone called Kebich. He said the same thing to me. "Do it". For instance there had been the incident at Narovlia, where the workers in a factory had decided to go on strike. Kebich said "Go and talk to them, evaluate the situation". I took agronomists, physicists, doctors, forestry workers with me; it was a multi-disciplinary expedition. We went there. The government thought we were going there to reassure the people. I came back and I said "We need to evacuate another 7 or 8 villages there". We didn't only visit Narovlia, but the whole district. This is the sort of work I began to do. At the same time, we began making equipment, dosimeters.

But when I made the decision—because I was still at the institute—to stop working on atomic energy and to work only on radioprotection, I came up against the same problem I had encountered in 1986. You can say "Halva! Halva!" [49] a hundred times but you can't taste the sugar in your mouth. Because we have no sensory organs to detect radiation, we needed to give the people a machine to do this for them. So we had developed, and tested a dosimeter called Sosna (the Pine). It was at this point that the obstacles began to be put in our way. I was told that this research was not fundamentally a worthy academic subject, that it should be left to the Ministry of Health and that I should be working on something more serious. But I think we have a duty towards the victims of radiation, we have the expertise and we must help them. This is when Sakharov, Adamovich and Karpov suggested that I set up an independent institute. As I could not leave on my own, I put the proposal to some of the people I worked with: there was Devoino, my assistant deputy, and my brother Volodia. I asked the people in the select group who were working on this machine. Actually there were many people who asked to come. They preferred to come with me probably because they were interested in my ideas and they shared my point of view.

—They weren't afraid of compromising their careers?

—They were Soviet citizens, all used to the idea that the government provides the money. Whereas I proposed "Come on lads, but I'm warning you: we won't get anything from the government, you'll be living off only what you can earn. It might work but it might go very badly". About thirty people came with me.

Returning to the subject of the dosimeters—in 1989, I went to Kiev to see two scientists: Paton, who was president of the Academy, and Trefilov, who was working on Chernobyl. They were convinced of the need to produce an affordable dosimeter for the public. We already had one, it was ready in 1986, developed by my colleagues. But we hadn't been allowed to produce it. Gorbachev had come and we talked to him about this dosimeter and he said "Yes, we need to produce a people's dosimeter". That was what he called it. The Belarusian government placed an order and we set up this temporary work collective. I was the director, my brother Volodia was the production manager… That was when we decided to get out, to leave the institute. I went into town. I looked for some premises. I managed to find 180 square metres in Plekhanov Street. That was where we set up production. We had it tested by the State in 1990. Since 1991, the three factories in Belarus, at Borissov, Gomel and Rechitsa, have produced about 100,000 dosimeters.

---

[49] Oriental dessert made from flour, sesame oil, honey, fruit and almonds.

My analysis of the government's monitoring system had shown that it worked in the capital and the towns, but in the villages, where people ate food that they grew themselves—if not what is the point of living in a village?—only a few families were monitored on the rare occasions when a doctor made a visit. There was no systematic monitoring of radiation in food products, and a severe shortage of radiometers for measuring. Only the veterinary services had them, and yet there were 3,200 contaminated villages that needed one. Our institute made 50 of them for the collective farms producing meat. But it was a drop in the ocean.

The whole system for radiological monitoring of food products, that had been created long before, for a state with a planned economy, no longer corresponded with decentralised production and the post-accident situation. The location of radiometric laboratories in district centres or in food processing plants, their remoteness from the villages (15–20 km), the administrative orientation of their work, the crisis in transport, the particularities of diet among villagers with local food products made effective monitoring impossible.

That was when we created our own system of local radiological monitoring centres (LRMC) for food products. It was Adamovich again who suggested it to me. He said: "Let's talk to Anatoli Karpov, the chess champion, about it, he'll help us". A Telethon had been held for Chernobyl around that time and some of the money collected was given to us, half a million roubles, which was a considerable sum.

I approached Ivan Smoliar, a good man, president of the permanent commission of the Supreme Soviet of Belarus for Chernobyl. At a meeting of the Joint Regional Soviets of Gomel, Mogilev and Brest, he put forward our idea to create 15 to 20 centres, as a start. The idea was approved.

We came to an agreement with them, received the first amount of money, and then Karpov gave us money from the Peace Foundation that he directed. And we started this work. We attracted teachers and nurses into the centres and we trained them. Fairly quickly we were able to gather together data. We found that the levels of contamination were between 5 and 10 times higher than those reported by the Minister of Health.

We measured contamination in milk and in the main food products containing the highest levels of radioactivity. More than 60% of the radioactivity was coming from the milk. The results do not vary: between 15% and 25% of the milk brought to us by local people for testing cannot be consumed. It is not fit for children, or even for adults.

But when the Ministry began to say that this level of contamination of food was not dangerous, and that children had very active metabolisms, we thought we had better start measuring the level of accumulated caesium in the bodies of the inhabitants. Because there was a direct link and it was clear that the inhabitants were going to be eating contaminated food over a number of decades. So the idea came to me that we should buy a spectrometer for the radiation emanating from the contaminated human body. It is a machine, an armchair, in which the person sits, and it has a receptor in the back (a counter that registers radiation and elementary particles). After three minutes, you can see the level of radiation in the body. It is, in fact, the most objective way to test the effectiveness of the protective measures put in place by the government or by the individual himself. The effectiveness can only really be evaluated on the basis of the radionuclides contained within the organism. We began to measure the inhabitants in 1994–1995. This is when I went to Gomel, which is the most contaminated region. And it was here that I met Yury Bandazhevsky. The first time our work brought us together, was in a little town called Vetka. This was in the years 1995, 1996, 1997. We gave him our results with the names next to them and he examined alterations in the health of those with the highest accumulated levels. This is when he discovered, with his wife Galina—he is an anatomopathologist and she is a paediatric cardiologist—that there was a correlation between dose and the alterations detected in the heart by an electrocardiogram.

Speaking as a specialist, I consider the most important scientific information is coming from Gomel, unfortunately not from Minsk and, not from the institute of the Ministry of Health in charge of radiological medicine. Bandazhevsky is the first, not just here but in the world, to have said that once caesium has penetrated the organism, it accumulates unevenly. His measurements of organs during autopsy showed that for an average of 100 Bq/kg incorporated in the organism, there were 2,500 Bq/kg in the heart, 1,000 Bq/kg in the kidneys…He said that a dose of 30–50 Bq/kg in a child will result in pathologies in the vital organs and systems. This is extremely significant in my view. I was then able, thanks to our 4 mobile spectrometry laboratories (eventually I hope we will have 12) to begin to inform the inhabitants. There are 1,100 villages where the conditions of life are dangerous for children.

—How much higher is 30–50 Bq/kg than normal radiation levels?

—Normally there shouldn't be any at all.

—But normally we receive 0.114 microsievert/hour (0.114 μSv/h) from natural background radiation from the cosmos, rocks etc.

—You're talking about external radiation. I'm talking only about internal radiation ingested through food. Obviously I'm taking into account the radiation we get from the cosmos, and from the external pollution that was created by Chernobyl, but today, if we exclude natural background radiation, and include only the radiation from Chernobyl, it's worth remembering that 70–80% and in some cases up to 90% of the radiation dose to the body is coming from food. That is why I put such emphasis on food products and on internal radiation. Another important point: if adults and children eat the same food—and that is the case here in Belarus with the levels of poverty today—the children will receive a dose 4–5 times larger than the adults. So, the most vulnerable group, the critical group, is made up of children, pregnant women, and women who are breastfeeding. Children have less body weight and their metabolism works 20 to 30 times faster than an adult's; that's why the irradiation is so high.

The information given to doctors should not be obtained on the basis of indirect deductions, because today we determine the danger from milk and potatoes on the basis of their supposed consumption by children. An official register of dose has just been set up. In the previous register, as I said, there were 1,100 villages where we needed to intervene. They cheated with the milk: by fixing the admissible level at 100 Bq/litre, it was impossible to keep below the annual admissible level of 1 millisievert per year, so it was decided that children should only drink half a litre per day. This is how they established the dose in the new register. In reality, to establish norms in radioprotection, as a specialist in the subject, I would say that you need first to identify the most contaminated group. To do that you need to proscribe the indirect method. You arrive in the village, you measure the children in nursery, you measure the whole school and about fifty adults. You select the 15 most contaminated subjects from these 200–300 people and you have your critical group. You don't need to work it out mathematically.

For instance, we have already investigated Vetka, which is about 80 km from Gomel on the Russian border. The accumulation threshold of 30–50 Bq/kg in the body observed by Bandazhevsky—above which he found alterations in the different vital organs and system in the child, represents, for me, a safety test: the contamination should not exceed this limit.

Chapter 4
THE CAPTIVES OF KRASNOPOLIE

However, from the mental health point of view, the most satisfactory solution for the future peaceful use of atomic energy would be to see a new generation of people who would have learnt to accommodate ignorance and uncertainty...

Technical Report No 151, WHO,
Geneva, 1958, page 59.

The beginning of the 1990's corresponded to the brief "democratic springtime" introduced under *perestroika*. People had great hopes that a historical and scientific investigation into Chernobyl might take place. But the collapse of the Soviet state and the institutional crisis that removed all means from the civil society that had emerged during perestroika favoured the nuclear agencies and their auxiliaries who occupied key posts at the Ministry of Health and in scientific institutes and they were able to regain control of a situation that had looked, for a time, as if it had escaped them. However, under pressure from society, certain partial measures to help the population were taken. They would be withdrawn over the next few years during 'normalisation'. All the same, some truths were beginning to emerge, and serious research into radiological medicine was being undertaken on the margins of official science. Yury Bandazhevsky, a brilliant young Belarusian anatomopathologist, was making huge progress in the field. He would soon describe the correlation between low dose radiation incorporated into the body and the pathologies in the contaminated territories. He would end up in prison for revealing these findings and for denouncing the policies of the Ministry of Health in his own country.

The disaster at Chernobyl had destabilised the most fundamental certainties and added a profound existential crisis to the more general material and moral collapse, that accompanied the disintegration of the Soviet system. The following testimonies show the disarray of both the rural and urban population struggling with the scourge of radiation. They were left to themselves, while the local leaders fled and the government authorities made their pact with the enemy. The testimonies also demonstrate how people resisted in spite of the injustice and humiliation.

1. THE PEASANT COMMUNITIES

For the population, under attack from invisible contamination, the danger was not only invisible but insidious. Fallout from Chernobyl had dispersed in "leopard spots", dependent on wind and rainfall, adding to topographical contrasts, and the contradictions and lies from official sources. In the same region of Krasnopolie, where the radioactive cloud, blowing in a north easterly direction was artificially seeded with chemicals fired from rockets in a helicopter in order to protect Moscow, the main town is relatively clean, whereas the surrounding villages are as badly contaminated as the town of Pripyat, 200 km away.

It was only in 2001, when the map[50] showing contamination was published by Com Chernobyl (Committee of the Belarus government on the problems caused by the disaster at Chernobyl) that we would discover that dozens of villages, just like those from which the peasants that we interviewed that day came from, had been evacuated, as a result, after a delay of five, ten or even fifteen years. Hundreds of villages, contaminated by 5 to 40 curies per square kilometre, are still inhabited, twenty years after the disaster, in a huge area, 170 km long and 50 km wide, on both sides of the river Sozh, that runs parallel to the Russian border. Fifteen large areas (the smallest measuring 8 square kilometres and the largest, 300 square kilometres), distributed along the border in the area of Bryansk, are contaminated at levels of more than 40 Ci/km². The reality of this extremely dangerous situation is concealed by the UN agencies, whose latest reassuring statement, dated 5th September 2005, only deals with the 30 km exclusion zone around Chernobyl itself. Galina Bandazhevskaya, a paediatric cardiologist, who worked for nine years in this region, found that 80% of children living in areas where the contamination measured 5–15 Ci/km² suffered from heart problems.

When we got the camera out of the car on 17th November 1990, in the main square at Krasnopolie, twenty or so women and men gathered around us, timid and curious. They had no more idea than we did about the radiation levels. We realised very quickly how these peasants, whose knowledge of nature's cycles and laws is timeless, are confused and disoriented. Yury Shcherbak had described their bewilderment. "These are poor people, humiliated by fear. People who don't know what is going on—who have always been lied to. They are helpless. And then, many of them do not even know what neutrons are, or what leukocytes are. They lack the most basic information. You can't blame them for that. They're peasant farmers who used to live off the land. Now they sense a danger surrounding them on all sides.

---

[50] See map on page 359.

118

A danger that can't be seen, can't be heard. You can't see it, you can't touch it. They can't cope". When people's lives are threatened by a hidden menace, God and anger are their first recourse, before they submit.

—What's the situation here regarding radiation?

—Only God knows.

—Does anyone here understand the situation?

—If I was a doctor maybe I'd understand, but…What can we do? We know nothing.

—Do you feel in danger?

—Of course! But what can we do?

—Nothing.

—Exactly.

—Is your food clean? Do you trust it?

—Ah my dear, we just eat what we find,

—Have you always lived here in Krasnopolie?

—No, I come from a village where no-one lives any more, at Novaya Ielnia. Near Zavodok, Rovnishche…They were evacuated. A few people stayed but they've already got one foot in the grave.

—Did you understand the danger from the radionuclides?

—Because they said that it was more or less alright to live here, we stayed. Then they came back: "You can't live here! You've got to leave!"

—When did you realise that you had to leave?

—They started to evacuate people last year. Some left sooner, others later. Difficult to find housing. We lived there for three years after the disaster.

—The military had set about decontaminating the area, hadn't they? We don't know exactly what sort of work they did but I suppose that was why they weren't in a hurry to evacuate people.

—Do you understand anything now about the radiation?

—How would we understand anything about it?

—You'd have to be some kind of great mathematician.

—Whether you understand or not, you've got to eat.

—Everyone should have a dosimeter.

—I used to live in Starayia Buda, one of the villages that has been abandoned. You can't live there. They destroyed the village, now it's been invaded by cats. We left in August. I liked it there. I get headaches here.

—You're just nostalgic. You're not used to it.

—Don't be silly.

An angry peasant.—They say you can live at Krasnopolie. Yes, you can! But you have no idea what it's going to do to you … One person says one thing, another says something else… No-one says anything concrete. Nothing concrete. What you can and can't eat. I've got a vegetable garden, carrots, onions and I eat all of it. But I don't know if it's dangerous. When I feel like eating, I eat. That's it.

—Do you test the food products from time to time?

—I don't test anything. It's all in my cellar.

—But there's a laboratory at Krasnopolie!

—No-one tests anything. I keep pigs, for example. What should I test? I slaughter a pig, I cook it and I eat it.

—You could take the meat to the laboratory to test the levels of radioactivity.

-- What's the point?

—That way, you can see if it's contaminated or not. Measure the level of radioactivity.

—Measure … in any case, most of the food we eat comes from our vegetable gardens. Test what? We've got to eat something. I grow everything in my garden, cucumbers, tomatoes…And that's what we eat.

—Even if you take your onions, garlic, potatoes to the laboratory, and they tell you they're alright, we don't know if they're spinning us a line or telling us the truth. We just have to believe them. What else can we do?

A young woman, holding back tears—I have a three year old son. But no-one's examined him. I don't know if he's healthy or not. We were told we were going to be evacuated soon from our village, Bereziaki. But at the moment, those who could not afford to leave under their own steam are still living there.

—But you are provided with clean food, at least?

A woman.—They bring us milk … We go and get milk and meat. Yes, they provide that.

The angry peasant.—But the milk they deliver, it's from here! It comes from the same contaminated villages. We produce it ourselves. They mix it with who knows what at Mogilev. Then they deliver it here, mixed and radioactive. It's the same with the crème fraîche. That's what we're eating!

—Farming is still going on in this region?

The angry peasant.—Farming? Of course! It carries on just the same. We raise beef, pork, in the contaminated zone. And we eat all that. I'm a mechanic, for agricultural machinery. We're working as we did before, the tractor drivers, all of us. They're feeding us with contaminated food. They're taking the mickey out of us! They themselves have left. The council leaders, the deputies, they've all left, leaving us as hostages. The mayors, Maximov, Pshekhotny, the doctors, they've all gone. We're here like guinea pigs.

—Like rats.

—They're taking the mickey out of us, that's all. They swindled us with their 15 roubles "for funeral expenses" as they say.

—Yes, you're right.

—All the local politicians scarpered: the council leaders, the deputies, the presidents. They released all this radiation then they made a bolt for it. Leaving us to deal with it on our own.

A woman.—They took all the money too. Took everything. They filled their pocket and left. They were replaced and the next lot filled their pockets in turn and left. That's politicians for you!

A worried peasant—Have you come because you're sorry for us?

A woman.—You've come to show you care and then you'll be off.

The worried peasant—You'll leave and we're going to stay.

A woman.—That's for sure.

The worried peasant—And we'll die one after the other like rats. Why? Our villages have got 12 Ci, 40 Ci, 45 Ci, 38 Ci. And these curies are a terrible thing. After two or three years….leukaemia and then that's it.

—Do you know how many curies there are at Krasnopolie?

—I know how many there are at Paluzh, you can go and measure it. It's 42 curies. That's where I'm from, I'm still living there. There are lots of us at Paluzh. We haven't got anywhere else to go. Where can we go?

An elderly person.—And where would I go if I left Krasnopolie? If it carries on like this I won't last longer than six months. If I go to bed, I can't get up again. I pass out. I don't want their 15 roubles compensation. I want to live. Today I fell out of bed. I can't walk. Tell me, sweetheart, what is there to eat? What can I eat?

—You're not being evacuated?

The worried peasant—Where to, if there are no houses, there's nothing? Go where? We're going to die here! The food that's sent here from other places, we don't see any of it. It's all sold on the black market. Our shopkeepers sell it on, behind our backs. The food arrives, it should be distributed to people who are ill, and they, they sell it. We're just animals in a laboratory.

—Is that what you think?

—We know it, that's all, we're not inventing it. It's obvious. They should have explained certain things a long time ago, like what we're being shown now on the television, or that I saw in Mogilev myself. Every person needs to be examined, every worker, every child to make sure they're "clean". But no-one explained anything to us, no-one came to examine us. As for us, the old people, we're of no interest to anyone. Former soldiers, war wounded, we're of no interest to anyone. No houses for us.

A voice.—At least give the young people a life!

Distressed mother.—They're not interested in the young people either. My daughter's at school and my son is three. We live at Bereziaki. There isn't even a hospital there. We have to go to town to buy medicines.

The elegant appearance of Lyudmila Kozlova, head teacher at the school, contrasts with the peasants' clothing. Last summer she accompanied a group of children to

Switzerland for a period of decontamination. She is visibly moved as she listens to us. We ask her what she thinks of this discussion.

L. Kozlova.—Everyone's complaining. Everyone gets headaches. Dizzy spells, nose bleeds. Going anywhere, getting out of the contaminated zone, even if it is only for a month, would do them a lot of good. But where can these elderly people go? There are places for the children to go. They have been to stay in Switzerland. Quite recently, I accompanied children on a holiday to Moldavia for a month and a half's holiday. But where can the old people go? Retired people.....How can they leave their gardens, their houses? Who's interested in them? No-one's interested in them. We feel our lives are condemned.

2. TWO LOCAL OFFICIALS

The following morning we met up with Semyon Kozlov, Lyudmila's husband, and Alexei Lazuko, a Communist Party official, in the main square in Krasnopolie, and we set off for Paluzh, a very contaminated village. At the edge of the village, looking out over a vast expanse of fields covered in high grass, stretching away as far as the eye could see to the horizon, Lazuko described the situation.

Q.—How many curies per square kilometre are there here?

A. Lazuko.—On average, 58 curies per square kilometre. This is an obligatory evacuation zone. But there are still people living here. Above 40 Ci, nothing is being cultivated any more. It is land that has been planted with grass, there is no agricultural activity. Below 40 Ci, this year and the next year, the land will be flooded and then sown with grass. There won't be any agriculture there either. We are continuing to use soils for agriculture in areas of the zone that have between 1 to 15 Ci.

—You are Secretary of the Communist Party in Krasnopolie?

—Yes, of the local Committee.

—Yesterday we talked to people in the main square. They told us that the local leaders in Krasnopolie left in the first few years. Is that true?

—Many did leave. Yes, there is some truth in that.

—They compared them to the captain who abandons his ship in the middle of a storm.

—The local people think the officials should be the last to leave, like the captain. But there is also, how can I put it, the human point of view, according to which when a person does not want to take risks with their health, they have the right to leave. So...

—Of course, if it was possible for everyone...

—Everyone should be guaranteed that right, and they should have stayed until everyone could benefit from that right. That's also my point of view. But the people who left justified themselves in all sorts of ways, looked for all sorts of reasons. They claimed to be afraid for other people's children, but it was them that left. I don't approve of them. In the first two years, the directive was to organise... of course it was a mistake to try to manage the zone instead of evacuating it. But at the beginning the scientists claimed that we could live here permanently even on land contaminated as much as 70 to 80 Ci per square kilometre.

—Ilyin's famous theory?

—We know now that it was Ilyin's theory. But at the time, for us it was the official government theory. In other words, this scientific theory was put into practice by government organisations. It was the official theory, you could say, government policy. Today, we know it was a mistake. Now everyone's gone to the other extreme. Abandon the land completely, evacuate the whole population in areas of 1 Ci and above. Obviously that's a mistake too. We have to save the people, but we must save the land too. To make sure there aren't other mini Chernobyls. If we abandon this grassland and everything around here, it could catch fire and the radioactive dust could be carried by the wind as far as Mogilev or even to Minsk. That's why we have to create the optimal conditions for the protection against the danger of radiation for the people in centres like Krasnopolie, and in the zones where contamination is low, from 1 to 5 curies. We have to adapt our agriculture to it. Develop specialised technology, within scientific production associations. We need help in this area. Near here, in the Cherikov region, there is a scientific production co-operative, Nadejda (Hope). People there are doing research to try and save this area. They are developing special techniques for growing different plants, studying different ways to treat and decontaminate production and to eliminate the radionuclides from the soil. In other words, they are learning how to neutralise them.

—But the scientific world does not know whether it is possible to neutralise it or not! Medical research is still looking into the effects of low dose radiation on the body. All this is still to be discovered. The people who live here are right when they say they are guinea pigs, don't you think? Are you, yourself, confident about the food that you put on your table?

—I'm absolutely sure that the food we are provided with is clean, or at any rate within the contamination limits for food that have been adopted here.

—Yes, but what about locally grown produce?

—You'd need to ask the health services here about local produce. It might seem strange, but they say that the potatoes grown here are below these limits.

Q.—Semion Efimovich, you work mainly with children. What exactly is your role here in Krasnopolie?

S. Kozlov.—My role is to ensure that the schooling process runs smoothly and to do my utmost to limit the danger of contamination of children. In other words, we are concerned not only with education but with their recovery.

These children have been deprived of their childhood by the Chernobyl tragedy. They spend nearly all their time indoors. We limit the amount of time they spend outdoors. We don't allow them to go beyond the school perimeter, to reduce the probability of radionuclides entering the body. The school is open at all times. The children are at school for nearly twelve hours at a stretch. We've organised a refectory that supplies four free meals a day. In summer, during the holidays, and throughout the year, we take children away from Belarus, to clean areas. At the moment, some of our children are in Moldavia, others are in Anapa… We receive a considerable amount of aid from other countries in the West. I have to say that people there have reacted to Chernobyl as if it were their own tragedy. This year, a lot of our children have gone to recuperate in Germany. Switzerland, Austria, Norway. I would like to express our gratitude, on behalf of the children, the parents, all of us. The restrictions that we have placed on our children have almost certainly had a psychological effect of one sort or another, but we haven't noticed any effects on their health. There will have been some effects and there will be effects in the future but we can't link them directly to the effects of low dose radiation.

3. A FAMILY IN KRASNOPOLIE

The housewife.—The food they give us has improved. We get clean milk. Before, they always brought the milk in large cans. Now we get milk from Mogilev in cartons, crème fraîche too. Everything's packaged. It's the same with the meat. It has to be said that from the point of view of food provisions, we have almost been too well served, apart from this year when supply was a bit irregular. But it's the same everywhere. Even in the shops, there are shortages everywhere… What's going on in the Soviet Union? I can tell you, as a woman, I am worried about the future. My husband and I would like to have another child but we've decided to… not to have one… It's terrible, ultimately, when a child dies before your very eyes. It's unthinkable.

Q.—Has anyone explained clearly to you the degree of danger here?
The husband.—No

The housewife.—That's the worst thing about it… When the accident happened, if someone had told us straight away that we mustn't go outdoors, that we should do this or that…Up to now, there has only been the odd short newspaper article here or there. At the moment I couldn't tell you what levels of radioactivity there are in Krasnopolie. We haven't got any machines to measure it with. They gave three dosimeters to the Town Hall, that's all. Where I work, there are none. Why? Everyone should have one. We live with uncertainty.

Q.—You don't trust the news?

The housewife.—They contradict each other all the time. To make things simple, I just go with the flow. This year, in our region we have changed president three times. 90% of the politicians have changed. They've left. You want to know why? Who would want to live in the middle of this radiation? Anyone who could leave has left. They make a lot of promises and then they leave. The local politicians take advantage of the situation to improve their standard of life. Why would they bother with anyone else? When our children visit uncontaminated areas of the country, even in Belarus, near here, they are treated like lepers. People are afraid to touch them; they make them take their clothes off. They won't let them come in the house, they take away their sweets.

The husband.—At school the children refuse to sit on the same bench with them. They call them "glow worms" or "fireflies" Often they have to return home.

Q.—Have you had any health problems following the disaster?

The wife.—For the first two years, I had absolutely no energy. I just wanted to sleep all the time. Terrible headaches. Then gradually, it seems that you get used to it; you learn to live with it. But if I ever go away for a month or more to a clean area, when I return to the Krasnopolie zone, I start to feel the same way again. The terrible headaches come back. Then it passes. Maybe the organism adapts.

Where I work, everyone complains of aching bones, in their arms, and in their legs, in their joints. Maybe we are just imagining it, I don't know. But then why does everyone ache in the same places? What causes it, I don't know.

Q.—Have your ideas about life, about the future, changed?

The wife.—I don't feel we have a future at all. I'm frightened. And not just because of the radiation. If it was only the radiation, I would have just abandoned everything, Krasnopolie, the house... I would have gone to another area. And I would have lived there happily. I've got family all over the place in the Soviet Union. But in the present situation, where would I go? What would I find?

Q.—The same conditions everywhere in the country.

The wife.—Of course, of course... I'm frightened. I am so indignant about the position Russian women find themselves in. How humiliated they are. They are supposed to provide everything when everything is in short supply. Apart from that, women really want to look nice sometimes, to wear nice clothes, to go out... We can't allow ourselves even the basics! How did we get to this? We're working just the same as before. Who brought the country to this state? I don't know. The radiation, of course, we understand. Maybe it wasn't adequately monitored, like they say. Everything is bureaucratised. No-one has to answer for their actions, for what they do in their job. We have become irresponsible.

Chapter V

DISINFORMATION IN KIEV

The solution [...] would be to see a new generation of people who would have learnt to accommodate ignorance and uncertainty...

Technical report, No 151, WHO, Geneva, 1958, page 59.

1. THE MARKET IN KIEV

On this Saturday in November 1990, crowds of citizens across all the eleven time zones in the USSR are trying to survive within the free market introduced by Gorbachev. Perestroika is in full swing. In an imitation of Lenin's 1921 new economic policy (the NEP), the new leader of the Kremlin has opened its doors to new forms of capitalist economics which co-exist with the empty shops of state socialism. Denationalising small business, allowing private enterprise is an attempt to get the blood circulating in the body of the country that had stagnated under Brezhnev. That day in Kiev, the historic capital of Holy Russia, where, ten centuries before, Prince Vladimir, who had converted to the orthodox Byzantine church, baptised the Russian people in the Dnieper River, the huge market was bustling.

With the murmur of the crowd in the background, two old women are selling their onions to passers-by. A young woman at a stall on one side is selling potatoes.

Q.—How do they monitor your produce?

—These potatoes are from my vegetable garden. If they had been too contaminated I would not have been given scales to sell them. They measure the levels of radioactivity and if they are "clean" they give me a certificate that I have to present before they will give me scales. Then I can sell them. And there are also inspectors going round, monitoring the stalls.

—Your area was not contaminated?

—It seems that where we live is clean.

—Do you know what radionucleides are?

—Invisible particles, which are bad for you. Everybody knows that.

—Do you feel safe, or do you feel that there is a strange danger around you?

—Everyone feels unsafe. Even now.

—Are you absolutely confident about the food that you eat? The meat, the milk?

—Me personally, I'm not really confident. But what can we do, there's nothing else to eat.

A queue has formed in front of the laboratory that measures food produce destined to be sold. Everyone is presenting food that has come from their own vegetable gardens. A woman is making notes behind a counter. An old man, poorly dressed, holds out some apples in front of the counter. The woman scolds him.

—They should be in a bag, grandad!

—A bag… Can you give me one, please?

—You need a bag … OK, I'll get you one. How many apples have you got?

—About ten kilos.

—Where are they from?

—My house, over there, under the Patton Bridge, next to the river. I've got some apple trees.

—Are you retired?

—I'm a war invalid.

After the test has been done, the old man goes off with his certificate to get some scales. A customer has been watching.

Q.—Do you think it's safe?

—I don't know…We trust her. She doesn't have any reason to mislead us. But perhaps they are not monitoring everything?

A smartly dressed woman has just bought some carrots. They are covered in soil. She is young, beautiful and seems tense. When we suggest that they should be cleaned, the earth removed and then washed, she becomes aggressive.

—So according to you, the carrot itself isn't radioactive? Really? You think radioactivity can be washed off, is that what you think? You wash the food and then it's no longer radioactive? Come on… you can't wash radioactivity off with water!

Q.—I'm talking about the radioactivity deposited on the outside.

—You can't wash radioactivity off with water! It doesn't wash off. You can't wash radioactivity off with water because radioactivity, it's like that… *(she makes a fluttering gesture with her hand in the air).*

Q.—The radioactivity is carried in particles of matter, in dust.

—I'm telling you, you can't remove radioactivity with water! I don't believe it. You can say what you want. Radiation isn't dirt, it isn't dust, it's made up of ions. You must know that.

Q.—But they exist in physical matter ...You must have been told that. You don't believe it?

—Yes but... particles... alpha... I don't believe it. Those particles can't be washed. It's obvious.

I remembered that in Italy, Switzerland, Austria, Germany, everywhere in Europe apart from France, people were warned of the danger and told to wash vegetables, especially leafy varieties. But I too, like this woman, I could not conceive of radioactivity except in its immaterial form. Yet the evening before, in the laboratory at the Scientific Research and Radiological Medicine Institute in Kiev, the good witches, who were analysing the liquidators' clothing, had cut them up into tiny pieces, and had showed me, in material form, radionuclides of caesium-137 that they had extracted. A thin silvery red film in the little metal flat cups, 2 or 3 cm in diameter. I had put my Geiger counter next to it and it had started to beep and the needle moved straight away from 0.114 μSv/h (natural background radiation) to 30 or 40...

2. SCIENCE IN THE GOOD WITCHES' LAIR

In the laboratory at the Institute for Scientific Research and Radiological Medicine in Kiev, an elderly woman in a white lab coat turns over some grass in an electric oven. It is a sample brought in by the physicist Volodymyr Tykhyy from a site where a Greenpeace medical centre is going to be built. The grass is not contaminated. In a pan are some carbonised mushrooms. There are many samples of all sorts piled up on the floor in the next room; they have been collected from the environment, including from vegetable plots. There are several women bustling about around the pans and the ovens.

Q.—Can you explain what you are doing here with all these instruments? It looks like the witch's scene in the Shakespeare play?

—We are like witches actually. I'll tell you why. The samples need to be carbonised in stainless steel containers. Ordinary pans do the job very well. That container is for sterilising surgical instruments. The oven is just an ordinary kitchen oven. It's the only way to do it in the first stage of carbonising. Suppose we were doing it on a gas stove. It's impossible to maintain a steady temperature: any increase in the size of the flame might cause the caesium to vaporise. It needs to just simmer very gently, so that the radionuclides don't vaporise.

—So it is actually the best scientific equipment?

—I can't say it's the best, but for the first stage of the process, we haven't found anything better. Naturally it would be better if we had hermetically sealed containers but that's not possible. It needs good ventilation. We would very much like to have modern equipment but we don't even think about it.

—Is it dangerous using this equipment?

—It's not dangerous if the samples are not very radioactive. But if we are burning samples like the ones we showed you, if we are burning the liquidators' clothes or very contaminated samples, then yes, it can be dangerous. You have to do it extremely carefully.

—How do you get on with the local people? Do you go into the contaminated villages? Do they trust you?

—The local people trust us. Maybe because I go with my team. I'm not young any more and I try to explain everything as clearly as possible. They are more likely to believe me, looking like someone's grandmother, than an institution. We go there with our measuring equipment and we show them what it reveals. After I've measured the milk, I do the calculation in front of them on a piece of paper and I say "If you drink all this milk you will receive this much contamination. You shouldn't drink more than one glass a day. It's better not to give this to your children".

—Are the people here suffering from "radiophobia" as the authorities say?

—No, I don't think there has ever been widespread "radiophobia". It's exaggerated. But they are frightened. Especially the people in Kiev. About 80% of the population are very frightened, they think the town has very high levels of radioactivity. Which is not true. It's because they have never been told the truth. Until very recently, our work was kept absolutely secret and the level of training in health was very low. Everything was secret. Anyway our people know very little about these sorts of things. Nothing. Even specialists, such as physicists, know nothing about radiological risks. They hear the word radioactivity and there's an outcry. But they know nothing about levels, effects or how to protect themselves from it... These things are not easy to explain to people. And, even teachers, academics with scientific qualifications, understand very little.

—Because we have no experience of the way in which it acts on the organism?

—Yes, that's it. We have no experience and our knowledge is insufficient.

3. THE MAIN SQUARE IN KIEV

A meeting has just taken place in the centre of Kiev and a lively discussion is going on among the demonstrators.

—They hide everything from us. They say that it's all within acceptable levels. In reality, Kiev is contaminated. Especially when it's sunny, you can see layers of contamination in the atmosphere, in the air. Everywhere in Kiev is contaminated with radioactvity, even though they say everything's fine. They've shown us a map, they say that all the areas around Kiev are not normal, but strangely in Kiev, everything's fine. They're lying; the Party is lying to us all the time.

—Tell me, you can see the radioactivity with the naked eye?

—Well no, you can't see it at all.

—So why do you say that everywhere is contaminated and no-one knows anything?

—Because the radiological laboratories take measurements, but they are never allowed to publish the results. There is a cover-up. The democratic press itself talks about it. They say in the press that the city is contaminated.

—Exactly. They say that 3000 patches have been identified in the town, but that doesn't mean they're lying to us all the time, or that you can see radioactivity on the rooftops. That's rubbish. I think that since the accident and its immediate effects, the level of radioactivity in Kiev is within normal limits. In any case, I personally believe the official information. I don't know about the food products that arrive in the shops because I don't know who is monitoring that.

—The situation is very alarming because suddenly a lot of people are experiencing health problems. The situation is very, very bad. I know because my wife works at the Kiev Institute of Medicine. They aren't publishing their figures, and every month they hide the number of deaths caused by the radiation. They are hiding it.

—Do you believe the information put out by Green World?[51]

—Yes, we trust what they say. Because it was set up by people who suffered the effects of the accident, not by those in power who take all the medicines for themselves. The people in power evacuated their own children a few days after the accident, but they forced the children of the town to take part in the May Day processions. None of their children were there.

[51] Green World was an association set up in 1987, in response to the Chernobyl disaster, which then became the Green Party in Ukraine

Chapter VI

SWISS *PERESTROIKA* IN A SOVIET HOSPITAL

To the great dishonour of the UN agencies and of the Western governments who failed to act, civil society came to the aid of the victims. Their initiatives were sometimes extremely effective though insufficient given the scale of the disaster. Before learning about one of these initiatives, we needed first of all to distance ourselves from another source of disinformation.

On 6th March 1991, the editors at Swiss Italian Radio-Television received some propaganda in the form of a cassette tape, accompanied by the following instructions:

SWISS ATOMIC ENERGY ASSOCIATION—ASPEA

The Soviet Union Atomic Energy Agency has supplied us with an English copy of the new film *Chernobyl as viewed from the 90's* which was produced by experts at the Kourtchatov Institute of Atomic Energy in Moscow. The film is of particular interest and value given the context of the 5th anniversary of the accident in Ukraine, and we are sending you the enclosed tape for broadcast. French and German versions, as well as the English version enclosed will be available soon in U-matic format for Pal, Secam and NTSC systems.

In our opinion, this unique document provides good material for television programmes about the accident at Chernobyl, which is why we would like to draw it to the attention of the editors concerned. There are no copyright problems as the Soviet Union had relinquished these rights to the European Nuclear Society (ENS) for viewing by the general public.

SWISS ATOMIC ENERGY ASSOCIATION
Administration Dept

Our documentary was ready to be broadcast. We replied with the following press release:

This officially authorised Soviet information describes a radiologically controlled situation, no longer presenting a danger from a scientific, medical and social point of view.

The documentary *Noi di Chernobyl* (We of Chernobyl), the result of an investigation undertaken in Ukraine and Belarus in November 1990, paints a different picture from that of the Soviet experts: it contradicts in particular the claims made concerning the safety and protection of those exposed to radioactivity. The conditions of *imprisonment* in which hundreds of thousands of people live, disoriented by years of lies on the part of officials and manipulated by local authorities who have no interest in evacuating them; the tragedy of children who are becoming more and more ill and are dying in the contaminated territories; the scandal of the 600,000 liquidators that the Soviet state refuses to care for medically and to help, after they were employed in useless decontamination work over enormous areas that are still radioactive today. These facts, reported by correspondents of the RTSI (Radio Television de la Suisse italienne) are as much an indictment as a call for assistance: a plea for our attention and for our help from the citizens of the Soviet Union who denounce the role and the responsibilities of the International Atomic Energy Agency of the United Nations.

We also sent out a second press release under the title "Swiss *perestroika* in a Soviet hospital. How we can save the children of Chernobyl", which said:

Last year, when the world learnt from the Soviet authorities that the disaster at Chernobyl had been much worse than they had wanted us to believe for the past four years, the Chief Medical Officer at the Oncology Hospital for Children in Minsk, Olga Aleinikova, finally managed to launch her appeal to the West. On the initiative of the Chaine du Bonheur (Happiness Chain) and a group of Swiss oncological paediatricians, a team of doctors and journalists was sent to Belarus to investigate and document the health catastrophe: the radioactivity was destroying the children's immune systems. The children were dying every day, not only from leukaemia, but from hepatitis and chicken pox, from a lack of syringes, minimal hygienic conditions, essential medicines. Thanks to the money collected from viewers who responded to the appeal, an aid programme was implemented, exemplary in its rational, practical effectiveness and a further documentary, entitled "Survivre à Tchernobyl" (Surviving in Chernobyl), broadcast the following year, on the 26th April, described the problems and the first results.

1. DIE IN JUNE, SURVIVE IN NOVEMBER

In June 1990 the Chief Medical officer, Olga Aleinikova, shows us round her empty and dilapidated hospital.

O. Aleikinova.—As you can see, we have no equipment. Bare walls and beds. All we have is our brains and our hands, that's it. If the children feel unwell, if they faint, if they have complications, we have practically nothing except this electric mucus suction pump and this oxygen cylinder on the wall; and The Austrian Red

Cross gave us 2000 indwelling catheters at no cost, three months ago. This is the last box. When that's finished, we will have to give the children an injection whenever they feel they need it and that could be up to five times a day.

We don't have any disposable infusion and transfusion systems either. We only have a few boxes left. As we cannot interrupt their care, we have to sterilise the same equipment several times, which increases the risk of infections. From Hepatitis B, among other things...

Katia is four and a half years old, and has lost all her hair. She is from the Mogilev region which is badly contaminated, and she has been ill for more than a year.

Katia's mother (in a resigned tone).—We have tried every possible treatment, radiotherapy, chemotherapy, everything... with no positive result. My daughter has a tumour in her back. Her bone marrow is compromised. She hasn't walked for five months. She had such beautiful hair...

Artiom's mother cradles him in her arms.

Artiom's mother.—When he was born, he was a really good baby... But he fell ill on New Year's Eve, and six weeks after, he was diagnosed with leukaemia.
Q.—Do you sleep with him?
—Yes, in the same bed. That way, I am at least close to him.

NOVEMBER 1990
O. Aleikinova.—Artiom completed a course of treatment and we hoped that he would survive. But he caught chicken pox, which might seem a mild childhood illness. For our children, it's dreadful, because they have no immune defences. To prevent the virus multiplying inside his body, we had to inject him with Varitect, an immunoglobin against the chicken pox virus. But we didn't have any and he died of generalised chicken pox. His intestines were covered in vesicles. The vesicles had invaded the lungs. He died of emphysema. But I knew how to save him ...

We talked to the despairing parents of another child called Dima Pranovich.

The mother.—We haven't got the necessary drugs. He could be treated abroad, but here, we have very little chance. I so much want our child to live! We have already written twenty letters to different countries, without any result, without any reply. I would do anything to save my child. My husband is also ill himself. He was treated for early stage cancer in December, and in January, it was the little one... what a tragedy!

The father.—They tell us here at the hospital that he has a 10% to 15% chance of recovery. I haven't got the money to pay for his treatment but if someone wanted one of my organs, I would sell it to them…

O. Aleinikova.—Many children are dying but over the last two months we've noticed a slight decrease. Last year, 21 children died in our centre and about the same number at home, because the parents sometimes take their children home before they die to avoid an autopsy. At present, the new law says that we cannot proceed with an autopsy without the parents' permission. This year we have already had 33 deaths at the hospital. That's an enormous number. I couldn't take it any more and last year, I approached a journalist from the television programme *Vzgliad* (Look), Alexandre Poliatkovsky, who was in Minsk, and we decided to make a programme right here in the children's haematological hospital to show people the conditions in which these ill children were being cared for.

If you had come here with your cameras a year ago, you would probably have wanted to cry. It was frightening: battered old camp beds, torn sheets, no equipment, no sanitary materials… When people saw that on screen, it was a real shock. Children who were so ill, in conditions that gave them no chance of survival. We didn't even have a private room for the children who were dying; they died in the communal room, in plain view of the others.

You can't imagine the suffering the doctors and nurses go through and how the mothers who stay here lose hope… Because, straight away, they think: "It could be my child next". And I think that too: "Who will be the next?" It's terrible. The feeling of powerlessness, that's the worst thing for a doctor, when he knows how to help but can't, through a simple lack of equipment.

On 22nd June a Telethon was organised in Switzerland that collected 1,300,000 Swiss francs for us. That money allowed us to buy equipment, as you have seen…Before we were supplied with the equipment, a group of specialists from Lausanne came here under the direction of Doctor Beck, who organised the medical side of the aid programme. He stayed here for three days and worked out what we needed: the aid that came was exactly what we needed. Not a single franc was spent in vain. As well as equipment, they sent us specialists. Speaking as a doctor, I have to say that it was the first time in fifteen years that I was able to work under normal conditions.

So, it was as a result of this programme, *Vzgliad*, that public opinion was alerted, first in the USSR and then in the West, and now we are able to administer the same treatment as in Western clinics.

From October 1990 to March 1991 the departments of Pediatric Oncology from Lausanne, Bern, Aarau, Basle, St. Gallen and Zurich sent teams of specialists to Minsk and, over a six month period, they installed technical equipment and trained their Soviet colleagues in a treatment system that raised the percentage of cases cured from 10% initially, to 60%, the average in Switzerland. This targeted approach, under the direction of the Swiss doctors, went far beyond the usual relief operations. The Western technology and protocols were more or less grafted on to the obsolete health care structures of the Soviet Union.

Dr Imbach, who has been researching childhood leukaemia for fifteen years, has his own secret weapon against the disease. It relies on very precise dose calculations of the powerful anti-leukaemia drug called Methotrexat. It needs to be administered using extremely expensive analytical instruments so that the treatment can be adapted according to how each organism reacts. This is why the treatment needs to be extremely rigorous and the doctors and nurses must be disciplined, in order to monitor and record the biochemical reactions of each patient according to a precise protocol.

Dr Imbach.—With the other Swiss specialists, we decided that a team would come first accompanied by a doctor to install everything we had brought and to train the local staff to use it. That was the task of the previous team. My task now is to organise the treatment, so that it is administered in the same way as in Switzerland.

We lacked the most basic equipment for controlling infection. Everyone here, including the children and the doctors still has a lot to learn. We need receptacles to be available in every room to dispose correctly of paper, hand wipes, and towels. We need disinfectant soap so that people can wash their hands properly. Then there are all the instructions about how to care for a child, in a sufficiently sterile manner. But there are still other problems. We have to ask, diplomatically, for example, that the nurses and doctors keep their nails short, because this is exactly the sort of place where infection can be transmitted.

We think that two out of three children are infected with Hepatitis B through blood products or through inadequate hygiene procedures. We need to take tests, change our blood transfusion procedures and improve hygiene in general.

My impression is that there is still a lot to do. Doctors here receive a different training than in Switzerland. The nurses are not very motivated. There are many improvements still to be made, and I am still not very optimistic.

O. Aleinikova.—The fact that we have introduced this method which has up to a 60% to 80% probability of curing leukaemia, allows me to hope that the children we have treated will live.

Dr Imbach.—We would also like to introduce our system of data collection and processing. First of all we need to train the doctors to use the system. Then we need to find nurses who can familiarise themselves with the procedures. That is what we are doing at the moment.

Then we need to help the chief medical officer to play a bigger role as director. We've noticed that she works completely alone. I am going to introduce her to our way of working. To do that she needs to be visiting patients on a regular basis and writing daily reports. We would also like the doctors to begin to visit the patients daily, accompanied by the nurses and improve the dialogue between the latter and the doctors.

We have advised the chief medical officer to train heads of department, who receive specialist training abroad, to ensure their younger colleagues are trained in their turn, so that oncology will spread from here all over the Soviet Union.

2. EIGHT YEARS LATER (JUNE 1998)

Eight years later, I met up with Olga Aleinikova, in her magnificent new hospital for the treatment of children with cancer, at Borovliany, on the outskirts of Minsk.

Q.—How did you manage to get this centre built?

O. Aleinikova.—With God's help, of course (*she laughs*), but also the help of a great many charitable people, both here and in the West. The construction of the centre was financed jointly by the Austrian State, an Austrian charity, Germany and the Belarusian government. But the biggest contributor was Switzerland.

1,800,000 Swiss francs had already been invested in equipment and medicines when we were still at the old centre. If we hadn't already started on the new treatment using this new technology, I don't think this centre would ever have seen the light of day. We were already prepared by the time we got here. As soon as this appeal had been launched on Swiss television, Doctor Daniel Beck came here. He looked around, he asked questions, then he sat down and drew up a plan: how to help as quickly as possible. And there had been this unforgettable contribution from the Swiss specialists, who worked here in relays for six months, to help us to introduce all this very quickly. That's how it all started, the source of it all.

We had a 12% probability of a cure. Today, this has increased to 70%. We have 70% survival rates for acute lymphoblastic leukaemia. For acute myeloblastic leukaemia the figure is 50%. For Hodgkin disease, the survival rate is 92%. For non-Hodgkin lymphoma, around 75%. For Wilm's tumour, or nephroblastoma, the figure is 65%. Of course, there are certain forms of cancer that are hard to cure even in the West, and we have problems with these too. But today we are able to treat most tumours and most melanoma.

The nucleus of the team that started with me in 1990 is still here. At that time, they were between 25 and 28 years old. Now they are 35. They have had a lot of experience, they have had the opportunity to train in the West and to put their skills into practice. They are very highly qualified professionals.

Q.—I remember that Dr Imbach predicted a great deal of work to introduce this very precise regime, and to allow the treatment technology to be used effectively. How easy was it to achieve?

O.A.—It was an enormous amount of work, but I think the secret lies in our enormous desire to succeed. We worked sixteen or eighteen hour days, with no days off, and maybe also, I was able to transmit my obsession to young people, because the young are still capable of being trained in something...On the other hand our task was made easier by the fact that we were benefiting from this same Dr Imbach's experience, and the experience of paediatric childhood oncology and haematology in the European community. Their experience had been built up over several years. In substance, the experience we had received and that had already been verified, had been being researched by others for twenty five to thirty years. We were simply adopting procedures that were already in place and so it was much easier for us. All we had to do was introduce them a disciplined manner.

Q.—As a citizen of your country, how do you view the role of the public authorities, both in your own country and also the global community, as regards this tragedy at Chernobyl?

O.A.—The tragedy had a direct impact on my own family. My children had to take part with their school in the May Day procession, as everyone did in the old Soviet Union. Every day, from 26th April to 1st May, the children rehearsed for the parade, taking part in the street marches... With no iodine...

But I myself had already started giving them iodine from the 28th April. It was a real stroke of luck. There had been a few lines in Pravda on 28th saying simply that there had been an incident at Chernobyl, and that four firemen had perished. That was all. But one of the nurses who worked with me had parents living in Bragin. They told her over the telephone that the power station was on fire. That was enough to alert me to the danger. I remember diluting iodine drops in milk. But even so, I wasn't sure. Was it really necessary? The lack of serious information, the silence from central government and the local authorities about the risks, and about the protection measures that should have been put in place immediately, constitutes a crime against humanity! As for the International Atomic Energy Agency—the IAEA or in Russian, MAGATE, the very name is used as an insult today. At any rate, among the people I know, among my friends.

Q.—Do you share that opinion?

O.A.—I share the opinion of our scientists, the academicians, Konoplia and Nesterenko, who have researched the matter, and have reached very different conclusions from those of the IAEA. Chernobyl has affected all of us, big and small, wherever we live. Children as well as adults. Including those children who weren't even born. It continues to have an effect through the mothers who grew up at that time with these high levels of radioactivity, and who, contaminated themselves, give birth today after a pregnancy in a radioactive environment. That's why I think that all our children are at high risk from radiation, the risk of radioactive effects on the organism.

A human being is a grain of sand in today's world and no-one cares about them. Governments of every country claim they are working exclusively for the people. But "the people" is an abstract concept. You and me, we are also "the people", aren't we?

At the hospital, we care about each of our cases; every patient is a human being that is suffering. And if we manage to save someone's life ... What a shame you didn't meet him! Yesterday, one of my patients, from 1990, came to see me. He was one of the first patients to be treated with the new method. He is 27 now. He is a professional soldier in the Belarus army, and he has a five-year old daughter. He brought me a photo of her. He came to see me just to say hello because he was passing by. It made me so happy!

3. DOCTORS WHO TELL THE TRUTH. THE MOTHER OF DIMA TUROVETZ—NOVEMBER 1990 AND JUNE 1998

We had filmed Dima eight years before, when his mother was carrying him up the stairs of the dilapidated hospital, at the old centre, for a painful lumbar puncture, undertaken without anaesthetic because none was available.[52] We had exchanged a few words with her before her child, who was not yet 3 years old, underwent his ordeal. The young woman was going out of her mind: she was personally going to help the nurse during the puncture operation. She was going to relieve his suffering, by whispering encouraging words and tenderness.

> **NOVEMBER 1990**
> Q.—Has he already had a lumbar puncture?
> The mother.—Yes
> Q.—Does he know what it is?
> The mother.—Yes
> Q.—Does he know he's going to have one now?
> The mother.—No, I haven't told him.
> Q.—Is it very painful?
> The mother.—Of course.
> Q.—How long has he been ill?
> The mother.—We've been here for six months. We're still hoping. They are using the new treatment from abroad. We'll see. Very few of the children who were hospitalised with him are still with us.
> Q.—Where are they?
> The mother.—Dead. They weren't able to treat them.
> Q.—Where were you during the accident?
> The mother.—In Minsk. We didn't leave the town. We were told Minsk was not contaminated, that in Minsk everything was fine. He was born after the accident. We have no idea how much radiation he received.
> Q.—Maybe it was the food? Radioactive milk?

---

[52] Documentary *Nous de Tchernobyl*, Swiss Italian Television (TSI), 1991.

The mother.—Maybe. No-one knows.

Q.—Are you confident now about the food you are eating, do you think it's uncontaminated?

The mother.—No, unfortunately.

JUNE 1998

The father of Dima.—We are still not confident about the food but we still eat it.

Q.—It looks as if you're out of the woods?

The mother.—We're hopeful but we're not absolutely certain. When Dima was taken into hospital, the mothers were not allowed to sleep with the children at the hospital. I was only able to stay with Dima because he was under three years of age. I slept in a ward with six children. I kept an eye on all of them including my own. It was only when Dr Aleinikova came back from Switzerland with this new procedure—she had simply observed the practice of hospitals abroad—where the children slept with their mothers—she gave permission for the mothers to sleep with their children. The first time my husband came in to the ward wearing a white coat, it was a real event: "This can't be happening, someone from outside coming into the ward, with the children!" Now everyone is allowed in. But before, it was impossible!

As a mother myself, I can tell you that, in general, the mothers are left in ignorance, left to cope on their own. I didn't know what this illness was. I didn't even know why the children were dying. Why? I really didn't understand! Someone should have explained it all to us, to tell us what we needed to do.

At the centre, Dima had a good general practitioner and I came straight to the point. I went to see him in his office and I said "I want to know everything. Can I ask you questions?" "Go ahead!" I bombarded him with questions, to the point at which he had to admit that he did not know the answers. These are things we don't know". It's great when a doctor talks to you honestly. It's rare. I am really grateful that right from the beginning he said, "We don't know". Instead of reassuring me that everything would be fine when we could see that nothing was right.

If I'm honest, I still don't really believe he's cured. No-one can give you guarantees, not even Doctor Olga Aleinikov. Especially as Dima was the first to receive this new type of treatment. They have perfected the treatment now and adapted it for the children of Belarus.

Q.—Children from Belarus are different from Swiss children?

The father.—According to Doctor Aleinikova, Belarusian children react differently to chemotherapy because of their diet and their way of life. The reactions of a child from Belarus and a child from a rich Western country differ at the cellular level. Without taking into account psychological particularities. The ignorance and incredulity of the parents have an effect on the child and the relationship between doctors, parents and children here is less open. We don't ask what tomorrow will bring any more, we live each day as it comes.

Chapter VII

THE ORDEAL OF THE LIQUIDATORS

Dr Michel Fernex has commented on the medical aspects of the documentary films we made about the deteriorating health of the Belarusian liquidators, that we met for the first time in 1990, then in 1998 and in 2001. The film "The Sacrifice", by Emanuela Andreoli was based on these films and was presented as documentary evidence at the International Symposium on "The health of the liquidators (decontaminators), twenty years after the explosion of the Chernobyl reactor", organised in Bern by the Swiss branch of the PSR/IPPNW[53], on the 12th November 2005.

Michel Fernex, Emeritus Professor at the Faculty of Medicine in Basle:

What we see and what we hear in the film provide answers to many questions and controversial subjects. It appears, for example, that the doses recorded for each individual depended on orders received by the officers in command who co-ordinated the operation. These orders resulted in the systematic underestimation of the doses received by the soldiers, who were working in unacceptable conditions around the shattered reactor.

The expert studies, which do not take into account the internal dose, resulting from the inhalation of dust, rich in radionuclides, will exclude many of these people who are ill or who have died because the dose noted in the register only takes into account what was written down at the time, a more than doubtful record as it relates only to external radiation.

The film explains the work undertaken by these "volunteers" and the magnitude of the risk of their inhaling radioactive gases and dust containing radioactive particles derived from burnt uranium or plutonium. The internal doses, of which the most serious relates to alpha particles, which give off enormous amounts of energy but do not penetrate the skin—damage nearby cells, and especially their genetic material, (causing cancers, hereditary illness and birth defects in the liquidators' children). It is the inhalation of dust containing high levels of radionuclides, and

---

[53] Doctors for Social Responsibility—International Physicians for the Prevention of Nuclear War.

in particular uranium derivatives released in the form of smoke or invisible dust, that results in deposits of nanoparticles leading to cancers.

The film shows that the work consisted of, among other things, scraping off the top-soil and using spades to throw it onto lorries, to be buried in enormous pits. It would have been impossible to avoid inhalation of the different radionuclides which would then settle in the lungs or circulate through other organs, irradiating and altering nearby cells for decades.

The liquidators describe their accelerated ageing. This symptom, premature ageing, previously recognised as a symptom of radiation, has been removed from the list of illnesses that can be attributed to radiation after Chernobyl.

The general practitioners seem unaware of the illnesses that result from this chronic irradiation. As the situation worsens, even the professors who were consulted seem to be out of their depth, powerless in the face of these unknown illnesses, while Belarus is renowned for its high standard of medical training. It is, for example, in this country that the reality of the epidemic of thyroid cancer was established, an epidemic that Western specialists refused to recognise for between five and eight years.

1. PIOTR Shashkov

1998: EIGHT YEARS HAVE PASSED SINCE OUR LAST MEETING

Q.—How have you been since we last saw each other?

P.S.—Worse and worse. I have an enlarged liver. I have problems with my spleen, pancreas and thyroid. I am diabetic. Last year, they should have operated on my thyroid but they had to see to my leg first. They removed a malignant tumour. I have four scars. I have to use a walking stick. I have no control over my left leg—it's as if it's dead.

Q.—Do you remember someone called Anatoli Borovsky, that we also met last time? Where could I find him?

P.S.—He's dead. I'm sorry. (*He is overcome and I look at his profile as he gazes into the distance with tears in his eyes*) Many of them are no longer with us…So many have died already. I have lists of all my men. Many have died. I buried a guy scarcely a week ago.

Q.—Another! But they're so young.

P.S.—Well yes, they'd be 35 or 36 now.

2001

Q.—Last time we came to see you was three years ago. How is your health now?

P.S.—It's worse. I've had another operation on my leg. The flesh started to detach from the bone and decay.

Q.—Was that side of you more exposed to radiation than the other? Was there something in your pocket?

P.S.—No, I was in an armoured vehicle converted into a bulldozer, my left side against the plating. No protective clothing, just an ordinary soldier's uniform treated with some sort of fluid. Nothing else. The whole of my left side was irradiated. The flesh on my leg comes away from the bone and suppurates.

Q.—As a liquidator, do you get any sort of help?

P.S.—No, none. Even for the fifteen years since Chernobyl they've given us absolutely nothing. They even suspended our pensions. I was supposed to be going into hospital in April but I was told there were no beds. They're reserved now for those that can pay. You have to pay for everything.

Q.—And what was it you had to go into hospital for?

P.S.—I go in every year for an extended cure. They treat my liver, stomach, heart....legs. Footbaths bring me a lot of relief. The worst thing are the abscesses, first on one finger, then on another. Same with my feet. They heal over and then start to suppurate again. The doctors don't know what it is. You see? Same thing. I had one there and now there's one here. There's always somewhere that's rotting. I apply plantain leaves to draw out the pus, that's all. My nails fall off all the time.

Q.—What do the doctors say?

P.S.—That all of it is because of the radiation. Before the explosion, I was never ill. I was a steel worker in a foundry. I was a weight-lifter. I could lift my wife with one hand.

Q.—How old are you?

P.S.—I was 50 in November. I was invalided at 37. First degree invalidity.

Q.—First degree, meaning what?

P.S.—It's the last. For those on the brink of death. But I've become… an optimist. Do you understand? I ignore the pain; I get up in the morning. Straight away I feel sick. I vomit. Bile. It's the same every day. They don't want to operate on me... Because of the diabetes. But I'm just carrying on as if I'm well. It'll be the mushroom season soon. I'll go to the forest, for sure.

Q.—You go out walking?

P.S.—I walk two or three hours a day even though my legs hurt. It's probably something to do with my circulation only working at 45% in my lower limbs. I've no circulation… I've even taken a needle and pricked a vein to see: there's no blood. I also have heart problems—cardiac stenosis. The left ventricle is not functioning.

Q.—That's why the blood is not reaching the extremities?

P.S.—No, it's because of the radiation.

Q.—Yes exactly, the radiation affects the heart.

P.S.—It must be that. Hiroshima and Nagasaki were a drop in the ocean compared to what we got.

Q.—The comparison makes no sense.

P.S.—It's true. The fallout was much greater here. The whole of Belarus is contaminated. Measure any product and the dosimeter starts clicking. But people have to eat, don't they?

Q.—Is your wife also being monitored?

P.S.—No, they don't take the wives of the liquidators into account. They have no rights. Children born after Chernobyl are on the lists... But children born before Chernobyl are not taken into account either.

2. THE ILLNESS AND HUMILIATION OF ALEXANDER GRUDINO

1998

A. Grudino.—Chernobyl happened just as we were thinking of having children. After the disaster, we were forbidden to have children for 3 years. As for those who did, their children are all sick. Thyroid and the rest. All the children are ill, every last one. I can't have any children myself now. Apart from the fact that I am not capable of having children, the Medical Commission at work also wrote down "hypoplasia" and explained that it was a congenital condition. But, how can it be congenital if I served in the army and was perfectly healthy? My ulcer too, they said that was hereditary. How come? Neither my father nor my mother had ulcers. All of that, it's to hide the real reason, to act as if Chernobyl has nothing to do with it. But I'm not the only one; there are hundreds of us lads like that, hundreds.

Q.—We saw each other eight years ago. How has life changed for you since then?

A.G.—At first it was just about alright. Then all the illnesses started to get progressively worse. Looking at me on the outside, I look as if I'm alright. But on the inside, nothing is healthy. Before 1986, I didn't know what illness was. I just had colds like everyone else. Whereas now I have a stomach ulcer, I have pains in my bones, in my vertebrae, permanent pain in my joints. When the weather turns, I walk like an old man, and I hurt everywhere. Where have you seen a man of 40 with pain everywhere? I have constant dizzy spells, especially in the sun. I can't tolerate sunlight. I always wear a hat and look for the shade. I can pass out if I stay in the sun.

As for the children, I'm not in a position to quote precise figures but those I've seen are all ill. For us, the old, as they say, it's one thing, but for children, it's a tragedy. They will not live to our age.

And then the injustice never ends. I was summoned before the recruitment board, twelve years after Chernobyl. When I showed them my liquidator's certificate which exempted me, they marked "valid until 2000" at the top of it. Why? Up to 2000, I'm a liquidator and after that I'm nobody anymore? War veterans are still war veterans even forty years after the war. What does that mean? What do they take us for?

—They want to liquidate you?

—Probably. To forget Chernobyl.

—Liquidate the liquidators.

—That's it.

3. THE ILLNESS AND HUMILIATION OF VICTOR KULIKOVSKY.

1998

V. Kulikovsky.—Briefly, I suffer from all sorts of diseases of old age: sclerosis of the brain with oedema of the brain, alteration of the brain cortex and of the circulation. Not just of the brain, of the whole body. I have problems with my memory, headaches, failing eyesight. Since 1995–1996 my blood composition started to stabilise but it was completely altered. In 1990- 1991 I started to have major problems with my legs. They wouldn't do what I wanted them to do. I stopped going out. The doctors say sclerosis but in fact, they don't understand it. It's getting better, but yesterday, I walked a little , in the evening my legs swelled up and were still swollen this morning. Mainly the left leg, you can see it's more swollen than the right. I've got other problems too... ulcer, gastritis, what's the point of enumerating them all? My arm hurts, my shoulder joint is frozen. If I try and move my arm, it is very painful.

It's hard to say how much radiation I received, in reality. The fact is that I worked on Reactor No. 4, in other words right next to the holes caused by the explosion. The 11.92 rem noted on the certificate, it's laughable. After our hunger strike in 1990, the Institute of Radiological Medicine tried to reconstruct the dose I received. It was about 100 rem. That's a lot. Nobody who had worked, like me, on the reactor in 1986, had the right to 25 rem on their certificate. Because at 25 rem, they had to pay them special compensation. One day Georgi Lepin, the scientist who created the Chernobyl Union in 1991, after he also, had been a liquidator, asked me to describe the places where I had been working, in order to reconstruct approximately my dose. When he'd read the twenty or so pages I'd written, he asked me "Is all this true?" I'd gone into great detail including sketches. Even today, with the ground concreted over, and the scattered fuel removed, there are certain places where the background radiation is up to 1.5 roentgens! How much would there have been before? And how much radioactivity would a man working there have received? He said that even with my detailed recollections they could never accept the extent to which I'd been irradiated because it would be too high. Of the order of ... less than 1000 but more than 500. Over 500. The men I worked with... Two years ago I tried to get in contact because to obtain the documents, you have to have confirmation from several people who worked in the same area as you. You need two, three people, at least. I tried to find the other men I had worked with there.

144

I can honestly say I didn't find one. Not one. All the requests I sent, whether to the police or to relatives in different towns came back marked "Deceased". There were about a dozen in our team.

Before Chernobyl I played a lot of sport. I did my military service with the airborne divisions. I was strong. Now I can't climb the stairs. I can hardly walk even on the flat. I can't get enough air. My heart...

Even so, in 1995, they accorded me third degree invalidity, which gives the person the right to the minimum pension, which they were then ordered to withdraw because I'd taken part in the hunger strike. If you want to humiliate a man, crush them; it's very easy. The hospital called me in for examinations. When I arrived there was the professor and a female doctor in the office. They closed the door and said "Why do you need a walking stick? You're cured! You can throw it away!" And the woman doctor asked me "Do you drink?" "No". "Even at a party?" Maybe a small glass, no more". "And your father?" "What about my father? He does what he likes". So they wrote "Father chronic alcoholic". I did not know how they came to this agreement between the two of them. "That will be all, you can go now". They gave me the extract and I read "The father is a chronic alcoholic. On the basis of the data noted above, the illness is due to chronic hereditary alcoholism". "I shall go to the tribunal with this diagnosis". "You can go where you like". I went as far as the regional prosecutor with this diagnosis. He read it and said. "Keep this diagnosis until things are better. The time will come when we will deal with the facts of this case". That was in 1993, at Hospital Number Five in Minsk. After that, in 1995, this is what I hear being said at the Institute of Radiation Medicine Polyclinic: "Kulikovsky? Are you not dead yet? Why have you come? You'd have been better off staying at home!" "I don't understand why you're speaking to me in that tone". She examined me, wrote down her diagnosis, then snapped: "Everything has to be paid for". "It's because of the hunger strike that you are behaving like this towards me?" "Yes, there's always a price to pay. It's for the strike".

4. THE DECLINE AND AGONY OF ANATOLI SARAGOVETS

1998
One hot June day we found Anatoli Saragovets in his wheelchair stripped to the waist. Eight years had gone by since our last visit. He is a different man. He has difficulty breathing, apparently because of the partial paralysis of his ribcage. He cannot entirely control his hands and his arms, and therefore his movements. This is no longer the smiling young man of eight years ago. He welcomed us without embarrassment. His simple conversation has a strange tone, painful and humorous at the same time.

A. Saragovets.—All I did was fall down, fall down...My wife said: "You need to use a wheelchair", so I did. That's it. Now I'm in a wheelchair. If I had my legs

Anatoli Borovsky

Piotr Shashkov

Alexandre Grudino

Victor Kulikovsky

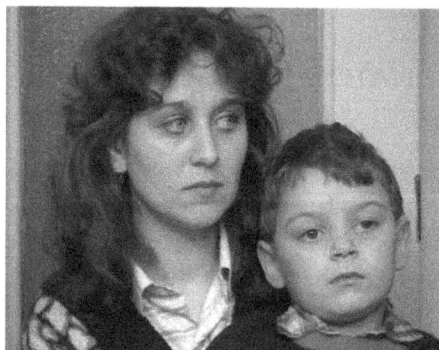
The wife of Anatoli Saragovets

Anatoli Saragovets

Anatoli

The nuclear power station

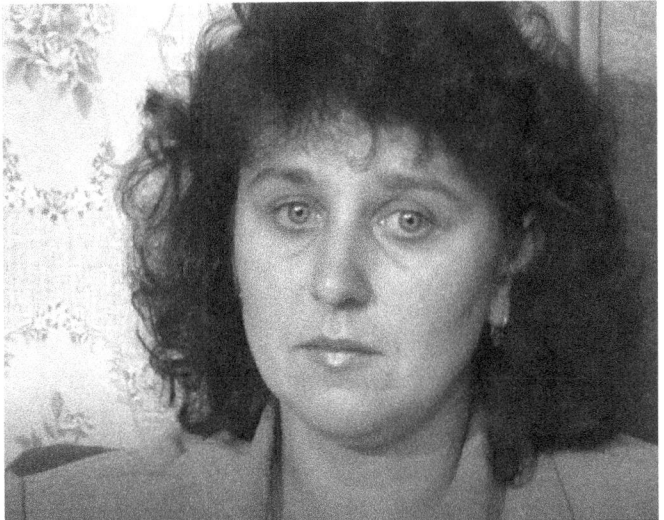

Anatoli's widow

Every day a team of specialists monitors the state of the building, the levels of radioactivity and humidity to prevent a chain reaction.

it would be alright. My legs don't work anymore, so that means I'm fucked. The diagnosis says: not curable neither here nor abroad. Nowhere. "Multiple sclerosis" is what they wrote. That's all. This year I'm due to appear before the Medical Board who have promised me a car. But with this diagnosis I'd be amazed if they gave me one. At Chernobyl, it was my job to drive in front of the column of liquidators with my water tanker, so that they wouldn't swallow the radioactive dust... We drove straight up to the reactor... but it's too painful to remember. It's better not to remember. The sun is shining outside, it's... a lovely day. If not, if you start remembering, it's a nightmare. It was a long time ago and it's not true. We say it a lot here: "All that happened a long time ago and it's not true". It's best not to recall those times... Before, I was a man. Before, I could walk: Before, I could drive. Now I can't do either. A nightmare.

I knew Vodolazhsky. He's dead. Migorek Klimovich is dead. Lionka Zaturanov is dead. In short there's only Kolka Verbitsky and myself left... Of the five of us, I'm the only one left... A white crow. I don't know. None of my friends are left.

I would just like to ask if there is anyone from abroad, who would like to... I don't know...help me find a car... even an old one... second hand... so that I could get out, go out into the countryside. Because like this, without seeing nature, it's difficult ... A nightmare... I sit at the window and watch. People go past. That's all. What else have I got left? An early death? I will live as long as God allows me. Man is finished, that's all. The boredom, ohhh! ... you could go mad. I just have to resign myself to it all. I'm still young, but...

Q.—How old are you?

A.S.—I'll be 38 in October. I might as well be 60, what difference is there? I'm resigned to it, after all these years. One day I was in the front room and my dog comes in and looks at me. "What are you looking at me for? " I said: "Woof!" He must have thought this man has gone mad... and who knows? And he went into the kitchen. He stayed there for a while, then he came back. "What's the matter? Woof!" He went off again, then he came back a third time. I said "Woof!" And he said "Woof!" So there you are, the two of us had a little chat. Nightmare. It was funny.

The dog was always trying to catch the parrot, when it taunted him. The dog slipped and fell over. The parrot who was running away slipped and fell over too. Both of them. What do I know, it was funny. Whereas now, of course, it's not the same.

When I was completely paralysed I was lying like a plank on the bed. The dog came up, put his paws on my hand and looked at me with tears in his eyes. He was ill as well. "You're not well either, little brother, go and have a walk". He died sometime later. He was on the sofa but he got down to die. He didn't die on the sofa. He was a good dog.

There must be something up there... I don't know. I didn't think I had sinned so badly against God... I don't know... normally. A nightmare.

Before we left, Anatoli commented one by one on the photos of his colleagues that we had taken during our last visit.

A.S.—I have no news of Anatoli Borovsky. I don't know where he is, what he's doing... (*Nobody told him he had died.*) Petia Shashkov...

Q.—Do you see each other?

A.S.—Shashkov, from time to time. He used to come but not now. Vit'ka Kulik came a couple of times. But I've lost touch with him completely. Same with Sachka Grudino That one, on the other hand, I don't know him. (*He looks at a picture of himself taken eight years ago.*) Hello darling! (*He laughs.*) A nightmare.

2001
When we went to his home in June 2001, it was his widow that greeted us. He had died on the 14th July 1999

Mrs Saragovets.—His health went downhill suddenly. He couldn't move his arms or legs anymore, he couldn't eat or drink unaided, he couldn't do anything at all. His legs were covered with eczema. The doctor said it was a result of the bone marrow decaying and that it was the end. He was a condemned man. He lay here for six months We didn't hospitalise him because we didn't want him to be a guinea pig for their experiments. The effects of radiation are practically incurable and they test all their new methods on the patients. He was bed-ridden for six months and then... he was practically rotting alive. All his tissues started to decompose, to the point where his hip bones were visible. I took care of him myself according to instructions from the doctor until the day his heart stopped. We tried to alleviate his suffering with injections, tablets... But there was nowhere left on his body where you could inject... The bones were bare. His whole body was disappearing. His back entirely, you could touch his hip bones with your hand. I put my hand in with gloves on to disinfect it and I pulled out...the remains of bone that had come away. Decomposed bone, rotting. He was fully conscious. He asked only to die quickly, so the suffering would stop... It was so painful... When I had to turn him from one side to the other, he'd grit his teeth, other times he'd groan. He never cried out, he endured it all. He had great strength of will.

We questioned the doctors, consulted a top specialist, we approached everyone we could to ask them about it. They said they knew nothing about this illness. The decomposition of the bone marrow left them dumbfounded. They were powerless.

I'm grateful to our local doctor Gula. She became our friend and I consulted her continually. I must also thank the English Fireman's Association who helped us. One of them, whom my husband knew, helped us financially, from England. Every three months he would send us something, as he might to a friend. But we have never received any help from the Ministry of Health or other government organisations.

We were married in 1983 and in 1986 he found himself at Chernobyl. That's where all our troubles started. He was always ill. Then the left side of his body started to go numb. The doctors said "You're playing the fool, you've just caught a cold from being in a draught". But in fact it was a completely different illness. His immune system had been destroyed. I think Chernobyl was a tragedy, not only for us, but for the whole of Belarus... so many innocent victims. To die like that, for nothing. To think those men who sacrificed themselves there are completely forgotten. Even this apartment where we live was taken over during the hunger strike that my husband and the other liquidators went on. Because when they were recruited they were given great promises of apartments, crèches for the children. And then it all came to nothing. It was the only way to get themselves heard.

Q.—Was he a believer?

Mrs S.—Not really but... he did say that "God allowed me to live thirteen years after Chernobyl" so that means he felt that it was a big thing. If not, how... how can you explain how he resisted so long? Here, the men died straight away. Vodolazhsky, a good friend of ours, a colonel, a helicopter pilot, died straightaway and exactly like him. Exactly the same phenomenon of decomposition of the organism like my husband. He flew over the reactor; he protected his soldiers by preventing them from flying, he piloted himself. He understood what it would lead to.

It hurts to think about it, to look at all this. Very painful. We don't understand why.

(A silence) He could talk about anything to anyone.. He could be funny and serious. He was always good fun. It was easy living with someone like him...who understood everything. And who lived life to the full.

You understand, there are people who are quite happy, who just live. "I have this and I have that" and that's enough. But he, he needed something else. He was reaching out for something. Always looking further, further further... He couldn't stay in one place. He was impatient for life.

Chapter VIII

THE CRIME COMMITTED BY THE UNITED NATIONS AGENCIES

A tragedy, enacted by a few individuals, who, day after day, decide the fate of millions of people.

YVES LENOIR, *op. cit.*

It is upon this radiological, health, medical and social reality of the abandoned liquidators and contaminated populations that the International Chernobyl Project—comprising 200 experts from 25 countries and representing, among others, the IAEA, UNSCEAR, the FAO and the Commission of the European community—imposed its verdict, during the IAEA conference, in Vienna, from 21st to 24th May 1991.

What research had been undertaken under its auspices, and what, in substance, did the International Chernobyl Project say?

According to experts, the radiation has had no effect on the health of the population. Both external and the internal doses of radiation were overestimated by the Soviet authorities.

The scientists from the countries concerned, Ukraine and Belarus lacked competence in the subject of radiation when they attributed to it the health problems they were encountering, when it was likely that these were caused by psychological and stress factors.

No attempt was made by experts from the International Chernobyl Project to estimate the doses received in the acute phase, the first weeks following the disaster.

In the opinion of the experts, the numbers of people re-housed and the food restrictions imposed need not have been so extensive. In applying a lifetime dose as the criterion for re-housing, there was no point in taking into account the dose accumulated over the three preceding years.

Thus, for strictly economic considerations, the seriousness of the concept of the lifetime dose, proclaimed as a radioprotection measure, has been eliminated. Manipulated in this way, this dose limit no longer has any objective value scientifically or medically.

> All the children examined were found to be in good health. There was no significant difference between children of the same age from contaminated localities and "control" localities.
>
> The data collected does not reveal any marked increase in leukaemia or in thyroid tumours since the accident. The only information available regarding these illnesses was based on hearsay.

"Hearsay!" exclaimed an astonished Bella Belbéoch "When Professor Demidchik, in Minsk, had already operated on 29 children in 1990 and 59 children in 1991 for thyroid cancer. In 1990, the incidence of thyroid cancer was already 20 times higher than before Chernobyl and this figure would continue to increase in the following years".

No mention in the International Chernobyl Project of immunity problems in children, nor of the increase in chromosomal abnormalities. No significant increase in birth defects, which is astonishing in the light of the studies published by G. Laziuk.[54]

Nothing about the health of the 800,000 liquidators. Cancelled out. For the international experts, they never existed.

This official untruth was put forward without the slightest serious scientific foundation. Using the "Hiroshima dogma" as a basis, it was preceded by no specific studies in the field. In the report Bella Belbéoch came across some very strange findings: Professor Pellerin had supplied 8,000 dosimeter films that had been worn

---

[54] In 1996, G.Laziuk wrote: "One of the unsolved problems of the Chernobyl catastrophe is the increase in the proportion of children born with congenital malformations, recognized at birth, and corresponding to the most common forms of hereditary damage (. . .). The appearance of this problem in our country deeply concerns the population; and weighs heavily on them. The radionuclides emitted from the power plant (Cs-137 and Sr-90) are damaging our genetic heritage (mutagenic effects) and interfering with the normal development of organs (teratogenic effects)". G.I. Laziuk, D.L. Nikolajew and U.W. Nowikowa. "Dynamik der angeborenen und vererbten Pathologien in Folge der Katastrophe von Tschernobyl", in Gesundheitszustand der Bevölkerung, die auf dem durch die Tschernobyl-Katastrophe verseuchtem Territorium der Republik Belarus lebt. Die wichtigsten wissenschaftlichen Referate, International Congress "The world after Chernobyl", Minsk, 23–29 March 1996.

for two months by the inhabitants of certain *selected* contaminated villages.[55] 90% of these films were below the detectable levels of natural background radiation. Bella Belbéoch then wonders whether "there is no additional radiation discernible above ambient "background" levels in these *selected* contaminated areas" (See the "political" explanation for this put forward by Nesterenko further on and for the technical explanation, see the footnote.)

Denouncing "Western responsibility in the health consequences of the Chernobyl disaster in Belarus, Ukraine and Russia", Bella Belbéoch concluded:

> The support given to the central Soviet power by the WHO and other international organisations, nullified the efforts being made by Ukrainian and Belarusian scientists to protect their people in the contaminated zones. *And for that we are responsible.* [...]
> This behaviour of our experts scarcely raised an eyebrow in the scientific community, or intermediary organisations (medical professionals, unions, civil society associations), or in the media. *Our responsibility for the health consequences of the Chernobyl accident is therefore total.* What is more, the actions taken by our experts, following Chernobyl, set a precedent for the use of purely economic criteria in the management of crises, with daunting implications should a nuclear accident occur in our own countries, an eventuality that cannot be ruled out. [56]

V. Nesterenko explains how the Soviets *selected* the subjects they would study for submission to their Western colleagues. The two parties worked hand in hand: the *tacit complicity* predicted by Bella Belbéoch, five days after the disaster.

This is what he told us in April 2000.

V. Nesterenko.—In 1989, various documents that had been kept secret were allowed into the public domain. Gorbachev and the president of the Council of Ministers, Ryzhkov, wrote to the IAEA to ask them to send specialists to answer the question of whether "we have taken all necessary measures to protect the population and what consequences should they fear". The experts arrived, three million dollars

---

[55] Bella Belbéoch's question is ironic because the statement is absurd. It would probably have been more effective to use thermoluminescent dosimeters rather than film dosimeters because the former are a lot more sensitive. These film badges, which are used as dosimeters in the nuclear industry, (they were developed for use in detectors and fixators, their sensitivity depends on the size of the silver bromide grains) have a threshold detection level of around 0.2 mSv. This threshold is roughly the same as two month's accumulated dose of natural radiation. In fact the dose debit before Chernobyl must have been about 10–15 μrem/h and two months represents about 1,440 hours. If you take 14 μrem/h (0.14 μSv/h) the accumulated dose through natural radiation will be around 200 μSv, i.e. 0.2 mSv. So there is no detectable radiation above natural background radiation in contaminated areas close to the area that was evacuated in 1986 and "selected" by the International Chernobyl Project.

[56] Bella Belbéoch, *op.cit.*

were allocated to the Soviet Union. The specialists worked at Gomel, in Ukraine and mainly in Russia. When I heard about it, I suggested to the Ministry of Health that I could give them the information I had in my possession. I was told that it served no purpose, that the specialists from the Ministry of Health in Moscow were already involved and working with them. So, they were knowingly disregarding anyone who might give them any different information. Given the expense of providing the daily needs of these 200 Western specialists, they only stayed for short periods at a time. Then there was the language barrier and secondly…they were filling their heads with rubbish. At Bragin, our people had decontaminated four times on a run. The experts arrive, do their measurements: "Oh, it's not so bad here!" They did not tell them that the soil had been removed four times.

Then, in September, they came to examine the children. They brought spectrometers, they measured the children: "You see, their accumulation is minimal". But they forgot to tell them that these children had spent the previous three or four months in the Naroch district in other words, one of the clean areas in Belarus. They had been eating clean food, and as children's bodies purify themselves very quickly, their levels of accumulation were very low. In other words… if, at least, this work had been undertaken on a competitive basis, if it had been announced that "Brussels has proposed a project, they are recruiting specialists…" But that's not what happened. In Belarus, it is the government and the government alone that put forward names to work with the Western experts. My opinion on this is very clear. It's obvious that there is a nuclear lobby, very powerful, a worldwide lobby. They pay. They give money for their projects. Here, they choose who to give the order to and in what way. Some years later, I was told by our specialists that, during discussions at the IAEA conference in Vienna in 1996, where there was a large delegation from Belarus, trained by the government so we were not there, our people were told: "Only thyroid cancer has been officially recognised in your country. Say nothing about any other illness". Each of them was told (some of my former students worked for the IAEA): "If you leave those questions alone, you will receive lucrative contracts from the IAEA. New contracts. If not, you will get nothing". Some people will starve before they sell their souls. Unfortunately not everyone is like that and it's no secret to anyone that scientists lived better during the Soviet era. Therefore, it becomes clear that the scientific research is intentionally organised so that correlations between illnesses and radiation dose received do not show up.

Professor Rose Goncharova, Member of the Institute of Genetics at the Academy of Sciences in Belarus, has studied genetic anomalies in fish and rodents, which have increased from generation to generation in areas with relatively low levels of caesium-137 contamination, 200 km from Chernobyl. We interviewed her in April 2000.

R. Goncharova.—In 1989, following a request from the former Soviet Union, research was undertaken as part of the International Chernobyl Project: they were asked to evaluate the radiological situation and its possible influence on the health of the population. The research was undertaken by well paid, highly qualified international experts. They were not able to demonstrate an increased frequency of chromosomal aberrations in somatic human cells, in peripheral blood in the inhabitants of the contaminated areas. Nothing was found. Three years ago, American researchers published an article describing a research project on emigrants from the former Soviet Union[57]. They had left Belarus between 1986 and 1989, from Kiev, Babruisk and Mozyr, where levels of contamination are low: at Babrouisk, less than 1 curie, and Kiev and Mozyr between 1 and 5 curies per square kilometre, where, in principle, levels of radiation were so low, "it could have no effect".

They emigrated during those years and soon after their arrival in the United States, they found themselves part of a medical follow up. Using a Human Radiation Spectrometer, they measured the concentration of radionuclides in their bodies. They studied cytogenetic lesions in the lymphocytes of peripheral blood, in other words, chromosomal aberrations and genetic mutations. And so these American researchers, quite simply, showed the effects of low level radiation. Very low levels of radiation. But it wasn't published until about three years ago (in 1997). In my articles, I explore the reasons why in 1989, in more contaminated areas, the international experts did not find any lesions in the hereditary apparatus of the somatic cells of the inhabitants, nor any chromosomal aberrations, nor genetic mutations when, among the Soviet emigrants who, before they left, were living in areas with lower levels of contamination, these effects were found one or two years after they arrived.

Q.—What is the significance of this? Were the first people lying or do we need to look for another explanation?

R. Goncharova.—The first people are "so good" at their work, in quotes, that they do not uncover effects that were obviously there.

Q.—It was their job not to discover them?

R. Goncharova.—I don't know what their job was. I am simply recounting the facts.

V. Nesterenko.—The International Chernobyl Project concluded: there is nothing to fear, the government has taken all necessary measures. Gorbachev got a positive response to the two questions he had posed. A lot of money changed hands and they got the answers they wanted. I don't contest the expertise or the qualifications of the experts.

R. Goncharova.—I said the same thing myself: these are highly qualified professionals. You have to be extremely "capable" to hide evidence of effects that in my opinion was leaping off the page. But of course, we are paying now, and will carry on paying, over the next decades for this criminal policy.

---

[57] G.K. Livingston, R.H. Jensen, E.B. Silberstein, J.D. Hinnefeld, G. Pratt, W.L. Bigbee, R.G. Langlois, S.G. Grant, R. Shukla, *Int. J. Radiat. Biol.*, 1997, 72, No 6, p. 703–713.

Chapter IX

ALTERNATIVE EXPERTISE

Valery Legasov, crushed by Ilyin's Moscow mafia, committed suicide a year too early. He could have been elected to the Supreme Soviet of the USSR along with Yury Shcherbak, Anatoli Volkov, like many other dissident scientists and experts. From 1st September, he could have participated in the work of the commission set up by the Supreme Soviet to analyse the causes of the accident at Chernobyl and to investigate the responsibilities of those who had not taken the necessary protective measures in the post-accidental period.

It was under threat from this investigation—which would have found them guilty—that the Moscow nucleocrats asked the IAEA and the WHO for the expertise of the International Chernobyl Project. The two commissions, one overseen by the UN and the other by independent Soviet citizens, worked simultaneously. The first concluded its task after a few weeks spent in the field, (published in May 1991). The second worked between 1990 and 1993, and survived the collapse of the USSR (21st August 1991) thanks to the financial support of a private entrepreneur, under *perestroika*.

1. PROVIDENTIAL MEETINGS AND SUPPORT

A few weeks before the putsch against Gorbachev, Vassili Nesterenko was writing a report while on the aeroplane taking him from Minsk to a meeting of the Supreme Soviet Commission on Chernobyl, in Moscow. He was chairing the permanent experts and specialists group which was analysing the situation in Belarus, the country that had been most contaminated by the disaster (Belarus, 23% of its territory contaminated, Ukraine 4.8% and Russia 0.5%). Absorbed in his work, Nesterenko did not notice the curiosity of the fellow passenger beside him. He introduced himself and said that he owned an oil company and was president of an NGO, the International Community for the Restauration of the Habitat for Humankind (SENMURV). His name was Afanasi Kim. Having collected money to help the victims of the Chernobyl disaster, he did not know who to give it to, to ensure that it would not end up in the wrong hands. He had observed that Nesterenko's work related to Chernobyl and asked his advice. Nesterenko invited him to sit in on the meeting of the commission. It was short of funds: some of the work was being undertaken voluntarily by its members. The oil man was pleased to have been able to attend this meeting and they were to meet again to come up with some proposals. Soon after that in August, there was the putsch, the end of communism, the Supreme Soviet was dissolved, and the members of the commission were in despair. They decided to complete their work using their own resources. But there was no money to publish their conclusions. The meeting with A.M. Kim proved to be providential.

The International Community "SENMURV" considered that it was its civic and moral duty to lend practical and financial support to the group of experts so that it could complete its report about the Chernobyl disaster.

Together with "SENMURV" the permanent group of experts from the former Supreme Soviet of the USSR was reorganised into a committee of experts (CUE-ОЭК), and can be seen as its successor. Nesterenko presided over its work until it was completed in 1993.

Significant financial support was given to the group of experts by the humanitarian NGO "Aide-Tchernobyl" (Aid for Chernobyl) set up by A.E. Karpov and R.S. Tilles.

The International Community "SENMURV" published 4 volumes of conclusions from the expertise of parliamentarians and independent scientists in Russian: "The Chernobyl disaster—causes and consequences".

The fact that the text was only published in Russian limited the dissemination of the independent experts' conclusions across the international community. To fill the gap, a Swiss journalist Susan Boos, asked Nesterenko to write an extensive "press release" that would sum up the essentials and could be published in German and English. The press release was published under the title: "Scale and consequences of the Chernobyl disaster in Belarus, Ukraine and Russia", Minsk, 1996.

In 1998, Susan Boos had convinced her colleagues at the weekly journal WoZ that the international community should not rely solely on the IAEA for its information about Chernobyl, but should examine the studies undertaken by independent scientists in Belarus, Ukraine and Russia, especially since the conclusions from these two bodies of expertise were diametrically opposed. The journalists collected enough money to publish 500 copies, in English, of the conclusions of the Supreme Soviet experts. The same year, they published Nesterenko's manifesto *The Chernobyl Disaster. Radioprotection of the Population* (Minsk, 1997), also in English, in which he openly committed himself to undertake the work that should have been done by the Ministry of Health in Belarus. Although they were not distributed widely, these documents had at least been rescued from media oblivion. They have a symbolic and historical value and thanks to the solidarity of a few individuals, the wall of deceit and ignorance has been breached. The intention was to send these documents to universities in America, Great Britain, Germany, France, Switzerland, Italy and Japan.

Paradoxically it turned out that the expertise instituted by the Supreme Soviet of the USSR was the only moment in history when the emerging civil society in the communist state was able to proceed with the independent study of the consequences of the Chernobyl tragedy. The conclusions of the government's expert group, alas, were never divulged. But testimony from the contaminated territories around Chernobyl and the search for scientific truth have not ceased since that time. This book, which chronicles the events from one year to another, is proof of that.

I am briefly jumping ahead in the account of events to quote an extract from the book that Nesterenko published in 1997. It reveals the way in which experts from the West denied the real levels of contamination following the accident at Chernobyl, which then had grave consequences for the health of hundreds of thousands of people contaminated by radionuclides.

2. DECEPTIVE AID[58]

THE DISENGAGEMENT OF THE STATE

The desire to reduce the burden of Chernobyl on the national budget has manifested, over the last two years, in declarations about it being quite safe to live in areas contaminated at levels between 1 and 5 curies per square kilometre (Ci/km^2) and the possibility of producing food here that is ecologically clean.

This policy direction has been based on the results of misleading measurements of the contamination of the inhabitants of Belarus, Ukraine and Russia, made between 1991 and 1993 by German scientists using a Human Radiation Spectrometer (HRS; also known as a Whole Body Counter—WBC). The National Commission on Radiological Protection (NCRP) lent its support to this idea, in April 1995, and adopted the "concept of radioprotection measures for inhabitants in the post accident rehabilitation phase". The philosophy behind this concept, developed by a working group under Professor E.P. Petriaiev, was that the emergency phase of the accident was over, the rehabilitation phase had begun and people needed to "learn to live with radioactivity". According to this concept, no further radioprotection measures were necessary where levels were below 1 millisievert per year (mSv/y). On this basis, in autumn 1995, the government of Belarus reduced radiological and social protection measures for the population in these regions.

DECEPTVE AID FROM GERMAN EXPERTS

In 1991, in response to the request for aid from the government of the former Soviet Union, the German Minister for the Environment gave 13 million marks for a programme to measure the radioactive charge in the body of people living in the contaminated areas of Russia, Ukraine and Belarus, following the accident at Chernobyl. The programme was entrusted to the Centre for Nuclear Research in *Jülich*

---

[58] Extracts from V. Nesterenko, The Chernobyl Catastrophe. Radioprotection of the Population, Minsk, 1997

(Federal Republic of Germany). In May 1991–1993, 317,000 people were examined in the three Republics using HRS.

These measurements were only taken in the towns because, according to the directors of the programme, the country roads were not suitable for the heavy vehicles carrying the measuring equipment. In this way, no measurements were taken of the rural population of Belarus (with the exception of two villages, Kirov and Svetilovichi: 1,651 measurements out of 41,785, less than 4% of the total). Yet we know that it is precisely in these rural areas that the Belarusian population receives 90% of the collective radiation dose, through the consumption of local contaminated food products. According to the German experts, only 1.4% of people measured in Belarus had received a dose higher than 1 mSv/y; 6.8% had received a dose of 1 mSv/y and the great majority, 91.8% of the population, had received the minimal dose of 0.3 mSv/y.

However, a register of radiation dose already existed for the population living in villages in Belarus at the time of the German programme (Minsk 1991, 1992). Why were the inhabitants of the villages listed below, who had very high levels of radioactivity in their bodies, not included in the FRG programme?

Brest area: Vulka—1.8 mSv/y; Zastenok—3.9 mSv/y; Dobraya Volia—2.5 mSv/y; Pare—1.1 mSv/y; Zhitkovichi—1.6 mSv/y; Gorodnaya—1.2 mSv/y; Derevnaya—1.3 mSv/y; Colonia—3.3 mSv/y; Otverzhichi—1.3 mSv/y; Olmany—3.0 mSv/y.

Gomel area : Komanov—2.9 mSv/y; Negliubka—1.2 mSv/y; Zhelezniki—1.6 mSv/y; Valavsk—1.3 mSv/y; Glazki—5.9 mSv/y; Kuzmichi—2.4 mSv/y; Skorodnoie—2.6 mSv/y; Buda—1.8 mSv/y; Grichinovichi—2.0 mSv/y; Korchevatka—1.5 mSv/y; Beriozovka—3.4 mSv/y; Lenino—2.5 mSv/y; Obukhovshchina—2.0 mSv/y; Slobodka—1.5 mSv/y; Shareiki—1.6 mSv/y; Volyntsy—1.1 mSv/y; Novaya Zenkovina—1.2 mSv/y; Staraya Zenkovina—1.4 mSv/y; Borovka—1.5 mSv/y; Markovskoye—1.4 mSv/y; Rudnishche—3.6 mSv/y; Pervomaisk—1.3 mSv/y; Viazovoye—2.8 mSv/y; Dzerzhinsk—2.0 mSv/y; Danilevichi—1.4 mSv/y; Zabolotie—1.6 mSv/y; Chiane—1.4 mSv/y; Manchitsy—1.5 mSv/y; Verbovichi—1.4 mSv/y; Grushevka—1.6 mSv/y; Konotop—1.6 mSv/y; Buda Golovchitskaya—1.2 mSv/y; Demidov—1.3 mSv/y; Zavoit—2.8 mSv/y; Smolegov—2.6 mSv/y; Khilchikha—2.4 mSv/y; Khomenki—4.7 mSv/y; Dukhanovka—1.7 mSv/y; Dubrova—1.4 mSv/y; Borisovshchina—1.6 mSv/y; Slabozhanka—1.2 mSv/y; Partizanskaya—2.9 mSv/y; Pikulikha—2.3 mSv/y; Krasnyi Bereg—2.0 mSv/y; Pokat—1.7 mSv/y; Krutoie—1.5 mSv/y; Selianin—1.5 mSv/y; Budishche—3.2 mSv/y; Novozakharpolie—2.8 mSv/y; Sapriki—2.4 mSv/y.

Many more villages like these, inhabited by hundreds of thousands of people with high levels of internal radiation, could be cited. It is hard to understand why the authors of the German programme did not investigate them.

The German measurement programme was undertaken without the participation of Belarusian scientists and the characteristics of the local diet were not taken into account. In this way, the inhabitants of villages, in particular those in Polessie, were not included as important subjects in the measurement campaign.

The error made by the German team in their choice of subjects to measure resulted in an underestimation, by several orders of magnitude, of the dose received by the population of Gomel, *via* food. The data bank at the Belrad Institute of Radioprotection contains more than 200,000 measurements of radioactively contaminated food products. Over the last five years, between 100 and 200 HRS annual measurement campaigns of the inhabitants from the above mentioned villagers, have been undertaken. Though the surface radioactivity is not high in Polessie, the characteristics of the soil are such that the coefficient of migration of the radionuclides of caesium-137 from the soil into the plants is higher here than in the fertile soils of the Ukrainian "chernoziom", by a factor of 30–50. The radioactive load in the body is 2–5 times higher in these areas than that reported by the FRG measurement programme.

It is very regrettable that the authors failed to accomplish the task they were set in the German measurement programme, that they neglected the main contingent of contaminated inhabitants in Belarus and have underestimated by a factor of between 2 and 5 the doses received by the Republic's population.

This information misled the President of Belarus and the new government of the Republic. On the basis of this information, it was decided that living in areas contaminated at levels between 1–5 Ci/km2 posed no danger to health, the mistaken concept of "protective measures in the post accident rehabilitation phase [...]" was adopted and the distribution of free medicines and vitamins to children living in the 1–5 Ci/km2 zone was withdrawn, leaving the inhabitants without radiological protection. 1,500,000 people, including 400,000 children, out of a total of 2,200,000 victims of the Chernobyl disaster, live in this zone.

The ideology behind this project is very reminiscent of the errors made in the International Chernobyl Project. The IAEA experts, having excluded from their statistics, the 800,000 liquidators and the 130,000 people who were evacuated from the 30 km zone around the nuclear power station, concluded that there were no health consequences following the Chernobyl disaster. In the same way, by omitting the inhabitants of the villages in the areas contaminated by Chernobyl, the FRG measurement programme underestimated the radioactive charge of the Belarusian people.

PART THREE

SCIENCE
BEHIND BARS

Chapter One

YURY BANDAZHEVSKY:
A SCIENTIST BEYOND CONTROL

In 1994, Vassili Nesterenko met the founder and rector of the Gomel Medical Institute, the pathologist and doctor, Yury Bandazhevsky who, since 1991, has been investigating the causes of new diseases appearing among people living in the contaminated territories. With his wife Galina, a paediatrician and cardiologist, Bandazhevsky has discovered that the frequency and severity of the structural and functional alterations of the heart increase in proportion to the amount of radioactive caesium incorporated in the body. He calls this "caesium cardiomyopathy": heart problems in young children, adolescents and adults, with myocardial degeneration. Death can occur suddenly at any age. Bandazhevsky and his team describe the "interrelated pathological processes in the heart, liver, kidney, endocrine organs as well as in the immune system". All these injuries arise from the same pathological process that researchers have called "the syndrome of incorporated long-lived radionuclides". These are the findings of a rigorous study into the health of thousands of adults and children, conducted by the staff at the research institute. For nine years, 25 medical professors have been working on the same subject, in three research areas: clinical, experimental (on animals in the laboratory), and anatomopathological. The Gomel Institute of Medicine has 200 teaching staff, 300 support staff and 1,500 students.

Since 1996, the Belrad institute and the Gomel Institute of Medicine have been working in parallel. Nesterenko travelled through the villages, measuring the levels of internal contamination by caesium-137 among the inhabitants, using spectrometers provided by Western NGOs. For the histological study of the effects of caesium-37 on tissue, he provided the researchers at Gomel with automatic gamma-radiometers that he had designed himself, to be used during autopsy, to measure the level of caesium-137 per kilogram in various organs. The two institutes were able to show, in both laboratory animals and children, that by reducing the level of caesium-137 in the diet, it was possible to avoid irreversible damage to the organs. This opened up an entirely new line of research.

In April 1999, the two scientists were invited by the Belarusian parliament to join a commission whose task was to check the register of radiation dose received and the use of funds provided by the State to the Institute of Radiological Medicine at the Ministry of Health. Their findings were not welcomed by those committee members who had close links to the Ministry. Bandazhevsky, Nesterenko and A. Stozharov, former director of the Institute at the Ministry, signed a separate report and sent it to the Security Council of Belarus, which was responsible for the health of the population. The Security Council rejected the old register of doses produced by the Ministry of Health and invited them to amend it as a matter of urgency, in line with the recommendations made by the three writers of the report. Bandazhevsky himself sent a report to President Lukashenko, in which he severely criticised the work of the Institute of Radiological Medicine, and claimed that out of the seventeen billion roubles spent by the institute, by 1998 only one billion had been spent effectively. In retaliation, over the next few weeks, the Gomel Medical Institute was inspected three times in quick succession, although no problems were found. During the night of 13th July 1999, Bandazhevsky was arrested on the basis of a decree against terrorism issued by Lukashenko. On 18th June 2001, he was sentenced by the military tribunal of the Supreme Court of Belarus, to eight years in prison for corruption, without any evidence being put forward. The new rector at the Gomel Institute of Medicine abandoned the research programme, declaring that it was not worthy of a higher education establishment.

Bandazhevsky represented a nightmare for the "experts" of the nuclear lobby, an unforeseen obstacle to their strategy of ignorance. Here was an anatomical pathologist restoring scientific research to its rightful place at the heart of the open-air laboratory, which had resulted from the most serious technological disaster in history. He had been preparing for this mission since adolescence, and he had devoted himself passionately to studying "the influence of different environmental factors (physical, chemical, biological) on gestation, foetal development and on the formation of the body's vital organs and systems". In other words, his training equipped him to understand perfectly the mechanisms and implications of the effects of proximity of radionuclides on vital organs at a cellular level, an area of study that official science currently refuses to recognise or even to discuss. It is at this microscopic level of his work that authentic scientific research will blow wide open the whole subject of toxic and radiological phenomena of low dose incorporated radionuclides in the human organism that the ICRP, UNSCEAR, IAEA, WHO, CEA, the UN Security Council and the Pentagon have kept under lock and key.

Bandazhevsky's quest for knowledge of these secret and forbidden mechanisms simply had to be stopped. His discoveries would be catastrophic for the nuclear lobby. Paradoxically, their attempt to silence him ended in defeat because by arresting him and throwing him into jail in such a brutal way, his case became known throughout the world.

The day after the scientist was sentenced to eight years in prison, Nesterenko, behind the wheel of his car, was driving us to meet Galina Bandazhevskaya. Despondent, anguished, she had just asked "How do I carry on with my life?" He was in shock, as we all were, and expressed his thoughts out loud as he drove: "It's terrible. He won't be able to continue his scientific research for so many years. I even wonder if there is any point in continuing my work. I'm not a doctor. I'm a physicist. It is very important for me that the doctors tell me the point at which I, as a radioprotection specialist, must intervene to protect the child against serious illness and death. To have this information, we needed to do more research, we needed to keep going…"

These two scientists were obviously not alone in understanding what needed to be done, and not the only scientists pursuing this research. But they were the only ones who found themselves physically at the heart of the contaminated territories, in the midst of the health, political and humanitarian problems posed by the disaster at Chernobyl. They were alone in their determination to honour their role as scientists and to honour science itself, in the face of their people's suffering. They had resisted extreme pressure—Vassili Nesterenko for fifteen years and Yury Bandazhevsky for ten years (in 2001). As well as the vilification and the continual obstacles put in their way by those who serve the lobby in the East and in the West to prevent their activities and muzzle the press, the European Commission programme TACIS (Technical Assistance to the Commonwealth of Independent States) had systematically refused requests to finance radioprotection projects for children, submitted repeatedly by Nesterenko.

The real and symbolic significance of certain moments in history appears to be inversely proportional to the apparent fragility of the individuals involved. This is one of those moments, perhaps a chance that should not be lost, at the heart of the Chernobyl tragedy. In supporting these two outstanding scientists, humanely, politically, and financially, civil society in the West, in Europe and in the United States have the opportunity to come together in a truly humanitarian effort, and to confront the *inner sanctum* of our governments whose suicidal policies are out of control and put at risk the whole of humanity. Their real aim is not to bring about the downfall of the nuclear industry as such. The industry is already condemned following the Chernobyl disaster and can only defend itself about what happened there with secrecy and deceit. The real objective of these two scientists and their supporters is scientific truth, properly funded and freely shared. More than ever, in the nuclear domain, if humanity is to survive, we must have independent research and knowledge. "Every human being has the right to know everything that relates to her health, to the health of her children and to the health of those closest to her, what to avoid and why.[59]" The fact that these two men find themselves in the same

---

[59] John W. Gofman, *Chernobyl Accident, Radiation Consequences for this and Future Generations*, 1993.

place, involved in the same struggle, is a fragile chance at this particular moment in history.

I heard about Bandazhevsky's imprisonment purely by chance in September 1999. He had already been in prison since July, and had been through some brutal experiences. No-one in the West knew. I might not have been able to attend the projection of my films at the festival " East-West " in the town of Die (France), where Svetlana Alexieich mentioned it to me in passing at the end of our first conversation between "two Russians abroad". I asked her "How is Nesterenko?"- I knew that Svetlana knew him, because he is one of the "voices" in her book "Voices from Chernobyl". "Not very good at the moment. One of his friends has been arrested". "Who?" "A doctor from Gomel". "You mean Bandazhevsky!" "Yes". "And you just tell me like that?!" Suddenly I was at an impasse. What should I do? I had not met him the previous year but Nesterenko had talked of him as an outstanding research scientist, who understood what was happening, and whose data were turning the official dogma upside down. I was struck not only by the brutal clarity of the message sent out by his arrest but also by the silence and resignation of the friends I had made in Belarus. Two months had passed and no-one in the West knew anything. I had just described the rebel Nesterenko, and his radioprotection work with children in the contaminated villages in Belarus in a television documentary[60], but he had not thought to pick up a phone to let me know. The Iron Curtain, no longer physically there but an enduring presence, still separated the minds of citizens of Eastern Europe from the victors of the Cold War. Even Svetlana Alexievich had only told me what had happened after I had asked routinely for news of a mutual acquaintance.

Something had to be done. But what? I didn't know anyone either in France, Switzerland or Italy with whom I could share the weight and the full significance of this apparently small news item. "You don't say! Another arrest somewhere in Belarus—not even in the Ukraine—with everything else that's going on in the world!" While preparing my programme I had read the name of an Emeritus Professor at the University of Basle in Switzerland, Dr Michel Fernex, in the book *Permanent People's Tribunal on Chernobyl*[61]. And now a fragile chain of coincidences (and opportunities) began to take shape: Professor Fernex knew Nesterenko's work very well. He admired Bandazhevsky's work and had been very impressed when he had visited the Institute of Medicine at Gomel. "Have you heard the news?" "What news?" "Bandazhevsky is in prison". There was a moment of stunned silence at the other end of the line, and then the fireworks began. Solange Fernex, Michel's wife, had been a Deputy at the European

---

[60] Le Piège atomique (The atomic trap), TSI (Swiss Italian Television), May 1999

[61] Rosalie Bertell, Permanent People's Tribunal, International Medical Commission on Chernobyl and International Peace Bureau. *Chernobyl: environmental, health and human implications*, Vienna, Austria 12–15 April 1996.

Parliament, and a long time anti-nuclear activist, with a vast network of contacts and Internet links to organisations all over the world. Michel Fernex had been a member of the Steering Committee on Tropical Diseases Research at WHO, and a critical observer of the 1995 WHO conference and the 1996 IAEA conference on the consequences of Chernobyl. He followed the international debate about the effects of radiation on health very closely. He was very familiar with the way the pseudo-scientists "accredited" by the lobby used subterfuges to falsify statistics and produce flawed epidemiological studies, containing epistemological errors to suit their purpose[62].

Appeals were made immediately to mobilise public opinion, political figures and institutions in the "free world" about the fate of the prisoner. This was the beginning of the campaign, led by individuals and organisations in the West, to free Bandazhevsky so that he could continue with his scientific research, and to support Nesterenko in his campaign to measure internal radiation in children and to provide them with some radio-protection in the contaminated villages.

At the beginning, fearing that telephone conversations could be tapped, I did not dare ask Nesterenko about it except by allusion and in veiled terms so as not to create problems for him. But he spoke openly about it and this allowed me to speak freely also: his wisdom and restraint, his reputation as a respected scientist in his own country did not prevent him, however, from speaking bluntly and criticising the Ministry of Health policies. The risks he took were carefully calculated and based on scientific understanding. He banished fear because he felt it blunted his judgement and paralysed his ability to act. He told me that Bandazhevsky had been arrested as a result of an organised smear campaign from a group of doctors and civil servants within the Ministry of Health, who wanted both to block his research, which was beginning to pose a danger to them, and to intimidate independent scientists. He faxed me an article by Irina Makovetskaya, a brave journalist who, later on, would follow up the story of Bandazhevsky's torments in prison in the pages of the opposition newspaper BDG *(Bielorousskaya Delovaya Gazeta)*. The articles written by Irina Makovetskaya and Lara Nevmenova helped us to understand the real context and allowed us to bring the affair to public notice internationally and to keep it in the spotlight. I quote the first two paragraphs of this series of articles that I posted on the internet:

Professor Bandazhevsky has been removed from his post as Rector of the Gomel Institute of Medicine.

On July 12th, Vladimir Ravkov, lieutenant-colonel in the medical service and holder of the Chair of Military Medicine at the Institute, was arrested in the garage of the Institute, taken by force by staff from the Directorate of the Committee against organised crime,

---

[62] M. Fernex, "La catastrophe de Tchernobyl et la sante", in *Chroniques sur la Bielorussie contemporaine*, L'Harmattan, 2001, Appendix I.

put in prison and interrogated for several hours. He remembers having been given a glass of water to drink in the judge's office, after which he could not formulate his thoughts clearly, and everything passed by in a fog. His wife claims that he was drugged because, in his normal state, he would never have agreed to slander Bandazhevsky.

Professor Bandazhevsky was arrested twenty four hours later, on 13th July, late at night. That day a ministerial commission was in the process of monitoring the entrance examinations for the Gomel Institute of Medicine, and Yury Bandazhevsky, in his role as rector, had asked the members of the commission to defend the Institute against "damaging insinuations". The call was heeded: a few hours later, the rector's flat was searched and two television sets, a video recording machine, a computer, a bunch of keys, and four diaries were removed in order to establish that no bribes had been taken. The commission staff also searched Bandazhevsky's office at the Institute, and the garage and his mother's apartment at Grodno, but they found no "compromising material". Nevertheless, Bandazhevsky was jailed.

(I. Makovetskaya—BDG. Bielorousskaya Delovaya Gazeta—8th September 1999)

Over the next months and years, Amnesty International adopted Bandazhevsky as a prisoner of conscience. The European Parliament awarded him a "Passport for Freedom", demanding that he be allowed to pursue his research. The Organisation for Security and Co-operation in Europe (OSCE) demanded that the decision of the court should be overturned due to eight separate infringements of the Belarusian Criminal Code.

The first to intervene with both NGOs (non-governmental organisations) and with political and scientific institutions in Europe, were Solange and Michel Fernex, Bella and Roger Belbéoch, the Groupement de scientifiques pour l'information sur l'énergie nucléaire, (GSIEN)[63], Danielle Mitterand (Fondation France-Libertés), Abraham Béhar (President of IPPNW France)[64], Women's International League for Peace and Freedom, (Solange Fernex was the President of the French section)[65]. CRIIRAD became involved in the campaign in February 2001. CRIIRAD organised a demonstration with a number of other organisations on 25th May 2002, in front of the United Nations and the World Health Organisation in Geneva. Many letters were sent to the authorities in Belarus[66].

At the same time, support came from organisations from across the Atlantic. The New York Academy of Sciences, of which Bandazhevsky had been an active member since 1996, approached the Belarusian authorities directly. In France,

---

[63] Later the International Human Rights Network at the Academy of Sciences would intervene.

[64] International Physicians for the Prevention of Nuclear War. Yury Bandazhevsky was awarded the XIV congress medal by its President, Abraham Behar

[65] Women's International League for Peace and Freedom

[66] Film *Youri et Galina Bandajevsky*, de W. Tchertkoff, Feldat Film 2000 at *http://enfants-tchernobylbelarus.org/zippy/nous_de_tchernobyl.flv.zip*

the *Collectif des centres de documentation en histoire ouvrière et sociale* (Collective of Documentation Centres in Workers and Social History) intervened.

Following his first arrest, Bandazhevsky was in prison for five and a half months. On 27th December 1999, under pressure from international opinion, he was released, on condition that he live in Minsk and not leave the country before his trial. Nesterenko was there to greet him when he was released from prison and welcomed him formally to Belrad as a member of the institute's staff. He stayed with his brother-in-law in Minsk temporarily but was able to visit his family in Gomel from time to time. In April 2000, we were able to interview him, with his wife Galina, about the dramatic effects his discoveries had had on the family.

Chapter II

BANDAZHEVSKY'S SCIENTIFIC DISCOVERIES

Since 1988, Bandazhevsky had put forward, through the official channels, a number of far reaching proposals for scientific research to the Academy of Sciences and the Minister of Health in Belarus. "I considered it my duty as a doctor to help to solve the problems related to the disaster. It seemed to me that what had been done so far had not resolved the existing problems. Above all there seemed to be no clear understanding of the mechanisms influencing the incorporation of radionuclides in the body, and how they act on the structure and function of cells and tissues, and on the metabolism". His passion for this area of research—"the influence of different factors (physical, chemical, biological) on gestation, foetal development, and the development of various vital organs and systems"—impelled him quite naturally to work in the Chernobyl territories.

In 1990, Bandazhevsky left the Grodno region, which had been spared any radioactive fallout. At 33, he already had a brilliant career, as director of the central laboratory for scientific research, but he left for Gomel to help the people living in the most highly contaminated territories in the South of Belarus. He was appointed rector of an Institute of Medicine, which did not yet exist and which he had to set up himself. Most doctors had abandoned the area.

Energetic, stubborn, driven by his vocation for scientific research, he worked relentlessly to set up the institute, train 1000 doctors and through three distinct, complementary lines of research, he discovered that caesium-137, incorporated at low doses in contaminated food, concentrates unequally in different organs of the body, leading to much higher doses in some organs than the average level over the whole body, and progressively destroys vital organs. With his wife Galina, a paediatrician and cardiologist, Yury Bandazhevsky described *caesium cardiomyopathy,* a new illness which some Western scientists believe will be named after him : beyond a certain threshold of prolonged intoxication from caesium, heart failure becomes irreversible.

1. FIRST INTERVIEW WITH BANDAZHEVSKY IN MINSK

We met Bandazhevsky at Professor Nesterenko's home in April 2000. Nesterenko showed us the official publication recently issued by the government about the consequences of the disaster at Chernobyl. It contained eighteen reports written by scientists, academics, government ministers, specialists and medical doctors, all highly qualified people, in positions of responsibility, including the two taking part in this interview. We talked about the Belarusian paradox: the publication contains Bandazhevsky's critical report to the government and President Lukashenko about the way the Ministry of Health had squandered public funds.

V. Nesterenko.—For a long time, I wasn't sure if they would publish it at all. They delayed it for a long time and then I heard by chance that it was about to come out. These are the official reports that were published on 21st April 1999. On the 5th or 7th January this year, the book was officially presented to the Belarusian Press Club. The president of the Commission on Chernobyl dedicated the book to Yury and myself. The indictment has still not been lifted. Yury is still being prosecuted, and even his movements are limited so that he cannot, for instance, visit Gomel for fear of destabilising the situation at the institute where he worked.

All the reasons for Yury's arrest can be found in this book. They reprinted his report in its entirety. It is a unique event. Contained within it are the results of his research under the title "Pathological processes in the body in the presence of incorporated radionuclides".

Bandazhevsky.—Read the conclusion: "Any amount… "

V. Nesterenko—"Any amount of incorporated radioactive caesium triggers pathological processes in the body. This contradicts radically the conclusions of the Republic of Belarus' National Commission on Radiation Protection according to which people can safely consume tens and hundreds of becquerels in their daily meals. On the basis of the information presented here, it is clearly necessary to develop a coherent programme of measures to protect those people whose bodies have been subjected to the action of these radioactive elements over a long period, and to organise restorative cures immediately."

Q.—The State publishes his report officially while continuing to persecute him. It is a country of paradoxes.

V. Nesterenko.—Yury is a public servant. He works at the heart of a government institution, and it's the government that he's criticising. I work in an independent institute and make constructive criticisms of Ministry of Health policies. I often criticise them severely. At the moment they are forced to tolerate these criticisms because I am a Member of the Academy and in the past, I had a high level post. Tomorrow they could start treating me differently. Yury began to obtain this information and was able to show that for all these years the government has been misusing the money that was destined to address the problems caused by Chernobyl and that the medical establishment in Belarus is misleading the government, stating that everything is good, everything is fine apparently, and there is no need to spend any more money. In reality the damage caused by Chernobyl has cost 235 billion dollars, about the equivalent of 70 annual budgets of the Republic in 1986. What should the government of Belarus have done? Demand reparations from the Ukraine, from Russia? Gone to the United Nations, set up an insurance fund to compensate the victims? Because the accident we had will not be the last, there will be others. But they prefer to deny the problem. They don't want to quarrel with Russia. So they say nothing. The government wants to reduce expenditure, and is quite happy when the doctors at the Ministry of Health say that everything's fine, they don't need to do anything, that everything's OK. That's their position. And then Yury comes along and upsets all that! He claims that an accumulation in the body of 50 becquerels/kg poses a danger to children.

Q.—How long have you been working on this problem?

Y. Bandazhevsky.—I was rector for a little over nine years. Here are the books that we published with Nesterenko that contradict the official position of the Ministry of Health. Look at this one[67]. It contains photomicrography of the pathological changes and the processes that occur in the organism. Look at the lesions on the glomeruli of the kidney, and, in particular, at this myocardium—it's terrifying. If a specialist saw that, he would say that it is no longer a heart. But it is. A man's heart. I don't know how it could possibly have functioned. It's unbelievable. This was our first publication. There is also a version in English. It was my wife, Galina, who first suspected there might be a link between heart disease and the contamination revealed in cardiograms of children. The discovery caused us a lot of anxiety, and my wife tried to dissuade me from following up this line of research. She knew it would cause us problems.

Here's how it happened. My wife had left a great pile of children's cardiograms on the table for her research project. I was a professor and she was still an assistant, and that evening I wondered how we could begin. It was quite late in the evening. She went to bed and I was left alone to think. Then I had the idea of noting on each child's ECG, the amount of incorporated radionuclides in the body. I had these results for the children because our institute had measured each child using a human radiation spectrometer or HRS. I wrote the results of the HRS on each ECG. And I started to sort the ECG's into piles according to the amount of incorporated radionuclides, and the degree of alteration in the heart. It was like a game of solitaire.

I ended up with four groups:

---

[67] *Pathologie du rayonnement radioactive incorporé.* (Pathology of incorporated radiation) Minsk, 1999.

1) children with between 11 and 26 becquerels per kilo of body weight—64% of them had altered ECG patterns;

2) those with between 26 to 37 Bq/kg–67% of them had altered ECG patterns;

3) those with between 37 and 74 Bq/kg–78% had altered ECG patterns and

4) those with between 74 and 100 Bq/kg in their body—88% of this last group had altered ECG patterns.

I soon noticed that the percentage of children suffering from heart problems increased with the amount of incorporated radionuclides. The difference between the first and the third group was striking. From that point on, we examined the hearts of people who had been contaminated, as a matter of course. I involved health professionals in the work and a number of doctorates were devoted to the subject. We began to make a closer study of other organs. On the basis of this research, we put forward a series of proposals, about how to proceed, and this is described in *Study of medical and biological effects in relation to the quantity of radionuclides incorporated into the body*. It was a completely new approach. It showed that you could live in an area that was contaminated by radionuclides without being affected or live in a clean area and be very badly affected by radionuclides. Illness resulted from the amount of radioactive elements *incorporated into the body*, not from the low level external radiation. It's very significant. When we started, no-one had ever talked about this. I had never found any reference to this correlation in any scientific publication before. Then having understood that it was necessary to deal not only with the contamination in food, mushrooms and berries, but above all in the human body, Nesterenko began to measure the population very thoroughly using an HRS at his institute. This engendered a whole line of research. It was really important, you see, quite crucial.

We have been interested for a long time—since 1992—in the incorporation of various elements. In general, I had chosen to do research on quantifiable phenomena. My preferred methodology was experimental simulation of the pathological processes under study. And this was totally new to us as well. Before, no-one did this. In our country, you were either a research scientist or a pathologist. Whereas we had set up models of the situation, where we were feeding animals with contaminated grain to simulate the food humans were eating. We had a group of rats who were eating wheat and bread, contaminated at 400 Bq/kg, which was the admissible level in 1996! And a control group with 40 Bq/kg. Absolute zero was impossible, unfortunately, given the situation. The results of this experiment are in my book *Clinical and experimental aspects of the effect of incorporated radionuclides on the organism,* published in Gomel in 1995. It contains a lot of data. For example, the metabolic effects of incorporated radionuclides, showing alterations in the biogenic amines in the brain structure. The study shows for example, that after ten days of eating this kind of bread, the animals accumulate 60 Bq/kg and their neurological mediating processes are abruptly disrupted. In other words, their behaviour is modified. After ten to twenty days of the diet, the radionuclides cause a slowing down of the serotonin system, and premature reactivation of other important biological systems. To cause such alterations, the animals would have had to be subjected to enormous levels of external radiation. This is the mistake that radiobiologists make, whether consciously or not, regarding external radiation. They bombard the animal with ionising radiation and they study its metabolism. Whereas we only give the animal a very small quantity of radioactive caesium but it is incorporated into the body and causes absolutely astonishing effects.

Q.—Could someone object that the effect on animals might not be the same as on humans?

Y. Bandazhevsky.—That's why Vassili Nesterenko and I began to look at the internal contamination of humans…

Q.—The dose in food that you gave to the animals to simulate the food being consumed by humans could perhaps have a different effect on a rat—which is a very small animal—than on a human being?

Y. Bandazhevsky.— Concentration…concentration. We calculate our quantities in proportion to mass. Though there is something very interesting here. This is emphasised by Vassili in our research. I am very grateful to him, as a great scientist, for understanding the significance. Our experimental research, on animals, white rats, rabbits etc, showed that the different organs accumulate caesium in different quantities. So, for example, if the overall load within the body is about 100 Bq/kg, the heart will contain about 2,500 Bq/kg, the kidneys about 1,500 Bq/kg. There is a little less in the spleen and in the liver. This shows that caesium penetrates selectively the cells of the most important, the most active of the vital organs. The changes in the metabolic processes in the central nervous system in animals with such a low level of accumulation signal a catastrophe developing within the organism.

Q.—Can you explain why these radionuclides behave in this selective way within the organs?

Y. Bandazhevsky.—Based on what I have observed, given that caesium penetrates intensively in the most highly energetic cells, in other words in cells that have a higher metabolic rate, and concentrate there, I am supposing that what traps the caesium are the mitochondria, the powerhouse of the cells. Although it's difficult to demonstrate this at present, but we'll get there. Electromagnetic photography shows that it is the mitochondria that react first. They undergo important alterations even though it is still very difficult to talk about apparent symptoms. In heart cells, they are more vulnerable to attack than anywhere else. These are organelles that provide potential energy for the work of the cell. They consume a lot of potassium, an element that plays an important role in high energy metabolic processes. And potassium and caesium are very close chemically; they belong to the same family.

V. Nesterenko.—The cardiac muscle contains 20 times as much potassium as any other muscle in the body. The body cannot distinguish between caesium and potassium. If it needs potassium, it will incorporate the radioactive caesium that has been spread so abundantly in the contaminated territories.

Y. Bandazhevsky.—And caesium can block the channels for potassium in the cells. But going back to what we were discussing, if we had only conducted experimental research, we would not have obtained these results. If we had only conducted clinical examinations of the children, we would not have obtained these results. But, when we started to put the two aspects together, and also the study of organs during autopsy, dissection…Can you believe I had 25 senior academics working on the same subject, each with his own specialty? You asked me: "How do you prepare a thesis?" Well, this is how people did it. And it didn't require enormous amounts of money. It was quite simply a matter of organisation. We had a Human Radiation Spectrometer. Later, Vassili Nesterenko devoted himself to the anthropogammametric measurements of children and we made use of his data. The specialists had everything they needed. I had paediatric professors, professors of cardiology, anatomopathologists who dealt professionally with the histology, and I am a professor myself. No-one can say that my research lacks good specialists or has not been undertaken with sufficient rigour. I also had opthalmologists… Anyway, here are the facts: from 20% to 25% of the children in Vetka have an accumulation of 50 Bq/kg…That may seem nothing, 50 Bq/kg, it's very little…But these children have cataracts! An alteration of the crystalline lens…

Q.—You could say that what you had in your hands was a unique instrument…

Y. Bandazhevsky.—Yes, a unique instrument, in a unique situation; you're right. They're in the process of destroying it. I try not to think about it. They are destroying something that has taken me ten years to create, with virtually no funding. I would never have got there if I had not been so obstinate. A government official advised me to avoid use of the word "radiation" in my scientific programmes, and to devote my time to any other subject if I wanted funding. Completely cynical. But I did not come to Gomel, to the contaminated zone, brought my family here, where they are exposed to enormous risk, to work on some other subject.

Q.—You chose deliberately to work at Gomel?

Y. Bandazhevsky.—To tell you the truth I was always passionate about medicine from a very young age, probably because of my father, who had a huge admiration for doctors even though he could never have become one because of the war. But it's possible he fulfilled his vocation through me. When I was 16, I began to study theoretical medicine and at the same time to breed guinea pigs. I bred my own laboratory animals. I completed my studies at the institute of medicine and received high marks, but I chose anatomical pathology as my speciality, which was the least prestigious and the least popular here. I was actually really interested in it. I really wanted to understand the mechanisms of disease. After a year and a half, my thesis was ready. In 1988, at the age of 26, I applied to do a doctorate in science, and after five years, I presented my doctorate on a completely different subject, but still linked to my experiments on animals. Then in 1990, out of the blue, I was offered the post of rector at the institute of medicine at Gomel, which did not yet exist—it needed to be set up. I was already a professor and had been awarded the Lenin prize from Komsomol.

V. Nesterenko.—This area was contaminated and people had fled. Doctors and educated people had already left and the government tried to send people here from Minsk and Grodno. But no-one stayed long, they left quite quickly. They needed to train professional staff on the spot and so they decided to set up a medical school, or, as we call it here, an institute of medicine.

Y. Bandazhevsky.—It was a good idea, but none of the government officials had much faith in my proposals. They did not need what I was proposing at all. From 1988 to 1989, I sent these proposals to the president of the Academy of Sciences and to the Minister of Health. I was proposing medical research programmes to help with Chernobyl. The response was always the same: "We're already doing everything necessary". Of course, they were doing nothing. Then suddenly they made a decision. They even joked about it at some banquet: they were going to set up an institute specially for that crank, Bandazhevsky, who wants to work in the contaminated territories. But they never provided any serious funding. The Soviet Union was collapsing; the Party that had promised to support us no longer existed. We took advantage of this, in fact, to occupy the regional Party committee building. I could tell you about the struggle we had to take it over, but that's another story. But it's true that I received several threatening phone calls saying things like "We're going to get you". After that the Ministry of Health tried again and again, directly or indirectly, to shut the institute down, "given that it is serving no purpose". They argued that it was difficult to cater for the needs of the institute in an area where no-one wanted to work and where trained personnel were lacking. So I developed a scientific programme myself and started to train personnel to put it into practice. Of 37 candidates who applied for doctorates during this period, 30 were my pupils. I trained them personally one by one. I didn't just sign the forms. In any case, we had recognition from Moscow. All the medical schools in the Russian capital, all the pathologists, without exception, supported the work we were doing. Even now, they support me and defend my case.

Q.—How many people were working with you before your arrest?

Y. Bandazhevsky.—There were 200 lecturers, 1,500 students, and about 300 ancillary workers. It really was a good university.

Q.—And all that is about to disappear?

Y. Bandazhevsky.—It exists formally but the research programmes linked to Chernobyl have been cancelled. They are destroying the laboratories, they have abolished chairs. The young people I brought here are leaving, saying: "There is nothing for me to do here anymore"

Our work always irritated the authorities and they spent years trying to find legal reasons to close us down. Even so, we managed to achieve a lot. We showed how important it was to offer radioprotection not only to the inhabitants of Chernobyl but to anyone living in areas of the Republic that had been contaminated, by preventing them from ingesting radionuclides through food. I could serve you mushrooms today that would make you very ill tomorrow. And thanks to the enormous amount of work we did, scientific knowledge has increased. It was pioneering work. In time, people will smile about it, and perhaps they'll find that it lacked rigour and precision. But we grasped the principles that underlie the effects of radiation. It's very important. Thanks to our experimental methodology, we were able to show also that caesium-137, even in small quantities, has a very toxic effect on the body apart from its radiological effect. I could put a small concentration of radioactive caesium in this glass, for example, and within a few days you would experience some toxic effects.

It might interest you to know that we have just began to look at exactly this subject with some French scientists. In the 1990's, I was invited to a centre (not open to the public) of the Commissariat de l'Energie Atomique near Paris to present this book[68]. We had numerous discussions. Then they broke off all contact with us. But I think my research must have touched a raw nerve, because they always had my book with them. They even came out to Belarus to see me. They questioned me about it many times.

Q.—So your ideas were new to them?

Y. Bandazhevsky.—It's difficult to say. They didn't give their feelings away. Relations between us were excellent and then suddenly it all came to an end.

He searches among his papers and gets out a map.

Y. Bandazhevsky.—Here we are. I've found the name. It was the Institut de Protection et de Sûreté Nucléaire—IPSN—in Paris. They gave me this map showing atmospheric contamination above Moscow. I gave them the scientific study that I had done in 1996. And here is the Russian translation of the agreement that we drew up together: "Activities…Specific recommendations"…We agreed that "…given our common interest in conducting research together in the area of incorporated dose, the two institutes will collaborate in the following areas: effects

---

[68] *Aspects cliniques et experimentaux de l'action des radionucleides incorpores dans l'organisme,* (Clinical and experimental aspects of the action of radionuclides incorporated into the body), Gomel, 1995.

of radionuclides on the human body; experimental research, clinical research", etc. Then, suddenly it all came to an end. I don't think I'll ever know the exact reason, it's obvious that you're going to come up against obstacles in this area, but, nevertheless, I managed to see a lot with my own eyes. This enormous building with all these people directing all of that world wide. They only showed me one part of the centre; there are other buildings, an incredible number of laboratories, and fantastic equipment.

At our first meeting we talked about the chemical action of caesium. At Gomel, we had some very lively discussions about radioprotection. We took part in a number of working groups and seminars together.

2. THE FRENCH CONTEXT

It would be interesting to understand the reasons for the comings and goings of the French between Paris and Gomel in 1996 and 1997, and also, to know what they have done or not done since then in their numerous laboratories. It was not long before Nesterenko and Bandazhevsky, in 1998, would severely criticise the register of doses presented by the Minister of Health in Belarus and the work that it oversaw at the Institute of Radiological Medicine. More importantly, it was not long before the police took action against the heretical Bandazhevsky, in 1999, and his French colleagues, who had been so interested in his research, had shown little interest in his fate since then.

Various reforms had taken place within the French nuclear establishment since these brief encounters between Bandazhevsky and the French nuclear scientists, and Bandazhevsky could not have had any idea what was going on behind the scenes. At the beginning of 2005, tension was building between the ASN (Autorité de Sûreté Nucléaire) the "nuclear watchdog" in France, and some officials at the IRSN (Institut de Radioprotection et de Sûreté Nucléaire) and this explains in part the apparently contradictory behaviour of some of the people who showed an interest in Bandazhevsky's work. As professional researchers their interest had been genuine but fell away quite suddenly when plans for collaboration were drawn up.

Originally the IPSN (Institut de Protection et de Sûreté Nucléaire), formed part of the CEA (Commissariat à l'Énergie Atomique), which was created on 18th October 1945 with the aim of "pursuing scientific and technical research in the use of nuclear energy in the field of science, industry and national defence". The 1945 statute brought together the civil and military wings of nuclear energy.

Since then, some restructuring had taken place: at the beginning of the 1990's, the IPSN formally separated from the mother ship, the CEA, and a number of people, with no connection to the CEA, were brought in to the scientific committee, to give the Institute a semblance of independence. In 2002, two new influential bodies were set up: the Autorité de Sûreté Nucléaire (ASN) and the Institut de

Radioprotection et de Sûreté Nucléaire (IRSN) which replaced the IPSN. The ASN's role was to issue warnings or give authorisation to companies like EDF and COGEMA. The ASN would be jointly overseen by three ministries, the Ministry of Industry, Environment and Health.

The IRSN itself was formed when the OPRI (Office de Protection contre les Rayonnement Ionisants), predecessor of SCPRI (Service Central de Protection contre les Rayonnements Ionisants), headed by the famous Professor Pellerin, joined forces with the IPSN, the organisation with whom Bandazhevsky had been in contact.

The IRSN is the expert body on which the ASN depends to evaluate the safety of nuclear installations, and the public health risks associated with exposure to radiation. Its work is no longer overseen, as it has been historically, by the Atomic Energy Commission (CEA) but by the five Ministries: Industry, Environment, Health, Research and Defence. The tensions that arose in early 2005 are most likely an indication of the dichotomy that Bandazhevsky came up against between the subjugation of scientific researchers to the military-industrial complex and the principle of independence, which is inseparable from science.

Here is an explanation of the conflict as it was perceived in a newspaper article entitled:

THE NUCLEAR INDUSTRY
Research organisation believes its independence is threatened
by the Nuclear Safety Authority.

Experts at the IRSN are worried

In fact it is the very purpose of the organisation that was set up only three years ago to reform the nuclear industry that is being challenged. One of the aims of this reform was to clarify, precisely, the respective roles of the scientists and the decisions makers, and for this reason, the IRSN was granted a certain level of independence.

But in drawing up the contract of objectives that the IRSN must sign with the government [...], the ASN seems to be trying to rein in the IRSN. "The IRSN was created in order to achieve a separation between those doing the monitoring and those being monitored", writes the ASN director, Andre-Claude Lacoste, in a letter to the directors of the IRSN..."but its objective is not to separate scientific expertise from management and should not have the effect of distancing one from the other: they should be partners in a supplier-client relationship..."

"This goes against the spirit of the reform!" retorts Francois Rollinger, an official of the CFDT (Confédération Française Démocratique du Travail). Furthermore, the nuclear watchdog claims that "it would be useful if the IRSN's findings were made available to the public, with certain reservations,[...] that the content does not lend itself to erroneous or malicious interpretation, that it contains only objective facts". These "reservations" are interpreted by some as "avoiding making public any contentious issues", of which there are many surrounding the nuclear industry. It is unlikely that this concept of transparency would be shared by all those working in the organisation. [...]

Does this mean that the IRSN is unable to undertake investigations on its own initiative, or respond to demands for information from other organisations? [...]

The director general of the IRSN, Jacques Repussard has distanced himself from some of the ASN director's statements: he does not accept that "the IRSN was not created with the aim of separating scientific expertise from inspection" and confirms that "our research is made available to the public and will continue to be so". When asked whether the IRSN today has the means to be autonomous, his answer is clear: "That remains to be seen".

Caroline de Malet, Le Figaro, 2nd February 2005

These two faces of the French scientific establishment offer some hope of its independence of spirit, which, if not apparent in their written statements or in the ability to take on board new information, can be seen in the differing responses of the two scientists and senior officials interviewed by Jean-Michel Jacqemin-Raffestin[69], when he sought their opinion about Bandazhevsky's report on "The role of radioactive caesium in pathologies of the thyroid gland". Here are their replies.

Jean-Francois Lacronique, a cancer specialist, and president of OPRI at the time of the interview, in 2001:

As a matter of fact, it is a really interesting report and should be taken very seriously! It is embarrassing in so far as it is extremely difficult to refute. My colleagues at the IPSN are already familiar with Bandazhevsky's theses. We have all spoken about the subject together, and they have been to Gomel and to Minsk to meet Bandazhevsky.

It seems that this man wanted to set up an international centre and was looking for funding. Sometimes the search for funds involves venturing beyond strict ethical boundaries. I think that's what got him into difficulties with his government. And besides, what he writes doesn't go down that well with the international scientific community; he has put himself forward as an original thinker. He deliberately chose to publish his work himself.

The documents that I have read are remarkably well done and conform to Western standards. There is the matter of his *curriculum vitae*, so the problems remain.

His thesis that if someone dies and high levels of caesium are found in the organs it can be concluded that the cause of death was the caesium, is a valid scientific hypothesis because, although a cause-effect relationship has not been established, it is a plausible explanation. I believe it should be taken seriously, and as with all scientific hypotheses, it should be looked at again and confirmed by others. Personally, I take Bandazhevsky's work very seriously and rather than saying "it is not recognised

---

[69] *Tchernobyl, aujourd'hui les français malades* (Chernobyl, today the French are ill), Editions du Rocher, June 2001.

181

scientifically", I would say that other teams should work on the same hypothesis in order to reproduce the result or to refute it.

I would like to make the following comments. It seems to me that for a *scientist*, the result of an experiment, if correctly undertaken, should not be "embarrassing" All credit to Lacronique for suggesting that other teams repeat the experiment in order to confirm or refute the result, though unfortunately, as far as I know, no such experiment has been undertaken by his organisation, unless it has taken place in secret. Contrary to what Lacronique says, a causal relationship has been established by Bandazhevsky, and is the most significant part of this "expertly written" research. As far as his other comments are concerned, Lacronique seems to know very little about the political context in Belarus and about the state of the judicial system there, if he can conclude *even before the verdict was announced* on 18th June 2001[70], that Bandazhevsky is guilty. As a citizen of the country that first introduced human rights, it is hardly the response one might expect to the plight of a fellow scientist. As for "originality"—according to M. Lacronique, Bandazhevsky "deliberately chose to publish his own work himself"—I hope that this book will show quite clearly that Bandazhevsky's isolation from the scientific community is very easily explained by the national and international context, rather than by some peculiarity in his character.

And here is André Aurengo, chief medical officer in nuclear medicine for the Pitié-Salpetrière group of hospitals and a French representative at UNSCEAR[71]:

Bandazhevsky's reports are not written in the classical scientific style, and so it is extremely difficult to comment. His methodology is very obscure and it is hard to understand what he has observed, how he has observed it, and what is his control group, etc. Bandazhevsky's report simply poses a lot of questions, which need further analysis and verification and which cannot really be considered to have been answered. What he writes is not convincing, and, as it stands, no serious international journal would publish his report because of these methodological flaws.

On the other hand, whatever Bandazhevsky does, says or writes about the subject, it is unacceptable that he be treated in the way that is being reported. Persecuting a scientist (or anyone) for opinions that go against the grain is monstrous, and I would willingly sign any petition to allow him to work and to communicate as he wishes, though this is not to say I endorse his work from a scientific point of view.

---

[70] See Part Seven, Chapter 1, page 476.

[71] United Nations Scientific Committee on the Effects of Atomic Radiation, set up by the United Nations General Assembly to evaluate world wide *doses* of radiation, their effects and the risk they entail.

How odd. Did the two scientists have the same document in front of them?

Professor Aurengo speaks with authority as one of France's representatives at UNSCEAR. He dismisses Bandazhevsky's work, without explaining the *methodological weaknesses* in the research carried out by the Gomel institute. A spokesperson has an important responsibility. For scientists such as Aurengo, it means verifying observations that do not conform to the dogma derived from research undertaken following the Hiroshima and Nagasaki explosions.

Scientists should have the courage or curiosity to look at the facts as they appear at Chernobyl and to make use of the results that have been found by the researchers at Gomel, at least as a working hypothesis. What one would not expect is that these scientists ignore Bandazhevsky for ten years, stand by while he is put in prison, as his colleagues at the Ministry of Health in Minsk did, and then offer to sign a petition to free him after these humiliations.

In any case, the two French scientists have at least one thing in common: each in his own way abandoned a colleague who was being persecuted for dissident opinions. One was able to acknowledge Bandazhevsky's worth as a scientist and the excellence of his research but joined in with the slander about corruption. The other presents himself as a defender of Bandazhevsky's (or any other individual's) right to the free expression of his opinions, even if those opinions have no significance because they are apparently "obscure" and "unpublishable". He "does not understand" and in this way is able to avoid presenting the merits of the argument,—but does not give a very flattering image of his mental capacities, given that his colleague at OPRI had no problem understanding Bandazhevsky's work. Is it possible that, because he is not an anatomical pathologist, Aurengo really has not grasped the mechanisms described by Bandazhevsky about the effects of proximity of incorporated radionuclides on the body? The results of a serious scientific investigation that repeated Bandazhevsky's experiments would have contributed to his release and allowed him to present his research at an international level.

A final observation: Lacronique states that "Bandazhevsky's work was known to his colleagues at the IPSN". They visited him in Gomel and in Minsk. Bandazhevsky reveals that he was even invited to Paris, in the mid 1990s. So, it is clear that for ten years, the French scientists knew perfectly well what the average reader will already have understood. Basically, there is a simple but crucial difference between the *violent external* exposure to radiation from the bombs dropped at Hiroshima and Nagasaki, and the *low level internal* exposure by the ingestion or inhalation of microscopic radioactive particles at Chernobyl. The proximity effect did not need to be *extrapolated from mathematical calculations*, but could be measured directly, using a radiometer, during autopsy (for example, in the heart). The number of becquerels incorporated in the tissue, could be measured and then Bandazhevsky could link it to the histological sections of damaged tissue in the organ. He discovered a constant, linear and statistically significant correlation between the level of incorporated becquerels and the lesions observed. This is not Bandazhevsky's "thesis" but a simple observation. It is a classic experiment, verified and repeated in clinical examination of children and by laboratory experiments on animals. Given their professional expertise, the French scientists will have understood the real significance of these findings far quicker than the average reader, during the hours they spent with Bandazhevsky. A collaboration was proposed. And then it was quietly dropped.

And it was only on 28th November 2005 (*ten years after!*) that the IRSN, as the IPSN is now known, published its "Response to the ECRR report"[72] in which it recommends that "research should be initiated in response to questions concerning the population living in the contaminated territories in Eastern Europe". I am neither a doctor nor a physicist, and cannot comment on the specific criticisms the IRSN makes of the ECRR report (in which they do, however, recognise that the fundamental thesis is *valid and merits debate*"), but as a simple concerned European citizen, I am appalled by the absolute passivity, the negligence, the ignorance of the scientific establishment, concerning the explosion at the Chernobyl nuclear power station, a unique phenomenon in the history of science. Twenty years have gone by since the disaster, and more than ten, since Bandazhevsky, the only anatomical pathologist conducting research where it was needed, communicated his findings, and we are still no further on.

3. THE BEGINNINGS OF AN AMBIGUOUS TURNAROUND IN FRANCE ?

Extracts from the IRSN document followed by my comments.

> IRSN—ECRR does not question the radioprotection system as it applies to external radiation but has significant criticisms to make about its application in the case of internal contamination by radionuclides. The committee believes that current evaluations underestimate the risk from internal contamination and it bases its argument partly on work published in the scientific literature. [...] [it] tries to respond to this lacuna by proposing to change the ICRP radioprotection system and to reduce, arbitrarily, the annual exposure limits. The questions posed by ECRR are perfectly valid, but nevertheless the arguments put forward for changing the system are not convincing, because taken as a whole, the evidence does not fulfil the criteria for a rigorous and coherent scientific approach [...].

---

[72] The report published in 2003 by the European Committee on Radiation Risk severely criticises part of the recommendations of the International Commission on Radiological Protection, adopted by the European Directive of 13th May 1996 and then by the French government, in 2002.

Commentary—Entrenched in the routine and the need for consistency within the official doctrine that originated with the American bombs and not the fire at Chernobyl (science's *black boxes*[73], as the experts from the ECRR call them), the IRSN puts a brake on the urgency of recognising the limitations of this dogma and prevents the adoption of appropriate measures to protect the population

While admitting its own ignorance and acknowledging that the criticisms are valid, the IRSN quibbles about the form and presentation of these criticisms— anything to avoid facing the possibility that there is a fundamental error in "the doctrine" that might force them to rethink the whole issue. Aurengo displayed the same attitude, though Professor Lacronique did at least admit that it was "embarrassing".

> IRSN.—Internal contamination results from transcutaneous absorption of radionuclides deposited on skin, or incorporated through ingestion, inhalation or through a wound. One of the challenges of radioprotection is to predict the risks associated with this sort of exposure. This assessment is difficult because it requires knowledge of the correlation between the quantity of incorporated radionuclides and the emergence of pathologies.

—It's been done. The correlation has been established. Why not go and see Bandazhevsky and Nesterenko who have been studying the subject for the last fifteen years? At Chernobyl—the laboratory about which a certain French professor was so enthusiastic[74]—there is no need to predict the risk, but to observe the illness, study it and provide care of the patients. There are millions of human guinea pigs in perfect condition for the experiment.

> IRSN.—[…] Usable data to determine these specific risk coefficients is hard to come by because normally, it is very difficult to link observed pathologies with precise levels of exposure […].

—Not at all! There is a mass of data on internal contamination in the Chernobyl area, mostly collected by Western agencies,[75] though not for the purposes of making a correlation with illness. As for the scientists who have researched and established the *exact level for a lethal exposure*, the IRSN is well acquainted with them. It does everything it can to prevent their work.

---

[73] ECRR, op.cit. p.29 and following pages.

[74] See interview with Y. Shcherbak, Part One, Chapter IV, p.38.

[75] See Part Three, Chapter VI, p.245, and Part Four, Chapter II, p.264 and following pages.

IRSN.—In fact, the only data available for research concerns people exposed to radon 222, thorium 232 in the form of thorotrast, isotopes of radium and finally plutonium 239. The main pathologies that have been listed after exposure to these radionuclides are lung, liver and bone cancer, and leukaemia.

Incredible! Chernobyl is three hours away by plane. There you will find plenty of caesium-137, and strontium-90 and millions of *people exposed...* No, *contaminated* by chronic incorporation of these radioactive particles through their food. As for the *main pathologies*, they are not *listed* at Chernobyl because it is forbidden. It would be "embarrassing". For the people living there, every stage of their life, from gestation onwards, is accompanied by all kinds of somatic illness. Cancers are only the tip of the iceberg.

IRSN.—In the case of epidemiological studies, many people thought that it was possible to understand the effects of chronic internal exposure by extrapolating from our knowledge of Hiroshima and Nagasaki. The Chernobyl accident, that has played a revelatory role in this area, shows that it is not so simple.

The statements published in recent years could therefore be revised. One reason for this is that it is now evident that the toxicity of an element is a complex variable that depends in part on the product, its concentration in the organism and how long it has been there (effects of proximity, Author's note). On this point, there has long been a theory according to which the incorporation of 100 Bq in one day was equivalent to 1 Bq incorporated for 100 days. This is true mathematically but not biologically. The second reason is that reference is being made more and more frequently to research that shows that the ingestion of contaminated food in Belarus has led to a significant number of illnesses and birth defects of all sorts amongst the population (Bandazhevsky, 2001). Although this research is not accepted by the international scientific community, it serves to maintain doubt in the public mind and for that reason should be followed up.

—Who, in the international scientific community, decides to *accept* or *not accept* a study? After all, the studies undertaken by Yury Bandazhevsky, Galina Bandazhevskaya and Vassili Nesterenko were published in *Cardinale* and *Swiss Medical Weekly*[76].

IRSN.—The most publicised of these studies attests to the number of illnesses experienced by people living in Belarus, affecting the cardiovascular, the central

---

[76] Y.I. Bandazhevsky, "Cardiomyopathie au césium-137", *Cardinale*, tome XV, No 8, October 2003; Y.I. Bandazhevsky, "Chronic Cs-137 incorporation in children's organs", SMW, 2003, 133, p.488–490; G.S. Bandazhevskaya *et al.*, "Relationship between Caesium (Cs-137) load, cardiovascular symptoms, and source of food in "Chernobyl" children—preliminary observations after intake of oral apple pectin", SMW, 2004, 134, p.725–729; V.B. Nesterenko *et al.*, "Reducing the Cs-137 load in the organism of "Chernobyl" children with apple pectin", SMW, 2004, 134, p.24–27.

nervous, digestive, respiratory, immune, reproductive systems, as well as the thyroid and the kidneys (Bandazhevsky, 2001). The author links these illnesses to the continued exposure to caesium-137 present in the region. [...] The interest in all of these studies is that they show or seem to show effects that were not suspected before, when studies were based only on the experience of Hiroshima and Nagasaki, or on acute radiation experiments on animals. This seems to suggest, even if the available data is partial, that chronicity plays a part in the toxicity of radionuclides. Research should be done on the subject to understand these effects better and to further refine the radioprotection system.

—Yes, that's good, but what is striking about this late concession, what makes one feel uneasy while reading this document, is the absence of any mention of the need to offer urgent prophylactic radioprotective measures to the people who have been contaminated, living in the territories around Chernobyl. It is like conducting a sociological study of the Auschwitz survivors rather than liberating them. Everyone has their own profession, of course. Experts and scientists are not nurses. But the physicist Nesterenko and the anatomopathologist Bandazhevsky are not nurses either. What is the point of the IRSN? It could at least recommend immediate action, take a stand, get its hands dirty politically. Make a humanitarian gesture.

IRSN.—An appropriate effort needs to be made to support new studies that are being launched soon on other cohorts of workers in Europe, more specifically exposed to certain radioelements such as uranium. The aim of these studies is to target populations where internal contamination has been correctly estimated, to follow their cases over a period of twenty years and to record a number of health indicators (cancer, leukaemia, chronic kidney disease, pulmonary, cardio-vascular...). These studies will be undertaken within the framework of the European Union. They will eventually increase our knowledge of the risks to health from low level internal exposure.

This fine programme that brings the IRSN document to a close does nothing to relieve our uneasiness. It sends shivers down one's spine. How does one "target over twenty years" the contaminated population and "record a number of health indicators", if inadvertently, the human guinea pigs are decontaminated by effective prophylactic cures? But wisely they abstain from this. The text talks about "*other* cohorts of workers in Europe". Is the crime of Chernobyl, financed by European tax payers, an officially planned programme? All that is needed is *not* to "target" people who have been seriously contaminated with any effective radioprotection.[77]

---

[77] See CORE programme, Part Four, Chapter III, p. 277.

4. THE INDEPENDENCE OF SCIENTISTS

The difficulties surrounding the independence of scientists was discussed at the People's Permanent Tribunal (previously the Russell tribunal): "Chernobyl: Consequences on the Environment, Health and Human Rights" that took place in Vienna, Austria, between 12–15th April 1996.

Professor Peter Weish, Human Ecology, University of Vienna.—I would like to come back to some questions that we discussed this morning, and to comment on a remark made by my friend Wolfgang Kromp, who said that we should not accuse persons for their responsibility in this catastrophe.[...] I would like to object to this view.[...]

I would like to answer the question of whether it is possible to separate people from the system, Yes and no, it must be possible. We should create more sensitivity and awareness of the fact that people tend to be slaves of the system. We should stress also the need for them to act in responsible way.[...]

Everyone has some freedom in his actions, and therefore no excuse for an irresponsible behaviour. We must try to have an open discussion, beyond the borders of a specific situation, to address the IAEA and to discuss with those people personally, outside their respective organisations, to stress their personal responsibility. [...]

Therefore we have to stress the responsibility of individuals. Human rights should not be violated. One basic human right is to be able to influence the environment in which you are living. Another is to be able to act in a responsible ethical way. Thus to be enslaved by institutions, represents a violation of human rights. [...] Scientists in particular have to liberate themselves from a system which reduces them to mere tools in the hands of industry.

Dr Wolfgang Kromp.—Physicist, Nuclear Adviser to the Austrian Federal Chancellor:

I wanted much more to draw your attention on the system, standing behind those individuals. These workers are like Martians and once you remove them, others will come.[...]

We have to go to the roots, and we should not forget to take care of those people which are the victims of this cruel system.

Dr Rosalie Bertell, Co-ordinator of the International Medical Commission—Chernobyl (IMCC), Toronto, Canada:

In our economy, it is very clear that scientists are dependent people. Normally scientists are paid by government or universities, and the universities are paying them with government money, or by the industry.

The public needs information, expertise, but it is not able to afford to pay scientists to help them to understand what the argument is.[...]

We have created a very complex technological society, but we don't provide the people, who are on the front-line of risk, with experts to help them to understand what is happening.

In 1978 I started to publish research on the health effects of low-level radiation, by ordinary diagnostic X rays. I immediately experienced a cut-off of all my funding. My name was going out on a list "Do Not Fund". In fact we even got a paper from the US National Cancer Institute, that said that, if I would like to change my line of research, they would consider refunding. I was so angry at that because it went against everything.[...]

It is a very high penalty that scientists pay to speak out on these issues. They need to be protected by society. We need to recognise science working in the public interest. We need people who can speak out about a hazard, without economic and social penalties, which our society hands out very freely to them.

5. SECOND INTERVIEW WITH BANDAZHEVSKY

Bandazhevsky.—I would like to show you the doctoral thesis of the famous Dr Botkine, which he defended in 1888 in St Petersburg. It is entitled "Influence of rubidium and caesium salts on the heart and on blood circulation". It was the great Pavlov who carried out the experiments. This means that the quality of the work was very high. The experiments were carried out on dogs using rubidium and caesium salts. The author shows that with a certain concentration of caesium chloride—10% caesium chloride—there is an alteration in heart activity, problems with cardiac rhythm, and then the heart stops completely.

Q.—And all this has been forgotten?

Bandazhevsky.—As always. What seems new is just old but forgotten, as the saying goes. But it confirms the fact that radioactivity is accompanied by a very damaging chemical effect. We are seeing the consequences of this.

Q.—Would the toxic effect be more dangerous than the radioactivity?

Bandazhevsky.—In the situation we have in front of us now, it is more dangerous.

Nesterenko.—I would say that the two work in conjunction, a synergy.

Bandazhevsky.—I agree. What Botkine showed was that caesium as a chemical element, is very active in biological structures. We are wrong to think that at that time people did not know how to conduct research. Their way of working was admirable. Pavlov was in charge of the experimental part, because for experimentation with dogs, people only had confidence in Pavlov, and Botkin analysed the results.

Q.—When I talked to Doctor Tykhyy in Kiev, he told me about these cases that the doctors could not understand: the sudden death of men between 45–50 years old in apparently good health.

Bandazhevsky.—This is exactly what I am talking about. Sudden cardiac arrest. People often point out that there were similar cases long before Chernobyl. It's true but there is an explanation: we have discovered a series of official documents, from the Belarusian health services that show clearly that caesium has been present in the atmosphere, the biosphere and in food since the 1960s. Here is another little book, in fact, about the same thing: "*Global fallout of Caesium-137 and Man*" by Marei, 1974. It talks about the different concentrations of caesium present in food across the whole of the Soviet Union. Other studies were done on the same subject in 1975. We have even found a description of the effects of radioactive elements in a 1970 report by Professor Mascalev. But it was not about incorporated substances. He implies that there are some inside the body but concludes that they do not pose a danger. Now look at this map of

Polessie that shows the different levels of microcuries per litre of milk. It reflects the situation in the 1960s. It's in this book.

Q.—Why Polessie? Surely it was the same situation throughout the Soviet Union?

Bandazhevsky.—Of course. There are figures in this book for 1964, 1965...1969 for all the republics of the Soviet Union. Look: "Concentration of caesium-137 in the daily diet of the rural population".

Q.—This is due to the nuclear weapons tests?

Bandazhevsky.—I have no way of knowing that. I am just an observer of the situation...By the way, the map that I showed you yesterday was given to me by scientists from the CEA. When I talked to them about Chernobyl, they said: "Come on, after Chernobyl, the concentration in the atmosphere was a quarter of what it was after a nuclear test". In fact, on the map, you can see that the high point in the curve shows high levels of contamination in the atmosphere above Moscow in the 1960s and a much lower level after Chernobyl in 1986. I'm not sure what explosive yield the bomb had, it's not my field...I'm not a physicist. It's the consequences that I'm concerned with.

Q.—Were they trying to minimise the seriousness of the disaster at Chernobyl?

Bandazhevsky.—Probably. But they also helped me to understand the mechanisms better. I began to study mortality figures from cancer not only after 1986 but from 1974 and even from 1960. Almost coincidentally, our own Ministry of Health also began talking about the cancer registers, showing that tumours and cardiovascular disease had begun long before Chernobyl. It's true. It started in the 60s. On the other hand, I was able to tell them something they didn't know: after Scandinavian countries banned the sale of food products containing radioactive caesium, their mortality rates for cancer and for cardiovascular disease started to decrease. In our country, they are increasing.

Q.—Have heart attacks, which have been the subject of so much discussion over the last few decades, always been as common as they are today?

Bandazhevsky.—Interesting question...First of all, what is a heart attack? Myocardial infarction is a vascular necrosis that occurs when the vessel is blocked by an embolism and this prevents blood supply to the tissue. When I first began my career as an anatomical pathologist, I was asked to do autopsies on people who had died very suddenly, and this necrosis intrigued me. I never discovered it! I should have encountered the classic version though. Well, I have to tell you that I very rarely met classic cases of myocardial infarction, that is, necrosis of the heart muscle that would have caused death. But with caesium, it's very different. It is not necrosis but cardiomyopathy, *meaning a diffuse toxic effect* across the whole heart muscle. Here it is! Look at this microphotograph. There is no muscle left. Is it the transverse myocardial band? On this photo there is still something left resembling a heart, muscular fibres with contractions. It was a child's heart. Here, in an adult heart, you can see that there is no muscle left, just holes. You can see on that photo the toxic effect of caesium on an animal's heart: myocardial dystrophy. If I put this heart in a glass, it will take the form of the glass. I am getting to your question. This sort of disease, diffuse myocarditis, that has appeared since the 1960s is much more common today than the classic infarction, even if it is called by the same name. Of course, caesium plays the same damaging role in causing classic infarction because it affects the coagulation of the blood and leads to thrombosis. This aspect has been well researched. We know all about it. But this new type of heart disease, this is the main harm caused by cesium…

Statistics show that, over the last three years, the death rate has exceeded the birth rate by a factor of 1.6. What's going on? Before, we had infectious diseases, plague, cholera, smallpox, typhoid, all sorts of deadly diseases. We had epidemics. During wars people died of wounds, from septicaemia…But when millions of people are dying from heart attacks!.. In Europe and everywhere else in the world we are constantly being told: "It's because we smoke too much, we drink too much coffee, too much sugar…" But who can explain to me why these things are sending us to our graves? Their toxic effect is nothing…zero, in comparison to the effect of caesium with the two factors working in synergy: radioactivity multiplied by toxicity. But what it causes is not really an infarction. It is a new disease that I want to call "toxic cardiomyopathy from radioactive caesium".

To untangle the strands of this problem would have required the best brains available. Because unfortunately, our understanding of what's going on here is being completely distorted. Knowing nothing about the real situation, we go from one extreme to the other. Some say the situation is absolutely terrifying, others that there is nothing to worry about. It's really alarming, but not just because of Chernobyl. It had begun long before. You know what I am talking about. It's terrible, there's no doubt about it, for the people here, in Polessie, who had already been subjected to radioactivity from caesium well before 1986, as is shown in the official statistics. Then on top of that, they suffer the effects of the disaster at Chernobyl. That is the cause of all these unexpected cancers. We expected the cancers to appear in the 1990s but they started well before that. In our research, based on autopsies, we have shown that the thyroid gland accumulates radioactive caesium in a very concentrated way.

Q.—So thyroid cancer is not caused only by radioactive iodine, as we are told?

Bandazhevsky.—It is the double action of caesium on many organs that is the principal explanation in my approach to the problem. Apart from the fact that caesium is distributed differently in highly differentiated structures, it also causes an *alteration of the whole organism*. The pathology needs to be studied as a whole. The thyroid gland cannot be treated in isolation as if the other organs of the body are not affected. It is the whole body that is ill. I don't agree with the diagnosis of "asthenic vegetative syndrome", as if the rest of the body is well. It is a much broader syndrome caused by caesium penetrating the whole organism: *incorporated radioactive elements syndrome*. It causes a variety of pathological changes and it manifests itself differently depending on the concentration. The early manifestation, caused by so called low dose radiation, is cardiovascular and nervous disease. Then, if the dose is heavier, the endocrine system is affected, then the liver, etc. To put it brutally, the organism begins to disintegrate, as evidenced by the reduced life expectancy. Humanity has become saturated with this poison for fifty years[78].

After impassioned discussion with his wife Galina, who knew the danger that his decision would pose to the family, Bandazhevsky published his findings and, in a television programme which caused a sensation, he denounced the government's policy of non-intervention.

In April 2000, in between his two arrests and before he was sentenced to eight years in prison, we talked to Rose Goncharova, a geneticist, about the value of Bandazhevsky's work.

6. NESTERENKO AND GONCHAROVA: THE SIGNIFICANCE OF BANDAZHEVSKY'S DISCOVERIES

Rose Goncharova, geneticist.—Shortly after the accident at Chernobyl, we began to notice an increase in somatic morbidity in Belarus.[79] That is, all the common illnesses of the cardiovascular system, the digestive system, gastritis, liver and kidney problems, were all increasing. This information came first from doctors in Ukraine and then from Russia. But the evidence that people exposed to the chronic effects of low dose radiation may have an increased incidence of somatic illness is not embedded in the usual postulates of radiation biology and radiation medicine. According to radiation medicine the general somatic illnesses cannot be caused by exposure to radiation, even over a prolonged period, particularly given the low levels of radiation to which the population of Belarus is exposed. Official radiological medicine recognises only those diseases caused by high levels of radiation (acute radiation sickness). At lower levels, but after prolonged exposure, it recognises cases of chronic radiation sickness in people who are involved professionally with

---

[78] The atmospheric nuclear weapons tests began in the 1950s.

[79] Somatic: concerning the body (rather than the mind). In radiological medicine, it is used to indicate organic illnesses other than cancer. All the illnesses described by Y. Bandazhevsky as a consequence of internal low dose contamination by Cs-137 are *somatic* pathologies. The official doctrine, based on the Hiroshima experience, where the victims studied had been exposed to very high levels of external radiation, does not recognise radioactivity as one of the causes of the somatic pathologies that have appeared in great numbers in the territories contaminated by the disaster at Chernobyl.

sources of ionising radiation. And of course, all sorts of cancers in various parts of the body. As was listed at Hiroshima in Japan.

But the facts tell a different story. In 1990, C. Shedlovsky, a doctor from Belarus, was able to show an increase in cardiovascular disease, digestive and respiratory problems in populations living in areas contaminated between 1 to 5 Curies. It is statistically significant in three districts of the Brest region—Luninets, Stolin and Pinsk. He showed that the illness in these districts was higher than in districts that were "cleaner". The higher rates of illness in the contaminated areas were confirmed by many other studies, in Belarus, the Ukraine and Russia. But the most crucial thing was to determine the cause. There is no consensus on this issue within the scientific community.

On this subject, I would like to point out that in Japan, after the bombs were dropped, certain groups of people had their health monitored over their lifetime: from what illnesses did these people suffer, what was the cause of death, what types of cancer did they develop, and in parallel to this, the radiation dose they had received was determined. Until that time, there was no evidence of physical morbidity in this category of the population. So it was possible to claim that radiation could not be the cause. The cause was stress, poor diet and emotional upheaval.

It is a purely scientific problem: is somatic illness a consequence of chronic exposure to radiation following an accident like Chernobyl, for example, or is it not? The fact is that we have been living for years in conditions of chronic irradiation. It is not really accurate to say things like "the population has been the victim…" The population is still being subjected to radiation. They are being subjected to it now. As a researcher I am very interested in this problem. I collect information about the subject wherever possible and analyse it. As a geneticist, I came to the conclusion that this increase in morbidity may indeed be the result of exposure to chronic irradiation from Chernobyl. I published these findings in the abstracts that I prepared for the international conference in Vienna in 1996. My attention was drawn to Bandazhevsky's work, which, in my opinion, was conducted correctly and rigorously from a scientific point of view. If you want to show the effect of a radiological factor on the body, you need to know the objective dose rates on the one hand and on the other, the particular *health effect* that you are studying within the population. With a well designed research protocol, if you can show a statistically significant correlation between the health effect under study and the concentration of radionuclides (and it is clear that the concentration of radionuclides determines the radiation exposure) you have proved that whatever health effect has been observed, has been caused by radiation. Bandazhevsky's approach was competent and correct. This is how I began to develop my ideas, using Bandazhevsky's work and that of his colleagues. They conducted their research perfectly correctly. They measured the concentration of radionuclides and related them to health effects. They were able to make objective measurements of the radiation in the body, not simply through autopsies, but also on patients using a human radiation spectrometer, the HRS. In Yahodvik's study, carried out at Bandazhevsky's institute, the concentration of radionuclides (primarily of caesium because it is the principal provider of the radiation dose) in the body of young women who had not yet given birth was determined, and this was correlated with indices of reproductive health. The results showed a close correlation between the pathological alterations observed and the concentration of caesium. This was proved. In our country, Bandazhevsky's work was the first rigorous scientific research in this field.

At the same time, similar work, but in a different context, was being done in Russia on children. This was the research done by the bio-chemist Neifach *et al.* They studied pathological alterations in children from a bio-chemical perspective. They were able to establish a correlation between the indices under consideration and radionuclide concentration.

In 1998, there was an international scientific conference held in Rotterdam to look at the problems of treating lesions caused by radiation injury. I presented my results there too, though I could not actually attend, because despite my qualifications, my monthly salary is only 60 dollars, and so it was not possible. But I received publications from there and read for the first time the theses of one of the Japanese researchers from the group studying the health status and the life expectancy of Japanese people who survived the atomic bombings. This is where I first read about a statistically significant correlation between illness in atom bomb survivors and the dose received. It is of paramount importance: for the first time, we had obtained data showing that radiation causes, not only cancer, but also other *common somatic illnesses*. It is worth remembering that illness amongst the Japanese population that had lived through the bombing only began to be recorded thirteen years after the bombs were dropped. So we have no epidemiological data for these illnesses. But in Belarus, we began recording these things much sooner. This is why our research came up against the views of the international scientific community. It contradicted their basic principles: what somatic illnesses could we be talking about at such low doses! I believe that the fact that we have shown this is a great achievement for Belarusian scientists. On the other hand, I think that because we have published these findings, sooner or later, Japanese and American scientists are going to have to reveal this data to the public. Namely, that even today, somatic illness in Japan is determined by radiation.

What is the explanation for this phenomenon? Up until now it was thought that radionuclides were distributed uniformly throughout the body, so that the dose seemed relatively low. But Bandazhevsky showed that the amount of caesium accumulated in the heart and in other vital organs was much higher than in ordinary muscles, so that the dose in these organs was much higher than the average. Bandazhevsky really seemed to have discovered something new.

V. Nesterenko.—It is important that he carries on with his work now. He is 43 years old. He has a very good brain. But over this last year, he has been living a precarious existence and his future looks very uncertain. All the time I've been working in Belarus, nobody ever asked me to make an instrument that could measure levels of radioactivity in tissue samples as small as 5 to 10g. Yury was the only one to say "I am doing biopsies of kidneys, liver and I need to know what radionuclides my samples contain". He was working with animals. He bred rats, and he fed them caesium and studied them. As a physicist, I know that I can make a crystal with a miniaturised well and I made him the instrument he needed. We can now determine the accumulation of caesium in a sample of 3g with a margin of error of 0.1 Bq/kg. It's possible. Of course, we need money. That sort of instrument would cost 2000 to 3000 dollars. I already know where I can get this crystal in Kiev or in Kharkov. The crystal alone costs about 300 dollars. Then the system needs to be designed and assembled electronically. He really needs this instrument. He also needs a good microscope. But more importantly, we need to find premises. He has been forbidden from working in Gomel. He should really have been working in Minsk in one of the departments at the academy. But Platonov, president of the Academy at the time said: "Chernobyl is not an academic subject". Then he confiscated all my equipment and I was forced to leave and set up my own non-governmental institute where I work today. I approached the current president, the new one, and I said: "I am a member of the Academy. I worked at the Academy. Give me back my institute; I won't ask you for any money, I've been earning my own money for the last ten years". His answer? "We've got plenty of institutes working on Chernobyl, we aren't going to duplicate work already being done by the Ministries". I think Bandazhevsky would get the same response. The Ministry of Health would prevent him from working. They would offer him alternative areas of research.

Q.—How can an academic say that Chernobyl is not a subject of research? Does it make sense to you?

V. Nesterenko.—Of course not! But the Party Central Committee had given the order to do whatever was needed to make sure "Nesterenko doesn't get involved with Chernobyl". Because, afterwards, "he will start writing letters to the Central Committee". And if there is a letter, they will have to reply. Ask Rose Goncharova. She will tell you how generous they are with their finances for this sort of research... Yet, her findings have provided hugely important information, just like Bandazhevsky. New information that is impossible to ignore. If you look at his reports, he presents evidence of direct correlations; it's easy to verify. His discovery is consistent with the laws of science. No-one had talked about it before. We suspected that radiation had certain effects, but Bandazhevsky is the first to produce quantifiable results. And those who are not convinced have only to verify them. No-one is preventing them.

Chapter III

CHRONICLE OF EVENTS

The conflict between the two dissident scientists and the Ministry of Health came to a head in April 1999. After he had criticised the Ministry's policy harshly during the television programme that had so terrified his wife, Yury Bandazhevsky gave an interview to Irina Makovetskaya, that was published on 26th April, the anniversary of the disaster, in the opposition newspaper BDG. Its headline was "Belarus has wasted too much time. We need to tell the truth to those in charge of our lives". This interview set the tone and the elements of the drama, which culminated, three months later, in the scientist's arrest.

1. THE INTERVIEW

I. Makovetskaya.—The Gomel Institute of Medicine is the youngest institute in Belarus. Presumably you deal with the effects of radiation...

Y. Bandazhevsky.—Obviously. A medical institute situated in the heart of the contaminated territories could hardly avoid being involved with the effects of radioactive matter on the human organism, on prevention methods and on the treatment of the illnesses that it can cause. For the moment we are only studying caesium.

—If you're going to make statements about the damaging effects of radioactive caesium, you would need to have solid proof, wouldn't you?

—We have that proof. Our research is devoted to the relationship that there might be between a person's state of health and the amount of radioactive substances accumulated in the body. To do that, we use the results of our experiments on animals in the laboratory, and clinical results from children and adults, including from autopsies. We are convinced now that radioactive caesium has a toxic effect on the body that causes very serious alterations in the structure of cells and tissues, leading to damaging lesions in the body and death. We have observed serious damage to cardiac muscle, which is the most frequent cause of death. The urinary system deteriorates because of the damage caused by caesium to the renal vascular system. These processes have serious consequences and lead to metabolic alterations. They will in time affect the liver, which will then lead to different sorts of hepatitis, cirrhosis, fatty liver disease. Almost everyone who comes into contact with radioactive caesium presents these sorts of alterations.

Radioactive caesium also accumulates in the thyroid, and therefore increases the severity of the damage. We are very worried about the quantity of malignant neoplasms that are appearing. It's a disaster. Radioactive caesium acts at a cellular level, and it does not cause a specific disease, caused by a specified agent. It causes the death of cells, followed by the death of the whole body. Doctors try to prevent cancer by early diagnosis, but no-one is trying to prevent the *genesis of these malignant processes*. And this is a fundamental difference.

—At what level does radioactive caesium become damaging to the body?

—My belief is that any amount of this poison—and it really is a poison at a cellular level—is dangerous to the human body. Low levels are sufficient: in a child's body, during pregnancy, when someone is ill, when someone is suffering from physical or psychological stress, it causes the slow death of the body.

—It seems that much time has been lost in the Gomel region...

—Time has been wasted throughout Belarus. And today, thirteen years after, we are still trying to explain this to the people who are in charge of our lives.

—Are you saying that the enormous sums of money that have been spent over the last thirteen years to reduce the consequences of the disaster have been wasted?

—The only way to judge how effectively the money has been spent is to look at people's health. And people's health is deteriorating. That's the answer to your question.

All measures should be aimed at reducing the presence of radioactive caesium in food. That is the only way to stop its destructive effect on people's health. The government needs to radically tighten up their monitoring of food and impose norms to reduce the amount of radioactive caesium in food as far as possible.

—What do you think of the food people in the republic are eating?

—We do not pay enough attention to this question. Here is the proof: the National Commission on Radioprotection has just adopted so-called acceptable levels of radioactive caesium incorporation which, according to our research, are capable of causing serious pathological processes in the body. Today, people living in the towns can consume up to 340 Bq of radioactive caesium every day, and in the country, up to 463 Bq. And these are the official norms!

—Is the government aware of your views?

—My views are known throughout the world.

2. THE MINISTRY IS CHECKMATED

V. Nesterenko.—Every year, in April, there are parliamentary hearings on Chernobyl. President Lukashenko, who had been much criticised over the way in which funds that had been provided to look into the consequences of the disaster had been used, ordered the president of the State Committee for Science and New Technologies (SCSNT) to look at the problem. The results of the work that had been undertaken by the Ministry of Health's institute on radiation and endocrinology needed to be verified. Between 1996 and 2000, seventeen billion roubles had been spent on a government programme to mitigate the consequences of the disaster. I had contested the validity of the 1998 register of contamination dose for inhabitants that had been produced by the Ministry of Health. Even though I am not a director of a State institute, the president of SCSNT had asked me to preside over the Commission, which as an academic I was qualified to do.

First I asked about the composition of the Commission and found out that among the members was a certain Professor Kenigsberg and others with whom I refused to work collaboratively. "Who would you recommend?" I replied that, as a medical expert, the first person that should be included was Yury Bandazhevsky, and he was then appointed. I named a few other people. But as they had not excluded Professor Kenigsberg and his partners, and I knew there would be no objective examination if they were present, I refused the presidency of the Commission although I agreed to take part.

Q.—It was while working for the Commission that Bandazhevsky wrote his own report separately and sent it to Lukashenko?

V. Nesterenko.—Yes. Every member of the Commission had to write their own report. In our group, Professor Stozharov, Professor Bandazhevsky and myself wrote our own final conclusions.

Y. Bandazhevsky .—Just to be clear, I knew that this commission would not present the true facts, that there would be a general conclusion, and I thought it was my duty to present my own conclusions. The second reason has to do with the fact that two months before, I had written to President Lukashenko to outline my views on the health problems that were facing the population, and I put the record straight. The report followed a letter I had already written. It had been included in the commission's conclusions.

V. Nesterenko.—The letter certainly played a role. I know that Bandazhevsky's name had been put forward for the post of Minister of Health in Belarus. It had been discussed several times. That was the result of his letter. But how had this commission ever come into being? There were a lot of journalists present during the parliamentary hearings on Chernobyl when I had criticised the new register of doses. I had said that the register was inaccurate because of the indirect method of determining accumulated dose in the inhabitants: instead of taking real measurements of the radioactivity in becquerels accumulated in the body of each inhabitant, they measured the level of radionuclides in 10 samples of milk and potatoes, from each village. It was not representative of the real levels of contamination in people. Our measurements, using a human radiation spectrometer, showed that the register had given annual doses 2 to 7 times lower than the real contamination. It came as no surprise then to find that, using this method, they had only found 128 villages (56,000 inhabitants) in which the people had annual doses that exceeded the limit above which the population needed to be given aid. In the previous register in 1992, the number of villages was 1,102. With a million inhabitants. According to this new register, levels in most of these villages were below 1 mSv/year, which meant they were no longer eligible for government aid. One of the members of the government challenged me aggressively. "All you do is criticise. Prove to us that they've got it wrong". Of course, he had been told that I had no proof. I replied: "OK I will, but only if I can choose ten scientists excluding those who compiled the present register. If you grant me this request, I will preside over the expert commission". That's how I became president. They let the genie out of the bottle. At the time I had five human radiation spectrometers (HRS) at my institute. We went to 45 villages that had been classified 'no risk' by the Ministry of Health and we measured the inhabitants. I printed up the results. I had the names of every person measured. It was impossible to refute.

I invited Professor Ostapenko, the new director of the Ministry's Institute, whose work we were monitoring and I put it to him: "You've spent 10 billion roubles on this project. If you withdraw the register, the government will finance any revision that they ask you for. If you allow us to prove that you did the work badly, you will have to refund the money". In his statement to the commission, he said that he agreed with our conclusions and that he would withdraw the register.

Professor Stozharov, Yury Bandazhevsky and I wrote a very critical report on the mismanagement of government money. They were so alarmed that Kenigsberg, the vice-president of the institute, and the IAEA's man in Minsk[80], telephoned me (it was a Friday and we hadn't finished writing the report) and asked me to come to the scientific council meeting to present our conclusions. I refused to go and told him that he would have our conclusions in good time. They complained to the Minister Zelenkevich, who wrote a 20-page letter to the president of SCSNT describing us as incompetents whose opinion could be ignored. We knew that our views would never be heard. With Yury Bandazhevsky's agreement, we sent our report, which both of us signed, directly to the Security Council. According to its statutes, the Security Council is responsible for the safety of the state, and therefore, for the health of the population. I thought that would be of concern to them. What followed is interesting. Ministers Kenik from ComChernobyl

---

[80] Mr Norman Gentner, a member of UNSCEAR, told us during the international conference in Kiev in June 2001: "Our source of information is the validated data supplied to us by the health services of the member States. We have official channels of information with Belarus, Ukraine and the Russian Federation. There are designated people whose job it is to transmit information to us. I could introduce you, for example, to Dr Kenigsberg, vice-director of the Belarus government register…" See Part Six, Chapter II, p.450.

and Zelenkevich from Minzdrav were invited to the Security Council and their document was sent back to the commission as non objective saying they needed to include our documents. Our five key points were included without exception along with two resolutions. The first, from Zametalin, the Vice-Minister of the Council of ministers, the other, Novitzki, Vice-President of the Ministry for Chernobyl. It was stated that the commission had found some obvious defects and the five ministers were invited to correct them as soon as possible on the basis of our conclusions. Re-establish the register etc…Everything that we had proposed was accepted and put forward as a directive to the Council of Ministers. Yury and I were listened to in meetings.

Y. Bandazhevsky.—Why did I write my own conclusions? I wanted to present each argument point by point, in detail. If I had been brief, it would have been easy to refute it. Vassili Nesterenko can testify to the way we were treated. When I arrived with my colleagues, they wouldn't give us the documents. They said: "They're over there; you'll have to look for them".

V. Nesterenko.—We were very interested in the method they used to compile their register of doses. They refused to show us the documents. They brought them to us on the last day of the commission.

Y. Bandazhevsky.—We weren't given tables and chairs. It's interesting, isn't it? "It's OK. I'll manage. Have you got a photocopier? I would like to copy some pages". "It's broken". "Too bad" I said to my colleagues: "We'll have to use pens and copy out these pages". We copied the whole thing by hand. I came several times, and in order to copy extracts from the reports, I had to sit on the window sill. I laid it all out in written form explaining how these funds had been spent and what effect this had on solving the problems of Chernobyl. I quoted directly from their reports and the sums that had been spent on their *in depth* research.

V. Nesterenko.—He claimed in his conclusion that out of 17 billion, 16 billion had been spent on nothing. I told him at the time that he was signing his own arrest warrant…

3. BANDAZHEVSKY'S REPORT

This report analyses, in particular, some studies carried out under the supervision of Professor Kenigsberg. One of these was a study by a certain V. Buglova, who was closely associated with the IAEA, and held in great esteem by Kenigsberg himself.

Extracts
The summary report on the projects listed above, signed by Professor Kenigsberg, analyses trends in the dynamics of dose formation for internal irradiation in relation to the establishment of admissible levels of caesium-137 and strontium-90 in food products in Belarus. [...]

The general scheme to update existing protection measures rejects any mandatory measures in the areas under consideration and proposes a transition to the adoption of voluntary and selective protection measures. In this regard, it makes it obligatory to take into account non-radiation factors (economic, social and psychological).

The recommendations made by the authors of this report amount to an abandonment of the principle of obligatory and universal respect for protection measures in the affected zones. Reading between the lines, we can detect the lifeboat ethic—"every man for himself", the state being freed of all responsibility towards the health of the population.

This is in flagrant disregard of the constitution of the Republic of Belarus and of the law regarding public health. Instead of explaining to people what radioprotection measures they should take, they are told to choose whatever measures seem right for them. Furthermore, the authors recommend that factors other than radiation be taken into account, though the part they play is not specified anywhere. We must point out that we were not able to familiarise ourselves with the method that was used to obtain the conclusions cited above. The study report directed by V. Buglova, medical doctor, was not made available to us.

General conclusions.—The vast sums given to the Scientific Institute of Clinical Research into the Medicine of Radiation and Endocrinology (more than 17 billion roubles in 1998 alone) produced no useful results for the national economy in terms of effective health protection measures for the affected population.

In our opinion, this is due to the complete absence on the part of the authorities of any effective control over the planning of research topics undertaken by the Scientific Institute of Clinical Research into the Medicine of Radiation and Endocrinology. These should have taken into account the urgency, the significance and the economic usefulness of the studies, as well as the possibility of implementing the results obtained in public health policy.

The irresponsible attitude shown by the directors of the Scientific Institute of Clinical Research into the Medicine of Radiation and Endocrinology, in the use of funds granted to them by the state to solve the problems linked to the disaster at Chernobyl, resulted in a large proportion of the money being spent uselessly, without achieving substantial results.

Meanwhile, funding of research in areas that are already well known has been favoured over research into new and unexplored areas. Because of this, in most of the reports, there are no real methodical recommendations for medical practice, nor any scientific conclusions of international import.

In connection with the above, we consider it necessary that:

1. Action should be taken immediately to revise the scientific programmes to mitigate the effects of the disaster at Chernobyl, by favouring those approaches that produce tangible results and play a significant role in the national economy. To achieve this, an expert commission should be set up involving representatives from all government bodies and from other organisations that have a real concern in the matter.

2. The Ministry of Health and the Committee for Chernobyl must assume greater responsibility for the use of funds for scientific programmes to mitigate the effects

of the disaster at Chernobyl. There must be effective oversight of the planning of scientific research from the point of view of their current relevance, their importance and their economic usefulness, as well as monitoring of the results obtained and of the possible applications in public health practice.

In order to implement these approaches, we propose the setting up of a scientific council on the problems of Chernobyl, made up of scientists from Belarus and from abroad.

The rector of the Gomel State Institute of Medicine
MD, Professor of Y.I. BANDAZHEVSKY

Knowing Bandazhevsky's work and the importance he attached to the *planning of research topics,* in other words, the selection of topics, it would not be hard to predict in which direction scientific research would have gone in the territories contaminated by Chernobyl, if he had been appointed Minister of Health in Belarus, as had been envisaged.

V. Nesterenko.—Of course, it was a very hard blow for them. But afterwards, the commission's work was not followed up by any concrete action. In reality, given the norms that were established under pressure from the IAEA, people will get ill, not this year, but in ten years. The people making decisions today won't be in power then. If you look at who has been in charge of the Chernobyl Committee, I can already name five different presidents. The country has changed its Minister of Health four times. They can always say, the scientists recommended it. Hide behind the scientists at all times.

This was the end of April 1999. The reprisals were about to begin.

4. REPRISALS

Y. Bandazhevsky.—In May, the Minister of Health sent three inspection commissions, one after the other, to my institute to monitor programmes, teaching, education and I don't know what else. They didn't find anything noteworthy. I was officially on holiday and I was busy with my scientific work…I had gone to Moscow to support some of my students who were defending their theses. It was at that time that I was awarded a medal from the Polish Academy for my contribution to the development of medicine. Then, it was June, the exam period…In July I found myself in prison.

V. Nesterenko.—When the three inspection teams came up with nothing, this was the situation: on the basis of the decision taken by the Security Council and of the Council of Ministers, the Ministry of Health needed to seriously reconsider everything—withdraw the register, recognise their mistakes, and change everything. It was impossible to contradict us on the scientific aspects of our work.

Impossible to refute it. So what do they do? They were left with only one option: discredit the people who had signed the report. It was probably around this time that the famous letter denouncing Bandazhevsky was sent to the government. I'm sure it had been organised from within the Ministry of Health and at the institute: they needed to blacken his name. If just one member of the commission that had criticised them came under investigation, the whole problem would be removed—you can do nothing.

Q.—What denunciation are you talking about?

V. Nesterenko.—They must have persuaded one of the staff at the Gomel Institute of Medicine to write to the authorities saying that Bandazhevsky was supporting the opposition financially. In the letter it states that he has a foreign currency account and that he supports the opposition. Faced with a government directive, what else could they do? They had to get rid of him. Because we could continue to cause trouble.

Q.—But who are "they"? Where are they?

V. Nesterenko.—At the Ministry of Health.

Q.—Do they have links with the international lobby?

Y. Bandazhevsky.—There's no doubt about it.

V. Nesterenko.—It became very clear. The woman whose optimistic report had been rejected by the Higher Commission when I was president works for the IAEA now. She's called Buglova. She worked with Kenigsberg—I have to mention names—who is the scientific deputy at the Institute of Radiological Medicine at the Ministry of Health that we had monitored. And in fact, it is he who directs this institute. There have already been several directors. But every time they appoint a new director, he manages to compromise their position in some way, the minister sacks them and there he is, once more, in charge. Anyway, Kenigsberg probably has close links to the IAEA. Buglova, who worked alongside him, and explained why enormous doses of radionuclides were admissible in foodstuffs, had to support his thesis. Some scientists signed a note criticising them, and saying it was not a serious study. And suddenly, I am asked to chair the conflicting experts' commission of the High Commission of Certification. A majority needed to be determined after votes were tied on the board of experts on her doctoral thesis As President, my vote was decisive. We listened, we looked at everything and we came to the conclusion that the work... in which it was written that *"due to the effective radioprotection measures taken to protect the inhabitants, Chernobyl has had no consequences in Belarus"* was clearly a political order to "conclude that there are no consequences". We concluded that the work was incorrect, that it did not reach the required scientific standard. Her work was rejected definitively with no right to present a thesis on the same subject again. She was angry. She left her job and works in Vienna now... for the IAEA as an expert scientist.

Q.—And you, Yury Bandazhevsky, were accused of being an "enemy of the people"?

Y. Bandazhevsky.—Yes, they wanted to label me as an enemy of the State, of the country. In reality, anyone who knows me at all knows that I work only for the State, and have done so throughout my career. I have never been involved in politics and hope I never am for the rest of my life. I have devoted my life to scientific research and to nothing else.

5. IRRESPONSIBILITY OF POLITICIANS

Y. Bandazhevsky.—The 92 scientists who signed the famous letter to Gorbachev on 14th September 1989, stating that Chernobyl had had no serious consequences, continue to work in their posts. The National Radioprotection Commission of Belarus continues its work with the same staff and the same president. These are the people who formulate this opinion and defend the official line. As for Nesterenko and myself, we are portrayed as cranks.

Q.—Is it true that there are plans to build nuclear power stations in Belarus?

V. Nesterenko—It's true.

Q.—Is it the IAEA that wants this?

V. Nesterenko—Obviously.

Q.—Who has the power to make that decision in Belarus?

V. Nesterenko—According to the Constitution, it is Parliament that decides whether or not to build a nuclear power station. Lukashenko dissolved Parliament and cut its numbers by half; there are now only 110 deputies. Eighty of them were willing to vote yes. The deputies who were opposed to the project knew this and asked me to intervene. I spoke at the parliamentary hearings on the question, and laid out my arguments. I said that it was a technology for rich countries. Chernobyl had shown that when an accident happens we are unable to protect the population. The government has no right to adopt a technology when it has shown itself incapable of protecting the population from its consequences; it's immoral. That was the first argument. Secondly, the cost. It would be much better to choose gas, which is two or three times cheaper. Third, we would need to spend billions of dollars. We would be indebted to the West and we would lose our independence. Parliament adjourned the vote until later and set up a commission made up of supporters and opponents of nuclear power stations to look at the question. They were to examine the question and report back to Parliament and the government. The commission was made up of 32 people. There were 7 against and all the others were in favour. It was people from the Ministries, from the Academy of Science etc. We started to "work" on them. I was glad that one of the deputies, the President of the Commission for Scientific Research, was on the Commission. There was also Mr Smoliar, who had been President of the Chernobyl Committee under two governments, Professor Lepin, a liquidator from Chernobyl and now an invalid, and myself. The others tried to show that there were no gas or oil reserves and we could not do without nuclear energy. We worked on each of them separately: interviews, discussions, presentations on the subject. On 30th December we voted. 8 people were in favour of building a nuclear power station, all the others were against. We had convinced them. In the end, they were all happy, relaxed: no more nuclear power stations. I gave an interview, I wrote an article for a German newspaper: "Belarus says no to nuclear", in which I explained everything. The next morning, the Minister for Energy and the President of the Academy of Sciences brought together the Praesidium of the Academy without inviting us and adopted a contrary resolution that was sent to President Lukashenko. We had heard about it by chance, and told the media to alert the public. Lukashenko was forced to say: "I cannot go against the will of the people". We proposed a ten-year moratorium before examining the question again, and we repeated what we had always said: "If you want nuclear energy, you need to call a referendum". They were frightened. The Prime Minister called up the Ministry, things calmed down for a bit before it all began again.

Chapter IV

THE LOBBY FIGHTS BACK

The nuclear lobby in Belarus is represented by people who gravitate around the Ministry of Health. No one tries to hide this.

In June 2001, at the International Conference in Kiev, Mr Gentner of UNSCEAR referred to Professor Kenigsberg as the trusted confidante of his own organisation and of the IAEA. He was commissioned to transmit information and official epidemiological data from Belarus to these two UN Agencies and they accepted the validity of this information, relying exclusively on it when they formulated their conclusions and recommendation for governments. Professor Kenigsberg refers to the IAEA every time he is in disagreement with Professor Nesterenko, and in particular, when he opposed the use of apple pectin for children. The European Commission dutifully rejected all Nesterenko's requests to TACIS to fund the use of this effective, natural adsorbent.

Understandably, the dynamism and energy of Nesterenko and of Banda-zhevsky, two highly qualified scientists, constituted a threat to the civil servants at the Ministry of Health. In concert with the Moscow 'nucleocrats' and the 'experts' from the IAEA these civil servants had been lying to the government and to the people since 1986. Their counterattack was not long in coming and was hastened by the arrival in power of Alexander Lukashenko. He had not dared to demand compensation for damages from the owners of the Chernobyl power station (Ukraine) nor from the designers and manufacturers (Russia). Lacking the financial means to deal with the problem openly, and protected politically by the pseudoscience emanating from the Ministry of Health and the UN Agencies, Lukashenko chose to bury his head in the sand: deny the scale of the disaster, hide the catastrophic deterioration in health among the inhabitants, and advocate the resettlement of evacuated territories. In 1993, towards the end of the "Belarusian Spring", the local radiological monitoring centres (LRMC), set up by Nesterenko and financed by ComChernobyl, numbered 370. The year after, there were 180. Within five years, from 1995 to 1999, they had been reduced to 82, of which 21 were financed by German NGOs, to whom Nesterenko, in the meantime, had appealed for help[81]. At the end of 2005, all the LRMCs financed by the state of Belarus (ComChernobyl) were closed; 12 were financed by German NGOs, 1 by the Association Enfants de Tchernobyl d'Alsace, 1 by CRIIRAD and 6 within the CORE programme, though the latter would not use the food additive pectin as a protective measure for children, for reasons which will become clear.

1. TORTURE

Yury Bandazhevsky does not like to remember the hell that he went through. His wife told us about the nightmare of his arrest and the torture which he endured during his first pre-trial detention, that lasted for a period of five and half months, and ended on 27th December 1999.

Galina Bandazhevskaya—I wasn't in Gomel on the night of 13th July 1999 when he was arrested. About fifteen armed police entered the house. They turned the place upside-down for hours. The search lasted from 11 in the evening until 4 in the morning. First in the apartment, then at his office at the institute, and at 4 a.m. they threw him in a cell, where he stayed until 4th August, completely cut off from the outside world. He slept on the floor. He was given a meal once a day. He lost 20 kg in 3 weeks. The lawyer was only able to see him twenty-two days later when he was informed of the charge. At this point, he should have been transferred to the normal prison. The lawyer warned me. I was behind the bars of the courtyard when they pushed him into a van. I didn't recognise him. He had a long grey beard. He was completely hallucinating. When he saw us he cried, "You won't abandon me the way everyone else has!" I was with our two daughters. We weren't allowed to even come close to him. He just had time to shout this one sentence to us. They put him in the van and took him away.

We thought they had taken him to the 'normal' prison in Gomel. When the lawyer went there the next morning, he was told he wasn't there. "Look for him, he must be somewhere". Afterwards, we realised that they had extended this period of solitary confinement to get him to confess. They had put him in an identical cell but in Mogilev, 120 km from Gomel.

---

[81] There have been two distinct responses to the Chernobyl disaster from Germany. The nuclear *establishment* influences the work of the independent institute Julich, which undertakes government contracts and is dependent on them financially. On the other hand, organisations from civil society, and especially the younger generation, sons and daughters of those who occupied Belarus during the Second World War, are very sensitive and supportive to the population on whom the nuclear scourge of Chernobyl hs been inflicted. The Julich Institute itself, some years after their initial "error" when they omitted to measure the contamination of people living in the rural areas, undertook a project with Nesterenko's institute, using an impeccable methodology, that confirmed the stability of plasmatic microelements after pectin treatment, thus contradicting Nesterenko's German detractors. (see Part Three, Chapter VI, 2, p. 239–240)

Before he was transferred, they read the charges against him in the presence of his lawyer, whom he was seeing for the first time. The lawyer told me that while it was being read, Yury came under heavy police pressure. They told him I was ill in hospital, and that his mother was close to death: "If you confess, you can go home. If not… " It was a very long interrogation. When he understood that they could keep him in prison for an indeterminate period, the lawyer saw Yury go white, press himself against the back of the chair; he was very close to giving in. He felt that if Yury had been handed a pen, he would have signed anything. The lawyer reacted. "What are you doing? You know it's illegal, what you're doing? Do you realise that?" Because his lawyer had dared to say that, Yury regained control. It was the first time since he had been arrested that he felt someone was on his side.

The cell in Gomel measured 2 metres by 2 metres. He shared it with another detainee who did not stay long. It was a "dungeon", a holding-cell where petty criminals were held for a maximum of 2–3 days. The floor was painted entirely in red, who knows why? Afterwards, his clothes stayed this colour because he was sleeping on the floor. He didn't even have a toothbrush, a razor or a towel.

During these twenty-two days, I asked if I could give him some clothes. The prosecutor said, "I'm not running a laundry..". When they realised they hadn't broken him, they hadn't got what they wanted, instead of transferring him to the ordinary prison as the law prescribes, they put him in the cell at Mogilev.

During this time, we looked everywhere for him, we didn't know where he was. I wrote a telegram to Lukashenko, telling him my husband had disappeared, that after being taken into custody, he should have been transferred to a prison but he had never arrived and I was frightened for his life. Perhaps he was dead? I begged him to find my husband, who was Rector of the Gomel Medical Institute. After the lawyer had appealed to the Gomel prison, he finally got a reply that they had decided to put Bandazhevsky in preventive detention, which was why he had been transferred to the detention centre of the regional executive committee of the Mogilev region.

Yury was very ill there, completely exhausted. To avoid complications, they ended up taking him to hospital. The doctor in charge was a young woman. She said: "I can't hospitalise you because you are a prisoner. Our hospital is not designed for this…" I learnt afterwards that Yury was on his knees in front of this young intern, who could have been his pupil, saying: "Listen, take me, because I'm about to die". She called the surgeon, and they decided to give him an endoscopy. They discovered two haemorrhagic ulcers and the surgeon ordered his immediate hospitalisation. They put him in a ward. They put two policemen at the door and they handcuffed his foot to the bed! The policemen wanted to take him straightaway but the consultant intervened: "When I have a patient in this state, it's me who decides when he can go. He is absolutely not capable of leaving at the moment".

After eight days, they took him to Minsk where he was hospitalised for three weeks. After that, he was transferred to a remand prison. It was there, for the first time, that I was able officially to see my husband. Fifty days after his arrest. They took me into his cell with a priest. The meetings took place in a room in the prison. He was already wearing a black prisoner's uniform that hung off him like a clothes-hanger. Close up, I could see that he was completely disoriented. He didn't understand why I was there, why the priest was there. We didn't manage to have a conversation. He cried the whole time. He had a handkerchief that he folded and re-folded at least twenty times.

2. AFTER-EFFECTS

G. Bandazhevskaya.—He was released on 27th December 1999, almost a broken man, terrorised. He would repeat at least a hundred times a day the same words: "Are they going to put me in that hole again?" On the street he thought he was being followed all the time. In the house, he would speak in a low voice. If he had something really important to say, he would find a corner somewhere in the house and write it on a piece of paper. He didn't want to set foot outside. He would take the car. Enclosed within it, it felt like a little home. Our daughter, the youngest, who went out with him a few times, said that when they went out on foot, if he saw someone he knew from afar, he would make a big detour to avoid meeting them. He tried to avoid any contact with people he knew. He thought these people had betrayed him. He told me "I don't want their pity. I can hear their words already "We're on your side. What a terrible thing to have happened! You, with the job you had, what a terrible thing to have happened!"

He was totally immersed in the situation. He had not distanced himself from it. He was absolutely incapable of escaping his experience. It was very trying for me and for the children. I understood eventually that it was absolutely useless to get angry with him. He wasn't being self-indulgent. He repeated a hundred times a day "Are they going to put me back in that hole?" For example, in the evening when he went to bed, he would say: "I close my eyes and I can see this narrow cell. There were eight of us in there. There was nothing to do the whole day. We stayed there the whole time". Here's an example: he spent a lot of time staring at his hand in front of him. He didn't have anything else to look at. It became a kind of image association, an obsession when he went to bed, and this image haunts him today. "I spent hours looking at my hand when I was there. I'm sick of the sight of it. I want to cut it off. When I close my eyes, it's always my hand that comes back to me and then I think of the prison". It's because he had to put his hand somewhere, given how narrow the cell was, with so many other prisoners in it. It became his horizon.

We asked that he be put in another cell where people weren't smoking, but there wasn't one. "That's the best cell" they said; "you can have up to forty people in the same space". There was one bed for three people. They slept in shifts. They passed infections onto one another all the time. They constantly itched. The bacteria and the humidity causes furunculosis. When he was released at the end of five and a half months, his body was covered in pustules.

He was so terrified by the experience of being thrown in gaol and everything else I've just described, that all he could do was keep repeating "Are they going to put me back in that hole?" He didn't say "I'm going to end it". He didn't talk about suicide, although he did say "I won't be able to stand it".

Three or four months after his release, he began to regain some normality. He was less afraid. He started to do some scientific work. He started to breed some Syrian hamsters at home, in the bath, for experimentation purposes. He actually completed a study which was published. But he still kept his distance from people. He had no contact with the institute. On the other hand, he took great pleasure and was very grateful for the visits he received from journalists, visitors and doctors from abroad.

Once, the judge came by. I think he wanted to get to know this person whose case he would be dealing with: not to interrogate him, but to allow him to tell the story in his own words, to get to know him better. I think this judge would have made a good psychologist. He said to me "Bandazhevsky has spent four hours telling me his life story and all he talked about was science, and his research. If they wanted to get a confession out of him, they went about it the wrong way". He didn't tell me what the right way would have been. Our police had been very brutal with him, whereas he is someone that would have had to be manipulated in a different way, if they wanted to get something out of him. He didn't say all this literally, but that's what I understood from what he said.

The subtler method would come later, during the long period of detention, and still the prisoner would not bend.

3. THE PRISONER'S TESTIMONY

Yury Bandazhevsky was reticent and evasive at the beginning but finally one evening he opened up to and gave us an account of the events that had turned his world upside down. Only his wife knows, and even she does not know everything, the innermost secrets of the life he led behind bars.

Yury Bandazhevsky.—You had to live it. Imagine coming home one summer evening, in a white shirt, tired, from a meeting of the regional executive committee to decide on admissions to the institute. And suddenly fifteen armed men have broken into your apartment, when you're alone, have just taken a shower, before the pandemonium begins. At five thirty in the morning, after all this searching and

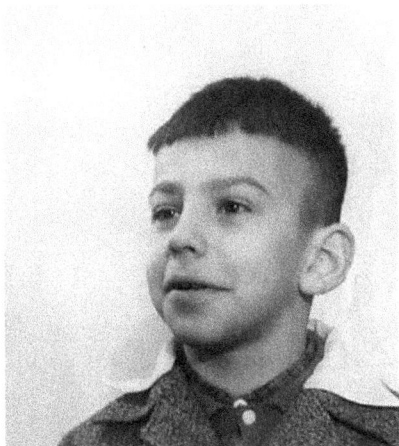

The young Bandazhevsky

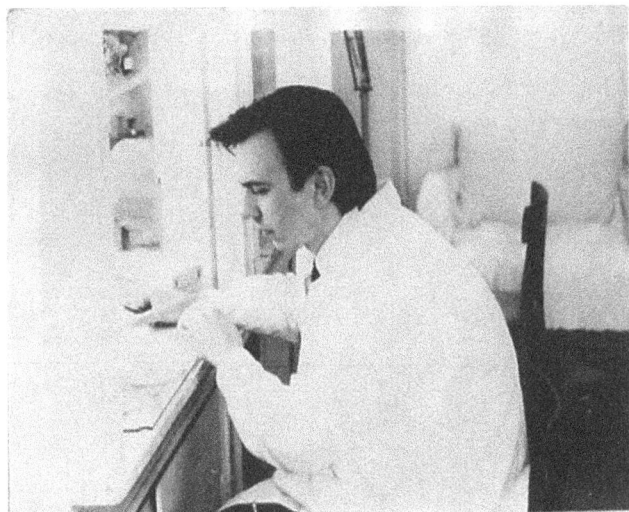

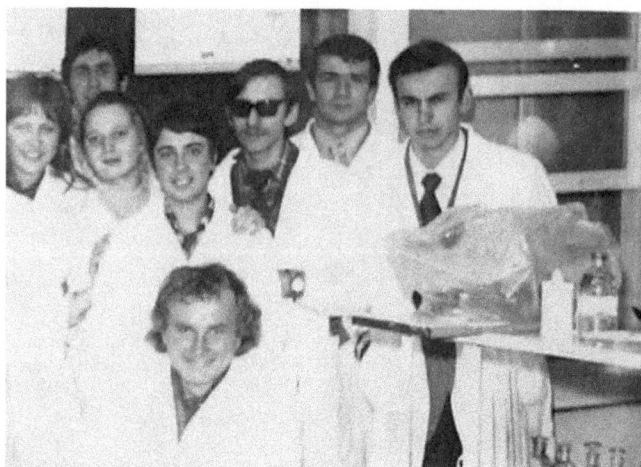

Y. Bandazhevsky, first on the right, at the Grodno Institute

Galina and Yury
Bandazhevsky

18th June 2001

The Volodarka prison in Minsk

Galina Bandazhevskaya

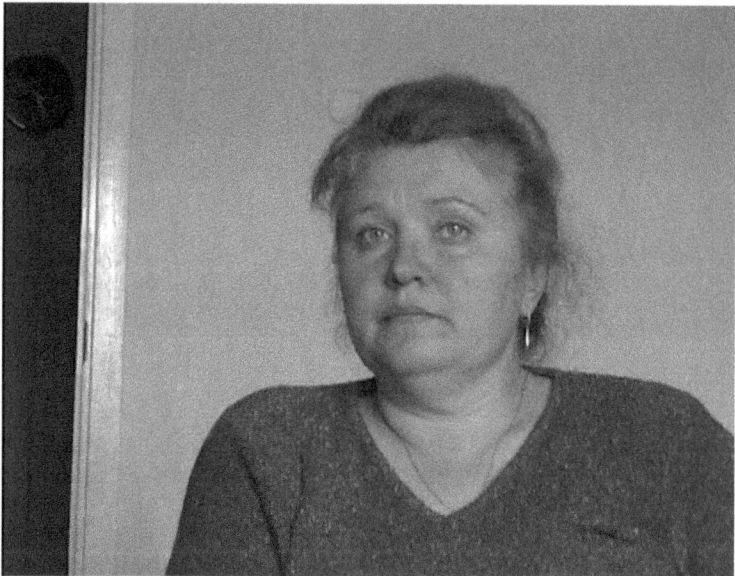

Yury and Galina Bandzhevsky

ransacking, I find myself in a cell, in the basement, you know? After a night like that, can you imagine? I would have understood if they'd handed me some kind of arrest warrant and said "There you are, you've been accused of this". Things could have been done differently, there are correct procedures. But I was literally just grabbed. And now they're presenting me with a paper saying I'm the leader of some criminal gang. Criminal!!! It was in writing. That explained the way I'd been handled, you see. It was terrible… Twenty two days in prison in inhuman conditions. I wouldn't say I was frightened. I wasn't frightened at all. In fact, that annoyed some of them, the fact that I didn't cry, that I didn't get in a state…But it was awful.

At Gomel, they tried to break me. I had no idea what was going on around me, or where I was. Later on, I learnt a lot from the other prisoners that were put in my cell from time to time. They had up to date news, they had links to the outside world. They told me: "Everyone's talking about you. The whole business has caused a big outcry". I was given newspapers and there was a lot of information about me in them. They had arrived in mid summer with pullovers to lie on, but I was sleeping on the floor with nothing. I put the newspapers under my head. It was cold at night, so I covered myself with newspaper. As a matter of fact, they really keep the heat in. If you ever need to keep warm some day, try newspapers.

When I was transferred to Mogilev, I was put in a holding cell again. I was put into a cell where there was a hardened criminal, whose job it was to look after the other detainees. Huge, terrifying, his face all burnt. It was very late in the evening. I came in. My face must have expressed all the agony I'd gone through. I was feeling really bad. And this man—they were expecting a different reaction from him, he holds out a metal cup of warm water, with little bits of mandarin peel in it. "Take it, it's cracknel", that's the word he uses for a biscuit. I'm so touched, I'm near tears. Imagine the man I am at that moment… First, I've been locked away, without charge, then transported in handcuffs across town in broad daylight. Accused, interrogated. Then I'm transferred again. It all starts again. I'm transferred to another town. I don't know anything. I see my wife running towards me: my family has seized the opportunity to see me, they're crying, banging on the car with their hands when I'm taken away. My daughter crying "Be strong, Daddy!" You see, I can't even talk about it now. It just makes me cry. It was hard… And then they put me into a cell with a guy like that. And he gives me…I began to cry. Nothing else I could do. They realised their little manoeuvre had backfired and they took him away the next day.

But the third day, I couldn't get up. I had already lost so…. You only get one meal a day there. And my ulcer had probably opened up again… I couldn't get up. My cell mate started to bang on the door. I must have fainted or something, because the next thing I knew, I was outside. A woman had died in there recently. They were worried. They got me out into the fresh air. Uniformed police were talking amongst themselves: "We need to get him to hospital. If he dies here, there'll be problems". They brought me some more food and they called an ambulance. The doctor at the emergency department in Mogilev thought I couldn't stay there. They took me to the regional hospital in Mogilev. There, people's attitudes varied, different doctors, different attitudes. But in the majority of cases, I was treated humanely. They were really kind, really friendly. I stayed there a little over a week, and they looked after me very well. Then, the same members of staff from Gomel came to get me and took me to the central hospital in Minsk, where they continued to care for me.

—Did they know who you were, in Mogilev?

—Everyone knew, obviously

—Through the newspapers?

—Of course. Everyone knew. The regional hospital in Minsk is part of the same public health system. "Rector of the institute, professor", everyone knew perfectly well who I was. I was touched by the attention they gave me, by the simple humanitarian care for my health, for my fate. Even if they were just following the procedures regarding prisoners… That's what I was, after all—a prisoner. Perhaps it was because I was so seriously ill. No-one beat me there. The people who transferred me on these various visits treated me completely normally…

—But they put people in your cell who offered you advice?

—Not just advice. All this is well known. I don't want to talk about it. What's the point of talking about it? But I just want to say that there's no doubt about it, I was being monitored very closely right from the start. I have no doubt about that.

— You told me that you always understood why you'd been arrested and why, out of the blue, despite your high position within society, you found yourself in prison

—I know now and I knew then that it was a very well planned operation, I understood that perfectly. Because I had already seen the denunciation against me…

—Even before?

—I had seen the denunciation a few hours before I was arrested. People who respected me and knew that I was an honest person showed it to me. A written denunciation. I know what it says word for word. I was described in the worst possible terms. They wanted to classify me as a declared enemy of the state. They tried to show that I was a criminal. They tried to persuade me and even said: "Confess!" At different levels…they placed me with people whose job it was to try to convince me. But when I was put in the Volodarka prison in Minsk, I have to say that they followed all procedures correctly. They respected me there…

There are two distinct worlds. At Gomel I was a prisoner. They handcuffed me, during transfers across town, in broad daylight, so that everyone would see me wearing handcuffs. They push you, push in front of you, push you into the prison van. Whereas in Minsk, I have no complaints. The conditions…of course, a prison is a prison. But their behaviour was very correct and they respected my psychological wellbeing.

—Did the guards know about you, did they know who you were?

In Minsk, no-one hid the fact that they were sympathetic to my situation. All these lads who accompanied me showed me great respect. They even said: "This is the Professor's cell". Or when we went out, they would ask "How is your health, Yury Ivanovich?" "Do you know me?" "Everyone knows you!"

And when I left, I said goodbye as if they were... maybe not family, but... In any case, I understood: they were doing their job, I understood completely. They wished me luck, all the best. They waved to me as I left. In general, wherever I went during this period, I found a common bond with everyone.

—What explains the hatred for you at Gomel?

—I think it was encouraged by certain people. Even though there were people who could have... helped me. They could have come forward and said: "We've known him for many years, he came here when he was 33..."

—Is it because these people have been terrorised for decades?

—No. I'm talking about high ranking officials. They invited me there, they promised to support me. They thanked my parents, my mother, for the fact that I had come to Gomel, for setting up the institute... My point of view was this: as long as these very young children and other people are here, they are going to need continual medical attention. I believe this absolutely. A doctor should be concerned with people's health at all times. I can't accept any other viewpoint. That's why I came...Certainly not in order to become a high ranking official. For the love of God! I've never been a part of the hierarchy. I am a typical research scientist... I threw myself into it. But quite obviously, I had a lot to learn about life, at the time—from a tactical point of view. I had no instinct for self-preservation, and even now, it's not particularly well-developed. The form of self-preservation I have at the moment is almost pathological.

—It saved the scientist in you.

—You're right, if I had any instinct for self-preservation, I would not be a scientist.

Chapter V

MINZDRAV SENDS
AN ULTIMATUM TO NESTERENKO

Minzdrav is the Russian acronym for the Ministry of Health (*ministerstvo zdravookhranenia*).

In July 2000, worried about the increasingly bad news that has been arriving from Minsk over the previous three months, I write to Madame Daillant, who works for Amnesty International in France and has responsibility for Belarus.

> There is bad news from Belarus.
>
> The Minister of Health, Zelenkevich, has sent an ultimatum to Nesterenko: the independent institute, Belrad, will have to cease operations if it does not receive authorisation from the very people who want it closed. The Minister has invented the notion that measuring the levels of radionuclides accumulated within the bodies of children is a medical act, and as such needs authorisation from him. It is false and it is absurd, as has been shown by the international scientific experts whose opinion was sought by the Minister in charge of Emergency Situations, who, unlike Zelenkevich, supports Nesterenko. It makes no difference. His situation is precarious. The question of whether or not Belrad may continue to work freely, is not yet settled.
>
> It is the civil servants and the doctors at the Ministry of Health, supporters of the IAEA's position in Vienna, who are hostile to Nesterenko and Bandazhevsky. They deny the seriousness of the health consequences caused by Chernobyl, do not recognise the correlation between radioactive contamination and the various illnesses described by Professor Bandazhevsky, and they publish reassuring data, that is systematically proved wrong by the bulletins sent by Nesterenko, three times a year, to all national and local government bodies. Bandazhevsky and Nesterenko are a threat to their policy of non-assistance to populations in danger, a policy that in another context would lead to prosecutions. It is these people that Bandazhevsky criticised in his famous report about the 17 billion roubles of government funds, in which he claimed that the money had been misspent, and this is probably the reason for his arrest. Professor Kenigsberg, the vice-director of the institute that was in charge of the work, was incriminated by Bandazhevsky and named personally as the director of projects that were allocated 3.35 billion roubles. He is a member of the group that has ordered Nesterenko to cease his activities. Kenigsberg refers frequently and quite openly to the IAEA to back up his theses.

Nesterenko is faced with a choice: either he complies, closes his institute and throws all his colleagues out into the street, or he continues with his work and fight back. Conscious of his rights and knowing the vital importance of his work, he will fight back. In fact, he has no choice: if he submitted, he would be sanctioning the ban on research and on informing the population (surely this is a question of human rights which the Organisation for Security and Cooperation in Europe (OSCE) and Amnesty International also defend?)

Nesterenko has no financial reserves to draw on if he suspends his work temporarily. He works under contract for charitable organisations in the West. The American MacArthur foundation, that has supported him for several years, has made a payment of 100,000 dollars for an ongoing project. Nesterenko does not touch this money other than when work has been completed, and sends them bills on a regular basis. He has to continue in order to pay his staff. At present he is working at Minsk airport: he is measuring levels of radioactivity in children's bodies before they leave to go on holiday, invited by charities abroad and he measures the levels again when they return to calculate the amount of radionuclides that have been eliminated from their bodies and to continue preventive treatment. He is on a knife edge. The police could be called in and equipment seized... Is he going the same route as Bandazhevsky?

Nesterenko has written to the Prime Minister and to President Lukashenko, with whom he has requested an audience. Tomorrow, he will meet the Ambassador Hans Georg Wieck, director of OSCE in Belarus, who has been alerted by Professor Michel Fernex *via* the director (a Swiss national) from the general headquarters of OSCE in Warsaw.

Given what is at stake, this case should be followed closely. These two scientists are alone in having the courage to divulge what those who should know, and in fact do know perfectly well, deny. (They may not understand every precise detail but they know enough to know that they are lying). All kinds of human rights are being violated here: the right to information, to health, to dignity, to life and to death... and it is not only the peasant population of Belarus whose rights are being denied.

I am in contact with Professor Nesterenko and I will keep you up to date with the situation.

12th July 2000

1. THE FACTS

Following the expert opinion offered by Nesterenko, Bandazhevsky and Stozharov, the Ministry of Health in Belarus is obliged to withdraw the government's 1998 proposed register and rewrite it. The three scientists recommend that the register be established on the basis of direct measurements of the inhabitants using a human radiation spectrometer (HRS), rather than on mathematical deductions based on a dozen samples of milk and potatoes from the contaminated villages.

On 20th April 2000, in an interview on Belarusian television with the president of the parliamentary commission on the problems following the disaster at Chernobyl, Nesterenko criticises the dangerously high levels of caesium-137 permitted in food products (Republic of Belarus admissible levels of contamination—ALC 99). He also criticises the method used to establish the new 1998–1999 register of doses as *"indirect, archaic and fallacious, using a limited non-representative selection of milk and potato samples instead of direct human body measurements which results in an underestimation by a factor of 3 to 16 of the true radioactive load of inhabitants of contaminated areas".*

The next morning, 21st April, a three-person commission from the Ministry of Health arrives at the Belrad institute with an injunction from the Minister, Zelenkevich to ban the HRS measurements, on the grounds that it is, supposedly, a medical procedure, and Belrad does not possess the requisite licence from the Minister of Health. Nesterenko, whose work is actually in physics, not medicine, has the required licence from the Minister of Emergency Situations (MES). Nesterenko writes an indignant letter to President Lukashenko complaining that the Ministry of Health is acting illegally. The MES is asked to intervene and sort out the conflict.

On 16th May, the MES convenes an international group of experts including Professor Michel Fernex, Bella Belbéoch (France), Alexei Yablokov (Russia), Ludmila Porokhniak-Ganovskaya (Ukraine), and Natalie Kolomietz (Belarus). Professor Fernex and the physicist Belbéoch would also write to President Lukashenko. The scientists confirm unanimously that the use of a Human Radiation Spectrometer (HRS) for measuring, constitutes a physical and not a medical procedure. However, an extremely tense correspondence continues between the Ministry of Health and Nesterenko, about both administrative and scientific matters.

On 5th July, the Ministry of Health, in reply to Nesterenko's objections, sends him an ultimatum, and orders departmental directors of local health authorities to stop any radioprotection programmes that were being carried out with the Belrad institute on the basis of agreements that had been signed previously.

MINUTES
From the June 2000 meeting "The medical nature
of the work being undertaken by the Belrad institute using
human radiation spectrometers"

Agenda: The medical nature of the work undertaken by the Belrad institute using human radiation spectrometers.

1. Deputy Minister's report A.S. Kurchenkov.

2. Contributions from Y.E. Kenigsberg, V.F. Minenko, L.S. Meleshko, V.I. Trusilo, G.V. Godovalnikov, V.I. Ternov.

All the contributors expressed the opinion that the work involving the measurement of the population using HRS machines undertaken by Belrad constitutes a *medical intervention*.

In the execution of a directive from the Minister of Health of the Republic of Belarus I.B.Zelenkevich, the commission, set up within the Republican Unitary Enterprise "Center for Expertise and Testing in Health Care", in May this year, monitored the use of HRS machines by the "Belrad" institute:

Belrad does not have the requisite documents from the Minister of Health of Belarus for the use of Human Radiation Spectrometers—in other words for a medical intervention, nor documents certifying that these machines have been registered with the Minister of Health.

3. After deliberation, we have decided:

3.1 Belrad must cease these examinations of the population until it has obtained authorisation from the Minister of Health.
3.2. Directors of public health departments of the regional executive committees and the executive committee of public health in Minsk must cease the work they currently undertake that is based on agreements made with Belrad.

Meeting secretary
Y.P. Platunov

On 17th July 2000, Nesterenko reminds the Minister of the existing rules and regulations:

May I remind you that, in conformity with the statute of the Academy of Sciences of which I am a member—a statute ratified by the Republic of Belarus Council of Ministers—all ministers and administrations are obliged to respond in depth to questions posed by members of the Academy.

And he responds by refusing to obey the illegal injunction to cease Belrad's activities:

The assertion that HRS measurements constitute a medical intervention has no basis in law. It is simply the opinion expressed by those present at the meeting and has no legal force.
There is no mention of the use of HRS machines in the law "Judicial regime in the areas contaminated by radionuclides from the Chernobyl disaster", nor in the Council of Minister's decree No. 386 of 16th October 1991, nor in the list of activities for which a licence from the Ministry of Health is required (ratified by the Republic of Belarus Council of Ministers No. 456 of 21st August 1995)
In consequence, the legislation currently in force does not view the use of HRS machines as an activity requiring a licence from the Minister of Health of the Republic of Belarus.
The Belrad institute does not accept the decision to cease measurements taken using a HRS machine.

The order that has been given to the regional public health authorities to break their contracts with Belrad is illegal.

On 19th July, Nesterenko sends a document to the organisations in the West that support him. It is partly an appeal to the international community and partly an accusation: "The Ministry of Health in Belarus is covering up the truth about the consequences of the Chernobyl disaster on children's health".

On 31st July, the ambassador Hans Georg Wieck, director of the Organisation for Security and Co-operation in Europe in the Republic of Belarus, intervenes. He writes to the Prime Minister of Belarus:

In accordance with the mandate of 18th September 1997, the Consultative Observation Group (COG) of OSCE monitors human rights in the Republic of Belarus, to ensure that it fulfils its obligations towards the OSCE.

We have received a letter from Professor Nesterenko saying that the activities of the Belrad institute, of which he is the director, have been terminated by the Minister of Health on the grounds that they do not have a licence.

His letter has been studied by our group.

In our opinion [...] the injunction to stop work at the institute has no basis in law. Nor does the Minister of Health have the right to outlaw work undertaken through contracts with this institute, as this represents an illegal interference in its activities: an abuse of power.

I would ask that the Minister of Health be advised not to put obstacles in the way of the work of the Belrad institute.

On 7th August, the Council of Ministers of Belarus requests the Presidency of the Academy of Sciences to set up a commission composed of scientists and experts from the Academy, the Ministry of Health, the Ministry of Emergency Situations, the government homologation committee, the Belrad institute and other "interested parties" to study the question and to report back to the Council of Ministers before the 1st September, in order to inform the Security Council of Belarus. Deputy Prime Minister Demchuk is in charge of the commission.

There are several Deputy Prime Ministers of government. This peculiarity would be pivotal in the subsequent manoeuvring that took place later at the Ministry of Health, which used Deputy Prime Minister B. Batura to get round Demchuk.

On 14th August the presidency of the Academy invited participants to a meeting on 22nd August 2000.

On 16th August, the Ministry of Health, knowing that general opinion is not in its favour, writes directly to another "friendly" Deputy Prime Minister, Batura, a long letter in which he reaffirms his position. The same letter is sent separately to Nesterenko, but he is unaware that other people had received the letter.

On 21st August, i.e. the evening before the commission meeting, Deputy Prime Minister Batura gives the Ministry of Health's letter, of the 16th August the force of a decree, by writing in the margin of the letter, in his own hand, the resolution: "For information and guidance". This resolution *that reiterates the obligation for Belrad to obtain a licence for its medical activities,* would be transmitted on 7th September 2000 by Deputy Prime Minister Demchuk himself (wolves don't eat each other) to the members of the commission who, not knowing about the manoeuvre, would be meeting on 22nd August for absolutely no purpose.

Thus the presidency of the Council of Ministers dispensed with a procedure that it had set up itself, before it had even taken place, showing utter contempt for those involved.

On 22nd August, the meeting was to take place, as always, at 2 pm in the presidency room at the Academy of Sciences. The "interested parties", in other words, scientists, specialists and journalists that Nesterenko had invited are all excluded, whereas several representatives of the Ministry of Health are present. As he leaves for the meeting, Nesterenko discovers that one of the tyres on his car had been slashed twice with a knife. He changes the wheel as quickly as he can and arrives at the Academy, out of breath, two minutes before the meeting started. He was supposed to speak for about 7 minutes. That did not happen. Instead there was a heated and chaotic exchange of views during which a lot of pressure is put on him by representatives from the Ministry of Health, to submit: he must not publish any more data from his HRS measurements, outside the Ministry. However, the majority, that is to say everyone except the Ministry of Health, agrees that Belrad's activities does not constitute a medical intervention and that a second licence makes no sense. He already has a licence from the Ministry of Emergency Situations.

At the end of the meeting a member of the Academy, a lawyer by profession is asked to summarise the findings which would then be approved by those who had taken part in the meeting and sent, before the 1st September to the Council of Ministers. This text was agreed by most participants but the Ministry of Health has refused to sign it (no-one knew yet about the letter to Batura, sent on 16th August). The president of the commission, is embarrassed, and asks one of the vice presidents to rewrite another "compromise" text.

On 1st September, the compromise text, signed by the president of the Academy of Sciences, A. Voitovich, is sent to Deputy Prime Minister Demchuk, who has been given responsibility for the task by the Council of Ministers. The central message, exonerating Nesterenko, is contained in the following sentence:

> The legislation on public health of the Republic of Belarus does not define the concept "medical activity". One can therefore conclude that the Institute of Radiological Protection, Belrad, is acting in accordance with the licence from the Ministry of Emergency Situations, which it already possesses.

On 13th September, Nesterenko receives the "directive" from Batura, in other words from the government.

In the meantime, unaware of what had been going on behind the scenes, he responded to the letter sent on 16th August. The controversy with the Ministry, whose central arguments are presented below through extracts from Nesterenko's reply to the letter that Orekhovsky sent to him and to Batura, clearly shows the fundamental incompatibility in their positions. On the one hand, we have a bureaucracy that distorts reality by trying to make it fit a predetermined scheme, and on the other, the objective approach of a scientific dealing with concrete reality.

2. TRENCH WARFARE

Minzdrav, 16th August.—Regarding Nesterenko's statement that the register of dose, produced in 1998–1999, was inaccurate, we would like to point out that one of the errors made by the corresponding Member of the Belarus Academy of Science, Doctor of Technical Science, Professor V. Nesterenko, is that he does not recognise the essential point: *radiological protection measures should be based on the values of projected average effective annual dose (EAD), and not on the dose of the critical group.* This procedure conforms fully with the internationally recognised principles of radioprotection[82] according to which it is standard practice to estimate the need for radioprotective measures and their optimisation on the basis of the significance of the damage to health, as determined by collective dose.

Nesterenko, 28th August.—International safety standards for protection against ionising radiation and safety in the management of sources of radiation (IAEA, Vienna, 1997) state that the averted dose and the number of projected case of illness *should be calculated* according to the collective dose and the values of the projected average effective annual dose, *but* radioprotection measures should be based on *the values of the dose for the critical group in the population.*

In this distorted interpretation of the regulations, Minzdrav refuses to differentiate between *calculated epidemiological predictions* (based on what can be deduced from average dose) and the *absolute criterion for obligatory radioprotection* in the villages (based on the highest levels of contamination). In so doing, the Ministry is not only sacrificing the villages that are most at risk (by omitting the critical group) by excluding them from radioprotection, but going against the "basic safety standards" established by the IAEA, to which it refers. The IAEA itself, recommends[83] that *radioprotection*

---

[82] IAEA, Vienna, 1997. Safety Standards Series No 115, "Radiation Protection and Safety of Radiation Sources: International Basic Safety Standards".

[83] *Idem*

should be based on the dose to the *critical group* within a population. The error made by Orekhovsky explains the sentence at the beginning of Nesterenko's response where he says: "Could I suggest that you find someone more qualified to respond to my letters". There is a joke in Belarus (people in the ex-Soviet Union maintain their ironic and slightly black sense of humour) about how the average temperature in hospital would prove that the patients are doing very well.

V. Nesterenko.- I need to underline the main omission on the part of the Ministry of Health specialists, when they underestimate the significance of direct measurement using a human radiation spectrometer (HRS) of all patients undergoing medical examination at the clinics. It is only by combining medical examination of the population and HRS measurement of incorporated caesium-137 in the body that the causal relationship can be determined between the incorporation of radionuclides within the human organism and the increase in illness.

Information about this phenomenon can only be obtained in the areas of Belarus, Ukraine and Russia affected by the accident at Chernobyl. It is an important argument for the organisation of radioprotection and in medical treatment, in convincing the international community of the need for aid for Belarus and for a reduction in the radiation exposure of the inhabitants, as well as for an understanding of the consequences of the accident at the Chernobyl nuclear power station (at the moment the only causal link that is recognised is between the Chernobyl disaster and thyroid cancer).

Minzdrav.- Nesterenko's comments and his conclusions that "Methodological errors of calculation and errors in the 1998 dose register led to a substantial underestimation of radioactive load following exposure to radiation, both external and internal" reveal his lack of understanding of the procedure used to calculate annual effective dose, an erroneous interpretation of the parameters used and an ignorance of the empirical data available. It is a question of estimates of coefficients used to calculate radiation dose, both external and internal, of insufficient information about the people's diet in Belarus and his elementary ignorance of the application of statistical methods to estimate the indeterminacy of the data obtained.

V. Nesterenko—As a member of the Academy of Sciences of Belarus, in conformity with its statutes, I am responsible for assessing the radiological situation and organising radiological protection for the population of the Republic. My professional career in radioprotection started in 1958, when I was still working at the Academy of Sciences in the USSR (Obninsk). I am, therefore, surprised at your assessment of my "elementary ignorance".

The Ministry of Health's grievances against the 1999 expert commission for the 1998 dose register are incomprehensible. The Ministry of Health's institutes established a register of dose in 1992, according to which of the 3,324 villages affected (more than 2 million people) 1,120 villages, had levels of radioactivity that exceeded 1mSv/year. In the 1998 register, following calculations based on indirect and archaic methods, there were only 169 villages with levels measuring more than 1mSv/year, with 55,181 inhabitants.

The director of the Institute of Radiological Medicine of the Ministry of Health, Professor V.A. Ostapenko, accepted the claims made by the commission of the State Committee for Science and New Technologies concerning the underestimation of internal radiation dose to the population by a factor of 3–8, and the errors made in calculating the external dose. As far as I know, the 1998 register of dose was withdrawn in order to be reviewed.

In January 2000, you invited me to a meeting to examine the new 1999 register where new versions of the register, in which coefficients said to be "in reserve" with a quantile of 0.95%, were proposed. But this approach excluded precisely the critical group within the population (5–7%), who would not then receive the necessary radioprotection.

When we measured incorporated caesium-137 in the bodies of inhabitants (mainly children) in 45 villages in the Gomel region, with a human radiation spectrometer (HRS), and calculated the dose for the critical group, we found that the real dose had been minimised by a factor of 3–8, as with the 1999 register, that was based on selected samples of milk and potatoes.

It is quite staggering to read in your letter the following sentences: "We cannot exclude the possibility that in particular villages the internal radiation dose, measured with a human radiation spectrometer (HRS), could exceed the calculated values. Such cases may arise following the consumption of fruits and agricultural products, mainly milk, which is contaminated above the admissible levels of contamination in Belarus (ALC) But this does not mean it is a "bad" method, because it was never intended to register cases where food contaminated above the admissible levels (ALC) had been consumed" Is it possible, Professor, that you have only just found out that in 550 villages in Belarus, according to data from the Ministry of Health itself, people are drinking milk that contains inadmissible levels of caesium-137?

Finally, it has to be said: we need to abandon these useless attempts to improve an indirect method of calculating internal radiation dose. We need to allocate all our scientific resources to creating a register of real dose, based on direct HRS measurements of the incorporation of radionuclides in the organisms of inhabitants of the regions contaminated by Chernobyl.

Minzdrav.—Analysis of the changes that have occurred over the last five years show an increase in both general and primary illness among victims of all ages. However, the rhythm of this increase in illness in adults and especially in children has diminished in 1999 in comparison with 1994, and is no higher than 1.4% in adults.

V. Nesterenko.—On 21st April 1999, we all listened to the Minister of Health, I.B. Zelenkevich, at the parliamentary hearings at the National Assembly of Belarus. He stated that "1,940,000 people, including 414,000 children, are at risk. In the period following the accident, 1,100 children were operated on for thyroid cancer. Illness among victims of the disaster was higher than in the population that was not affected, and there were indications that this was increasing each year".

On 25th April 2000, during parliamentary hearings at the National Assembly of Belarus, the deputy Anatoli Volkov stated that among 85,900 children examined by doctors in the Stolin, Luninets and Pinsk districts, 27,000 had diseases of the thyroid gland, 23% had heart problems and 22% digestive tract problems. Only 13.7% of the children could be described as "healthy".

Given this information, I cannot accept your statement about the decrease in rates of illness in children between 1998 and 1999.

Minzdrav.—There is no convincing scientific data concerning the pathological effects of various levels of radioactive caesium incorporated into the vital organs and systems of the body, as cited by V.B. Nesterenko, referring to the work undertaken by Y.I. Bandazhevsky.

V. Nesterenko.—This statement does not reflect the real picture either. The work undertaken by Bandazhevsky was accepted previously by the Ministry of Health in Belarus. As for the author himself, he received the Albert-Schweitzer Gold Medal in 1998 and an international prize in 2000 in Paris (IPPNW). His work shows that "pathological states are recorded in the vital organ systems of children when levels of incorporated caesium-137 exceed 50 Bq/kg". It is extremely regrettable that his work at the Gomel Institute has been interrupted.

Minzdrav.—In accordance with the Ministry of Health directive No. 282 of 13th September 1999, radiological diagnosis is subject to a licence from the Minister of Health, and therefore measurements using a human radiation spectrometer (HRS) constitute a medical activity and, according to the regulations, require a licence from the Minister of Health.

There is no doubt that measuring gamma rays in the environment is a physical procedure. However, if the equipment is used for diagnosis of a patient, either for treatment or prevention, as is the case under review, the use of this technology constitutes a medical intervention and according to the regulations, needs to be registered with the Minister of Health.

V. Nesterenko.—It is clear that you are *substituting one concept for another* here. The introduction of radionuclides—iodine 131, technetium—into a patient's body, for the purposes of treatment or radiodiagnosis is clearly a medical intervention, but the measurement of gamma radiation emanating from a human body, using a human radiation spectrometer, is a normal physical procedure, similar to measuring someone's height or weight.

Minzdrav.—Using the food additive Vitapect to aid the elimination of radioactive caesium from the human body has been discussed a number of times in a series of meetings, with specialists from the Ministry of Health, the Ministry for Emergency Situations, the consortium Belbiofarm, the Ministry of Agriculture, etc. The Ministry of Health believes that apple pectin products and polyvitamins have a part to play in the prevention of negative effects from internal radiation. However, they need to be prescribed under medical supervision. [...] It is unacceptable that they should be produced by random individuals with no medical or pharmaceutical qualification. Moreover, pectin-based products do not specifically eliminate radioactive caesium from the body. The statement made by Nesterenko, who is a doctor in technical sciences, that "there is a 60–80% purification effect in the children's bodies" is a simplistic approach to a very complex process.

V. Nesterenko.—According to eminent doctors in Ukraine (Professor, Doctor of Medicine, M.I. Rudnev), and in Belarus (Professor, Doctor of Medicine, N.D. Kolomietz), "pectin promotes the elimination of radionuclides, heavy metals and other toxic substances in the body".

In April 2000, the Belrad Institute received a licence to produce and use the pectin-based food additive Vitapect. Given that pectin is a natural substance, there is no real necessity for a dose limit. It is a dry concentrate of apples enriched with vitamins, whose content in the final product is ten times less than the admissible dose. Therefore the Vitapect recommended by Belrad constitutes a dietary food product and not a medicine.

Several thousand children have been given the food additive, and this has decreased the burden of caesium-137 in their bodies by 40–80%. Your assessment of the effectiveness of using this pectin-based product as "a simplistic approach" is surprising. The HRS is an excellent machine for evaluating the effectiveness of different food additives in the diet.

It is no coincidence that in all the sanatoriums in Ukraine, or in places where children from the Chernobyl area are sent for periods of rehabilitation, Human Radiation Spectrometers are used. Measurements of the incorporation of gamma radiation taken at the start and at the end of a period of rehabilitation provide objective information about the effectiveness of different protective measures.

Minzdrav.—The Ministry of Health has asked us to look into the activities of the corresponding member of the Academy of Sciences in Belarus, doctor of technical science, Professor V.B. Nesterenko. The commercial business, of which he is director, the Institute of Radiation Safety "Belrad", rather than contributing to finding the most effective solution to the complex problem of reducing the consequences of the accident at the Chernobyl nuclear power station, is exacerbating an already difficult social and psychological situation in the contaminated territories, through activities that are ill-conceived from a scientific and practical point of view. Bypassing the official channels of the Ministry of health and ComChernobyl, V.B.Nesterenko continues to publish his own bulletin, *The Chernobyl disaster*, a publication which, in the light of our current knowledge about radiological medicine, radioprotection and safety, does not stand up to scrutiny.

Unfortunately, Professor Nesterenko's approach with its disregard for universally recognised principles of radiological protection, still finds support among a few politicians and scientists, poorly versed in matters of radiation protection. For example, when he was asked to preside on an independent commission of scientists by the Chernobyl committee the outcome was a completely unnecessary delay in attributing dose in the official 1998 register.

As for the scientists, including those from other countries, co-opted by Nesterenko to take part in the monitoring of the "Belrad" Institute's activities, none are recognised general practitioners, or specialists in clinical medicine or ecopathology.

V. Nesterenko.—Have we really learnt nothing from the enormous damage done, at the time of the Chernobyl disaster, when the Ministry of Health of the USSR had a monopoly on information, and imposed secrecy on the consequences of the accident?

Displaying utter contempt for the laws of the nation, the Ministry of Health defends its monopoly on information that it presents to the public and to the authorities on the radioactive contamination of the population.

I would like to draw your attention to the fact that I undertake this activity as a member of the National Academy of Sciences, and that I publish the results of the measurements of the incorporation of caesium-137 in the bodies of the inhabitants of Belarus, that have been obtained not only by the Belrad institute, but also by other organisations in Belarus. The institute is authorised to publish these figures, with the names of the children's families and the levels of incorporation of caesium-137 in their bodies.

Your criticisms of my publications contradict Article 22 of the law in Belarus "On radiological safety of the population" where it says "civil society associations serving the public interest are entitled, according to current legislation, to exercise control over the observance of the rules, regulations and hygiene standards in the area of radiological safety". Your censorship is illegal and I do not accept it.

Your decision of 27th June 2000 to suspend Belrad's HRS measurement activities has no legal basis, and your order to the directors of local public health authorities to break contracts with Belrad is illegal.

Collecting data in Belarus to show the correlation between radioactivity and the frequency of heart problems, kidney disease, diabetes, cataracts etc, is the only way to convince international organisations of the existence of negative consequences of the Chernobyl disaster, not only for the thyroid, but for other vital organs.

It would be helpful if this scientific research could be conducted with specialists from the West. It would allow us to obtain aid from charitable organisations in the West to help with decontamination and with medical treatment for children.

The Belrad Institute is willing to collaborate with Ministry of Health institutes in monitoring the health of children using HRS, and in their radioprotection, so as to reduce the health effects of the Chernobyl disaster.

On 14th September 2000, Nesterenko writes for the second time to President Lukashenko, asking him for his help and protection against the unfounded accusations concerning Belrad's activities that were being made against him by the Ministry of Health. He reminds him that:

> On 6th June 2000 the United Nations Scientific Committee on the Effects of Atomic Radiation (UNSCEAR) published a new scientific report (UNSCEAR-2000) in which new estimates about the consequences of the Chernobyl disaster were made. [...] The Committee bases its calculations and estimates on standard international criteria in nuclear safety, adopted by the IAEA. The authors of the report state: "Although the personnel at the nuclear power station were subjected to the biggest risk from radiation, *the majority of the population will probably not suffer any serious health consequences resulting from radiation from the Chernobyl accident*".

> These conclusions are only possible because scientific information from Belarus linking illness with radiation exposure hardly ever reaches the UN.

> Remaining silent about the extent of illness among children and the lack of information in the West on the subject, has resulted in a situation in which most foreign countries believe that there has been no significant impact on the people of Belarus from the biggest disaster of the century.
> Today, only a few compassionate "fanatics" are helping the victims of the Chernobyl disaster, providing them with medicines etc, which cannot possibly meet the real needs of the people.
> Basically, many errors and false calculations have been made in the radioprotection of the population in the fourteen year since the disaster at Chernobyl.

3. A LETTER TO NON-GOVERNMENTAL ORGANISATIONS

On 1st November 2000, I sent a letter to the members of our network updating them on the situation in Belarus:

> This morning Deputy Prime Minister Batura convened a meeting to discuss the differences of opinion between Nesterenko and the Ministry of Health in Belarus, after Nesterenko's last letter written to President Lukashenka asking him for protection. Representatives from those ministries and governmental committees who support

Nesterenko were present at the meeting. They could not express their opinions., as they are subordinate to Batura.

Nesterenko's adversaries are furious about the fact that Nesterenko communicates with the West and receives aid from foreign organisations: "Foreigners support you because you are in conflict with the Ministry!" "I have been resisting the illegal claims made by the Ministry of Health since April 2000, and anyway I have been receiving help from various charitable organisations since 1995!"

Having failed to establish that the use of human radiation spectrometers (HRS) to measure the incorporation of radioactivity in the body was a medical intervention requiring a licence from the Ministry of Health, Nesterenko's adversaries are now saying that he needs a licence to distribute the pectin-based product, when he already has a licence from the Ministry of Health to distribute it as a *food additive*.

Nesterenko is coming under enormous pressure, to force him to submit and to prevent him from disseminating information about the real consequences of the Chernobyl disaster on health, which the Ministry of Health has been minimising since 1986.

Nesterenko is standing firm. He is continuing with his activities, and is working within the law, but the only aid he now receives comes from non-governmental organisations. The Ministry of Health could interrupt his work at any time, even though it would be acting illegally.

It should be noted also that the Ministry of Health's accomplices in the West continue to spread slander about both Professor Nesterenko and Professor Bandazhevsky. The latter still has a court case hanging over him like the Sword of Damocles.

Legal and financial aid from abroad is urgently needed to defend and to reinforce the position of these two independent scientists.

On 5th November, V. Nesterenko writes, in an insistent manner, to President Lukashenko again:

Once again, I find myself having to write to you to ask for your support in allowing me to continue with my radioprotection work with children from Belarus at the Belrad institute.[...]

On 1st November 2000, during a discussion with vice Prime Minister Batura, I was reproached for having asked for the support of scientists from Russia, Ukraine and from international public opinion. [...] I was forced to take this action because my letters to the government, to the Security Council, to you yourself, about the dispute with the Ministry of Health, are blocked: they never reach their destination and each time, it is the Ministry of Health that arbitrates in the dispute.

During the meeting on 1st November, about Belrad's activities, B. Batura, while admitting that he was not an expert in either medicine or physics, and that using a Human Radiation Spectrometer is a physical procedure, nevertheless concluded that the Belrad institute must obtain a licence from the Minister of Health, if it wanted to carry out measurements using HRS.

I was forced to consult the Ministry of Justice who concluded that examination of the population using a human radiation spectrometer is not mentioned in the list of interventions that required a licence. [...]

The Belrad institute does not carry out medical interventions. It monitors the incorporation of caesium-137 in children's bodies using HRS, and then offers them radioprotection in accordance with its statutory objectives.

Staff at the Institute hope that their work will be supported by the Head of State.

Following the response from the Ministry of Justice and no longer having the President's ear, the Ministry of Health finally gave up on the "licence". The struggle took up almost a whole year. "Our work around Chernobyl certainly comes up against some real obstacles" was Nesterenko's laconic observation.

But there now began an apparently senseless attack, both from inside the country and abroad, on the distribution of pectin to children in the contaminated areas. Unfortunately, it was not only officials from the Belarusian government that were putting obstacles in Nesterenko's way. A campaign of hostile disinformation was being disseminated by certain Western "benefactors", who had no hesitation in slandering him. Nesterenko complained to the ambassador Wieck at the OSCE in a letter dated 29th September 2000.

> I am surprised and upset by the fact that while a dispute on principles is going on between the Belrad institute and the Ministry of Health, the German association Aid to Chernobyl (Tschernobyl-Hilfe DVTH, Professor Lengfelder), in its comments on my article "The Minister of Health in the Republic of Belarus is covering up the truth about the consequences of the Chernobyl disaster on the health of children", uses false information about me personally and about my work at the Institute of Atomic Energy at the Academy of Sciences, as well as our collaboration with medical practitioners from Belarus, and in effect is lending support to the illegal actions of the Ministry of Health against our institute.
>
> I hope that Mrs Frenzel, as befits any honest individual, will retract the false information that she has published. She will have to answer to the law.

Mrs Frenzel, the author of the slander dated 29th August 2000 retracted nothing. Who were these German "benefactors"? Why, and in whose interest, were they acting in this way?

Chapter VI
RADIOPROTECTION SLANDERED

Professor Edmund Lengfelder is the director of the very wealthy Otto Hug institute in Munich and president of DVTH (Deutscher Verbande für Tschernobyl-Hilfe), and Mrs Frenzel is the vice-president. In an email to ZDF, written on 28th December 2004, Professor Lengfelder describes the institute as "a federation of 70 organisations and corporations that since 1990 have invested a total of 75 million euros in medical and humanitarian aid projects for the population around Chernobyl in Belarus. After visiting Belarus more than 150 times, I know the country and its political and social situation well".

We do not know where the institute gets its funding, but it is able to provide a constant supply of money that far outweighs the amount gathered together by the other German NGO's—barely enough to cover the low salaries paid to the staff of the 16 remaining LRMCs set up by Nesterenko in the most contaminated villages in Belarus. No less surprising is the aggression shown by this organisation towards Nesterenko and Bandazhevsky. They have no hesitation in slandering and vilifying these two independent scientists, whose only crime is to criticise the Belarusian Ministry of Health's radioprotection policy for its servile attitude towards the negationist policy of the lobby in Vienna..

1. SLANDER

Mrs Frenzel's report is dishonest and creates an entirely false image of Professor Nesterenko, both personally and in his professional life. Among other things, she reproaches him for his timidity, or even ambiguity, towards the nuclear lobby, whereas, she claims, the Otto Hug Institute is quite open in its criticisms.

> In his article, Professor Nesterenko criticises the registers, the directives and estimates made by the Minister of Health in Belarus concerning the radioactive load in the population. [...] However Professor Nesterenko does not explain in his article the main reason behind this attitude. He should have known that when government authorities evaluate health risks, they rely on the recommendations of Western organisations such as the IAEA, Euratom, the US Department of Energy and the United Nations. Before you criticise Belarus for its attitude in the face of these radiation problems,

you should first of all direct your criticisms openly at the attitude of the international pro-nuclear organisations mentioned above.

After this little lesson in consistency, Mrs Frenzel gives some information about Nesterenko that bears no relation to the truth.

Nesterenko's attitude to the policies adopted by Belarus on the consequences of the Chernobyl disaster and the part he has played within the situation are problematical from various points of view. Professor Nesterenko is a physicist who was the director of the centre for research and atomic production at Sosny, near Minsk, during the Soviet era. Following certain incidents at the centre, he was sacked. This was years before the Chernobyl disaster. He then set up his company Belrad, which contained the word institute in its title. His company produced radiometers. After Chernobyl, he spent many years on the measurement of radioactivity in food, amongst other things, on behalf of the State. In the meantime, the authorities in Belarus began to do their own measurements, and Professor Nesterenko received practically no official contracts.

This text might have been dictated by the Minister of Health in Belarus in its campaign against Belrad, and Nesterenko replied in a long letter addressed to Mrs Frenzel, after he found what she had written on the official site *www.chernobyl.info*. He began by expressing his surprise at finding these falsehoods in a document produced by an institute directed by a professor at the University of Munich, then replied in detail to the two points cited above.

CONSISTENCY TOWARDS THE LOBBY

My criticisms of the Minister of Health do not date from yesterday but from 30th April 1986, when the Ministry, supported by the Moscow Professor L. Ilyin ignored my request that children should be given, as a matter of urgency, preventive iodine tablets and should be evacuated immediately from the Southern areas of Belarus.

In 1989, six doctors from the Ministry of Health in Belarus joined a group of 92 doctors from the USSR and signed a letter to President Gorbachev stating that there were no health consequences for the inhabitants of Belarus from the Chernobyl disaster, and accusing scientists from the Academy of Sciences in Belarus of incompetence and radiophobia. From 1991 to 1994, I presided over the Joint Committee of Experts (JCE) which contradicted the conclusion of the International Chernobyl Project (1991) and showed that the Chernobyl disaster had serious negative consequences for the health of the inhabitants.

The scientists from the Academy of Sciences of Belarus also stood out against the proposition of 35 rem over a lifetime. For their part, the doctors from the Belarus Ministry of Health solicited the support of experts from the IAEA, the WHO and the Russian Ministry of Health, to defend this inhuman concept.

I collaborate very closely with doctors working in the field and have nothing in common with the doctors working within the structures of government, who seem to have forgotten the Hippocratic oath and lost all real contact a long time ago, with medical practice and with the children who live around Chernobyl.

SACKED FOLLOWING "CERTAIN INCIDENTS"

I am surprised and sorry that Mrs Frenzel uses false information about my own and Belrad's activities. Quite obviously she has chosen to speak to unscrupulous informants from the Ministry of Health who have given her inaccurate information about my career and about the Institute.

I feel sorry for her because she will have to retract the false information she has published, as befits any honest individual.

The National Academy of Sciences of Belarus (NASB) issued an official statement regarding my professional life.

When the accident at Chernobyl happened, I was the director of the Institute for Nuclear Energy of the National Academy of Sciences of the Soviet Socialist Republic of Belarus (from July 1977 to July 1987) and straight away, I directed all my staff at the Institute towards the study of the radiological situation in Belarus. In May and June of 1986, maps showing the contamination of the areas of Gomel and Mogilev were completed by the Institute under my direction.

From July 1987 to October 1990, I worked as director of the laboratory for radioprotection at the Institute of Nuclear Energy at the Academy of Sciences of Belarus.

In 1990, I was appointed director of the Centre for Scientific and Technical Radioprotection (CST "Radiometry") and I left the nuclear energy institute of my own volition.

In January 1992, following a proposal from the physicist, A. Sakharov, the Belarusian writer, Ales Adamovich and the chess champion, Anatoli Karpov, the CST "Radiometry" was reorganised and became the CST Institute of Radiation Safety Belrad.

2. THE WAR AGAINST PECTIN

For years, at conferences, seminars and during commissions held in Germany, Professor Lengfelder has claims in a peremptory manner, without providing scientific evidence to substantiate these claims, that the pectin distributed by Nesterenko is ineffective as an adsorbent of radionuclides, and dangerous because it eliminates from the body oligoelements necessary for life such as selenium, copper, zinc, magnesium etc. But the effectiveness of pectin in eliminating radionuclides from the body is proven and it has none of the secondary effects described by Lengfelder. It is an evil form of disinformation, that has succeeded in blocking funding from various charitable organisations that might have supported Professor Nesterenko's radioprotection work with children. Professor Lengfelder suggests instead that children should eat apples and avoid contaminated food. Derision, contempt or ignorance on the part of this man who boasts about having visited the country 150 times? It costs a lot of money to buy "clean" food—money that neither the Belarusian state nor the peasants in the contaminated villages can afford. As for eating apples, they would need to eat 4 kg a day to achieve the same dose as that contained in Nesterenko's food additive. In her comments Mrs Frenzel repeats the same untruths as Professor Lengfelder.

> Professor Nesterenko has never provided any scientific basis to prove that his pectin cure achieves a noticeable decrease in the annual radioactive dose among the children. This comes as no surprise to most scientists who know that the biochemistry and medical properties of pectin have been studied for a long time. No "decorporation" effects have ever been attributable to pectin.
>
> One of the biggest manufacturers of pectin in Germany, the firm Herbstreith and Fox, undertook a large study with the Institute of Biophysics in Moscow (under Professor Ilyin,—*author's note.*) to see if pectin could be used for the decorporation of radioactive caesium: the results were negative. Russia was very interested in the subject because there were always cases of radionuclide incorporation from their nuclear installations not only among their workers but also among the population. They were looking for a product that could be taken for relatively long periods of time without causing negative side effects. Pectin was not found to be effective.

What audacity from someone who claims to be a scientist to publish such gross disinformation. But Professor Lengfelder has a lot of influence in medical circles in Germany, and up till now, things had gone his way. There must be some powerful interests behind this absurd war on pectin, for them to leave themselves open to such an obvious refutation to which I now refer. On 12th February 2003, the directors in charge of medico-biological problems and emergencies at the Ministry of Health of the Russian Federation distributed the following official document:

> *To heads of central health and medical units and local health and medical units, to Chief Medical Officers of the Government Central Health Monitoring and Epidemiology and to directors of companies:*

> I am sending a document to you entitled "Methodological recommendations for the use of the pectin Zosterine-Ultra as a mass prophylactic in the nuclear industry and other industrial enterprises working with radioactive substances, heavy and polyvalent metals, as well as in territories contaminated by radioactive and other poisonous substances". [...]
>
> Medico-biological tests and other experimental studies of the product, as well as clinical tests and observations of patients from the centre of research Institute of Biophysics (Professor Ilyin N.D.A.) and in other clinical centres have demonstrated this product's effectiveness in eliminating from the body the toxic components of lead,

mercury, cadmium, zinc, manganese and other heavy metals as well as radionuclides including plutonium.

Zosterine-Ultra is effective in the treatment of a number of illnesses, is not toxic, is perfectly tolerated by patients and has no contra indications; it can be kept for five years. The product is registered with the Minister of Public Health of the Federation of Russia as a biologically active food additive (Registration Certificate No 004963. P643.11.2002)

Deputy-director A.V. Sorokin

The recommendations specify:

Over the last few years and in particular after the Chernobyl disaster, great emphasis is being placed on the development and practical application of specialised methods to prevent chemical and radiological damage, lowering accumulation levels and the concentration of toxic substances in the body, and reinforcing the body's own defence mechanisms[…].

Among food additives, fibres, especially pectins—of citrus fruits, apples and of Zosterine—are being used, that possess remarkable adsorbent and fortifying properties and a positive influence on carbo-load and cholesterol metabolism.

Since the 1960s it has been shown that pectins aid the elimination of radionuclides such as strontium and caesium from the body, and that they lower the absorption of lead. This has led to their use as a mass prophylactic measure by professionals and the public living in areas contaminated by the fallout from Chernobyl. These properties of pectin have been confirmed in work undertaken by scientists at the State Scientific Centre (GNC), the Institute of Biophysics and other research centres. [..]

Zosterine-Ultra was approved as a therapeutic and prophylactic food additive by various medical research institutes, hospitals and clinics, including the State Scientific Centre, Institute of Biophysics, the Institute of Research of the Academy of Medical Science of Russia, the Kirov Academy of Military Medicine, the Institute of Toxicology at the Ministry for Public Health in Russia, the Academy of Continuing Medical Education (St Petersburg).

We therefore recommend the food additive Zosterine-Ultra for its curative and prophylactic properties for:

—people employed in the nuclear industry and in other industries that use radioactive materials, and in areas that have been contaminated by radioactive or other toxic substances, to prevent the accumulation in various organs of the body, of radionuclides or heavy metals and to decrease the likelihood of undesirable clinical effects;

—the general treatment of various illnesses contracted in the workplace.

In order to get a clearer picture of who Edmund Lengfelder was, I approached Professor E.P. Demidchik, a thyroid specialist in Belarus, who has already operated in 10,000 cases of thyroid cancer. Lengfelder had collaborated with him, and provided equipment and freezers for the conservation and morphological analysis of thyroid tissue, which seemed to be of great interest to him.

According to Professor Demidchik, Lengfelder, who has significant funds at his disposal, undertakes no medical work in Belarus, but instead collects data on thyroid disease, in exactly the same way as the Americans do, as described by Sebastian Pflugbeil in an article that I quote further on. The equipment that Lengfelder provided to Demidchik was not for medical use. On the other hand Professor Lengfelder also finances radiotherapy in Gomel. In this sense, he is helping, indirectly, the victims of the Chernobyl disaster, if only the victims of thyroid cancer.

In November 1999, following a series of conferences in Germany, during which Professor Nesterenko presented his data, Dr Michel Fernex summed up the current situation in Belarus:

> The disaster is of unequalled proportion, in its human and social impacts, as well as in health and the economy. There is a progressive deterioration in health, a highly significant increase in birth deformities and an increase that has never been encountered before in mutations of children born to parents living in the contaminated zone (250 km from the nuclear power station). The illnesses being described by people living in the region are being reproduced in animals at the Institute of Pathology under Professor Bandazhevsky. There is almost no organ of the body that is not damaged by incorporated radionuclides, which often concentrates in certain tissues.

In the face of this tragedy, one has to wonder at the underlying motives of Mrs Frenzel and Professor Lengfelder, to have waged this lengthy campaign of disinformation, the main result of which is to put obstacles in the way of a preventative treatment for children.

On 12th May 2000, during the dispute between the Ministry of Health and Nesterenko, a meeting was held with Astapov, the Minister for Emergency Situations, one of Nesterenko's supporters. A number of representatives from the Ministry of Health and from the National Commission for Radioprotection of Belarus were at the meeting. All were opposed to Nesterenko apart from Astapov and Dr Kolomietz, who was not able to speak and left the meeting angrily. Professsor Kenigsberg quoted Lengfelder and the IAEA in his criticisms of Nesterenko. He referred to a "scientific" discussion he had had with Lengfelder, who had told him again that pectin was ineffective and could damage health. Kenigsberg even repeated the same contemptuous phrase uttered by Lengfelder to describe Nesterenko's thesis—in Russian "profanatsia", a " parody" or "fake science". Kenisberg concluded by saying:" I propose that we ban the use of pectin". He quoted the IAEA and the ICPR to support 1mSv/year as the limit above which children must be given radioprotection, instead of the 0.3mSv/year recommended by Nesterenko. Following this meeting, Minister

Astapov sought advice from five international scientists, who supported Nesterenko against the Minister.

On 7th February 2002, during a conference in Germany, Nesterenko met Professor Hans Ulrich Endress, director of Herbstreith and Fox, who had undertaken "the large study with the Institute of Biophysics in Moscow" (of Professor Ilyin, *author's note*.) that according to Mrs Frenzel, had shown pectin to be ineffective in the decorporation of radioactive caesium. Endress was very interested in the work being done by Belrad. When he was asked "Is it true, as Professor Lengfelder claims, that pectin does not adsorb radionuclides?" Endress said that the study had not looked at radionuclides. They had only been looking at the effect of pectin on *heavy metals*. So, Lengfelder and Frenzel are not telling the truth. Despite the obvious falseness of their thesis, they persist with it, doubtless relying on the effectiveness of the repression in Belarus and on the ignorance of the international scientific community, dominated as it is by the nuclear lobby. Nesterenko is a lone voice and he is powerless against the rich medical *establishment* in the West.

In the spring of 2005, the two Germans contributed to blocking the funding from the European Commission (the TACIS aid programme) for the project "Reducing the dose in the population…" presented jointly by the foundation Maison de Belrad, the Association Les Enfants de Tchernobyl Belarus[84] and the centre for nuclear research, Julich (Germany). V. Nesterenko told us:

> At the beginning of March, TACIS told us that our project had been approved. TACIS projects need to be approved by the European programme CORE[85], in order to be free from the main presidential taxes.
> Initially there had been some negative response from Belarusian scientists, but on 19th April 2005 our project was discussed again at a meeting of the decision board in Braguine (I was at the meeting and presented the project). Professor Lengfelder's representative, Mrs Christine Frenzel[86], spoke against distributing pectin to children. Dr Dimitri Mikhnyuk (head of the Chernobyl medical programme for technical aid

---

[84] On 27th April 2001, five people met in Paris for a constitutive general assembly and set up, at Nesterenko's request, the association (under French Law 1901) Enfants de Tchernobyl Belarus to support the Belrad Institute in its aid work with 500,000 contaminated children: Solange Fernex (President), Vassili Nesterenko (vice-President), Galia Ackerman and Wladimir Tchertkoff (Secretaries), Michel Fernex (Treasurer). In accordance with Belarusian law concerning humanitarian aid, this association, which must have one or more Belarusian citizens among its members, facilitates the transfer of funds for the preventive protection and medical treatment of children *(http://enfants-tchernobyl-belarus.org).* In 3.12.2014, the Association was recognized as an establishment of public utility.

[85] European programme that was being set up in Belarus. See Part Four, Chapter III, p. 277.

[86] CORE members.

from Switzerland) actively opposed her and was in favour of our proposal. At that meeting our proposal was accepted.

The next day at a meeting of the Approval Board of CORE which is attended only by the directors of the organisations involved (I was not present, because at this level only the directors of partner organisations of CORE participate) , Lengfelder came out categorically against the use of pectin. He said that in 1985, (before Chernobyl) he had studied the effectiveness of pectin in the elimination of Cs 137 from the body, in co-operation with the Institute of Biophysics (under Professor Ilyin), and had found no effect.

The French and German ambassadors, Stephane Chmelewsky and Martin Hecker, had come out against Lengfelder and the Ministry of Health, and suggested that a multi disciplinary study be conducted on the use of pectin in the radioprotection of children. I came to know about this through Jacques Lochard (CEPN) and Jean-Claude Autret (ACRO) who were at the meeting. Following this proposition from the ambassadors, the decision on our project was adjourned till the autumn when it would be put before a panel of experts.

On 20th June, Solange Fernex writes to Stephane Chmelewsky, the ambassador of France in Minsk. After enumerating the slanders against Professor Nesterenko and Professor Bandazhevsky, and the numerous obstacles put in their way by Professor Edmund Lengfelder, a highly respected German figure, she wrote:

As president of an association that uses funds donated by hundreds of our members to finance anthropogammametric and radiometric measures as well as pectin cures, I cannot accept these gratuitous statements from Professor Lengfelder against the effectiveness of pectin. If he were right, our donors would be justified in asking us to account for our misuse of donations.

Yet, to my knowledge, Professor Lengfelder has not provided any scientific basis for the assertions he makes against pectin. For the first time, on 20th April, he has mentioned research that he undertook, before Chernobyl, with Professor Ilyin of Moscow. As you know, Ilyin is the principal architect of Soviet disinformation about Chernobyl. Our association would like to know what scientific proof there is against the use of pectin. If none exists, then we are left with a gratuitous slander that we will not tolerate.

You have the right, Ambassador, as a participant in the CORE project, to demand that Professor Lengfelder back up his assertions and provide scientific references confirming those assertions at the CORE meeting due to take place on 20th April 2005. From our side, we base our view on the research published by Belrad with Julich and on the double blind study of pectin *versus* a placebo and published in English in the Readers Committee Review of the publication Swiss Medical Weekly.

Perhaps you and your colleagues, in particular the Ambassador of Germany, could explain to us who Professor Lengfelder is, and why he is spending so much time attacking a fellow professor, who is defenceless, imprisoned, has been removed from his post and relegated to the status of a criminal (Professor Bandazhevsky). What could motivate him to crush the Belrad Institute, and for no reason, condemn its work, and in particular its pectin cures?

I thank you most sincerely, Ambassador, for helping me to clarify this matter which is causing great anxiety among our donors who want only to continue helping the children of Chernobyl in an effective and useful way. This question does not simply affect Bandazhevsky, Nesterenko and Belrad, but our donors in France who need to know the truth.

Solange Fernex
Honorary European Deputy
President of Enfants de Tchernobyl Belarus

On 13th July, the ambassador replies that the French authorities, on the one hand, and the European authorities on the other, fully supported the cause of Yury Bandazhevsky, whom he has visited several times with his German colleague. On the subject of Lengfelder, he writes:

> As regards pectin, this controversial subject came up at a meeting of the Approval Board of the programme CORE. While making no judgement about its effectiveness, I can only say that its use is not unanimously approved, particularly because, to date, no research into the subject has succeeded in bringing together all those involved in the area of radioprotection.
>
> These disagreements were preventing the implementation of various projects including Professor Nesterenko's, and so, during the meeting, I proposed that a multidisciplinary evaluation of pectin be undertaken, involving supporters and detractors of the product. The objective was to obtain definitive conclusions on its effectiveness and conditions of use, conclusions which could not then be contested by its supporters or detractors. It was decided that the IRSN should organise a pilot study and in June, they had already contacted Nesterenko about it. The latter was naturally going to take part in the study, which would begin after the summer.
>
> Lengfelder does indeed seem to have reservations about Professor Bandazhevsky, and about the use of pectin. However, Bandazhevsky's fate is not in the hands of the German Professor and far more people are working for his release and for the renewal of his scientific work. The conclusions of the multidisciplinary evaluation of pectin will bring this controversy to an end.

During our stay in Belarus in 2000, I asked Nesterenko if he thought Western scientists had a clear view of these questions and were acting in good faith?

V. Nesterenko.—There is a total lack of understanding about our situation in the international scientific community. Let me use Lengfelder as an example. Here is what he wrote in response to my proposition to distribute pectin-based food additives. "They would be better off eating fresh apples. In general it would be better if families were brought into the schools and it was explained to them that it is dangerous to eat contaminated produce. If they were not eating contaminated produce, they would not need to be given pectin, which in any case, would cost more money". I told Lengfelder that if the peasants received the same monthly salary as he did, they and their families would be able to eat clean food. But since their salary is not 10,000 marks, nor 1,000, nor 100, but only 30 dollars, it's obvious that they are going to be eating what they grow in their gardens and what they can gather from the forest. The international community needs to realise this. The countries that use nuclear energy should set up an insurance fund to help these people. It would not be a charity, I consider it a duty of the international community to help the people here survive in these conditions.

Q.—But what explains this hostility to pectin?

—It was Lengfelder who suddenly began to oppose its use, and I've no idea why. Personally, I think, and it may sound a bit harsh and brutal, that they are interested in conducting an experiment. This is how these pseudo-scientists work: they make observations, they diagnose, and they record the number of illnesses. But they do nothing to reduce these illnesses. I think our work with pectin prevents the collection of pure scientific data.

It's true that Lengfelder set up a laboratory here in Minsk and another in Gomel. But all they do is diagnosis. They do nothing to tackle the real cause of the illnesses—the contamination. Let's be clear, I don't think that pectin is a universal panacea. But it's an effective product, which works, given that the population is not going to be evacuated from the contaminated territories. We have seen Bandazhevsky's work and we are convinced that children must not have more than 50 Bq/kg in their body. Yesterday we were measuring children: all had levels above this. Every child had a range of different illnesses, affecting different organs of the body. These are all children born after Chernobyl. It means that if we do not decontaminate them, a whole generation will grow up ill.

—But why this war on pectin?

I kept asking, even though Nesterenko had already answered me. The terrifying explanation, that Svetlana Savrasova had hardly dared believe, came back to me. The human guinea pigs needed to be experimentally pure, unchanged. Hannah Arendt, describing Eichman, talked of the banality of evil: stupid, bereft of intelligence, of moral conscience and emotion, as flat and calculated as a train timetable, and it seems to be the guiding force now around Chernobyl. Because the petty evil men who carry out their orders in the nuclear gulag, are covered by the official scientific truth sanctioned by the highest political authorities in the world. Humanity caught in its own trap. Not one government dissenting.

V. Nesterenko.—You want to know why? Because if pectin is administered three or four times a year, it really can lower, by a factor of two or three, the annual concentration of radionuclides in the child, in other words they will be less ill. Our food is contaminated. I think, I hope, that if France or Germany had the misfortune of experiencing such an accident, contaminated produce would be banned and everyone would eat clean food. But here, the government can't provide it, and the people can't afford to buy it. They eat what they grow.

I wrote to the Minister of Health. I can understand why he is against it. But their job was to guarantee the safety of the inhabitants. They should have organised the production of pectin. Tackled the problem and found ways to obtain it. And they should have done it in 1986. It's 2000, fourteen years have gone by[87].

Deputy Minister Orekhovsky told me: "Pectin does nothing. It only eliminates heavy metals. As regards radionuclides, no-one knows. It's better to eat apples".

I think they didn't want to recognise that there had been mass contamination. We encountered enormous opposition to the use of Human Radiation Spectrometers (HRS). But thanks to the international appeal that we launched, we received the help we needed. Ireland is a small country but, following Adi Roche's initiative, it supplied us with 5 mini buses and spectrometers. Germany gave money too. America also. But it was only NGO's who helped us. It allowed us to measure 50,000 children. We have already distributed pectin to 15,000 children. With the financial support of the Assistance Committee "Children of Chernobyl" ELFI and Walter Meusburger (Austria), we measured 1000 children last year, and gave them pectin for a year. The train has left the station...Lengfelder is too late to stop it.

That was in 2000. At the beginning of 2006, when I was finalising the French edition of this book, the picture was not as rosy as Nesterenko hoped when we had that conversation. Lengfelder continued his attacks against the use of pectin and finally got his way. Following an initiative from the German Green Party deputy, Gila Altman, who, in 2003, was the secretary of state at the Ministry of Environment and of Nuclear Safety, the German government provided funding for a three-stage radioprotection project for the children in Belarus, to be jointly undertaken between 2002 and 2004, by the Research Centre "Julich" (Germany) and the Institute "Belrad". The third stage, which consisted of monitoring the effectiveness of the pectin cure in families following a stay in a sanatorium, was cancelled. According to Gila Altman, Lengfelder threatened to provoke a scandal in German scientific circles if the government gave money for this follow up in the villages. The "train" that we thought had left the station, was moving very slowly with frequent stops. Lack of money was a constant anxiety.

The United States (MacArthur Foundation),[88] whose considerable financial support had provided some long term security for Nesterenko's institute, abandoned

---

[87] 2006, twenty years have gone by... (Author's note)

[88] It is understandable that America disagrees with the Lukashenko regime, but it should not punish the population.

them five years ago. Also ELFI and England. The Swiss Green Cross backed out of a project that had been put together under M.Wiederkehr, who was president at the time, following an attack on the use of pectin by Mrs Frenzel at a symposium at Soleure in 1998.The Italians made huge promises but nothing ever came of them: they preferred to deal with the Belarusian authorities, while the health catastrophe grew worse year on year. An important project with the Julich institute, which was to be financed with 80,000 euros from the German government in the autumn, fell through, apparently because of the failure of the Green Party at the elections. Not to mention the eight or ten projects that had been put forward, after enormous expenditure in time and effort, to various international financial bodies, five of them at the European Commission. All were rejected for reasons that were never satisfactorily explained, or not explained at all... There remained a few faithful organisations from France, Belgium, Germany, Austria, Ireland and a new recruit from Spain, that were able to offer sporadic and limited funds to help with Nesterenko's projects.[89] The salaries that he is able to pay to his forty or so colleagues at the institute are among the lowest in Minsk. Some of his staff had to leave. From one month to the next, Belrad, the independent institute of radioprotection, risks closure from lack of funds. As I write, in February 2006, funding from the NGOs has dried up, signatures on various projects are late in arriving and Nesterenko tells me that he does not know how he is going to pay salaries in March or April...

The article published by Dr Pflugbeil, in the journal *Zeit-Fragen*, No. 10 of 17th March 2003, could explain the real motivations of a seemingly senseless war against pectin waged by some very powerful interests—a war with criminal consequences.

3. HUMAN GUINEA PIGS

---

[89] FRANCE: Enfants de Tchernobyl Belarus—ETB, 65, quai Mayaud F-49400 SAUMUR *etb@enfants-tchernobyl-belarus.org*, *http://enfants-tchernobyl-belarus.org* ; France-Libertés-Fondation Danielle Mitterand, 22 rue de Milan, 75009 Paris, *contact@france-libertes.fr*; Association Solidarité de Biélorussie et de Tchernobyl, 74 rue de Falaise, 14000 Caen, *assbelarus@hotmail.fr*; Les Enfants de Tchernobyl, 1 A, rue de Lorraine, F-68840 PULVERSHEIM, *lesenfantsdetchernobyl@gmail.com*, *www.lesenfantsdetchernobyl.fr*.
  BELGIUM: Association belgo-biélorusse pour les enfants de Tchernobyl (ASBL), 16 rue Marache, 5031 Grand-Leez, *bel.asbl@belgique.com*
  GERMANY: Jugends Aktions Netzwerk Umwelt und Naturchutz e.V. (JANUN) Gr.Barlinge 58a, 30171 Hannover; NIKOBELA, Grosse Drakenburger Strasse, 3, 31582 Nienburg
  AUSTRIA: Tirol hilft den Kinder von Tschernobyl, A-6521 Fliess 111a, Tirol.
  IRELAND: Chernobyl Children's Project, 8 Sidneyville, Bellevue Park, St Luke's, Cork, *adiroche@indigo.ie.*

ARE THE CHILDREN OF CHERNOBYL BEING USED AS GUINEA PIGS BY THE UNITED STATES?

Doctor Sebastian Pflugbeil, president of the Radioprotection Association , Berlin.

Are the authorities in Minsk aware of the background to the United States thirty year diagnostic programme monitoring the Belarusian victims of Chernobyl?[...]

In 1994, the United States Ministry for Energy approached the Ministry of Health in Belarus with a proposal for a joint scientific project, called BelAm, that was to last at least thirty years. The aim of the project was to undertake a long term study to establish how many cases of cancer and other thyroid illnesses would appear among groups of people living in Belarus, who had received different levels of radioactive iodine from the Chernobyl reactor. About 13,000 people were selected, who were to receive regular examinations over the next three decades. [...]

A number of institutes and individuals were chosen in Belarus to co-operate: Highly placed scientists from the Ministry of Health, under the direction of the Minister himself, the Institute of Radiological Medicine and Endocrinology attached to it, its affiliates in Minsk and Gomel, the specialist dispensary in Gomel and other clinics. On the American side, the partner organisation was the National Cancer Institute of the United States government. Financial arrangements were made and the contract signed. The project has already been going for six years. [...]

According to the agreement, the BelAm project is only concerned with the diagnosis, screening and epidemiology of thyroid disease. Once the cancer has been discovered, America is no longer interested: it becomes Belarus's problem. It's hard to believe, but Belarus on behalf of its own patients, never asked the United States to offer appropriate treatment or even finance for that treatment, for the thyroid disease that was discovered, as an integral part of the project. The people of Belarus are quite simply the guinea pigs in an enormous radiological laboratory, from which the United States are gathering their scientific data. [...]

On 1st April 1996, after a long inquiry, British television showed the film "Chernobyl Ten Years After", in which it is shown that of all the countries taking part in research around Chernobyl, it is the United States that is most vigorously opposed to the establishment of a correlation between radioactive iodine and thyroid cancer. "The American government has particular reasons for its defensiveness. In the 1950s, the United States Ministry of Energy deliberately released a cloud of radioactive iodine into the atmosphere in order to verify whether it was possible to observe the trace of a radioactive cloud. This increased the contamination that had already been caused by the nuclear weapons testing. [...]

Today, the victims are demanding compensation from the government for damages. The government is contesting these claims based on the grounds that there is insufficient evidence linking radioactive iodine with an increase in thyroid cancer.

The victims in the United States argue that the dramatic increase in thyroid cancer after Chernobyl constitutes the definitive proof of the correlation between the damaging effect of iodine liberated into the atmosphere. Scientific experts from the American government—the same people with whom Belarus is collaborating on the BelAm project—are contesting this argument also and say that long term research needs to be done within the CIS before an opinion can be formed with any certainty. Under current American legislation, the government in Washington expects compensation to run into billions of dollars, whereas the BelAm project is costing them 1% of that figure (10 million dollars)! In all likelihood, this is a political and legal stratagem devised by the US, in which the BelAm project (UkrAm in Ukraine) will unfold over decades in order to achieve some "clarification of the issue". How many of the American victims will still be alive in thirty years to claim their rights when the research is finally completed after 30 years? It becomes very clear then, that the United States' principal interest is to have absolute control over the science, the data and the information concerning thyroid cancer that has resulted from Chernobyl. Why does Belarus lend a hand in these machinations?

From the point of view of medical ethics in the United States and in Europe, scientific research on patients who are ill is not admissible unless, right from the outset, the treatment of the patient is guaranteed. Why should this be different, just because United States research is being conducted in Belarus? [...]

Western countries which depend on nuclear power, would like to establish, with the aid of another scientific project in Belarus, absolute control over the medical information on thyroid cancer. The United States, Japan and the European Community Commission—a kind of nuclear alliance—have imposed on the Ministry of Health in the Republic of Belarus (and the Ministries of Health in the Ukraine and Russia) a project for the creation of a tissue bank and data about thyroid cancer following Chernobyl. The West, with a majority of representatives, decides which scientific group in the world will be authorised to conduct research on tissue taken during operations, in order to avoid "cross publications" [...] and to reach "successful" conclusions, as it states in the project description.

The promoters of the project would be concerned, above all, with the correlation between thyroid cancer following Chernobyl and not only radioactive iodine but also with the genetic component of the disease and with environmental pollution that might cause cancer[90], so that in the final analysis, it might be the faulty genes of people in Belarus, or the polluted environment, and not radioactive iodine that are to blame for the dramatic increase in thyroid cancer. Although no contract has been signed with Belarus, the project has been going on already for several years, with agreement from Minsk. And once again, the alliance is only interested in scientific data, while patients in Belarus are quite free to look after themselves.

---

[90] Unfortunately, the author ignores one essential radiological factor that is jointly responsible for a number of pathologies at Chernobyl: the chronic accumulation of caesium especially in the thyroid gland. It is not only Bandazhevsky's work that has proved that, particularly in young children, caesium accumulates more in this gland than in most of the organs or tissues, but also, that isotopes of caesium were used, in scanners, in Western Europe, in the 1960's to locate nodules in the thyroid. As a general rule, the IAEA and experts would prefer us to forget the role of Cs-137 as a cause of illness, in particular in paediatrics. The FDA (Food and Drug Administration), on the other hand, is extremely worried about Cs-137 and has called on pharmaceutical firms to develop and register products to protect people and the army in case of an accident like Chernobyl, or following the use of a dirty bomb that might release Cs-137.

But let us now return to the year 2000 and take up the story of what is on the point of disappearing.

V. Nesterenko.—In September 1998, we went to the village of Sivitsa near Minsk. The children had levels of caesium-137 measuring 340 Bq/kg. We had distributed pectin four times over the year. We had taken measurements before and after each three week cure. Now all the children had levels below 40 Bq/kg. During the year, we had succeeded in lowering their levels by a factor of 8. In another village, Polessie, situated in the Chernobyl zone, where conditions were even worse, we had succeeded in halving the levels of contamination. Now we had irrefutable proof. The really important fact was that doctors from Bandazhevsky's institute were examining the children in parallel to our measurements. They examined electrocardiograms, the state of the immune system, and they had shown that the product acted as an antioxidant, eliminating allergens from the system, eliminating heavy metals, like lead. People say: "It's not possible that a simple apple can do that". But we all know the saying "An apple a day keeps the doctor away". Because pectin really is a very unique natural substance. The way it works is very interesting. The pectin that you take is not digested in the stomach. It swells and passes into the digestive tract, in the large intestine, where the food is transformed into amino acids. Soluble radioactive caesium (Cs-137) is present in this liquid. Pectin cannot be assimilated into the organism. It fixes the caesium and eliminates it through the digestive tract.

There is another mechanism that is even more important. We all have deposits—that's what they're called. It is in the muscles, in tissues, that caesium forms a sediment. Caesium is the exact chemical equivalent of potassium. We know that the heart muscle contains 20 times more potassium than any other muscle in the body. It also contains much more caesium, given that it is the chemical equivalent of potassium. The more active the muscle, the more radioactive caesium it will accumulate. Pectin powder is very fine, 50–70 microns. It can pass through vessel walls and gets into the blood. It penetrates these deposits, capturing the caesium and eliminates it through the kidneys. These two mechanisms have been studied in clinical experiments in Russia and in Ukraine at Krivoi Rog[91]. We no longer have any doubt about the capacity of this product to adsorb lead, cadmium, mercury and radionuclides. But it has no effect on the oligoelements that the metabolism

---

[91] Otchet a naoutchno-issledovatelskoi raboté po klinitcheskim ispyta-niam pektinosoderjachégo préparata iz iablotchnogo chrota "Yablopekt" XD.14.02.002.97, (Report on the scientific research and clinical tests on the apple pectin-based product "Yablopekt" XD.14.02.002.97), Research Institute Oukrpommed, Krivoy Rog (Ukraine), 1997.

needs: copper, zinc, iron, manganese, selenium, potassium etc. Thank God, and his creation.

Dr Michel Fernex informs us further—The effectiveness of pectin as a food additive resides in the fact that it blocks the absorption of radioactive caesium and strontium in an animal that is eating radioactively contaminated food (as shown by Korzun[92] in Kiev). In humans, pectin is also effective in patients who are contaminated but are eating radiologically clean food. This is explained mainly by the fixing of caesium (that is excreted by contaminated patients) in the bile, or in the intestinal lumen. In the absence of effective adsorbents, Cs-137 is reabsorbed into the small intestine and continues to cause damage to the noble organs (endocrine glands, heart and thymus). Adsorbed by oral pectin, the stock of caesium in the body is reduced, radiotoxicity decreases or disappears altogether.

This was demonstrated by Nesterenko and his colleagues in a randomised double-blind, placebo-controlled study of a homogenous group of contaminated children from villages close to Gomel. The children gave their verbal consent after a detailed explanation during a holiday in a sanatorium; their mothers had agreed in a letter. An ethical committee duly monitored and approved the experiment. The results were published in Swiss Medical Weekly and showed that eating clean food for three weeks had reduced the load of caesium-137 in the body by 14%, but levels never fell below 20 Bq/kg, while in the group who ate clean food and also had a teaspoon of pectin before each meal, the burden of Cs 137 was reduced by 63%, and all children had levels below 20 Bq/kg.

A complementary clinical study was undertaken by Nesterenko and Galina Bandazhevskaya, who is a cardiologist, and was the main author of the publication that followed. The real interest of this study was the statistically proven correlation between the Cs-137 load and the symptoms and anomalies found in ECGs. Pectin cures reduce this load, but they also significantly reduce certain clinical symptoms. This study was also published in the Swiss Medical Weekly, a journal that is renowned for the rigour of its peer review. These two studies can be accessed on the site *www.smw.ch*.

The work undertaken jointly with the German research institute Julich confirmed not only the effectiveness of Belrad's pectin product but also how well it is tolerated. Emphasis was placed on the stability of microelements in blood plasma, as it had been in a previous study, undertaken by Nika Gres and colleagues in Belarus, that produced similarly positive results.

---

[92] "Nutrition problems under wide-scale nuclear accident conditions and its consequences", *International Journal of Radiation Medecine*, 1999, 2, (2) : 75–91.

4. THE GERMAN PROFESSOR DENIGRATES BANDAZHEVSKY

On 6th December 1999, Bandazhevsky is still in prison, awaiting trial.[93] A petition has been sent to President Lukashenko from IPPNW Germany defending him. Professor Lengfelder, an influential member of IPPNW, sends a letter to them. He condemns this initiative, describing it as superficial, badly timed, and claims it was inspired by preconceived notions about post-Soviet Belarus. According to Lengfelder, the promoters of the petition, Professor Fernex (President of IPPNW Switzerland at the time) and Gottstein (President of IPPNW Germany at the time), knew nothing about Professor Bandazhevsky, his scientific work, his contacts with the faculty of medicine at Gomel, nor the underlying reasons (*Hintergrunde*) that led to his trial. Questions needed to be answered: how could this young man, at the age of only 33, during the Soviet epoch, have become professor and rector of a newly founded medical institute at Gomel? "In my view, the idea that it is his anti-nuclear views that have led him to be incarcerated is not realistic", he wrote. And he went further. The complaint lodged against Bandazhevsky by the State Prosecutor of the Republic said that "…during the selection process for students—the number of candidates being much higher than the places available—his faculty and its directors had accepted payment to take on extra students. The parents, some of whom (*etliche*) I know personally did not make a fuss, and were only too happy to pay a few hundred or a few thousand dollars, for their child to study there". One cannot help wondering if Professor Lengfelder—the only person who did not retract his statement—was the perfect witness that the military tribunal had searched for so desperately at the time, and never found.

Is it possible that the "wise men" from the West knew before the trial had even taken place, what accusations would be made against these researchers with too much scientific data at their disposal? One example. Lengfelder, at the canteen during a medical congress, revealed that Professor Okeanov had been guilty of misuse of funds. This accusation, made at the dinner table, came as a surprise to Professor Fernex[94]. Okeanov lost his post as director of his own institute after an epidemic of cancers among the liquidators and the inhabitants of contaminated areas in the south of Belarus[95]. These "experts" from the West like to pass themselves off as anti-nuclear in their own countries. They play a dual role: providing financial

---

[93] Y. Bandazhevsky was in prison from 13th July to 27th December 1999. He was under house arrest until his trial on 19th February 2001.

[94] M. Fernex was no less surprised to learn that Professor Lengfelder had told various colleagues that Dr Fernex was either addicted to drugs or to alcohol. Professor Lengfelder's vocation is quite clearly as a slanderer.

[95] Cf. "La catastrophe de Tchernobyl et la santé" (The Chernobyl disaster and health), by Michel Fernex

manna for the institutions for whom they work and conducting a kind of scientific tourism in which they glean data collected by others. It is a great publishing opportunity. Their career is assured. Many a scientist has been tempted in this way.

At the end of his letter, Professor Lengfelder arrogantly dictates his terms to IPPNW. "In my view, IPPNW's involvement in this affair, and in the campaign to gather signatures, is of such importance that I demand a formal binding declaration from the office (*volle Haftung*—full liability) as I propose here (contradicting the one that was foreseen):

—IPPNW will take no action, and make no comment about the proposal for a petition, without previous agreement from Professor Lengfelder. [...]" It would be funny if it were not so contemptible.

In autumn 2000, a German journalist, Alexandra Cavelius, wanted to undertake an objective investigation into the two "heretical" scientists in Belarus, using interviews with people, including Lengfelder[96]. She put questions to Bandazhevsky and to Nesterenko, using me as an intermediary. Here is Nesterenko's response which he asked me to pass on to her.

> Yakov Kenigsberg is deputy director of the Institute of Radiological Medicine at the Ministry of Health. I have no dealings with him. His position as regards the radioprotection of children in Belarus and the use of pectin is as inhuman as that of Lengfelder. On 12th May 2000, during a meeting with the Ministry of Emergency Situations, Kenigsberg argued strongly against the use of HRS measurements of children and against using the food additive pectin to protect them. He claimed that he was supported in this view by E. Lengfelder of Germany. All these people belong to the nuclear lobby and cannot be trusted with the protection of our children's health. It is surprising: what objective view of Bandazhevsky, Nesterenko or of our work could the journalist expect from these people?
>
> As regards the mortality figures in 1985 and today in the Gomel region, Y. Bandazhevsky could supply these. As a doctor, he has the official annual statistics of Belarus.

On 1st October 2000, the German journalist sent me a letter in Russian addressed to Bandazhevsky, that I sent on to him in a fax.

> I hope I am not putting you to any trouble but I need to take into account the views of critics from the West in my investigation (Mr Lengfelder and his friends). I am aware that some of their information is not accurate. It is therefore very important to do everything possible in order to refute it. In order to do this, I need more information.

---

[96] Article published on 26th April 2001 in *Sud-Deutschen Zeitung Magazin* entitled "Tod dem Teddybar".

1. There is this allegation: "Mr Bandazhevsky has got where he is today only because of his wife's good relationship with various politicians. His brilliant career, at this young age, owes more to these political connections than to his academic achievements". What can you say about this? Were you, at the time, a Party member (this has no importance to me)?

2. There is another allegation: "Bandazhevsky uses partially erroneous indices in measuring the parameters of radioactivity. He has been criticised for this by scientists in his own country". A second criticism: "His statistical results are false. Poor mathematical work. In this connection, could you tell me the name and the address of your French assistant? Are there other witnesses who are familiar with your original work who could give me a brief assessment? Could you give me their names and addresses (a fax number would be best)? Perhaps there are even some Japanese or American people (in connection with the award "Sun. The Golden Emblem" [97]).

3. Is it true that many of your colleagues have been dismissed as a result of the accusations made against you? If yes, how many are there? Is the person who retracted the false statement the vice-rector? Is this the case?

Y. Bandazhevsky replied on 2nd October 2000.

1. My wife, Galina Sergueievna Bandazhevskaya knew no politicians or government employees at the time that my career was taking off. Her parents were peasants who spent their entire lives in the same village. If you have any written information about relationships she had with politicians at that time, I would be very grateful if you would send it to me so that we can lodge a complaint...

Up until 1991, I undertook scientific research and achieved good results at the Institute of Medicine at Grodno, and there are several specialists at the central scientific research laboratory who know my work very well. During those years, 6 scientific doctorates were prepared under my direction and our team of scientists was awarded the Lenin/Komsomol prize from Belarus, for our work in experimental pathology. My scientific work was published widely in all the major journals in the USSR and is protected by copyright. My area of research was completely new at the time and had not really been studied: almost no-one in the USSR and even fewer in our Republic of Belarus had done research in this area of teratology and of experimental pathology.

In 1990, I was appointed rector of a further education centre in Gomel that did not yet exist, that I had to set up entirely on my own. I had no dealings with anyone else and I relied only and still rely only on my own scientific knowledge and expertise. This claim is borne out by the fact that when I found myself without any professional help at all, I carried out a complex technical experiment—the study of the effects of radioactive caesium on the development of the foetus—in circumstances that were utterly inadequate for such research. Over the last nine months, I have written three books and a series of articles and presentations[98].

I became a member of the Soviet Communist Party in 1988 and I left in 1991.

---

[97] International prize awarded by the United States for scientific work in the area of radiopathology.

[98] It was during the nine months of house arrest while awaiting trial that he undertook this work. He did not prepare himself for the trial because he was so sure that no case would be found against him.

2. In my research, I do not use any parameters based on complicated calculations of radioactivity, in other words I do not calculate the radioactive dose received by people. I rely solely on the quantity of accumulated radionuclides in the body, measured in becquerels.

Where statistical data is used, it is strictly defined according to criteria that have been reliably established in particular according to Student's t-distribution.

The Scientific Councils of the Institute of Stomatology in Moscow, the University, Amitie des Peuples, the University of Medicine in Moscow, the corresponding member of the Academy of Sciences in Russia, Alexei Yablokov and his colleagues, Professor Fernex and his colleagues, are all familiar with my scientific work.

I should point out that most of the scientific information obtained by myself and my colleagues has been presented in the form of doctoral theses that have been defended in various scientific institutes in Russia and in the Republic of Belarus, since most of them include a morphological section and their specimens strongly support our conclusions.

3. In my role as rector, I directed the institute and had to take responsibility for disciplining those who did not fulfil their obligations. Once, I had to dismiss the administrative vice-rector, who had seriously contravened his professional duties. He appealed to be returned to his post, but the tribunal did not find in his favour.

I am surprised that Professor Lengfelder has said these things. A week before my arrest, he visited me in my office and said very complimentary things about my research (he even gave me a film on the role of American scientists in the liquidation of the consequences of Chernobyl).

It is very odd that, after participating, as many scientists from our own country have done, in numerous seminars and symposiums held at our institute, he never contested the scientific information that I presented.

What is more, in 1998, during a meeting at the Academy of Sciences on medico-biological problems, the scientific material that I had presented was approved and a proposal was made to develop this line of research. Similar support was expressed by the eminent scientist John.W. Gofman, who is familiar with my work.

On his return from a conference in Germany, where he had learnt of the campaign being waged against the two Belarusian scientists, Professor Fernex expressed this hope:

The Republic of Belarus will eventually see where its interests lie, in spite of the advice being given from those who arrive from abroad with expensive equipment, or even with money and medicines. This country has its own researchers, and highly qualified doctors. They deserve to be listened to, and helped in the country that they serve so fervently. They also deserve worldwide recognition. Western universities should introduce them at conferences and symposiums : invite them to speak rather than speak for them as happens today; they should be helped to publish their work in the most prestigious scientific journals, even if this entails paying a scientific editor, as American academics do.

From the point of view of medical ethics, it is not acceptable to measure high levels of caesium-137 in food, and therefore a high level within children's bodies, without providing them with a pectin cure (given that these children are not likely to be evacuated permanently). It would be like discovering the Koch bacillus in a child's mucus and not treating the tuberculosis.[99]

[99] M.Fernex, "Pourquoi la pectine de pomme?" (Why apple pectin?), extract from a document sent to the Swiss and French embassies on 30th June and 1st July 2004.

PART FOUR

THE DEMOCRATIC GAOLERS OF THE GULAG

Chapter One
EUROPE'S DUPLICITY

Why was Professor Nesterenko never able to get any financial support from Europe to protect the health of children contaminated by the nuclear industry in Belarus?

The idea for the TEST project "Evaluation of the effectiveness of the food additive Vitapect in the elimination of radionuclides from children's bodies" was born following a proposal made by European deputy Yves Piétrasanta in October 2000, when Nesterenko found himself in Strasbourg, having been invited by Green MEPs to the European parliament.

Piétrasanta wrote to Barry McSweeney, Director-General of the Joint Research Centre of the European Commission (JRC) at Ispra (Italy). McSweeney, who was familiar with the properties of pectin as an adsorbent of heavy metals, was keen to look at its effects on caesium-137. A working group met at Ispra in June 2001, with Nesterenko, Michel Fernex—who had proposed the idea of a prevention campaign with Vitapect in twelve villages in the contaminated territories—and myself. Once the proposal was accepted, Michel Fernex took charge of designing the protocol for the double blind experiment.

We hoped, naively, that the funds would arrive by September 2001, knowing that Belrad would begin to experience financial problems around that time. But the Vitapect sample analysis and the writing up of JRC's *Technical Assessment* were only completed in January 2002.

On 22nd February 2002, Mr Sarigiannis, scientific assistant to the Director-General of JRC, sent us his conclusions and approval with the hope "[...] that our analysis will help to develop a project that will be even more useful to the children of Belarus".

I asked Nesterenko for his comments on the JRC report. He agreed to all the suggestions and modifications put forward by the Ispra experts, adding only a comment on the section in which JRC cites as an example, the work of the French group ETHOS in one of the villages where Belrad's teams were working.

JRC—"The project ETHOS, undertaken in the village of Olmany near Stolin, in the Brest region, is a good example showing how the education of local people about ways to reduce individual risk can be effective in lowering the dose and achieving a certain degree of individual risk control".

Nesterenko's comments.

"We have the following observations to make on the ETHOS Project (and other similar projects)

Too little attention is given in the ETHOS Project to working with the local population to teach them to make practical use of the various radioprotection methods.

Furthermore, for people to make use of the various procedures that would allow them to reduce their annual radioactive load, it is necessary to:

- raise funds so that inhabitants of contaminated regions can fertilise their land with mineral fertiliser and thus reduce the transfer of radionuclides from the soil into the food chain;
- ensure fertilisation of fields and pastures with mineral fertiliser (potassium, phosphorus and calcium) to obtain "clean" milk and meat;
- spread mineral fertiliser (potassium, lignite) on forest soil in a radius of 8–10 kilometres around settlements so that mushrooms and berries are "clean";
- work with the inhabitants to introduce the habit of regular radiological control of the fruit and vegetables grown in their gardens and the products that they consume from the forest;
- using HRS, take regular measurements of the quantities of caesium-137 incorporated into the body and ensure that the people take a pectin-based food additive regularly to reduce the level of accumulated radiation and annual radioactive load;
- train teachers, parents and schoolchildren in methods of food preparation in order to reduce their levels of radionuclides (separation of milk, soaking of mushrooms, meat, etc).

Since then, everything has stalled. Pierre Frigola, McSweeney's Assistant in the Directorate General for External Relations (DG RELEX) at the European Commission, confided to me at the end of the year, unofficially, that not everyone was in favour of the project, but that the directors of JRC had finally obtained from DG RELEX of the EC a million Euros for the programme TACIS for one or two "environmental" projects "without specifying the beneficiary". The JRC provides expert scientific assistance to the EC, but "other services" he told me "have a different opinion from the JRC".

A further request, no doubt more insistent, resulted in the following correspondence, transmitted on 22nd April to Solange Fernex from the office of Marie Anne Isler Béguin, a Green Deputy and a colleague of Piétrasanta. In substance, I was to advise Nesterenko to seek the support of a ministry (a procedural requirement for EC financial support) and to make contact as quickly as possible with Norbert Jousten, EC representative in Kiev, referred to below.

Dear Mr Piétrasanta,

I have just received a copy of an internal note addressed by our representative in Kiev to the director of finance of the TACIS nuclear programme.

This note specifies that a sum of 6 million euros is to be attributed in the 2003 budget to the "social consequences of the closure of down of Chernobyl", of which 1 million euros must be devoted to combating the effects of the accident in Ukraine and Belarus[...] It would be desirable if Professor Nesterenko made urgent contact with our Delegation in Kiev and I suggest that you write to our Ambassador, Mr Norbert Jousten.

P. Frigola

From Mr Yves Piétrasanta,
European Deputy
Vice-President of the Commission on Science
and Research of the European Parliament

To Mr Norbert Jousten
Head of Delegation of the
European Commission,
Kiev

Dear Sir,

As part of the European Union's management of the social and health consequences related to the Chernobyl reactor, a budget of 6 million euros has been granted for 2003, as a continuation of the regional project entitled TAREG 7/03/97, initiated in 1997.

Last year, under my direction, the Joint Research Centre at the European Commission approved a cooperative project with Belrad, the independent institute in Minsk directed by Professor Nesterenko, which specialises in the study and prevention of the health impact of radioactive contamination in Belarus. We were recently informed that a budget of 1 million euros had been allocated to fund this project.

Belrad's scientific and humanitarian activities, undertaken in the difficult political context of Belarus and in the face of an extremely alarming health, medical and economic environment resulting from the Chernobyl disaster, require the full assistance and support of the European Union and so I welcome the signing of this agreement. All the more so as Belrad is facing a critical financial situation today, resulting from the explicitly independent stance of its research and activities.

It is therefore of great importance to me to ask for all your support and diligence in implementing this cooperative project as soon as possible and ensuring the delivery of 1 million euros from the Community budget to Professor Nesterenko's institute.

This letter clarified a confusing situation. It stated explicitly that the million euros, from the "Community budget", had been obtained by the JRC for Belrad's TEST project. It seemed to me that the least we could do was take this statement at face value, partly out of respect for Piétrasanta, the politician who had promoted the project, and partly in the spirit of citizen control of public European funds destined to mitigate the consequences of the Chernobyl disaster.

On 22nd April 2002, I passed on to Nesterenko, Michel Fernex's suggestion that he should not be afraid to present to Mr Jousten, a series of sections of the overall budget, of which Ispra was only the central scientific part, the essential objective being assistance to children.

The Belrad institute prepared a project for around 1 million euros, organised in four chapters around TEST, which Professor Nesterenko sent to Jousten on 24th April. The progress of this dossier "through the corridors" should have been followed closely before it disappeared into thin air.

On 30th June 2002, Nesterenko wrote to Piétrasanta:

> Following your instructions, we sent the 4 projects on 27th May 2002 to the head of the European Commission delegation in Ukraine, Mr Norbert Jousten, at the Representative Office of TACIS in Kiev, through the Representative Office of TACIS in Minsk (Mr Raoul de Lusenberger).
>
> We do not know what action has been taken in their regard.
>
> I have learned that letters were addressed to Mrs Isler Béguin between January and April 2002, from the leader of the Spanish government, José Maria Aznar, the President of the European Commission, Mr Romani Prodi, and Commission member, Christopher Patten. All these letters expressed the wish to see TACIS provide support to programmes designed to mitigate the human consequences of the Chernobyl disaster and to encourage civil society through the support of NGO projects.
>
> In Belarus, the government and administration give absolute priority to international projects submitted by research institutes that form part of government structures. Because of this, our projects, which emanate from independent structures, are *a priori*, disadvantaged.
>
> Dear Sir, Member of Parliament, we draw your attention to the fact that our first TEST project, despite having been approved by the JRC of the European Commission, has still received no funding
>
> I would be extremely grateful if you could help to speed up the implementation of the 4 projects that we have submitted.

On 7th July 2002, the president of Comchernobyl, Vladimir Tsalko, sent a letter of support to the head of the European Commission delegation in Minsk, for the four projects:

> 1. "TEST"
> 2. "Rehabilitation": radioprotection of children and rehabilitation of the population's local environment, in Narovlia district in the Gomel region.
> 3. "Radioactivity in food products": "radiological monitoring of food products in the Belarus territories contaminated by Chernobyl through the establishment of

20 Local Radioprotection Monitoring Centres and provision of information to the population on required radioprotection measures.

4. "Children's health": "monitoring of children's health". Project prepared jointly with Professor Michel Fernex (France).

The Castle remained silent.

On 4th April 2003, after more than a year had gone by (almost two years since Ispra) I prepared an interim summary of the facts which I sent to Marie Anne Isler Béguin, the Green MP and a colleague of Yves Piétrasanta, the politician who was promoting the project.[100]

Following Yves Piétrasanta's initiative in October 2000, Belrad committed itself fully throughout the spring of 2002 to the preparation of the four projects, which were duly submitted, within the time frame, to Mr De Lusenberger in order to receive funding from TACIS in 2003. To date, these projects have still not been registered in Brussels.

This fact has only recently been discovered by Mr Charles Deleuse, who works with Belrad in Belgium welcoming contaminated children from Belarus. Nesterenko asked for his help in finding out why the expected funding had still not arrived. Mr Deleuse contacted Mr Kozyreff who, after some research, found nothing either among projects that had been rejected or in the register of mail received in Brussels. However, at a reception in December at the French Embassy in Minsk, in response to a question about the Belrad projects, Mr Jousten said that they had been in Brussels since September and that in his opinion, a reply should arrive shortly.

Following regulatory procedure, resumes, or *abstracts* as they are known, of the project were sent by Mr Mikhnevich to Mr Konstantinov in Kiev, who was supposed to send them to Brussels. After approval, a dossier number should have been attributed and communicated to Belrad, who would then send the projects in full directly to Brussels. As the *abstracts* never arrived at their final destination, Mr Jousten owes an explanation to Mr Piétrasanta. European funding is of interest to the three countries contaminated by the Chernobyl disaster and it is in Kiev that the progress of the dossier has been interrupted . . . An entire year has been lost. At this stage, the question is political.

Nesterenko confronts financial problems daily. Some of his scientific collaborators have had to leave because their salaries are too low. He was very much counting on this European funding which would have allowed him a breathing space. His health (he has a heart condition) is also affected by this continual worry and stress.

---

[100] To properly understand the chronology of what follows, note that Kozyreff works for TACIS in its Brussels office; Norbert Jousten is the head of the European Commission delegation in Kiev, and he represents TACIS in the three countries of Russia, Ukraine and Belarus. He has a Ukrainian colleague, a Mr Konstantinov; Raoul De Lusenberger represents TACIS in Minsk, and the name of his colleague in Belarus is Mikhnevich.

Jeff Rivalain, assistant to Marie Anne Isler Béguin, replied to Solange and Michel Fernex and myself, on 22nd April 2003.

I have just called V. Kozyreff, of EuropAide A3 TACIS at the European Commission in Brussels (Public Health group). He was aware of this issue concerning the four TACIS dossiers from Nesterenko. And indeed, he has never seen a trace of them. From our telephone conversation, I gather that it seemed to him that the projects submitted by Nesterenko had not respected the strict and transparent framework of TACIS. The Commission works with this organisation through a bidding process after it has decided its priorities in projects in such and such an area. According to Kozyroff, Nesterenko was relying more on his political support in the European Parliament (letter from Piétrasanta, conference at PE) but could not provide answers regarding the precise tender to which his TACIS projects would have corresponded.

Mr Kozyroff also emphasized that the Commission only uses TACIS in partnership with government authorities and not directly with civil society actors. Belrad is not on good terms with the government, and neither is Brussels with Minsk. The community budget for this country, already derisory, will be even more restricted. Few of the next tenders would suit Nesterenko's work. The most promising would be CORE (I have sent this to you by post), where Belrad is already a partner.

A meeting has been fixed with Kozyroff at the European Parliament, this Thursday, the 24th. He is Russo-Ukrainian by origin, born in Belgium and convinced of the importance of increasing exchanges and projects between the European Union and the countries of the former Soviet Union. But he is well aware that among the leaders of the European Commission, this feeling is not shared.

I replied by return email.

Dear Jeff,

Nesterenko has followed the instructions given to him by TACIS officers strictly and professionally. He mobilised his entire institute who worked intensively for a substantial period in order to complete procedures in time and in due form. No one at TACIS mentioned anything about tendering. Mr Jousten did not question the regularity or conformity of the four projects when he met Nesterenko at the French Embassy. On the contrary, he told him that the dossier had been sent by his office in Kiev to Brussels, where Kozyreff now says they never arrived. The person who needs to be questioned, in fact, is Mr Jousten who is responsible for TACIS in Kiev, for the three countries concerned—from whose office Nesterenko's dossiers disappeared. Nesterenko did not "rely on political support": he did what TACIS told him to do and he thought, indeed, that he had the political support of Mr Piétrasanta. But after all, it is not this political support, that explains the disappearance of the dossier in Kiev. Who are we kidding?

As for the fact that the Commission only uses TACIS in partnership with government authorities and not with civil society, I seem to remember that President Aznar had written exactly the opposite. No doubt this is a mistake as Aznar is only a "politician" and, no doubt, naïve. *"Coincido con Ud en utilizar el programa TACIS con el fin de paliar las consecuencias de la tragedia del accidente nuclear de Chernobyl. Estas acciones deberían realizarse siempre a través de la sociedad civil, ya que el estado actual no permite la puesta en practica de acciones o programas de cooperación que incluyan al Gobierno o la Administracion bielorusos"* ("*I agree that you should use the TACIS program in order to mitigate the consequences of the tragedy of the Chernobyl nuclear accident. These projects should always be undertaken by civil society, as the current situation does not allow any cooperative initiatives or programs that include the government or the administration in Belarus* "

(Jose Maria Aznar to Hon. Mrs Marie Anne Isler Beguin—Madrid, 30th April 2002).

Who are we kidding? Aznar, Prodi,[101] Patten, Piétrasanta, Marie Anne, all "politicians" and then Nesterenko.

What seems clear from this dismal story is that the interaction between politicians and the Commission is virtual and schizophrenic. They act and talk as if their actions and their words had any importance. In reality they are not the ones who determine policy. Are they unaware of this or do they just pretend? What comes to mind in relation to the European Union is De Gaulle's reference to the United Nations as "the thingamajig".

A study by Dr Raoul Marc Jennar, *The European Community: should the Bastille be stormed?*, published in The Ecologist, Vol. 1, No. 2, winter 2000, describes the impenetrability of this "new centre of power, disconnected from citizens" very well. I will end with some excerpts from his text, which shed light on the Kafkaesque forest through which Nesterenko is forced to march aimlessly with his troop of sick children:

> The 23 Directorate-Generals responsible for specific policies undertake their activities as if, in terms of the impact of those policies, there were no interaction. [...] It can be stated, without fear of contradiction, that what is done, *for example*, by the DG in charge of development aid is destroyed (and even if it was possible, several times over) by the activities of the DG responsible for commerce, that competes globally with the USA in its zealous support of TNCs. Despite protestations to the contrary, the political manipulation going on here has reached unprecedented heights. Is it surprising then, knowing that the EU is the largest funder of development aid—that hunger in the world and underdevelopment are not decreasing? [...] In fact, there is no formal structure, before the decision making process, to coordinate the proposals put forward by each DG. This makes it easy for pressure groups who are skilled at exploiting the divergent interests of different DGs.[...] Every one knows about the 3000 lobbyists working in Brussels, a majority of whom are working on behalf of

---

[101] Romano Prodi replies on 8th April 2002 to the letter sent by M.A. Isler Béguin on 7th February 2002: (...) I completely agree with you, Madame Deputy, that we must continue to support aid programmes whose aim is to mitigate the human consequences of the Chernobyl disaster, and to support civil society".

industry. They may extol the merits of texts submitted by the Commission or they may propose amendments that the Commission is only too willing to adopt.[...]

The real decision makers that determine the choices of the European Commission are hidden by the latter in a category where one would not expect to find them. Pressure groups representing the business world are categorized as NGOs, non-governmental organizations. Everyone knows that this label usually refers to humanitarian or development aid organizations, associative movements in general, where the pursuit of profit is the last thing on their agenda.

The CEPN is one of these NGOs and is the subject of the following chapter.

Chapter II

AN INCOMPETENT FRENCH TEAM OCCUPIES THE TERRITORY

In 1996 a group of French researchers, named ETHOS, entered the vast and unique laboratory provided by the contaminated territories of Chernobyl, with the purpose of collecting radiological data and to receive training in radioprotection, from the local Center of the Radiological Control (LRCC) [102] created by Professor Nesterenko in Olmany village. The ETHOS consortium is part of the CEPN (Study Centre for Evaluation of Nuclear Protection) established by Electricité de France (EDF) and the Atomic Energy Commission (CEA). The French nuclear lobby is actively and amply represented here! The health of the population does not figure in the statutes of this organisation, so what is it doing here?

One of ETHOS' objectives is to prepare a document for the European Union on the management of nuclear accidents and of regions that have been contaminated by long-lived radionuclides, by establishing "guidelines on sustainable management of radiological quality and public confidence".[103] For two years, from 1996 to 1998, ETHOS made full use of Olmany's LRMC data and of its personnel, trained and equipped by Nesterenko, to measure radioactivity in food, milk, etc—with no thought to reimbursing the technician for the extra work that this involved. It was a happy and fruitful collaboration until one day in January 2001, when ETHOS, through the Belarusian authorities, succeeded in expelling Nesterenko from Olmany and from four other villages in the Stolin district.

Today, even though it has no qualifications in the area of health, ETHOS presents itself as the scientific reference in radioprotection in the Chernobyl territories and coordinates CORE. The CORE program has been the subject of a critical review that we submitted to the European Commission and European parliament in June 2003[104].

---

[102] The creation of the LRMCs is mentioned in Part Two, Chapter III, "Citizens unbowed", p.109.

[103] ETHOS, minutes of the meeting held on 17th April 2001.

[104] See Part Four, Chapter III, p. 277.

The Belrad Institute struggles for survival in the midst of great financial difficulty. It gets its funding from ordinary European citizens, members of health and environmental NGOs. Following criticism of the activities of ETHOS by these associations, Nesterenko was included in the CORE programme as an expert in radioprotection. But he was prevented from distributing pectin as a prophylactic adsorbent to contaminated children because the programme follows the line taken by ETHOS on this question—even though ETHOS has no competence in this area. Mr Frigola (Sweeney's assistant at DG RELEX—Directorate-General for External Relations at the European Commission) confided to me that not everyone is in agreement in the corridors of power at Brussels, notwithstanding the fact that the ISPRA Common Research Centre is in favour.

The provisional budget of the programme is 4 million euros, funded among others by the Belarus government, the UNDP, UNESCO, OSCE, the European Commission, the Swiss Agency for Cooperation and Development, and in France by the Ministry for Foreign Affairs and the IRSN. The European Commission, for its part, allocated 2 million euros to CORE (information that was communicated to their Belarusian partners on 8th May 2003).

1. EXTRACTS FROM A PSR/IPPNW DOCUMENT[105], March 2001

The nuclear lobby, represented by the IAEA at the UN, systematically dismisses people who try to disseminate the truth about Chernobyl, discredits these researchers or dismisses them from their posts, and gives its support instead to those academics that are willing to undermine or contradict studies revealing the extent of the medical disaster resulting from the radioactive fallout from Chernobyl.

German, Swiss, French and other researchers and academics whose results conform to the wishes of the promoters of commercial nuclear power and discredit those who express any other point of view, sometimes militate in antinuclear groups where they are able to exert influence. They know how to adapt their discourse to the audience at hand. In the devastated areas of Chernobyl, where they gather material for their publications, these foreign researchers work locally in collaboration with those who are trying to disguise the indelible signs of the catastrophe.

In France, a group of academics, united in a consortium called ETHOS, with their own particular ethical concerns, are planning research in relation to Chernobyl. As these studies require considerable financial support (travel, salaries, equipment, publications etc) these academics and higher education teachers appeal to NGOs and submit projects that are likely to be considered favourably by funding

---

[105] Physicians for Social Responsibility/International Physicians for the Prevention of Nuclear War. Nobel Peace Prize 1985.

organisations. One such NGO is the CEPN, Centre d'étude sur l'Evaluation de la Protection dans le domaine Nucléaire (Study Centre for the Evaluation of Nuclear Protection).

Who is hiding behind CEPN?

The Centre d'etudes sur l'evaluation de la protection dans la domaine nucleaire (CEPN) is a non-profit association, (as defined by the French Law 1901). The CEPN funds multidisciplinary projects and on its site (www.cepn.asso.fr) we learn that its president and treasurer represent Electricité de France (EDF) and its vice-president represents COGEMA (Areva). The organisations behind this munificent NGO are among the most powerful in the nuclear lobby worldwide. The priority for these companies is that Chernobyl should be erased so that their business, that has been somewhat adversely affected by the suffering of the 9 million human beings who live in areas around the reactor, will prosper.

It is not surprising then, that having scarcely arrived in Belarus, certain teams funded by the CEPN (400 euros per day for mission expenses), take measures to get rid of those who are helping the people living in territories contaminated by radioactive fallout and who study the consequences of Chernobyl, in the field, in order to protect children, in particular, who are the most vulnerable to the radioactive contamination of the environment.

2. HOW I GOT TO KNOW ABOUT ETHOS

In April 2000, while filming the village of Olmany in the Stolin district, I learnt that a French team came to work there from time to time. It was the first time I had ever heard the name ETHOS. All I had been told by Nesterenko was that this team gave information to people and that Pasha Polikushko, the dosimetrist at his LRMC, provided them with measurements of various food products.

A few months later, I learnt in more detail about the consortium ETHOS from some friends—sociologists at the University of Caen, who were working on a project (ETHOS 2) initially planned for 2000–2001. This meeting took place on 20th January 2001 in Brussels, during a theatre presentation, film projection and public discussion around the theme of Chernobyl. As I viewed the colour photographs of farmers in Olmany exhibited on the walls of the theatre, my ears pricked up when one of the sociologists told me that Jacques Lochard, head of the ETHOS project, was from the CEA (Commissariat à l'énergie atomique) and that his definition of their task could be summarized in the following surprising way. "We need to occupy the territory". Although the people I was talking to assured me that Mr Lochard was not a nucleocrat and that he did useful work with the inhabitants of the contaminated territories, my curiosity was pîqued. I took the glossy, colour album presenting the work of ETHOS in Olmany 1996–1999, with me to read on the train. The photos of the peasant farmers were accompanied with short poems and a text which ended with this sentence: "living with Chernobyl means learning to live again, to live another way, integrate the presence of radioactivity into daily life as a new component of existence". This new version of the world order, decreed by the CEA, stuck in my throat. I did not know at that point how my feelings of discomfort would turn to revolt when I learned that the decision had been made, following a proposal from ETHOS, to expel Nesterenko from the villages in the Stolin district (including Olmany) where he distributed pectin so parsimoniously, only to the most contaminated children, without a penny from the CEA to help him.

The latter part of January was marked by a series of odd coincidences. On my return from Brussels, the TV channel ARTE was studying a proposal for a documentary on this precise subject—ETHOS' activities. They asked me what I knew about this charitable organisation of volunteers working in the contaminated territories of Chernobyl. In order to check my information, I called Nesterenko in Minsk, who gave me the latest news. He had just written to the President of ComChernobyl[106] to protest against the committee's decision to deprive him of five LRMCs of Stolin district, including Olmany, where he had been working for the past ten years, and this precisely at the request of ETHOS!

3. I WRITE TO ARTE ON 23 JANUARY 2001

ETHOS is not a voluntary organisation, but a multidisciplinary French organisation, funded by, among others, the European Commission's research program on radioprotection, headed by a representative of the CEA (Commissariat à l'Energie Atomique, the French nuclear lobby).

At first four, and later seven, research institutes were associated with the project, which has already undertaken activities in the village of Olmany. I made a visit and filmed there last April without seeing any positive results[107]. I interviewed peasant families abandoned to cope on their own, whose children had radioactive loads exceeding permissible levels (20 Bq/kg body weight). I filmed a small child who weighed 12 kilos and had 800 Bq/kg which means 10,000 atomic disintegrations per second in the organism. They had never met representatives of ETHOS—which does not mean of course, that other families have not been involved. But what are the criteria for selection? ETHOS operates through central and local power structures.

---

[106] Chernobyl Committee, an interministerial body in Belarus that coordinates policy to deal with the consequences of the Chernobyl disaster.

[107] Measurements taken in January 2001 on 31 samples of milk from the Olmany area revealed 22 samples where levels were well in excess (up to 2,600 Bq/l) of the maximum admissible limit of 100 Bq/l (a level that doctors already consider excessive for children especially if it continues over many years. In Russia the "norm" is 50 Bq/l. See Part Five, Chapter I, "The illusion of norms", p.300).

What is more, this French organisation provides a service to ComChernobyl in the eviction of independent Belarusian scientists and specialists from the area who publish the true figures on contamination every three months and provide real assistance to the population with very few resources and in the face of incredible obstacles. On this subject, in addition to the document attached, I will send you the letter of protest that Professor Nesterenko wrote to ComChernobyl just eight days ago, as soon as I have translated it.

4. I MAKE CONTACT WITH OUR ADVERSARIES

I found myself, once again, in sole possession of a shocking piece of information (the previous one was the arrest of Bandazhevsky) and under an obligation to take some form of serious action, beyond informing the TV channel ARTE. None of the NGOs or the people involved in supporting Nesterenko and Bandazhevsky knew about ETHOS. What could I do? Regardless of the lobby, I decided to write to the men and women of the consortium ETHOS themselves. When I had received and translated the correspondence between Nesterenko and ComChernobyl, I asked the sociologists from Caen University for the email addresses of members of the consortium and I wrote them the following letter:

> Letter written on 8 February 2001
> to the 16 directors and operators of ETHOS

> Dear Sir/Madam,

> The great writer, Albert Camus, winner of the Nobel Prize for Literature, was a master of succinct style. In one of his notebooks he wrote "To live is to verify". When someone asked him rather disingenuously "What is truth?" he said "The opposite of a lie". He had a tragic trust in humanity[108].

> It is in his spirit of trust and the quest for truth "in spite of everything", that I have decided to write to you.

> You have chosen a beautiful name—ETHOS—for your action. And I am appealing to your ethical sense as a human being. To find out whether you are aware of the damage you are doing, or whether it is corrupt government officials in a country that has been so wounded, who are doing it in your name, without your knowledge.

> Despite everything stated in the documents attached to this letter, I still find it hard to believe that the name ETHOS masks a conscious determination to do harm.

> The independent work undertaken by Nesterenko must be protected, supported and developed in the territories contaminated by Chernobyl. You are blocking his path and chasing him out.

---

[108] Vassili Nesterenko, who had been battling for fifteen years against the nuclear plague and the lies of the scientific bureaucracy reminded me very much of Dr Rieux, the hero of Albert Camus' novel *La Peste*.

With the financial, political and human means at your disposal, you could help him pursue his work, the most intelligent and effective response to the current situation. I am in contact with him daily and I know the difficulties he faces.

I look forward to your response. The future of this scientist—a man who is devoted to his country—depends on it. It is still within your power to stop this infamy.

I attached to this *email* the two letters to ARTE and the correspondence between Nesterenko with ComChernobyl. (They are reproduced in part in this chapter on pages 272–274). My letter unsettled Jacques Lochard, as he told me himself the next day over the telephone. He tried to convince me of his best intentions regarding Professor Nesterenko. He considered that the activities of ETHOS and of Belrad were complementary, at the same time making a puzzling and inaccurate distinction between a "national observatory managed by Professor Nesterenko and the ETHOS program which would undertake local and operational dosimetry, involving the population". Yet this was precisely the originality of Nesterenko's direct objective method that he had perfected and developed[109] alone in the face of inertia on the part of the Ministry of Health in the villages ignored by the German institute Julich.[110] What had really happened was that the consortium ETHOS had learnt from Nesterenko's experience, used his data and was now appropriating it in order to supplant him—a true case of plagiarism. But there was one fundamental flaw: ETHOS' mission, as defined by the French nuclear lobby, had an unbridgeable statutory limitation which excluded any work in the area of health. ETHOS was not qualified to deal with the health of the population. So what was it doing in Chernobyl if it was not qualified? At this point, the phrase about the foreign "occupation of the territory" to deal with this "new atomic element of our existence" began to take on a more sinister meaning. Supported politically and financially by the nuclear lobby, ETHOS, which gave the appearance of addressing problems caused by the Chernobyl catastrophe, was actually preventing the recognition of the health consequences, which independent scientists like Nesterenko and Bandazhevsky were revealing in the face of thousands of obstacles. Furthermore— for those who had an interest in it—see the US American programme BelAm mentioned by Pflugbeil[111]—the biological purity of the contaminated guinea pigs under observation remained intact because the radioactive load in their bodies was not reduced by an adsorbent that the European Commission had refused to finance. A puzzle of global dimensions made up of apparently unconnected compartments each closed off from the other. As for Jacques Lochard, he assured

---

[109] See the description given by Dr Fernex to Lukashenko that I quote further on in my detailed letter to Arte.

[110] See Part Two, Chapter IX, p. 160.

[111] See Part Three, Chapter VI, p. 245.

me that collaboration between the two bodies was important and to be encouraged, and that Nesterenko could benefit from European support: 1,500,000 euros were budgeted, of which 80% were to be spent in the country, he told me. (This "manna" has never materialised despite the submission of several projects). He also told me that the removal of Nesterenko from the villages followed a decision taken by the Belarusian authorities within the framework of ETHOS 2, which involved the French team and its Belarusian (governmental) partners in five villages of the district of Stolin. That it had not resulted from any request by ETHOS. However, the laconic response from ComChernobyl leaves us in no doubt:

> Committee on Problems Relating to Consequences of the Chernobyl Catastrophe (ComChernobyl).
>
> 25.01.2001
>
> To the Director of the
> Belrad Radioprotection Institute
> V. Nesterenko
>
> The Committee on problems relating to the consequences of the Chernobyl disaster has taken the decision to transfer five local radiological monitoring centres (LRMC) of Olmany, Gorodnya, Belousha, Rechitsa and Terebezhov of the district of Stolin, to the Radiological Institute in Brest.
>   This decision has been taken on the recommendation of ComChernobyl's working group, in collaboration with French scientists from the European Commission's project "ETHOS 2", in line with the proposal from the latter.
>
> Vice-President, V.E. Shevchuk[112].

Following our telephone conversation, I wrote this *email* to Jacques Lochard to explain my view precisely:

> 9th February 2001
>
> Following our telephone conversation yesterday, I need to make the following points, in order that matters be clear between us:
>   1.—I understood from our telephone conversation that you do not want conflict— quite the opposite—you are seeking to clarify the matter in a way that will be constructive and beneficial to the work of independent Belarusian scientists.
>   I visited your internet site yesterday www.cepn.asso.fr. I read there that ETHOS is a CEPN project. The founding members of CEPN are EDF, the CEA and COGEMA.

---

[112] Shevchuk is an *apparatchik* from the Ministry of Health in ComChernobyl.

The declared intention of ETHOS is to assist the victims of the Chernobyl disaster. Three of the founders of CEPN are promoters and users of the very industry which caused the disaster. Without free and independent information on the real consequences of the disaster, no assistance is possible. But this scientific information is a threat to the industry on which you depend. There is therefore a conflict of interest. Can you hold an objective view on these essential points?

Only your political actions will convince us of the reality of your intentions.

In addition, the link between those who are persecuting the independent Belarusian scientists, of whom we are speaking, and the CEPN has existed for several years.

I.V. Rolevich, mentioned in Nesterenko's letters, was the vice-president of ComChernobyl during those years when 285 of Nesterenko's 370 functioning centres, were eliminated. It is Rolevich personally who took that decision. And the name of Rolevich appears in CEPN publications between 1996–1999.

To an outside observer, whatever your intentions, ETHOS appears objectively to be in the service of the nuclear industry, whose deepest wish is to muzzle the scientific information relating to the Chernobyl disaster. Professor Fernex' text "The Chernobyl catastrophe and health" provides evidence for this. The operation to remove Nesterenko, undertaken in ETHOS' name, the subject of our discussion today, seems to confirm this.

If you approach ComChernobyl (which, you say, falsely implicated ETHOS) to reverse its decision, Shevchuk, the official in question will, in all likelihood, write another letter to correct his mistake: the decision to evict Nesterenko from the five villages in Stolin will have been taken independently and not on your suggestion.

No one will believe it, because you occupy the territory in its place, while nothing would prevent a collaboration.

But nothing is ever final. It is people who make history. The institutions and politicians that sponsor you carry a lot of weight and have sufficient prestige so that the Belarusians could put this story of the five villages behind them. This could be a credible start to a collaborative approach in developing a radioprotection policy worthy of the name.

2.—Radioprotection in the contaminated territories of Chernobyl is impossible without a scientific approach, applied to the organism of each child and of the food s/he absorbs. This is what the Ministry of Health of Belarus does not want to do, because it will not be able to continue publishing false generalised statistical data. It explains its opposition to the work undertaken by Nesterenko who uses spectrometers for human radiation, a concrete measure that reveals the true doses of contamination. These measurements are essential for the targeted prevention programme for each child and for the establishment of the correlation between radioactive load incorporated in the organism and the many illnesses studied by the anatomopathologist Bandazhevsky. But they also reveal the true scale of the Chernobyl disaster, which has only just begun. Failing to undertake scientific work and only providing education and social support, can become an alibi, which leaves everything just as it was—in a state of "ignorance and uncertainty".

I am really only just discovering ETHOS now but I have known about the levels of contamination in Olmany for three years. What you say and write bears no relation to what I have observed and for which I can provide the documentation. Giving the green light for the sale of milk from Olmany certainly helps the local economy and reinforces the power of local leaders—they will be very grateful to you—but it means the epidemic will continue to spread throughout the country. Because it is not the ETHOS team, who are present in the field for ten days now and then, that will guarantee that only clean milk will be exported.

You are right that the Belarusians should take their destiny into their own hands. The Belarusians—in the first place, this means scientists like Nesterenko. Either their expertise will be used to develop a genuine radioprotection policy or there will be no radioprotection at all. Without these scientists, the peasant farmers are trapped, with no money and no information, and they will never have the strength to deal with what confronts them. The 370 centres must be re-established.

I needed to clarify this as briefly as possible after our conversation; for myself, for you and for all those who read this.

Lochard sent me an *email* which crossed mine expressing his surprise at the way ETHOS was presented in the documents that I sent him; that there was a misunderstanding about the nature and objectives of the project and that "the ETHOS team was very willing to cooperate with Nesterenko in the future".

5. PROFESSOR NESTERENKO'S PROTEST

In his protest addressed to Vladimir Tsalko, president of ComChernobyl, on 15 January 2001, Nesterenko set out the facts:

By reducing financial support, ComChernobyl has caused, year on year, a systematic decrease in the number of local centres managed by our radioprotection institute.

The exclusion of the LRMC of Olmany, Gorodnya and Berezhnoie from the list of centres managed by the radioprotection institute will interrupt the continuity of information that is transmitted to ComChernobyl, on the annual contamination of food products in the Brest region. The three monthly comparisons from one year to the next which permit the observation of trends will be made impossible, and the development of general radioprotection recommendations will become more difficult.

Furthermore, Olmany, Gorodnya, Berezhnoie are large villages with populations of between 1,500 and 2,500 each. The loss of information on the contamination of food products in these villages will make it impossible to undertake the targeted monitoring of inhabitants, through the measurement of radioactive load with human spectrometry correlated with Cs-137 contamination of local food products, consumed by the different families.

I consider it necessary to inform you that the vice-president of the previous management of ComChernobyl, I.V. Rolevich, withdrew authorisation from Belrad for the management of twenty-four centres in Leltchitsy, one of the most contaminated districts of the Gomel region, entrusting them first to local authorities and then to Gomel's scientific radiological institute. In our opinion, this has merely resulted in the loss of important information on the contamination of food products.

In view of the above, I urge you not to take the decision to remove the local centres of Olmany, Gorodnya and Berezhnoie from the Radioprotection Institute of Belrad's management.

In fact, the decision was final. Today in 2006, not one LRMC is financed by the Belarusian government.

6. THE EXPLANATORY LETTER TO ARTE WHICH WORRIED ETHOS

For the attention of Thierry Garrel
Director of Documentaries
Arte, France

31st January 2001

I promised you a translation of Nesterenko's letter of protest that he sent recently to ComChernobyl in Belarus. It highlights the role of ETHOS in the management of the consequences of the Chernobyl disaster.

In order to understand his letter, which I am also sending to European parliamentarians who are interested in the fate of independent scientists, it is important to know that between 1991–1993, inspired by the democratic renewal of post-communist societies, ComChernobyl supported and financed 370 Local Radiological Monitoring Centres (LRMC) set up by Professor Nesterenko's independent institute Belrad. Located in the largest villages in the areas of Belarus that are contaminated by radionuclides, these centres provided information to the population and helped them deal with the dangers. The centres were equipped with machines to measure radiation and were managed locally by doctors, teacher and nurses. They provided advice to people on food hygiene and preparation to limit their ingestion of radionuclides.

Responding to a request for expert advice from the Belarusian Minister in charge of Emergencies, Professor Fernex had this to say, in a letter that he addressed to President Lukashenko:

"Today, there is a new epidemic which deserves as much attention as the plague or smallpox: radiological contamination affecting vast regions. People are mainly contaminated through food that contains radionuclides.

The Belrad Institute has measured hundreds of thousands of food samples, brought by people from villages in the contaminated zones, using remarkably precise mobile spectrometers. These are produced in Minsk and are therefore, relatively cheap. The people get the results of the measurements and at the same time they receive advice on nutrition and ways to prepare food which will reduce the content of radionuclides (soaking meat in salt water, skimming cream off milk etc). The institute is providing the public with information and a continuing education for people of all ages.

Professor Nesterenko has contributed to the discovery of the aetiology of new clinical syndromes in children who consume food that is contaminated with Cs-137.

In order to prioritise interventions, the children of the most contaminated regions must be identified and followed up. The Belrad Institute measures whole-body radioactive load with a very effective, mobile radiometric chair. They have undertaken more than 60,000 measurements. "

(*As I explained to Garrel:*) The extreme economic difficulties in Belarus and the absence of any interest from the rich countries have allowed the nuclear lobby to rapidly normalise this situation that threatened its monopoly on scientific information and could prevent it hiding the real dimensions of the catastrophe. It introduced its own advisers into various decision-making bodies and using research that produced false conclusions, it determined the direction taken by the Belarusian medical establishment.

Michel Fernex: "It took doctors and pathologists to demonstrate the correlation between levels of Cs-137 and Sr-90 in food and in the environment, and the appearance of new and known illnesses such as diabetes and atherosclerosis, where symptoms of illness such as hypertension and myocardial infarction are appearing earlier and earlier, even in young children.

It is thanks to Professor Yury Bandazhevsky and his team in Gomel, that knowledge has advanced, correlations have been established, and the necessary experimental proof provided to demonstrate the physiopathology of a whole range of illnesses, each of which can be correlated to an abnormal concentration of Cs-137 in the affected organ (heart, liver, kidney, digestive system etc)".

With the support of Western NGOs, Nesterenko and his team travel between contaminated villages in their minibus-laboratory, donated by the Irish. In order to reduce radioactive load in children's bodies, they provide periodic treatment with an effective adsorbent—apple pectin—which has positive effects on their health. Until the last few months, Belrad worked under contract with local health authorities of the contaminated territories. The Ministry of Health recently issued instructions to health authorities to discontinue their preventive programmes with the Belrad Institute, to the detriment only of the contaminated children. (Belrad is a non-governmental organization and does not report to the Ministry).

The support that this tiny island of civil society receives from the West has allowed it to stay afloat until now. If that support is reduced, the scientists will leave.

What does all this have to do with ETHOS?

ETHOS does not question the nuclear lobby's policy which, in practice, leaves the inhabitants exposed to the consequences of the disaster and, through every possible means, obstructs scientific research and the diffusion of reliable information. ETHOS undertakes no systematic individual measurements followed up with a preventive cure such as that provided by Belrad but it uses (to what end?) the data gathered by Belrad in the field. It provides no information to people about the way radioactivity in food causes an increase in radioactive load in the inhabitants. This data should be used to direct effective and achievable aid programmes from the international community.

In fact the help offered by ETHOS is a sham.

ETHOS, though it may employ some sincerely motivated people, shares the interests of the nuclear lobby and of the Belarusian government and contributes to stifling the information in the three monthly bulletin, published by Nesterenko, that provides a unique source of information about the consequences of the Chernobyl disaster on human health.

Knowing that this information was being read by people at Arte may have worried ETHOS. Needlessly. Two months later, Arte was to turn down my proposal for a documentary about the conflict of interest that was paralysing, at the highest levels of the UN, any action on the part of the World Health Organization and blocking medical and scientific research about Chernobyl. (The documentary was however made at a later date by a Swiss television company) Instead, Arte produced a faceless documentary on ETHOS[113].

7. ETHOS MANAGEMENT PROVIDES AN UPDATE TO ITS MEMBERS

Position of the ETHOS team regarding Professor Nesterenko's activities.

On 8th February, a report was sent to ETHOS' different partners, to various French and Swiss personalities in a number of NGOs and to French journalists by Wladimir Tchertkoff. The report is a protest against the removal of Professor Nesterenko from the Local Radiological Monitoring Centres (LRMC) in the villages involved in ETHOS Project 2. Tchertkoff, until then, was unknown to the ETHOS team who do not know the exact nature of his links with Professor Nesterenko. Tchertkoff's presentation of the ETHOS project is inaccurate and unacceptable to the team.

In response to my critique which was qualified as "virulent and based on imprecise and partial information and mistaken interpretations", the letter provides a historical account of the activities undertaken by ETHOS Project 1 (1996–1999), of its relationship with Professor Nesterenko which was more or less non-existent, and with the dosimetrist of Olmany which was highly productive, and sets out the development and the ideas behind the consortium which was to become ETHOS Project 2. In fact this new project was never to see the light of day, but became the programme CORE, financed by the European Commission and directed by the CEPN team. There was never any question of radioprotection but rather of "radiological quality", an obscure and ambiguous term (see Chapter III, page 284). For the moment, within this new hypothetical framework, as stated in the letter addressed to ETHOS members, "a discussion has been initiated" with partners in local government for the "development of indicators of radiological quality suitable for use by the population in the context of local management" and for the "local development of charts and tables for follow up work on the radiological situation at village level". In the context of these discussions, it was recognized that "the project submitted to the Chernobyl Committee did not take into account the effect that this option might have on Professor Nesterenko's work". But what, in any case, is the

---

[113] *Pouvons nous vivre ici?* (Can we live here?), Arte France, 2002.

purpose of this "option", these "quality indicators", these "charts" and "tables" if, as Mr Ollagnon, a team member said to Dr Fernex, "We are doing good work, but the children are more and more unwell"? The only thing that is clear from this letter is that ETHOS is occupying the territory and publicising its work *pro domo* through a substitution of roles in which it takes over the information and local training activities initiated by Professor Nesterenko, and leaves him with the "construction and management" of a mythical and non-existent "national observatory". A public relations and marketing exercise which satisfies those from whom they receive their orders, who in turn present ETHOS as the model and expert in the management of the CORE programme.

In concrete terms, the usefulness of ETHOS' presence among the inhabitants of the contaminated villages is incomprehensible since it is not involved in health protection and the contamination in food does not decrease but rather tends to increase after their visits according to measurements made by Nesterenko.

I refer the reader to Michel Fernex' analysis of the ethical aspects of this programme launched by the nuclear lobby in an article entitled "When the nuclear lobby attacks its victims". In the introduction entitled "The key lies", he writes:

> Teachers, PhD students in agronomy, sociology, technical studies and physics, were brought together to work on the project ETHOS in the contaminated zones. The role imposed on them by the lobby, of which they are probably unaware, is the elimination of existing structures that provide radioprotection for the population. In fact, any measures that draw attention to the seriousness of the radioactive contamination of the country and its impact on people's health are unacceptable to the nuclear lobby".

Chapter III

CORE: AN INEFFECTIVE AID PROGRAMME

On 6th December 2003, *Le Figaro* announced the launch of the program CORE in an article optimistically entitled "A turning point in international aid":

> There was much emotion at the meeting in Paris on Thursday which brought together Belarusian leaders of the Chernobyl Committee, representatives of UNDP[114] and UNESCO as well as French organisations and associations that are to participate in the program CORE (Cooperation for Rehabilitation). Seventeen years after the explosion of the Ukrainian reactor, this is the first time that an international development programme has been launched in Belarus. CORE is the successor to the ETHOS project which was set up at the end of the 1990s by a handful of experts in radiation risk from France.
>
> On 2nd December at the European Commission, France, Germany, Italy and Great Britain finally provided their support to the programme, after President Alexander Lukashenko, who governs the country with an iron fist, finally agreed to repeal a law which imposed taxation on humanitarian aid. CORE will be involved in the most contaminated districts: Bragin, Stolin, Slavgorod and Chechersk. Overseen by the Chernobyl Committee (its director, Vladimir Tsalko is a prefect) CORE is involved in a number of areas, including health, radiology, economic and social aspects, education and culture.
>
> The health component, directed by the IRSN, will consist of an evaluation of the health of children aged 0-15 years. This is an important step because unusual illnesses are regularly diagnosed in this age group. What illnesses do the children suffer from? "The Belarusians do not define the illnesses in the same way, so we do not really know" explains Catherine Luccioni of IRSN. Under the direction of European scientists, the evaluation will be undertaken by doctors at the Chechersk hospital, and paid for by IRSN, which is allocating 50,000 euros to this operation. In about a year we may finally know whether these illnesses are due to chronic contamination at low doses, or not. On its side, Médecins du Monde will undertake various activities with infants and pregnant women.

---

[114] United Nations Development Programme.

A victory for the imprisoned Professor Bandazhevsky.

"I had the privilege of visiting Professor Bandazhevsky in his prison cell. I can assure you that he would be more upset than anyone if the work he initiated on children's health in the contaminated territories, was not pursued" declared Stéphane Chmelewsky, French ambassador to Belarus, by way of encouragement to the CORE participants.

I was familiar with the CORE programme because I had read the draft document which had been sent to us *via* the European Parliament by the Office EuropAid Cooperation Office at the European Commission in Brussels.

On 18 June 2003, on behalf of the association Children of Chernobyl Belarus of which I was secretary, I sent a detailed critique of the programme to European parliamentarians and political and institutional authorities. In the covering note I explained that:

The program ignores health problems in a region where more than 80% of the children are ill as a result of the Chernobyl disaster. This figure was 20% before 1986. The memorandum of the programme CORE, to which thirteen international partners[115] have signed up, promises an independent audit in five years time to evaluate its effectiveness. In our view, the critical analysis that we present here must be taken into account from the start of the project, because the health catastrophe in the contaminated territories is worsening and reaching epidemic proportions. The contaminated populations, abandoned for seventeen years by the international community, cannot wait another five years for a project that offers no qualified medical intervention.

I have restricted excerpts from the report I wrote criticising the CORE programme to what is strictly necessary. Some of the people involved in the programme at the time I sent the report, have since been replaced. I added a paragraph to the 2003 text (see page 294) to include Belrad's assessment of the first stage (between 2004–2005) which demonstrates the ineffectiveness of the CORE strategy in terms of the radioprotection of children.

---

[115] The Belarus government Chernobyl Committee; the United Nations Development Programme (UNDP); the French Embassy; the German Embassy; the European Commission; The Swiss Directorate for Development and Cooperation; UNESCO; the World Bank; the European Committee for Partnership and Preparation of CORE; the executive committee of the Bragin district; the executive committee of Svetlogorsk district; the executive committee of Stolin; the executive committee of Chechersk district.

ETHOS IS REBORN FROM ITS ASHES
IN THE FORM OF CORE

DOCUMENT

CORE—*Cooperation for the rehabilitation of living conditions in the territories of Belarus contaminated by the Chernobyl accident.* (Project, Version of 15th November 2002).

While reading my comments on this EU financed program, I suggest that the reader bear in mind the famous report sent by Bandazhevsky to the Belarus government mentioned above (see p. 200).

Everything that CORE proposes in its text, was learned by ETHOS from Nesterenko, before they removed five of the Stolin district villages from under his control, where he had worked for more than ten years in the local radiological monitoring centres (LRMC) that he had founded. It was Nesterenko, in fact, who created, beginning in 1990, 370 local radiological monitoring centres (LRMC) for food in the most contaminated villages; he who trained professionals in the villages (doctors, nurses, teachers etc) in radioprotection; it was he who gave instructions to children who are more receptive than adults, so that they, in their turn, could teach their parents about ways to prepare food before cooking;. he was the first and the only person to undertake systematic measurement of the internal contamination in children's bodies and to provide this data to the Ministry of Health and to Professor Bandazhevsky, who correlated the data with various illnesses he was studying.

But here the parallels between Nesterenko and the authors of the ETHOS-CORE programme end. There is a major omission in the ETHOS-CORE programme and an ideology which totally changes the nature and renders inoperable the proven radioprotection measures advocated by Nesterenko and Bandazhevsky. The scientific contribution of these two scientists was the demonstration of the nocivity of the chronic incorporation of radioactive caesium to cells and tissues, as well as the possibility of eliminating this radioactive poison by administering pectin orally. This problem of eliminating caesium-137, when it has been incorporated through contaminated food is not taken into account within the CORE programme. On the contrary it is opposed. If, as Nesterenko has recommended, a policy of prevention through the administration of pectin-based adsorbents were officially adopted, it would radically change the nature of this European programme. Were pectin to be recognised as being genuinely effective and useful, were the health benefits of this natural product that accelerates the elimination of radionuclides from the body to be recognised, it would then follow that there really had been mass contamination, that it really was caused by caesium-137, dispersed by the fire at Chernobyl and not by "stress", and that it necessitates, if not the evacuation of all children from the contaminated territories, at least the urgent distribution of the food additive to slow the pace of this health catastrophe. But ETHOS-CORE refuses to finance pectin even while it rehabilitates the territories. In their statement of intention there is no explanation of how the *rehabilitation* that they recommend could guarantee the protection of the 80% of children of Belarus who are ill as a result of the Chernobyl disaster. Before 1986, only 20% of children were ill, according to official figures from the Ministry of Health[116].

SOME QUESTIONS AND COMMENTS ABOUT THE CORE PROJECT[117]

CORE.—Preamble: why the project CORE? (p. 3 in the original document)

"The choice of a cooperative programme for rehabilitation, on which CORE is based, derives from the partners' shared conviction that the situation experienced by the inhabitants of contaminated territories following *the Chernobyl accident* and the search for methods of rehabilitation is a major issue not just at a local and national level in Belarus but for the international community.

This view results from the shared feeling of great vulnerability of people and societies when faced with a situation of large scale radioactive pollution. It is an issue that confronts all countries that have chosen nuclear power as part of their energy production, as well as their neighbours. Knowledge of this situation and of what is at stake has become an inevitable dimension in developing energy policies which have an impact on people as well as the environment.

Beyond the issue of nuclear power, the programme partners are committed to the belief that this situation is central to the *ethical problems* posed by technical and scientific development and its effect on human beings".

Comments.—Right away, the real issue has been skirted, diluted.

How can an accident at a nuclear reactor be considered "beyond the issue of nuclear power"? The incoherence, the absence of any logic, so absurd that it renders it invisible, creates the illusion that this is a serious and pluralistic approach, but the entire project comes to a standstill: having removed nuclear power from the debate, the moral question relating to "its effects on people" is raised. But the "effects of what"? Is this ethical question about technical and scientific development being raised in relation to Chernobyl? In this text, you have to pay great attention to each word. As Orwell had already shown "it is words that determine action". A conflict that involves the lives of millions of people is made to depend on nuance, a semantic hide and seek, verbal evasiveness. According to the official dogma, internal low dose radiation has no significant effect on human health. CORE gears

---

[116] Statement from the President of the Academy of Sciences of Belarus, in December 1999, confirmed by the vice Minister of Health in Belarus at a parliamentary hearing on the consequences of the Chernobyl disaster, in April 2000.

[117] The commentaries refer to the key phrases that I have put in italics in the original.

up therefore to rehabilitate the area. You can live there perfectly well; you just have to follow the instructions of the "experts". Not only have they been subjected to the contamination but now the victims are subjected to authoritarian rule: they are crushed twice. If they fall ill, it is because they didn't follow instructions. It is their fault. They are both victims and guilty. In this scenario, pectin serves no purpose. It is just an embarrassment because it lays bare the radiological cause that needs to be concealed. This is followed by a sociological analysis of symbols and behaviour, the way people will live and what will become of them in this "new world".

> CORE.—"The implementation of *rehabilitation measures in no way interferes* with the individual and collective choices of those directly affected by contamination in the territories affected by Chernobyl".

What, concretely, does "rehabilitation" of an area, saturated in caesium 137, mean? What freedom of "choice" can there be when one is forced to continue living in such a place? A choice between what and what? The implementation of rehabilitation measures by CORE, which is funded by European taxpayers, predetermines and conditions any decision that might be made by the population themselves.

Censorship of the scientific studies undertaken by Professor Bandazhevsky, and of Nesterenko's radioprotection measures, combined with this supposed management of the consequences of the accident, financed by the EU, guarantees that people will remain in "ignorance and uncertainty" in a new world, in other words a world to which they are condemned to live and to which they will have to "accommodate" themselves.

> CORE.—"CORE aims to create conditions of choice and an *autonomous and informed way of living for the people living in these territories.* This is why the approach developed by CORE is above all based on people and their *individual and collective capacity for initiative.* The aim is to support local initiatives and implement educational measures".

What is this autonomy? What individual and collective initiative could possibly be commensurate with the scale of the disaster? "Every man for himself!" Bandazhevsky wrote, just before he was thrown into prison as the government abdicated all responsibility for people's health. Because it is the government—governments, Europe, the rich world—that is responsible. Nesterenko called for the immediate evacuation of children living in a radius of 70 km at first, and then of 100 km, around the reactor. And despite the fact that ETHOS promised to involve him, Nesterenko was not included in the preparation committee of CORE. He did in fact participate, (but not as a right) in the preparation phase of the project but only after the German and French ambassadors intervened.

CORE.—"With this perspective in mind, CORE aims to encourage the development of a *shared heritage* at local, national and international levels, through recognition of, *understanding and remembering the Chernobyl event* and the response made by humankind to the situation".

A museum to commemorate a past event? The new illnesses, the miscarriages, the abortions, the birth defects occurring now and in the future are not part of the "Chernobyl event"? In order to understand what is happening in the contaminated territories and organisms from a medical urgency point of view read Dr Michel Fernex's note[118].

In the chapters entitled "Preparation", "Projects", "Partners", there are lists of the people involved in the programme, some of whom are known to us

CORE.—Some of the contributors to the preparation phase (p.4 in the original document)

Preparation Committee

Valéry Berestov, President, Executive Committee of the District of Slavgorod; Neil Buhne, United Nations Development Programme (UNDP); *Vladimir Shevchuk*, Vice President, Chernobyl Committee, Belarus; *Gilles Hériard Dubreuil*, ETHOS; *Norbert Jousten*, Chief of Delegation, European Commission for Belarus and Ukraine; Sergei Kulyk, World Bank; Guillaume Kaspersky, French Embassy in Belarus; Sylvie Lemasson, French Embassy in Belarus; *Jacques Lochard*, ETHOS; Vassili Maximenko, President, Executive Committee of the District of Chechersk; *Henry Ollagnon*, ETHOS; Vladimir Pashkevich, President, Executive Committee of the District of Stolin; Richard Stefanovich, President of the Executive Committee of the District of Bragin; *Vladimir Tsalko*, President, Chernobyl Committee of Belarus; Matthias Weingart, Swiss Consul, Belarus, Coordinator Humanitarian Aid.

Details:

• Vladimir Shevchuk is the Ministry of Health's man, close to Tsalko in ComChernobyl. He is the one who informed Nesterenko that the villages in Stolin District had been removed from his project at the instigation of ETHOS.

• Gilles Hériard Dubreuil, ETHOS, Director of Mutadis. This study group, specialized in social management of risk, is responsible for coordinating the CORE project. As such, he may play an important role in the decision to accept or refuse financial support for pectin, requested by Nesterenko. For the moment, he is opposed.

---

[118] See Michel Fernex at the end of this chapter, p. 297.

From 2nd-4th May 2003, Nesterenko took part in a conference in Brühl, Germany, where the most interesting discussion took place during the session "Debates at the IAEA, at the European Commission and at the United Nations on the Consequences of Chernobyl". The presidents (and speakers) at this session were Dr Edmund Lengfelder and Hériard Dubreuil, CORE programme, Paris.

Hériard Dubreuil, who arrived from Paris accompanied by Professor Ollagnon, asked the Germans to participate in the CORE programme. After Ollagnon's presentation, Lengfelder used the occasion to show his hostility to the nuclear lobby publicly by posing the following question: "ETHOS was financed by Euratom, isn't CORE also a project of the nuclear lobby?"

Before the session, Hériard Dubreuil and Ollagnon told Nesterenko that the French Ambassador had sent them *Information Bulletin* No. 23 from Belrad, in which CORE was criticised for its refusal to protect the children with pectin cures. Hériard Dubreuil declared that "in future French scientists would give their approval for the use of pectin". Nesterenko replied that the children were ill today and they didn't have time to wait for the French scientists.

After our denunciation of Nesterenko's expulsion from the villages where he had worked for years, meetings took place between Nesterenko and the two ETHOS managers. The two men declared themselves willing to collaborate with Belrad in a joint project that would be presented to the European Commission (the TACIS programme)[119]. During one of these discussions in their hotel room in Minsk, they intimated to Nesterenko that he would have to cut off relations with Professor Fernex if he wanted to work in collaboration with them. They were forced to back down when Nesterenko interrupted the conversation and made for the door.

In reality, ETHOS, with CORE already in preparation, was shirking its responsibilities. Nesterenko had written to them, in fact, at the beginning of April 2001:

> At our meeting on 17 April, I propose that we examine the possibility of:
> 1. re-establishing the 20 LRMCs which were closed down in schools in the Stolin district, with the financial support of ETHOS (cost of equipment for one LRMC is 1800-2000 dollars and the cost of running one LRMC is about 900 dollars a year);
> 2. re-establishing the regular distribution of animal fodder mixed with adsorbents for the cattle of the inhabitants in the 30 villages of the Stolin district, throughout the whole period of lactation;
> 3. buying and installing cream separators in school kitchens and nurseries in the 30 villages;
> 4. with financial support from CORE, providing pectin prevention (4 times a year) for children in the Stolin district, and undertaking 5-6 annual whole body measurements of these children (annual cost per child, 25 dollars);

---

[119] Telephone conversation between W. Tchertkoff and J. Lochard on 8th February 2001 and during a meeting in Minsk in June 2001.

5. organising advanced radiometry courses for 80 teachers and health workers, training them for work in the LRMC and teaching them the principles of radioprotection in schools;

6. organising a month's convalescence every year for children of the Stolin district, in French families, as is being done currently in Germany, UK, Ireland, Spain and Italy, in order to restore the health of Belarusian children;

7. undertaking a three year follow-up of the health status of children in 36 secondary schools, 12 colleges and 12 primary schools of the Stolin district, within the framework of a special international programme.

I propose a discussion of this project for radioprotection of children who are victims of the Chernobyl disaster in the district of Stolin.

I want the children of the Stolin district to benefit from the genuine, effective radioprotection, that will result from our joint scientific work.

Nesterenko's proposals were not discussed at the meeting on 17th April "because the ETHOS team had no new funds to pursue its activities". In fact, ETHOS was preparing to relaunch its programme of *sustainable management of radiological quality and social confidence* under the new name CORE. This advertising gem, "*radiological quality*"—the leitmotif of the ETHOS strategy, given renewed prominence in CORE, serves to remove reality, to pass over the essential, to do things but not to do them, to say things but not to say them, and ultimately to cover up the reality. Let me explain.

An oxymoron, as the dictionary tells us, is a figure of speech in which two words with opposite meanings are put together for greater effect (for example, passive aggressive). But there are more subtle and less innocent operations. Take for example, the well known term *greenwash:* this is the name given to the clever marketing techniques that are designed to provide an attractive veneer to concepts that are a little jaded or have had a bad press. An image is created of something that is worthwhile and entirely new, which in fact remains fundamentally unchanged. The term "radiological quality" is a good example of this. What is it?

The radioactivity in the territories contaminated by the nuclear disaster can just as well be good quality or bad quality? A trivialisation, emptied of meaning. Thus artificial radioactivity, in contact with the human organism, poses no problems in itself, but by virtue of its supposed good or bad quality? What "radiological quality" can there be in the "effects of proximity"[120]? Is there some other meaning that has escaped me in this marketing operation, this *greenwash* invented by Mutadis-ETHOS-CORE?

"Language plays a critical role in the construction of political lies", as Jean-Philippe Jaccard observes. "It allows the powerful to name, define and describe the world, not as it really is but as they would like it be perceived by the population. It is

---

[120] Described in Part One, Chapter V, p. 54.

true of all systems and starts at the level of vocabulary: in the USSR, they used the term "camp for re-education through work" to designate a concentration camp of the worst kind[...]. Linguists called this opacity, the loss of any connection between words and objects, a language artificially created and therefore false[121]".

The archetype of the oxymoron is the famous *Arbeit macht frei* (work makes free) the sign placed by the Nazis over the entrance to Auschwitz. Financial support from the European Union for this propaganda about *radiological quality* in the Chernobyl gulag is essentially no different. Are European MPs aware of this or have they already been taken in by this scientific greenwash?

• During a contentious discussion (CORE meeting on 12th July 2002 in Minsk), Norbert Jousten disagreed with Professor Kenisberg, and argued that local initiatives and an independent system of radiological control must be developed. He was therefore, defending Nesterenko's position. Remember that Jousten was responsible for the four Belrad projects, which in 2002 were not under TACIS control, despite assertions to the contrary. During a meeting at the French embassy in Minsk, in December 2002, he said he was expecting a response any minute, from Brussels, which never arrived. Who is responsible?

• Jacques Lochard, director of CEPN. During the same discussion on 12th July 2002, Lochard arguing against representatives from the Belarusian Ministry of Health, said that "all projects should support the LRMCs and the whole body measurement of radioactive contamination with spectrometers. The population does not trust government information about contamination of local food products, nor in the official monitoring of the levels of caesium 137 in children's bodies. The inhabitants of these areas have expressed their wish for an independent, non-governmental system of radiological control to be established".

He stressed "the importance of the creation by Belrad of a network of LRMCs and of 8 minibuses, mobile laboratories for anthropogammametry". It seems therefore that he was willing to restore Nesterenko's functions and to make use of Belrad's expertise. But make use of them to what end? Because Lochard, along with Hériard Dubreuil, opposed the funding of pectin-based adsorbents, the one benefit that the children of contaminated areas could have gained from an international programme such as CORE.

Taking advantage of the work of the dosimetrist at the Olmany LRMC over several years, ETHOS appropriated Nesterenko's data and then used them in ways that were contrary to Belrad's objectives: to justify keeping inhabitants in contaminated territories by reassuring public opinion about the risks of radiation,

---

[121] Professor Jean-Philippe Jaccard, University of Geneva, *Le Courrier*, 7th June 2003.

and by minimising—and even ignoring—its direct effects on health. A cursory glance at CORE's budget and a comparison of the amounts allocated shows this quite clearly. 25% was allocated to medical and radiological monitoring of inhabitants and 75% to the socioeconomic and cultural context. The ratio of the amount destined for what constitutes a medical emergency for people who have lived with chronic contamination for 17 years and the amount destined for socioeconomic and cultural needs is the exact opposite of what it should be. The extent of the underestimation of the health component is indicative of a politically motivated decision, one which preconditions and determines all the individual and collective decisions of those concerned just as ETHOS-CORE hoped. If Lochard had shown the same determination to support the distribution of the adsorbent pectin as he showed for the re-establishment of the LRMCs, his stated desire to work with Belrad might have been credible. Was it not Lochard who used the phrase "We need to occupy the territory", reported to me by friends from the University of Caen. "What for?" one wants to ask him.

• Henry Ollagnon, National Institute of Agronomy, Paris-Grignon, ETHOS. At the conference in Stolin in November 2001, his statement of disillusion in relation to the health of children was dramatically reinforced by the pediatrician, who summarized the situation during this conference. Ollagnon approved sustainable development in contaminated territories to ensure their rehabilitation. Prior to 1986, and without "sustainable development" the children were full of joy, living in a beautiful natural environment. Today they are dejected, beaten down, undermined, as we have observed and as their parents, the old peasant woman of Skorodnoie[122] and the doctors that we interviewed testify.

• Vladimir Tsalko, President of the Chernobyl Committee of Belarus (ComChernobyl), member of the *nomenklatura*. He did not appear to share the position of the Minister of Health, supported by Professor Kenigsberg, which minimised the consequences of Chernobyl.

> CORE.—1. Presentation on the context of the preparations for the CORE program (p.7 in the original document).
>
> "Although 80% of the populations live in regions where the level of contamination with caesium-137 in soil is relatively low, for a variety of reasons, the possibility of continued radiological risk cannot be ignored. Indeed, *complex radiological processes of re-concentration can lead*, in certain situations, to *levels of exposure* of the people in these

---

[122] Film *Controverses nucléaires* of Wladimir Tchertkoff, Feldat Film, 6945 Origlio, Switzerland; email:eandreoli@vtx.ch. See Part Five, Chapter II, p. 318. (English version *Nuclear Controversies* https://www.youtube.com/watch?v=MZR_Fvp3RrQ)

regions, in particular the children, that may be problematic. A second argument is based on the precautionary principle, given the uncertainty about the long term consequences for human beings, associated with life in an environment that is characterized by *diffuse radioactive pollution*".

Contortions.

Unable to deny the increase in illness in children and wishing to avoid the hypothesis that food is the source of internal contamination, the CORE document explains "non-negligible radiological risk" in terms of levels of exposure (most probably external). But this is impossible according to the official dogma, because the levels are actually very low. Yes, but they suffer "complex radiological processes of re-concentration" (Where? In the air, in the soil, or in their organs?).

"Diffuse radiological pollution": this is not true. The radioactive pollution is not diffuse, as the studies of Nesterenko and his colleagues show. Radioactive pollution is not uniform either, particularly in the food chain and in vital organ systems, as Bandazhevsky and his colleagues show in their studies.

> CORE.—"Radioactive pollution has a significant impact on the economy in areas which depend heavily on *agriculture. It is a source of deep concern* for the population in terms of possible health effects, especially in children. The presence of radioactive contamination is also the source of depreciation in living condition, natural resources, and the *cultural, symbolic and aesthetic* values that are attached to these living conditions. Finally, the process of evacuating and re-housing populations since the Chernobyl accident has deeply disturbed the social and demographic equilibrium of these areas".

A deft sidestepping of the issue. We begin with a discussion about agriculture but are immediately diverted. It is a source of deep concern and not about internal contamination with Cs-137 through the ingestion of food. They do not explain why agriculture should be a source of worry. In order to avoid entering an area of discussion that is forbidden by the scientific establishment (IAEA, WHO, UNSCEAR), heartfelt concern is expressed about "cultural, symbolic and aesthetic values". ETHOS-CORE will teach the farming community a new culture: optimism, initiative, rules of behaviour but they will offer no help in relation to issues of real danger and urgency: illnesses of vital organ systems. Illnesses against which they refuse to take preventive measures, and which are then not treated properly because information relating to these illnesses is prohibited.

In reality, the people we interviewed and filmed in September 2002 in the villages east of the river Sozh, in the Slavgorod[123] district (the people are being monitored by a German organisation called Julich which carries out measurements but does

---

[123] See "Nesterenko's villages". Part Five, Chapters VII to XII, p. 357–403.

not protect them) have no idea what they are ingesting. In true "Soviet" style, they are used to submitting, to feeling more or less unwell and to dying prematurely. The situation for children is even worse than for adults. In the beginning they appear to be perfectly well, the harm being done is invisible, and then when illness develops, it is already too late, whether it is in the heart, the eyes, gastritis or some other illness. 80%, 90% or 100% of children in these regions are ill. The people have gone from submission to the Soviet system to submission to an elusive radioactive power. In both cases, there is resignation, powerlessness. Disinformation: we did not observe "deep concern" in these interviews, but ignorance, a lack of information and a total absence of monitoring and qualified medical help. Parents do not attach great importance to their children's symptoms, such as persistent headache, breathlessness or fatigue because they do not understand them. They do not know that the heart is affected.

"The people of Belarus are quite simply the guinea pigs in an enormous radiological laboratory, from which the United States are gathering their scientific data. [...] From the point of view of medical ethics in the United States and in Europe, scientific research on patients who are ill is not admissible unless, right from the outset, the treatment of the patient is guaranteed. Why should this be different, just because United States research is being conducted in Belarus? [124]" In the case of the ETHOS-CORE programme, no scientific research worthy of the name is even undertaken[125].

> CORE.—"Capitalising on the valuable experience gained by the ETHOS project over the last few years in this area, the aim is to develop initiatives in the Stolin district based on coordination and cooperation adapted to the *complexity* of the situation. The projects developed will be characterised by strong levels of integration within health, economic development and radiological protection. In addition, the projects must involve cooperation at different levels: local, national and international."

What complexity are we talking about?

Nothing is more complex than a living organism. But the CORE program turns its back on this in favour of promises of sustainable economic development, even if this is in contaminated terrain ("[...] integrate the presence of radioactivity into daily life as a new component of existence [...][126]" The word health is mentioned here in vain: the research undertaken by Professor Bandazhevsky at the Gomel

---

[124] *Zeit-Fragen*, 11, Jahrgang No 10, 17th March 2003; and Part Three, Chapter VI, p. 245.

[125] On this subject see the final note from Michel Fernex, p. 297.

[126] Quote from ETHOS' glossy brochure with colour photos of the peasants of Olmany, Cf Part Four Chapter Two, page 267. *Regards sur Olmany* ©ETHOS, Paris, 1998

Institute, showing a quantifiable linear correlation between levels of *Cs-137* radiocontamination measured in children and various vital organ and system pathologies, is ignored. Also in vain is mention of the word radioprotection, for what are we protecting and how, when we do not know and do not want to know what is happening inside the contaminated organism? The refusal to finance the use of pectin is negative proof of this wilful ignorance.

In a letter addressed to the ambassador of Belarus in Paris on 20th February 2002, Bella Belbéoch (GSIEN) summarises:

> At the Gomel Medical Institute, a great number of clinical examinations were undertaken at first hand, and the results of immunological, haematological and biochemical tests were analysed by Professor Bandazhevsky and his colleagues from the population of an area that was extremely contaminated by Chernobyl. The health status of thousands of adults and children was the subject of a rigorous investigation.
>
> From clinical data, laboratory testing, autopsies—of both adults and children—and animal experiments, Professor Bandazhevsky showed that chronic incorporation of long lived radionuclides such as caesium-137 plays a leading role in the pathological processes. Incorporated Cs-137 is pathogenic and leads to the deterioration of cellular membrane structures and dysfunction of metabolic processes, interdependent morphological and functional alterations, resulting in problems in all vital systems and organs (heart, liver, kidneys, endocrine glands). The severity of these disorders increases with the concentration of Cs-137 in organs that are affected simultaneously and chronically by this radiotoxin which is not distributed homogenously between different organs of the body, as measurements during autopsy have demonstrated.
>
> CORE.—2. Methodological options of the CORE approach
> *"Management of the radiological aspect of the situation: Radiological quality* is the aim of CORE's work, and will *result* from the dynamics of *sustainable development* (in health, food, environment, agriculture...) of the contaminated territories, and not in isolation."

So the Chernobyl territories could be decontaminated through sustainable development? The term "radiological quality"—in the context of the post Chernobyl health disaster—is an arrogant expression and an insult to intelligence (not to mention morality). When someone has power and money, he can say anything he likes. Like Gentner, from UNSCEAR, speaking at the Kiev conference (June 2001) on the medical consequences of the Chernobyl catastrophe: "For those who believe, no explanation is necessary: for those who do not believe, no explanation is possible" (sic)[127]. You buy them off or you crush them from a position of strength.

---

[127] See Part Six, Chapter II, p. 424.

An absurdity or a lie is presented as axiomatic and whatever conclusions are most convenient are then drawn (An axiom is a truth which cannot be demonstrated but is obvious to anyone who understands the meaning—and in this case, anyone who does not understand the meaning is sidelined.) The impossible is declared possible: post Chernobyl radioactivity may be of high quality on condition that it initiates sustainable development. If it fails the population will be responsible. In short, the scientific truth proposed in these two lines from CORE is as follows: If you increase GNP, the caesium-137 will disappear from the food chain. Belarus is on its knees. ComChernobyl has no funds. A contaminated and humiliated population is being sold this lie about radiological quality—for which it will be responsible—with a promise to finance economic development.

> CORE.—"Using experience gained during the ETHOS project, the CORE programme will encourage the development and *intergenerational transmission of a practical radiological culture* at the heart of the daily life of the inhabitants of the contaminated territories. It will also include projects relating to *memory* and its intergenerational and *international* transmission."

In this paragraph, CORE alludes, indirectly, and then skims over what is certainly the most worrying aspect of the Chernobyl catastrophe: the health consequences for the descendants of the population contaminated by radionuclides. Once more, this aid programme avoids confronting in any useful way the problem of health while showing that it is not completely unaware of it. Why concern oneself about future generations when there is no mention of what it is that threatens them, and no action planned to prevent whatever misfortune awaits them?

Why talk about "culture" and "intergenerational memory"?

In 1956, a group of brilliant genetic scientists, including the Nobel Prize winner H.J. Muller, raised the alarm in a well known statement: "As experts, we declare that the health of future generations is threatened by the increasing development of the nuclear industry and of sources of radiation".

After the Chernobyl catastrophe, many studies were undertaken on rodents in areas that were more or less contaminated. These animals are very representative of humans although their reproductive cycle is much faster. R.J. Baker and his colleagues, in a study of the DNA of a gene transmitted from mothers to babies in the field vole, observed a level of mutations, from one generation to another that was hundreds of times higher than any that had been observed before in the animal kingdom[128]. The environment in which these field voles live has seen its levels of radioactivity decrease because rainfall leeches the Cs-137 downwards

---

[128] R.J. Baker *et al.*, "High levels of genetic change in rodents of Chernobyl", *Nature*, No 380, p. 707–708, 25th April 1996.

through the soil. One would expect the animals to react positively to improved radiological conditions. However, mutations and genomic instability continue and increase over 22 generations. Goncharova and Ryabokon are seeing the opposite of an adaptation to radiation[129].

As humans and rodents are comparable genetically, these studies led Professor Hillis of the University of Texas to conclude in an editorial in *Nature* (25th April 1996): "Today we know that the mutagenic impact of a nuclear accident may be far more serious than previously suspected and that the eukaryotic genome may present rates of mutation that were never considered possible up to now[130]".

But the programme CORE, that has no interest in these challenging scientific questions and no plans for preventive measures for the contaminated population, limits itself to a fatalistic intergenerational transmission of a radiological culture. It is also concerned with memory because Chernobyl now belongs to the past. Radiocontamination and its health consequences are to be considered as facts of nature, to which people all over the world must learn to adapt (international transmission), thanks to the savoir faire (radiological culture) that the "experts" will teach them. "Every man for himself! The state no longer has any responsibility for the health of the population" (Y. Bandazhevsky)

Is it in conformity with the statutes of organisations such as UNDP, the European Commission, the Directorate for Development and Cooperation of Switzerland, UNESCO, the World Bank to finance a mock aid programme rather than demand from its promoters a targeted and commensurate programme of action to combat the evil that is ravaging the population and threatening their descendants?

> CORE.—"In this perspective, the approach will be implemented through the access of the population and local specialists to the *operational and effective means* of measurement and evaluation of the *radiological situation*".

What is the point of knowing the radiological situation without providing preventive measures and qualified medical care unless the aim is to gather just enough statistical data to fill publications justifying the expense[131] without any benefit to health. The effective operational methods to protect the population were, and are those used by the independent Institute of Radioprotection, Belrad, directed

---

[129] R.I. Goncharova and N.I. Ryabokon, "Dynamics of gamma-emitter content level in many generations of wild rodents in contaminated areas of Belarus", 2nd Intern, 25–26th October 1994, conference "Radiobiological Consequences of Nuclear Accidents".

[130] Extracts from a report by Michel Fernex, "La catastrophe de Tchernobyl et la sante" (The Chernobyl disaster and health), May 2000.

[131] Daily allowance for French specialists on mission for CORE. 400 euros is more than the monthly salary of a member of staff at Belrad.

by Professor Nesterenko and the Medical Institute of Gomel directed by Professor Bandazhevsky, that have shown, over ten years of work in the field, what can be done for people's health. Health which, apart from the mention of the word, is an invisible problem in this programme that is being financed by the European Commission.

> CORE.—"an important aspect of the programme CORE is to encourage rehabilitation based *above all* on initiatives taken by the inhabitants of the contaminated territories".

According to the rationale "Every man for himself" denounced by Bandazhevsky, it will fall to the peasant community itself to take responsibility (above all) for the secret wounds to their bodies caused by caesium-137 and strontium-90. But surely Nesterenko, at Belrad, and Bandazhevsky, with his research at the Gomel Institute, were precisely members of this contaminated community in Belarus who took their own initiative and responded to the disaster? Initiatives that CORE seeks to sabotage rather than to encourage or to support.

> CORE.—"The accent is put on prioritising *autonomy* among the local people and on allowing them to fulfil themselves personally. The aim is to encourage the abilities of local men, women and families in the contaminated territories to meet basic needs and to reinforce their capacity to take control over their own lives. CORE aims to encourage the emergence of a "critical mass" of projects that will bring about change within families and their communities to bring about a process of *sustainable development*".

In an area that is contaminated for generations to come . . .

> CORE.—3.1 Medical follow up and quality of health
> "People living in the contaminated territories are concerned about their *state of health*, especially *the health of their children*. Local health professionals are aware of the existence of these problems. However, *the information gathered is fragmented and incomplete*, and this reinforces the feeling of abandonment among the population, and leads it towards an attitude of fatalism and risky behaviours. It seems necessary therefore to implement a project on health problems".

The most effective use of their tax payers' money in the area of information about health, which is currently *fragmented and incomplete*, would be for the countries that fund CORE to ask the Belarusian authorities to re-establish the Gomel Institute of Medicine with Professor Bandazhevsky as its rector, in such as way as to avoid squandering the large quantity of information and knowledge—accumulated with scarce resources—over nine years of rigorous scientific research on the illnesses caused by the Chernobyl disaster.

CORE.—Individual health monitoring of children.

"The aim of this project to monitor the health of individual children is to make an objective assessment of their health to address the concerns of the population. This project will provide the means for local health professionals to undertake a *health assessment of children, in collaboration with experts from Europe*. The health assessment in the contaminated districts (Chechersk then Bragin) will be conducted in parallel with health assessments in an uncontaminated district sharing the same socioeconomic characteristics, which will serve as a reference (this region has not yet selected)."

Back to square one

Seventeen years after the disaster, with research at the Institute of Gomel interrupted and its rector in prison, a health assessment is to be undertaken by local health professionals, who dare not even mention the word radioactivity in relation to the illnesses that they see.

Galina Bandazhevskaya tells us that one of her husband's pupils, appointed chief medical officer in gynaecology in a contaminated area and appalled by the number of birth defects, abortions, and perinatal deaths she was coming across among her patients, began to measure radioactivity in placentas, and received the following warning from a local superior: "So, what happened to Bandazhevsky isn't enough for you? Describe these pathologies any way you like but don't link them to caesium-137". This young woman has a child to feed. She cannot risk losing her job. She gave up her analysis. *[These words were spoken in March 2003]*

What is more, these local professionals will be working under the supervision of European experts. What do these expensive European experts know about these problems? They have never undertaken medical research around Chernobyl and they will arrive with their mathematical formulae based on the Hiroshima model, according to which only external exposure to high doses of radioactivity can cause illness. Who pays for their travel, their conferences and other events? And who will benefit in the end? The comparison between the amounts allocated in the CORE budget for "indigenous" staff and for "experts" needs no comment.

The rule to apply when providing technical assistance to a poor country—and Belarus, following the accident at Chernobyl, is one of the poorest countries in Europe—is to strengthen local structures and not to replace them. The lobby has managed to get rid of Bandazhevsky's institute. Now it is Professor Nesterenko that they are trying to neutralise.

Two years have passed since *Le Figaro* wrote: " In about a year we may finally know whether these illnesses are due to chronic contamination at low doses, or not. Does Ms. Luccioni, who stated that "The Belarusians do not define the illnesses in the same way, so we do not really know", know anything more today?

First assessment, in October 2005—The ineffectiveness of CORE.

In his *Bulletin d'information* No. 28 of October 2005, sent to the Belarusian authorities, the ambassadors of France, Germany and Great Britain, the Directorate for Development and Cooperation of Switzerland, and to the IRSN, responsible for the expert evaluation of the efficacy of pectin-based adsorbents, Professor Nesterenko presents the results of CORE'S first year of activity in the district of Bragin.

A comparison of results obtained by the programme CORE, *without pectin* and of those obtained four years before by the Belrad Institute, *using Vitapect*, in the same villages in the district of Bragin, shows the failure of CORE's strategy and an unforgiveable loss of time for the contaminated children who are still being deprived of an effective preventive treatment because CORE refuses to finance it.

Proceedings of the Belrad Institute:

> "The medical programme of the CORE project limits itself at present to the collection of data on the health status of inhabitants and the health effects of low doses of radiation.
>
> It is highly regrettable that the proposal of Belarusian experts to implement the first steps of the project on practical measures for radiation protection of the population (mostly children), such as the introduction into the diet of a regular intake of pectin-based adsorbents, was not accepted as part of the CORE project.
>
> Analysis of the measurements made by the BELRAD Institute for the CORE project (under contract with the Agency for Development and Cooperation of Switzerland), shows that the rate of accumulation of Cs137 in children's bodies measured during examinations between spring 2004 and spring 2005 remained unchanged (eg. in the villages of Komarin, Mikulichi, Khrakovichi) or even increased (in the village of Burki) ... Whereas in 2000–2001 the BELRAD Institute monitored levels of Cs137 contamination in children from the same villages in the Bragin district where they had been given the pectin-based food additive (Vitapect) between examinations. In one month, the rate of Cs 137 in children's bodies had decreased on average by 27%. In the villages of Burki, Mikulichi and Khrakovichi, it had declined by 32% and in the village of Komarin, by 35%".

> V. Nesterenko, "The Chernobyl Catastrophe"
> *Bulletin d'information*, No 28, Minsk, 29th August 2005.

The question that has to be answered is this: do we have the right to leave untreated, contaminated children that are the subject of our observation, while an effective and safe treatment is available?

> CORE.—The health education of pregnant women
> "This project aims to reduce the radiological risk posed by the ingestion of food through teaching pregnant women who are being monitored at the hospital in Stolin to control the risk themselves To allow *pregnant women to reappropriate their environment,* this project will build on the approach developed and implemented in the program

ETHOS. This process of reappropriation will be based primarily on the integration of a module of practical radiological culture and of operational dosimetry that is already in place for pregnant women."

Chernobyl has severed the natural biological link between life and the environment that it has poisoned. Now life is summoned to reappropriate this environment under the direction of well-fed experts from Paris. Young women are to be taught to cope by themselves in this contaminated gulag. To their burden, is added the guilt of being fatally bad mothers, because inevitably they will be poor pupils taught by poor masters. The poisoned mother will be a source of poison to the new life inside her—new life that trustingly takes all its sustenance from her. This happy symbiosis has evolved into a monstrosity. What reappropriation are they talking about, these blind counterfeiters?

Young mothers should be evacuated from contaminated areas throughout the period of pregnancy and lactation. Levels of radionuclides in the placenta should be measured systematically. The correlation between these levels and congenital malformations and perinatal mortality needs to be studied scientifically[132].

CORE.—"A *retrospective study* of dosimetric data should be undertaken. For each child, *medical and dosimetric data will be correlated* to determine whether chronic contamination with caesium may have played a role in the pathologies observed".

Making this correlation is a step forward. But who will undertake the work? Using what methodology? Will Bandazhevsky's work be taken into account? In any study, it is essential to know who is behind it, what are the curriculum vitae of the authors and co-authors of the projects. As for retrospective dosimetry, this is a farce. Direct measurements should be taken today. The children are ill today. Caesium-137 is present today in their organs.

CORE.—3.2 Agriculture, rural development and management of the living environment.
   Project for the development of family income in rural areas in economic recession after Chernobyl (*English*)
   Summary of the translation.—"The radioactivity circulates mainly within the food chain, in other words from agricultural production, but it is also transmitted to the families through the exploitation of forests (wood, fire, fruit and game).
   The project involves teaching agricultural techniques to produce potatoes, vegetables, and dairy products containing fewer artificial radionuclides. This method of production is expensive."

Who will be given this aid?

---

[132] See Michel Fernex's note at the end of the chapter, p. 297.

These projects are difficult to manage. They will need a new bureaucratic structure.

The feasibility of such a project needs to be demonstrated in terms of safety, low prices, job creation. It makes use of the region's biomass. What remains will be used in industry.

Nothing is said about protecting workers, loggers, sawmill workers, and heating engineers, chimney sweeps, etc. It is not clear what kind of forestry work will be developed, in which all grades of wood are supposed to be "renewable". There is no mention of filters for Cs-137 and Sr-90 and for transuranics. Is there not a risk of transporting the radioactivity stored in the forest, into an urban environment, with dust and smoke that could contaminate people who have so far been relatively protected?

CORE—3.4. Project in radiological culture, intergenerational and international memory and education

This is the cultural section. Chernobyl will be turned into a museum of "intergenerational and international memory and education".

It is here that the West's investment in ideological pretence seems most significant.

The emptiness regarding the problems of health is at its most apparent here. A stupefying void!

This part of the project has a delusional character, as the reality of Chernobyl, and the only thing that justifies all this effort, is made an abstraction: the consequences of radiation, mainly internal, on children, on pregnant women, and on men whose descendants will suffer health and genomic effects.

CORE.—3.4.1. Context.

"The economic difficulties encountered in the contaminated territories of the Republic of Belarus result in the inhabitants being dependent on farming in small family plots, gathering produce from the forest, fishing and hunting to meet their daily needs. *All these products are likely to have very high levels of contamination according to the locality*. In everyday life, information about radioactivity and recommendations regarding food and the environment are not well disseminated and even where they are, not applied."

Does this mean that CORE recognizes that contamination at low levels of exposure is largely internal, *via* the food chain, and that sometimes levels of contamination are very high? If so, we have to ask why, not only nothing is being done to research and offer radioprotection against this internal radiation, but obstacles are put in the way of this research and radioprotection.

"This is an original initiative that is in marked contrast to previous approaches. It is a decentralized approach, based on the strong involvement of local people in the rehabilitation process. The initiative aims to create conditions that will allow *the residents of the contaminated territories themselves to recreate their quality of life* and, within this framework, *to manage the radiological risk generated by all their activities on the contaminated territories.* The reconstruction concerns all aspects of their daily lives that have been affected by contamination: health (in particular of children), the environment, safety at home, living standards, professional activities, social and cultural activities, individual and collective identity".

Incredible! The risk from radiation was not caused by the explosion at the nuclear power plant. The people themselves are guilty.

The text in italics speaks for itself. It is at least true, it tells the whole truth. From now on, the residents really are responsible themselves, trapped in their everyday activities—in their lives!—for the harm that will come to them. As for the contamination of their land, that is now simply a fact of nature according to CORE. With instructions from the *experts*, the inhabitants must learn how to accommodate it. It is they who will be responsible and it will be their fault alone if it does not work.

It has to be understood that CORE represents through and through the moral stance and the policy adopted by the European Union and all of us who belong to it. We have financed and continue to finance this lie and this arrogant deception; the complacency and wilful ignorance of the United Nations agencies responsible for nuclear power, using European citizens' money.

No-one is more qualified than Professor Fernex to conclude this chapter. He is a doctor, and he worked for the WHO. He has compiled the list of what needs to be done at Chernobyl:

"All research must have a clearly expressed purpose. We are beginning this research seventeen years after the disaster. For whom? To protect children who are ill? To protect the nuclear industry (and so neglect the victims of that industry)? If representatives from the nuclear industry are involved in projects for the victims, then we will have the same results that we got from WHO's IPHECA project, which was planned and prepared by the IAEA (just as studies of smoking related illness were planned by the tobacco industry that financed them). The IAEA gives priority to the study of dental caries and ignores genetic effects.

If the aim is to find out whether ionising radiation from artificial radionuclides from Chernobyl is playing a role in the deteriorating health of the population, we must formulate a set of hypotheses and a clear research objective. For each hypothesis, the protocol requires a double blind study, detailed questionnaires, finalised and available before the study begins. Statisticians need to specify how many subjects need to be included in the studies. The radiometrists must collect anthropogammametric measurements from the indigenous rural population and the data must be kept (secretly) on the computer.

Clinicians need to study the pathologies that relate to their own areas of expertise:
—Ophthalmology: Cataracts, number of opacities in the lens of children, retinitis.
—Cardiology: Medical history: ability to take part in sport, etc.. Pain, fatigue. Blood pressure. Children with hypertension, hypotension or normal. Complications. ECG. arrhythmias, Long QT syndrome (LQTS), repolarisation disorders. Clinical abnormalities. Heart defects.
—Gynaecology and Endocrinology: hormonal disorders at puberty in girls.
Collect the placenta for independent pathological examination to measure the amount of Caesium 137. Look at the fertility of local families over the last five years. Look at the levels of birth defects. Perinatal pathology.
Levels of autoantibodies against thyroid antigens or islets of Langerhans
Goiter, thyroid deficiency, Hashimoto's thyroiditis, type I diabetes:
—Infectious diseases: Number of children prescribed with antibiotic treatment, medical indications, hospitalisations. Recurrent diseases: chronic bronchitis, recurrent urinary infections.
—Surgery: fracture repair, quality and speed. Length of time for wounds to heal.
—Psychology: Achievement at school relative to chronological age. IQ. Absenteeism etc.
—Allergies: Bronchial asthma, age, severity, allergic dermatitis and others.
—Gastrointestinal: Gastritis, duodenitis, food allergies
—Oncology: Brain tumours, other malignancies, anaemia, lymphopenia, leukaemia.
Always based on the levels of caesium-137 load measured independently (and only known after the examination).
Only the statistician should receive both data records, when it is all in the computer, and will undertake the calculations based on the formulae mentioned in the plan, adopted before the study.

I have seen no proposals for scientific research into health from either the IRSN, from WHO's IPHECA project, set up by the IAEA, from the ETHOS project, or from CORE. It is wilful ignorance, a desire to cover up the central problem: people can see and they know that the children living in the rural areas around Chernobyl are ill and that food produced here or gathered from the forest poses a danger, and is very difficult to sell".

Dr. Michel Fernex

This is the nuclear gulag, financed by the West.

PART FIVE

NESTERENKO'S VILLAGES

Chapter One

A CLOSER LOOK AT THE CONTAMINATED VILLAGES

"Feeling pity for the plight of the victims or helping them, without engaging in a whole hearted denunciation of the strategy of the WHO/ IAEA lobby, will obviously result in the victims becoming habituated to the violence to which they are subjected and to "guarantee" the eventual repetition of the same scenario somewhere else in the world. Establishing radioprotection organisations that are independent of the nuclear industry is the only way to put an end to this scandal".

YVES LENOIR, *op.cit.*

Belarus is a landscape of forests, lakes, ponds and rivers. But 26% of the forest and more than half of the cultivated fields on the banks of the rivers Pripyat, Dnieper and Sozh are situated in the radioactive zone. For twenty two years, until his death in 2008, Nesterenko undertook a rearguard action[133], on his own, in hundreds of villages in these territories, publicising the true situation, and demanding at the very least, that the "norms", or "admissible levels of contamination" as they are called, are respected. (In Belarus: Admissible levels of contamination—ALC).

1. THE ILLUSION OF "NORMS"

Through a number of different biological and physiological processes, the human organism is engaged in a daily struggle against incorporated toxic and radioactive substances that it eliminates. It is more or less successful in this, if the immune

---

[133] Rearguard action; an attempt to prevent something from happening, even though it is probably too late to succeed, Collins English Dictionary.

system is not weakened. Human beings heal and protect their bodies by leading a healthy and hygienic life, and above all, if the quantity of toxic substances ingested and inhaled is minimal and of short duration. In the territories contaminated by Chernobyl, people's bodies are chronically incorporating radionuclides.

This is why Nesterenko's fight can be described as a *rearguard action*. The norms invoked protect no-one. They were invented in order to legitimise politically the radioactive emissions released into the atmosphere by nuclear power stations. All the nuclear power stations in operation all over the world (about 450) release these emissions continually. During the symbolic trial, organised by the Permanent People's Tribunal, of the management of the consequences of the Chernobyl disaster Rosalie Bertell explained these norms:

> Routinely, all of the gases are released and all of the liquids are released, so the only effort made is to retain the solid radionuclides. When we talk about radioactive waste, it's that solid portion only. By the design of the technology, we don't know any way to constantly collect and store a gas. We just keep getting more and more gases, because we can't contain them physically.
>
> So what happens with the industry, is that they try to set a standard, and they say that as long as you keep the releases of gases and liquids below this standard, we'll consider it acceptable to the population. If you go above that, it's unacceptable. When you do that, and give those numbers to the engineers, they design an industry in order to be legal—everybody assumes that if it's legal, it's safe.[...]
>
> If you deal with the engineers and physicists, they look at the standard and say, we stayed below it and therefore we are safe. If you talk to medical people, they'll say that the standards are not protecting health. Therefore to be in compliance with standards does not always protect the worker, nor the public health.
>
> [...] People are talking two different languages [...] The workers and the public are often deceived by standards, they expect that a standard is there to protect them. Therefore, if they're told their radiation exposure is within permissible limits, they think that they shouldn't be sick, or that it's their imagination...[134]

The UN agencies, subservient to the nuclear lobby, deprived of any rational scientific argument, have expressly adopted this all-purpose formula: the victims of Chernobyl are suffering from radiophobia and from stress, in fact, from "their imagination". The safety norms are a psychological exercise, a verbal crutch for these illusionists, who must get the population to accept these unavoidable becquerels, a permanent feature of our "modern world" in which WHO hoped to find "a generation which has learnt to accommodate ignorance and uncertainty". Nesterenko is a voice crying in the wilderness, not only because of the inherent deception in the very concept of a "norm", but also because the

[134] Rosalie Bertell, *op.cit.*

Belarusian authorities, backed up by the international "experts"[135], are advocating the resettlement of the evacuated territories. While what is really needed is a systematic government-run programme of information and prevention with pectin so that the inhabitants are warned of the danger and can learn to defend themselves against it.

The norm for life is zero artificial radionuclides in the human organism.

2. FIVE OF NESTERENKO'S VILLAGES

In 1990 and in 1998, with our cameraman Romano Cavazzoni and the sound engineer Nino Maranesi, we were sent to the contaminated territories of Ukraine and Belarus, to investigate a people abandoned to a nameless scourge that the scientific establishment refused to recognise. Then in 2000 and 2001, now retired, we went back there, but this time, with shared friendship, solidarity and indignation as professional journalists. Our job is to inform, and we could no longer stand idly by in the face of the persistent lies about the consequences of the Chernobyl disaster, which we now experienced as provocation.

We recorded the silent tragedy of the peasant farmers imprisoned in this environment that was once their friend and provider, where it was now dangerous to walk, to fish, to collect berries and mushrooms. An environment in which people fall ill or die simply because of what they eat, while an indifferent and avaricious world withholds the means that might be sufficient to save their fellow human beings. In Belarus, 2 million people including 500,000 children, eating food that is contaminated with caesium-137 and strontium-90, are living, humiliated and abandoned, in the centre of Europe. Condemned to a life of suffering, prematurely cut short.

In 1998, 2000 and 2001, accompanied by Vassili Nesterenko, we filmed the work of the teams from his radioprotection institute in five villages, among the hundreds ignored by the German scientists who conducted measurements in 1993. [136] The average radiation dose of the inhabitants, recorded in the 1991 and 1992 registers of the Republic of Belarus, was higher than 1 millisievert a year. All these villages are further from Chernobyl than the standard 15–30 km zone around nuclear power stations in the West, where the governments of these countries, supposedly, provide systems of radioprotection for the population. The villages we filmed are:

---

[135] See p. 71, "...so you will have to put up with 70 to 100 rem...", a statement sanctioned, on the French side, by Monsieur Pellerin, who was present and raised no objections to his colleague from the ICRP, who threw out this response to Nesterenko.

[136] In 1,100 villages in Belarus the caesium-137 contamination in milk exceeds Russia's accepted norm that has been set at 50 Bq/l. It should not be over 37 Bq/l. But even at this level it is still a "norm", a poison.

Skorodnoie—2.6 mSv/y, 5–15 Ci/km², 90 km from the nuclear power station at Chernobyl;

Rosa Luxemburg—2.2 mSv/y, 5–15 Ci/km², 72 km from the nuclear power station at Chernobyl;

Olmany—3.0 mSv/y, 5–15 Ci/km², 210 km from the nuclear power station at Chernobyl;

Slobodka—1.5 mSv/y, 5–15 Ci/km², 70 km from the nuclear power station at Chernobyl;

Valavsk—1.3 mSv/y, 5–15 Ci/km², 96 km from the nuclear power station at Chernobyl.

1998

On 29th May 1998, we left Minsk with Nesterenko for Polessie, a natural paradise in the south of the country, half in Ukraine and half in Belarus, at the very centre of the area in which Chernobyl exploded. We were returning there, after our first visit eight years earlier, with this extraordinary person as our guide. We wanted to gain a better understanding, from as close to the victims as possible, of the mechanisms that had reduced them to the status of guinea pigs and to document the radioprotection work undertaken by Nesterenko in the face of the Ministry of Health's inertia. After a night in Mozyr, on the very edge of the 100 km zone traced around Chernobyl on the map, we joined the teams from his independent institute in their work, measuring the children from the most contaminated villages using a Human Radiation Spectrometer (HRS), and at the same time teaching families new hygiene techniques.

Chapter II
SKORODNOIE

2.6 mSv/year, 5–15 Ci/km²; 10 km from an area contaminated
with 15–40 Ci/km², and 90 km from Chernobyl

1998

Nesterenko and the female assistant in charge of the radiological monitoring centre are looking at the measurements taken by Vladimir Babenko, chief engineer at the Belrad laboratory. Behind the door, in the school corridor, about fifty children are waiting to take their turn on the armchair that measures their radioactivity. The number of becquerels is not shown in figures, but the level of caesium is shown rising vertically on the screen. Nesterenko comments:

Nesterenko.—This boy has quite a high number of becquerels, I can see it in the curve. It's risen to a peak… According to the Ministry of Health, people are not interested in being measured. Look how many children have come.

Radioactivity is measured in becquerels. The presence of 50 becquerels per kilo bodyweight in the human body means, for example, that in a child who weighs 10 kilos, there are 500 artificial radioactive atoms disintegrating every second. The normal level should be zero becquerels of caesium-137 in the organism. This Human Radiation Spectrometer (HRS) is an automatic machine, that is very sensitive. If a person has only accumulated 200 becquerels in the whole of his body, this armchair will detect it. The back of the chair contains a crystal, in general of caesium iodide or of sodium iodide. When the person sits down and presses his back against the chair, the radiation emanating from his body strikes the crystal which emits a light beam. Behind the crystal there is a photomultiplier detector, a device that receives the beam and converts it into an electrical impulse. The signal is transmitted by an electronic spectroscopic system that determines which isotope is involved: whether it is potassium or caesium-134 or -137. The computer calculates the result. Normally, the majority of the machines provided by the administration detect levels between 800 and 1000 Bq in an adult body. Whereas this one detects levels at 200 Bq. This armchair was conceived specifically for measuring children. And, as you see, it can be moved around. It can be transported by car to any village in Belarus. It is this feature that makes it so suited to the unique situation that resulted from the Chernobyl accident. The government system is based around machines that are fixed. They have them in the capital, and in the provincial and regional towns but not in the villages, even though it is the inhabitants of the villages that receive the highest radiation doses. After three minutes, we know the exact level of internal radiation in the child's organism. Between the 8th and 10th May this year, we gave them pectin tablets. Twenty days have gone by, and they have taken this food additive twice a day. Among most of the children, about 30% of the radionuclides have already been eliminated. And today we have a child who is completely free of radionuclides. Given how little money we have, this is a very effective defence system. Besides, it is incredibly important that the parents are informed about the food products they are eating.

Q.—Is the government doing anything about all this?

Nesterenko.—You could ask the person in charge of the nursery when was the last time the health service did any measurements here.

The assistant.—Never.

Nesterenko.—There's your answer. They're not measuring the children. Even though this village should be a priority.

Q.—Do you publish your figures?

Nesterenko.—I publish them in a three-monthly bulletin at the institute, which I then send to the Ministry of Health, to the local authorities, to the office of the President of Belarus, to the Council of Ministers, to the Supreme Soviet, to the Ministry of Instruction, and to anyone who could help.

Q.—Are you able to publish it freely or do they put obstacles in your way?

Nesterenko.—This year, we have not had any problems with them yet.[137] But for some time now, they have started to limit our activities. In 1993 we had 370 local radiological monitoring centres (LRMC), and now there are only 79, and not all of them are financed by the government.

In February this year, the president of the Swiss Green Cross, Roland Wiederkehr, who is also a Deputy, visited our institute. We talked about the problems of funding so as not to have to depend on government aid. There are 567 villages where the milk is contaminated above the authorised levels[138], and I think if we had

---

[137] It is 1998, and the attacks on Nesterenko from the Ministry of Health would begin in spring 2000. See p. 218.

[138] These figures date from 1998. In 2005, the Minister published a figure of 111 villages. These official figures decrease every year: in 2001, 325 villages; in 2003, 216; in 2004, 165... Nesterenko contests the figures. The Ministry's figures are based on an annual measurement of contamination in milk in each village. It should be measured four to six times a year, according to the different seasons, and the "radiological quality" of the fodder which depends on where the animals have been grazing. The measurements taken by the Belrad Institute show that contamination in agricultural products, especially in the private sector, is not decreasing. Quite the contrary, as has been shown in the example of Olmany, where ETHOS works.

100 centres in these areas, going from one village to another, we would get a full picture of the situation and we would understand how best to help the people. Wiederkehr suggested that I go to Switzerland and give a series of talks at the universities to launch an appeal for funds for a programme of radioprotection for the children and to monitor their contamination. I went to Switzerland in April, I gave talks in Bern, Basle, Zurich, other universities, I went to Soleure, gave a talk in St. Gallen. We came to an agreement with the Green Cross, that they would collect money so that children could spend a period of convalescence in the summer in clean areas in Belarus. This programme has already been decided between us. About thirty LRMCs will be financed by the Green Cross, if they collect sufficient funds. It's one way to ensure that the monitoring of contamination is really independent.

Since then, unfortunately, Wiederkehr has been removed from his post and the Green Cross has done nothing more, while Nesterenko has received no explanation from them.

Q.—The Green Cross is an association. The lack of interest from governments is astonishing.

Nesterenko.—Yes, it's very unfortunate. That's why I've begun to approach international NGOs. For example, the programme set up by Adi Roche, that monitors contamination in children, has great support among the Irish people. I can tell you that there is not one government organisation that gives any concrete help towards the radioprotection of the people living here. I am hoping that these appeals will finally allow me to set up these laboratories. It's very important now that, in Ukraine, we have discovered an effective product like the pectin-based adsorbent Yablopect. In Norway, for example, a teacher wrote to all the schools and universities. She came up with this plan: three Norwegian schoolchildren would save one Belarusian schoolchild. She came in person to give us the money, we were able to buy pectin tablets for the children; we distributed them and we tested to see how effective they were. At Bremen, at the press club, during the course of one evening, journalists collected enough to buy tablets for 100 children.

This is a summary of what I learned from Nesterenko that day. In 2006, to continue to purify and to protect a child's organism, after some of the contamination has been eliminated following a visit abroad, would only cost about 100 euros a year: this is the cost of an effective, pectin-based heavy metal adsorbent and of anthropogammametric measurements of the contamination of the child's organism. But pectin on its own is not enough. A short stay abroad, or in a sanatorium in Belarus with clean food is necessary but not enough. By combining the two methods, it is possible to keep the child's level of internal contamination low enough to avoid the danger of serious damage to the organism. In Belarus alone, there are 500,000 children who need this treatment, and the cost of this would be 50 million euros. Nesterenko has not got this amount of money.

The cost to the rich countries of waging two days war against Belgrade (400 million dollars according to the American bank Merrill Lynch) would have covered eight years of treatment. Money is available for killing but not for living. I made this calculation during the bombardment of Serbia by NATO. Since then, with the war in Iraq, a few hours would suffice to obtain the same result.

1. THE FATHER WHO IS FOND OF FISH

Two little children look on in silence as the adults talk above their heads. Professor Nesterenko is reprimanding the father.

Nesterenko.—So, you ate some fish, did you?

The father.—A little bit.

—Did you have it tested?

—No, we didn't test it but they said it was good.

—What does that mean, "good"?

—They tested it and they said we could eat it. It was within the limits.

—But you know that children should not eat anything above 37 Bq. If they eat food with 37 Bq, they will have an accumulated level of close to 40 Bq. While your children have 70. We'll measure this one now. Are you a fisherman?

—Yes, amateur.

—It's all clear. We've finally understood why although the milk is clean, the children are contaminated.

—Before your previous visit, they weren't eating fish. They just had a little bit only a week ago.

—I can remember one mother who had two children. They went to collect blueberries. Just once, when I had come in June. The children's level of radioactivity did not go back to normal for a year. They had eaten roughly half a litre of berries. With fish, it's the same thing. Officially, the admissible level is 370 Bq/kg. If it's 350 Bq/kg they will say it's clean. But children should not go above 37 Bq/kg— in other words ten times less—whatever the food product. When you bring fish back home, at least put it in salt water (two spoons of salt for each litre of water) and let it soak for six hours. Even better: soak it for three hours, throw the water away and start again. I promise you, with salt water, you can get rid of 70 % of the radionuclides. That is, if you really can't go without fish. I understand that the family income today is low.

—They didn't eat very much, they just tasted it. I was eating and they said "Dad! Give us a little bit just to taste". That's all they ate.

2. THE WOMAN SELLING MUSHROOMS

This peasant woman had brought her 5 year old daughter from a village some distance away to be measured.

Nesterenko.—Of course, 61 Bq per kilo is much too much. If she had 20, I would say there was no problem. But…

The mother.—I've got four children, and each one has a different level.

Nesterenko.—What were the others?

The mother.—90, 80, and 116.

Nesterenko.—If you say the milk is clean, there must be something else. Do they eat mushrooms? We can assume that 30 of the 60 becquerels is coming from the milk. What else is there?

The mother.—*(she is justifying herself)* But someone buys them from us, our dried mushrooms!

Nesterenko.—The fact that other people buy them from you is their concern. But I am worried about your children. They are all contaminated above safe levels. According to the research done by Professor Bandazhevsky in Gomel, pathologies begin to appear in the vital organs and systems if the level is above 30–50 Bq/kg body weight. Why expose them to that? Give us the mushrooms to be tested! We will measure them. Do you use dried berries?

The mother.—We make jam from them.

Nesterenko.—Alright, then bring us the jam and we will measure that too. OK? We will wait for you. (*She goes off*). Why don't they bring the food to be tested? They could bring the mushrooms. If she knew…

The assistant.—They know they're contaminated.

Nesterenko.—So? Is that any reason to contaminate the children?

The assistant.—They've got to eat something.

Nesterenko.—So what are you saying? There's nothing else to eat?

The assistant.—They're just eating what they normally eat. They don't really understand the situation properly.

Nesterenko.—But if she had them measured, perhaps she would understand. The little girl is still young, but later on, it is she who will suffer. You can see how much she has accumulated. She has taken the tablets, and eliminated a little bit, but it will all start again. This is the blueberry season and her mother is going to give them to her to eat.

The assistant.—Yes, she will.

Nesterenko.—As for the girolles mushrooms, they contain several thousand becquerels per kilo. Several thousand.

The assistant.—Tens of thousands. In the area she lives, it's tens of thousands.

Nesterenko.—No really, if they eat the mushrooms, it's hopeless. 100 Bq/l in the milk, and then the mushrooms…

Q.—Why is the milk radioactive? Where are the cows grazing?

The assistant.—Their pasture is quite far away. They have to cross the woods.

Nesterenko.—And in the woods, they graze on the grass. In the woods, the grass is always radioactive.

3. THE NURSE'S SON IS NO LONGER CONTAMINATED

Q.—Is this the first time that your son has not been contaminated?

Elena Shulga.—The first time. I never even hoped. When they told me yesterday, I was relieved.

—The child does not feel too restricted because he isn't allowed to play in those areas that are dangerous?

—No, why? I let him go to the river.

—And in the forest?

—The forest, no… We don't eat berries, we don't eat mushrooms. Never. Not just the children, none of us eat them.

—Your life has changed?

—Of course. The forest for us is a support. There are berries, mushrooms. The forest supplies us with many things, but now we have to do without them. It's the same with the vegetable garden where there is radioactivity. It's everywhere. Wherever you go, there is radioactivity.

—So you've lost the forest and in part the vegetable garden?

—We still grow the things we used to grow before, like potatoes and vegetables.

—Do you get them tested, the potatoes?

—We were getting them tested before. This year, we haven't.

—And it's not dangerous?

—Of course it's dangerous.

—But why don't you have them tested? What stops you? Is it difficult to get them measured? You have a monitoring centre in your area!

—I can't explain it. We forget…we just get on with everyday life, we don't think about it any more. I say to myself: "Everyone else is eating it, so I will too".

—Does today's successful result chase away all the danger : "There, our child is no longer contaminated, we can live better"?

—No. We never stop worrying. We think about the children all the time. Especially in winter, when there are fewer vitamins. Over the whole winter the children drank Biophilite, a multivitamin product. I don't think they are deprived of anything, yoghurt, fruit juice, fruit…

—You're in the right profession too, being a nurse. You know more about these things.

—I understand a bit about what it can lead to.

—Amongst your friends, the people you know, have you noticed them taking more care about what the children eat?

—First, it depends if you can afford to buy clean food for the children, to buy them imported food. For example, my sister got herself measured. She had accumulated very high levels of radioactivity, up to 230 Bq. She works at the school. I told her "Buy adsorbents as well". She replied: "I can't afford it". In the end, she also took the tablets and her levels came down to 102 Bq. By more than a half. [139]

—The government should distribute them.

—Of course! It should be done as a matter of urgency.

—Why doesn't it? How do you explain it?

—I don't know. I'm not the government. If I was at the head of the government I would have made it a priority. Because above all we must protect the children's health.

4. A POOR FAMILY

Mrs Grits, a young mother, has two children, a boy of 7 and a little girl of 3 who has a glass eye. She is holding her little girl in her arms as she welcomes us.

Q.—What illness does she have?

Mrs Grits.—The scientific term is retinoblastoma. It is a malignant tumour of the eye. She is receiving chemotherapy.

—Is she in pain?

—Yes, a lot. She has lost her hair. After the second chemotherapy session she couldn't walk. She's a little better now. We are into the third session now. We need to do three more. There's no guarantee that she will regain normal health nor that she will live. They say: "No-one can give you that guarantee". They say that these tumours are unpredictable. It's terrible. They put poison into the organism. She lies down and can't get up, it's so strong. I say: "They could kill her this way". After the second session, she was in resuscitation for three days. The first day, she was unconscious. She regained consciousness on the second day.

The children have pains everywhere, in their legs, their joints…The adults? They have headaches all the time. There is definitely something going on here. My husband says: "As soon as I do a little work on the tractor, I come back to the house with pains in my head". He complains about it. But who can you go, to complain when almost everyone feels the same way?

—You too?

---

[139] Nesterenko can only afford to distribute pectin freely to children, as the money coming in from foreign NGOs is insufficient for wider distribution.

—Like most young people. I get the impression that everyone between 20 and 30 is becoming weaker. We have a cousin who works in radiation. He says there is a map showing all the levels of radioactivity. Here, near the hospital, where the road turns, where you came in, there's a place marked with very high levels of radioactivity. It fell with the rain in different places; there are patches here and there.

—Do you know where to walk and not to walk? Do you know which places are contaminated?

—No, no-one here knows anything. 10 km from here a village was evacuated, but we go there to collect berries and mushrooms, and we eat them. People have got to live, haven't they?

—You have a choice about what you eat, you can be disciplined and exclude the food that you should not be eating.

—But what choice do we have when salaries are not paid? The last salary I was paid was three months ago. It was 400,000 roubles *(the equivalent of 8 dollars)*. What choice is there? We produce everything. We have a cow, and some pigs, the potatoes are from our vegetable garden, we can't buy clean potatoes. It's all ours.

—Do you test the milk to see if it clean?

—They say we can drink it.

—OK, "they say" you can, but have you had it measured?

—No, we haven't had it measured.

—And the potatoes? Nothing?

—The potatoes haven't been measured either. It's clear that all that is radioactive to a certain extent. But the people here have got used to it, and in the end they have to eat something. We would die of hunger if we didn't eat something.

5. THE ERMAKOVS, A POOR FAMILY

The technique employed by Evgeni Ermakov and his friend Yury to catch fish for their dinner consists of entering the shallow water of the pond up to thigh level and armed with a net stretched out as a barrier in front of the reeds where the fish are hiding, give loud thumps with their feet in the water. The fish are frightened and swim away into the middle of the pond… and finish up on the table. We filmed them from a wooden footbridge over the narrow part of the pond, which divides it in two. In the past this place at the edge of the village was a major route through to Polessie.

Ermakov tells us that Catherine the Great had the road made to link the North of the country with Ukraine. There were six bridges along the route. All that remains are four wooden walkways for the animals to pass through: the road takes a different route now.

His ten year old son, covered from head to foot to keep the mosquitoes off the ecchymoses all over his body, is watching us. The fishermen come out of the water with three little fish, and his father talks to us.

Ermakov.—The boy is an invalid from Chernobyl. He has aplastic anaemia, a disease of the blood. His body is covered in ecchymoses. Blue marks. They wanted to give him a bone marrow transplant. We wanted to be donors, but there was no match. We were given a German medicine that he drank morning and night. He is monitored continually.

Q.—Why do you think he became ill? Was it because he eats radioactive food?

—Of course.

—That you give him?

—No. There is congenital anaemia, which can't be cured… It can change with age. For example, if I had congenital anaemia at the age of ten, by the age of twelve, it might have disappeared. But his is acquired. And he has a certificate from Chernobyl: "Due to the radioactive contamination of the territory". Because we live in this area. That's it. The radiation is too high here.

—Really, and what about your food?

—What food? The cow goes into the wood which is radioactive! I feed the pig? It's radioactive. I fish… Can I refuse to give him fish? Its radioactive too. I can go to the shop to buy him sausages. And where is the sausage from? It's all our meat. They accept our animals. What else is there? How else can I feed him? I can't buy food from Africa every day, pineapples, oranges, so that it's clean. Besides, I only get paid once every three months. I live on what I have… The vegetable garden, the cow, the pig and I go round in circles. You haven't got any money? You borrow until you get paid and then you go round in circles. It's good if you have relatives, like Yury; he has his father, and I've got an aunt who is retired. They receive a regular pension. For example, my son gets an invalidity pension. They pay him 400,000 roubles, *(8 dollars in 1998)*. Chernobyl was his punishment, he fell ill, and for that they pay him 400,000. But I have to take him to Minsk every ten days for his treatment and for tests: 100,000 to get there and 100,000 to get back. That's 200,000. And then, I have to feed him and I have to eat something myself. We go there for three or four days.

His legs were very painful, he could no longer walk because he felt dizzy and because of his legs. Now he is having chemotherapy. At the moment, his condition is normal…within normal limits. They say they may be able to postpone the sessions at some point. We don't know if it will get worse or what…

—Has your view of the world, of nature, changed?

—That's an understatement! *(He turns to his friend.)* Yury, which year was it that we worked together at Kuzmichi on the harvester?

Yury.—1991.

Ermakov.—That village has been buried now.[140] But we remember being at Valavsk also, which is 12 km from that area of contamination. The blood ran from

---

[140] On the map showing contamination, Kouzmichi is in the middle of an area of contamination with levels between 15–40 Ci/km², 8 km from Skorodnoie.

our noses, in a real fountain. I was driving the harvester, I had a whistling sound in my head, and blood pouring from my nose. A little rest on your back and then back to work. After that, perhaps the body gets used to it, little by little. But the first year was very difficult.

—You don't feel free like you were before?

—We go into the forest to collect mushrooms. Today I go somewhere, get back to the house, and everything is fine. I go somewhere else, come back and I have a fever and a headache. The radioactivity occurs in patches in the forest. Here it's clean and over there, there is a big patch. And in the evening you don't feel well: fever and headache. We don't have a dosimeter. When I go here or there, I don't know if it's radioactive or not. All I know is that I was walking in that part of the forest and I feel ill. Take game for example. What happens there? You kill an animal in the forest. The animal is normal. But if it comes from the exclusion zone, 30 km from the reactor, it's contaminated. They say, and I don't know if it's true, that the children of people who hunt game have more radioactivity than the others. The animal arrives here from the zone, they kill it and they eat it. They are more contaminated than we are. We have our son who is ill. Our neighbours, the Grits, have a little girl born without the crystalline lens in the eye, and she also has leukaemia. She's three years old. This is the situation we are in. This is how we live. We can't all be evacuated to a clean area. You can't accommodate all of us, as long as we are in Belarus, to be decontaminated in your areas.

He laughs, lifts his son up onto his shoulders and crosses the shallow water to reach the fields on the opposite bank.

2001

Three years later, in 2001, we pass by Skorodnoie again. We find Ermakov changed. He talks jerkily in sudden outbursts, nervously, then gets blocked, as if he can't find the right word or can't articulate it. His mind is perfectly clear and sometimes he looks a bit embarrassed, a bit ashamed of the problem. He almost shouts the words, when they finally come to him. So the three-way conversation we have with him, his wife and his son Kolia is full of eruptions and interruptions. Looking at him, we are suddenly reminded of what Nesterenko said about the effect of caesium-137 on the brain.

(*His wife greets us at the door*)

Q.—We telephoned a little while ago…

His wife.—Yes, come in please!

Q.—Has the boy run off?

Ermakov.—He took his bike and disappeared.

Q.—How is his health? He had anaemia, if I remember rightly.

The woman.—We were told…

Ermakov.—Once he reached…before he was 16… I don't know how to explain. The doctors told us… that before he had known a woman… *(He makes a circular gesture with his hands on his chest as if to indicate a hormonal change in the body).*

Q.—It would get better? Is that what they are saying?

Ermakov.—He could get over the illness. There he is, he's come back.

Q.—Hello, do you remember? Three years ago…at the pond?

The boy.—Oh yes!

Q.—What is your name?

The boy.—Kolia.

The woman.—You know, at the beginning, I didn't know he was ill. Specialists came from Gomel. I was teaching at the school. They came and they did a blood test, then they wanted to see me. They called me while I was teaching. I came to the laboratory and they told me: "Your son's haemoglobin level is too low. If you agree, we will refer him for further tests".

Ermakov.—He would play with a ball. He got bruises…We hadn't paid any attention: well, he was playing football and he got bruises! Then he said his legs were hurting him. He began to get tired. He would come back to the house and lay down to sleep. He felt weak like that. Then they tested him again at school… the haemoglobin levels had fallen.

Q.—Tell me, is it hard to study, or do you feel normal?

Kolia.—I feel normal.

Q.—Because they told me that at school he is very good…

Ermakov.—In mathematics. He is very good at mathematics. If he could…carry on with mathematics…he's better than me.

The woman.—Mathematics and physics, he is very good at these subjects. He solves his sister Ina's problems by himself. Even though Ina is in a class above him.

Q.—Do you like going to school?

Kolia.—Yes.

Q.—Are you interested in your studies?

Kolia.—Yes.

Ermakov.—It looks as if Kolia is on the way to recovery, but two children… have died.

The woman.—In front of Kolia, while they were in resuscitation, right in front of us.

Q.—Did they have a different illness?

Ermakov.—The same.

The woman.—The same. Aplastic anaemia.

Ermakov.—The doctor who treats him says the illness may disappear in a few years. When he is 16. He could get over it by himself.

Q.—In about two years then. Have you had enough of hospitals?

314

Kolia.—Pretty much, yes.

Q.—And what about your health, the parents, do you feel any effects from the radioactivity?

Ermakov.—Headaches.

The woman.—When I go in the forest, for example… At the moment, we are collecting girolle mushrooms, because we're short of money. We need money. We go and collect girolle mushrooms. When we come out of the forest, we have terrible headaches, I can't tell you what they're like. And we get a fever. Immediately. We get back to the house and take pills… Then it goes back to normal.

Q.—And you sell the mushrooms?

The woman.—Yes, we sell them.

Ermakov.—We sell them. We need the money.

Q.—Where do they go after, these mushrooms? Do you know?

Ermakov.—Probably to Brest (*She smiles*) Do you know, it's through Brest that you get into Poland. You get higher prices there.[141] Last year, we collected mushrooms and then we tested them for radioactivity. Kolia really likes mushrooms… girolles. Well, we had to stop collecting them for us to eat. Because, if we prepare them for us, he wants them too. So, we stopped collecting them. We threw them away. Apart from the girolles, which we collect to sell. But the others, that we call *babka, krasniutchok, belyi*, we don't collect them at all…so that he doesn't have any. How can I explain?… he sees us cooking them on the stove and he wants some. So, we don't collect them any more.

Q.—To protect him?

Ermakov.—Yes, of course!

Q.—We saw each other three years ago. Has your health changed in any way since then?

Ermakov.—I can tell you just one thing: fever. In the evening I have a fever.

The woman.—You didn't have that before.

Ermakov.—And now I always have a temperature.

The woman.—He has a temperature every evening.

Ermakov.—Three years ago, I could talk normally with you… I spoke normally… Now I have this fever… And headaches.

Q.—Have you seen a doctor for an examination?

The woman.—But what doctor? Tell us about them! At Borovliany, at Aleinikova, they only treat children. Kolia is the first to… respond to the treatment.

Ermakov.—…who responded to this… they gave him some German medicine and he responded to the treatment. And there was one boy there…it was Puzanski, Kolia?

---

[141] From Brest, they will go further afield, to the West.

The woman.—Puzanski.

Ermakov.—He didn't respond at all. Kolia followed this treatment and he improved.

6. SKORODNOIE'S OLDEST INHABITANT

Before leaving the village, we revisited areas that we had been to three years before. We wanted to find the elderly peasant woman that we had filmed discreetly without speaking a word to her.

A village in the distance beyond the fields. A man crossing the vegetable gardens carrying a scythe. A stork winging its way over the trees and disappearing. The pond at Skorodnoie. The rotting wooden platform where the pond had dried out is still there. Wooden houses close by. The electricity lines looking like a musical score running from an old wooden pylon from a bygone age. Kolia and his friend paddling through the water. An older boy crosses over the platform to join them. A courtyard behind a fence. Maria Kirilovna is scything the grass in the field behind the vegetable garden. She comes towards us. We stay on the road in front of the fence that surrounds her yard, just as we had last time, without going in. This is where we stand and talk, in front of these fence posts, washed a grey-green with age.

Maria Kirilovna is a sprightly woman, in spite of her advanced years. She has a steady voice, a lively mind, a twinkle in her eye, sometimes ironic, sometimes serious, as she recounts her life, as an escapee from Stalin's deportations to the Great North (Siberia).

Q.—Do you remember us? We were here three years ago?

Maria Kirilovna.—You filmed me that time.

—Yes.

—That was a long time ago. There have been others. They came last year… Well, we cut the hay; we work. I have goats in the stable…

—How is your health?

—Well, thank you for asking. What health can a person like me have? It's finished. It's time to go. When God decides… (*She gives the chickens something to eat*) I've got goats… For the moment, I look after them.

—The radioactivity doesn't bother you?

—Ah well, we can't see it, it's invisible.

—But how do you feel?

—Bah, I feel like everyone else: My legs hurt, my hands hurt, my fingers are deformed.

—Your heart?

—The heart too. When the weather changes, straight away, I have to take Carvolon, Validol. There's no cure any more. (*She laughs*)

—Don't you ever get your food products measured to find out if they're radioactive?

—The milk in the village is tested.

—And is it radioactive?

—Some of the cows are, some are not. Everyone drinks it. I drink my goats' milk too.

—Have you measured the goat's milk?

—No...The radio said it was OK.

—The radio?

—Yes, it seems that the white...the white goats' milk is completely clean. The white ones are, but the grey or piebald, they are radioactive.

—What are yours?

They are white. Shall I show you them? If I get them out, they will start to run. They are beautiful, my goats.

She goes to the stable. The dog attached to the fence begins to bark. The goats come out. She gives them some leaves from the trees. We can hear the goats bleating and the frogs croaking. She comes towards us. A white goat trots behind her.

Maria Kirilovna.—This is my Pasha following me.

Q.—Do you live here by yourself?

—By myself.

—Have you any family, any children?

—I have five, and they've all left. There's no-one. They've all gone. There's one in the Kaliningrad area...And here, in Belarus...

—You cut the hay by yourself?

—All by myself! I scythe the hay.... And I'm over 80 years old!

—Really! And you scythe the hay!

—The children don't want to scythe hay any more. I've got two grandchildren— no, I've got a lot more, but here, in this village, I've got two—and they work. One is a doctor in a hospital, the other is here. And one son is a forest warden who lives quite close.

—And you are 90 years old, is that it?

—Not quite, I'm nearing it.

—You're nearly 90 and you are scything. Radioactivity must be good for health.

—*(She laughs)*. For old people, maybe. That's what they say actually, it's the middle aged people who are dying at the moment. As for the old people, they are living. Only, when the weather changes, I can't get to the shops. Yesterday I went, another old woman came with me; My God, I was tired, I nearly died.

—So you have grandsons?

—I've already got five great-grandsons.

—And how is the health of these children? Is the radioactivity having any effect on them?

—One is being treated for something. He's got something here. *(She points to the throat)*. He has monthly examinations. He needs clean food products, vitamins. But there are no vitamins.

—They say the children here are tired.

—Yes, in this street there were a dozen children. The street used to ring out with the sound of their cries. That was before the radiation. Here, the neighbours had eight children. But now, they're all the same, sleepy. Basically, they're not how they should be… I would say these are children who no longer have anything about them that resembles childhood. They used to run about happily, went sledging in the winter, here on the ice. But now, they lie down all the time, sleep or stay sitting down. They are not well… And where you live, is there any radiation?

—No, not at this level. A little… a little.

—It hasn't spread to where you are?

—It isn't at the same level as you have here.

—We measure it here, but only in certain places, whereas at the place where it exploded, it disappeared into the air. The planes dispersed it. They say that the planes dispersed it far away from the towns. They dispersed it towards the villages. Then it fell down again on the villages. The land, we sow it, we work it and we swallow all that… With the air, in the dust, in everything.

—They damaged the natural environment. It is beautiful.

—Oh it's so beautiful here, near the river! In summer, in June and July, my God! Everyone used to go to the river, to sunbathe. Now that's all over. No-one comes any more. A few people, very few, come for a day or two then they leave. No-one sunbathes near the river, the way they used to. That's life. *(She addresses the camera for the first time)* You've already taken enough photographs of me to make a whole film…

—My name is Wladimir Sergeivitch. What is your name?

—Maria Kirilovna.

—It was a pleasure talking to you.

—Oh, if I told you everything!

—You have more to tell?

—What I have endured!

—Tell us then.

—Oh! We've only lived here since 1948. We were expelled from a village 12 km from here.

—Expelled where and by whom?

—*(In a very low voice)* By Stalin. It was Stalin.

—Where were you sent?

—To the Perm region.

318

—How long were you expelled for?

—Oh…maybe even for ever, I don't know for how long.

—And after, you came back?

—Mum…Dad was in prison in Mosyr. Afterwards, he was freed and later he joined us. He came after. As for us, we lived there. We did not die. People died on the way there. Children and adults died. But the rest of us, we carried on. We survived. We spent nineteen years there. And after we came back. Then after the war…I don't remember which year… was it in 1935?… we didn't have the right to vote. They took 25% of our wages.

—But why were you punished like that?

—Why? Because we had a lot of cows. Pigs and sheep.

—The so-called *kulaks*?

—Yes. There were the *podkulachniki* and there were the *kulaks*. But we… my father built a mill, he got together with three other men, he was the fourth, and they built a steam mill. Down at Rudnia. And it served the local community. So that was what displeased someone.

—Someone denounced him and said you were *kulaks* and that you needed to be got rid of?

—Yes: "You need to get rid of them, we don't need people like that". And we suffered for nineteen years there. That's where we grew up. And when we came back here, I already had two children. There, we worked in the forests. We improved the soil, we worked in the mines. On the Kama. The big river is the Kama, and the little river is called the Kossa. There were several little rivers… That's where we spent our youth. And we came here for our old age. I stayed here alone. I live here. That's how it is. Life… Do you know the North… Have you never been to the North?

—No.

—Ah… it's not everyone that can live up there! The cold, the taiga, snow storms…Oh it was a constant battle. It was very hard. They took all the men for five years. The women lived there alone in the kolkhozes. We did everything ourselves. Summers were short, three months at most. The first year, we hadn't cut the hay so there was nothing to feed the horses. They died of starvation. So what did we do? When it rained, we broke the branches. We made them into bundles and hung them all along the fence. And when the weather was good, we cut hay, and collected it…We got used to it, doing it like that, waiting. When a man appeared, who had lost an arm or was wounded… he would help us. Apart from that, all the work was done by women. They brought up the children too. The winters were hard. There were empty houses, like this one, with two entrances. We cleared snow from one house, we went into the forest with snow up to here, we cleared the roads. We cut small wood for burning… We kept the children's nursery warm, the school, the town hall and our houses too, our flats. We lived like that for five years, till the end of the war. When the war ended in 1947, the evacuees were allowed to leave, even to go to Moscow to live... But we came here, in 1948, just when the financial reforms took place.

—And did you find the other villagers, when you got here?

—Some people had already returned. My father got here before us. He stopped here, in this village.

—Little Father Stalin[142] tormented you?

—Who knows who tormented whom? He tormented us and we tormented him and he tormented us as well. In 1937, they took all the men, the young adults, and they never came back. They still haven't come back; we don't know where they are. Some were in prison for ten years... those ones came back. Some were in prison for a year... they let them go. But the first ones to be arrested in 1939... no, in autumn 1938... and in 1937, none of them came back. All those young men who were married. Oh, there was plenty of suffering! One person is guilty and all the rest pay... Those who worked over there stayed alive, they survived. But those who did not want to, who said "Let's go. Let's go back home", etc, well, they may have escaped, but then they were killed in the war. *(People go by on the path)* Such was our joy! We really suffered. And now, an old woman alone, carry on as best you can. You want to live? Then work, keep yourself fit and eat well...There's no milk at the shop. We only get crème fraîche... But if not, I've got the milk from my goats. I milk them and I drink it. Warm milk.

—Can you manage on your pension, with your vegetable garden and your goats?

—Yes, you can manage. It's my children who are short of money. Yes... I've told you enough to write a book!

(Some hens, the goat... Someone arrives to make a telephone call... She goes to the back of the farmyard, puts the goat in the stable. We leave.)

---

[142] The Tzar was known to the Russian people as "Little Father".

Chapter III

ROSA LUXEMBURG

2.2 mSv/y, 5–15 Ci/km², 72 km from Chernobyl

1. A HEADTEACHER LAMENTS THE SITUATION IN WHICH HE AND HIS PUPILS FIND THEMSELVES

After forty-six years teaching, the head teacher at the school in this village, named after the famous revolutionary, feels as if he is running a hospital.

The head teacher.—All of the children in our school are ill. They seem continually exhausted. They're lethargic… almost half asleep. I think it's the effect of the radioactivity. I've been a teacher since 1954. So it's easy to compare how the children were before the radioactivity and how they are now. Before, they were lively, they were happy.

Q.—What new illnesses have appeared?

The head teacher.—Diseases of the thyroid, the eyes, the heart. They often have headaches. Back pain. Their attention span and memory are not what they should be in children. It's definitely to do with the radiation.

Q.—Since Chernobyl, are you more lenient towards them, or do you carry on as normal?

The head teacher.—No I am no more lenient.

Q.—But, if it is difficult for them to learn…?

The head teacher.—In what way, difficult?

Q.—Well, their memory…

The head teacher.—First of all, we've had no directive from the minister. If they haven't achieved the required level, they have to do it again. They need to follow the curriculum… set out by the government.

Q.—So the children are under pressure from both sides. First, their health and then their schoolwork?

The head teacher.—Yes, there's pressure from both sides. The school text books are so theoretical, so far from their lived experience.

Q.—If you were free to choose, what would you teach them?

The head teacher.—I would concentrate on what will be useful to them in life. The most useful subject for them is agriculture. If the school is in a rural area, these subjects need to be taught. The child needs to know more about problems in agriculture, in animal rearing. Even more so now that we are changing over to a market economy. In time there may also be agrarian reform. The children need to be taught how to run a private agricultural enterprise. Basically, be educated for the workplace. It's even more important, given that we live in an area where the children are not allowed to go into the fields, not allowed to work there at all.

Q.—Because of Chernobyl?

The head teacher.—Yes, it's not allowed. Before, for example, we had a field where we did experiments growing different crops. Now, we're not allowed to do that.

Nesterenko.—Really? There is no clean land around the village for the children?

The head teacher.—It's not allowed.

Nesterenko.—I don't agree with this sort of blanket prohibition. If the land is not contaminated and the children won't be exposed to radiation there, why should you not work there?

The head teacher.—Only if I get permission… Yes, I can take them to work in this particular field, but I have to get permission from the health service. Then they have to come here, assess the situation, take measurements. Only then would I be able to take them to work there.

2. THE ILL-INFORMED COW HERDER

On the road, we stop to talk to a former teacher. She is now retired and is looking after two cows. We take measurements in the field with a Geiger counter.

Q.—The grass should be fine here. There is no radioactivity.

The cow herder.—So it isn't dangerous. Because we've got no idea, do you see? The authorities don't tell us anything. This field was allocated to us. Of course, our grass isn't that good. Last year, we were allowed on to the next field, but this year—I don't know why—we're not allowed. Yesterday, we put our cows on it and we were chased off. They are treading on the same ground. It's a big field but even so. Can you see what the grass is like?

—And in winter? What do you do for hay?

—We make provisions. The kolkhose *(collective farm)* helps us too; they give us their plot. They cut it for us, we rake it up and store it for the winter. Last year, we had good weather and we gathered enough hay for our three animals: a horse and two cows. I don't know what it will be like this year… At the beginning, there was a lot of panic. Children were evacuated but it was never ours that were taken.

—Were they measured?

—No, no-one measured them. The children are weak, you know, very sickly. There is a lot of illness. The medical examinations are very superficial, they're not very thorough.

—Many of the food products contain radionuclides. It's the same all over Belarus.

—Of course, but how can you find out? We've been going to the woods all of our lives. I used to love going there, for example, but now I no longer have any desire to go. We want to, and we go, but then we don't know what we've collected… Basically, we're afraid. If someone could just come and measure it with a dosimeter from time to time. There are services for that, after all.

—They don't do it?

—No, they don't. We don't know anything. We go and collect berries, we look at them… I look at maps of the area showing the contaminated areas. We check. It's mainly for the mushrooms. My children pick them, and I throw them out secretly. It's a shame, because nature is wonderful. But it's all been destroyed. Who is to blame?

Chapter IV

OLMANY

3 mSv/y, 5–15 Ci/km²; 210 km from Chernobyl

2000

Nesterenko.—Polessie is famous because it is one of the most beautiful areas in the country. The river Pripyat is not far from here, it crosses the country from West to East, parallel to the frontier with Ukraine, which is very close to where we are in the village of Olmany. The river flows into the Kiev basin near Chernobyl. The town of Pripyat, which was built 1.5 km from the nuclear power station, took its name from the river. Pripyat was abandoned permanently thirty-six hours after the disaster. After that it is the Dnieper River that drains the water and all the radioactive contamination in a southerly direction, as far as the Black Sea, along the series of dams that Ukraine has built. Near here, in Ukraine—we are about 3 km from the border—is the beautiful village of Perebrody, described by Alexander Kuprin in his novel *Olessia*. The book was made into a very good French film, with Marina Vlady.

Olmany is 200 km from Chernobyl. It's very far away. You see, in your countries, in Switzerland, Germany and especially in France, they say that beyond 30 km there is no danger. Here we are 200 km away and look… The government has invested a lot of money here, to improve the fields. The president visited this village and this building was constructed with a clinic. Why? Because there has been an enormous increase in illness in children. There have been some improvements in agriculture; the milk is a little less contaminated. When we started work here in 1990, the milk contained 5000 becquerels of caesium-137 per litre. Today, it contains 800. There has been some progress but 800 is too much. It's unacceptable. Children should not drink it when it contains more than 37 Bq/l.

The radioactivity was carried over these villages. The wind blew the clouds here, full of water. All that rain fell, and contaminated the soil. Significant amounts of radioactive iodine, caesium and strontium were deposited on the soil. I think that the harvests will be contaminated here for decades. That's why the children need to take an adsorbent food additive at least three times a year—four would be better—to eliminate more rapidly the radionuclides that they have absorbed.

324

—Are new born babies protected by their mother's milk?

—Quite the opposite. We know that when the mother consumes contaminated products, the radionuclides are eliminated from her body and are concentrated in her milk so that the child receives a contaminated food source. At 8–10 months, young babies can have already accumulated levels of 200–300 becquerels per kilo body weight. They are very badly contaminated by their mother's milk. She decontaminates her body through the child. The medical principle that mother's milk is best does not hold if the milk is "dirty"; it's better to give the baby artificial milk. The child will be healthier. Of course, the cow's milk needs to be clean. That's why we have an LRMC from our institute at this medical centre. The school nurse, Pasha, works here as a dosimetrist. Everyone who lives here can bring their milk and have it tested for free. But if it is contaminated, it needs to be purified with a cream separator. Unfortunately it all costs money and they haven't got any.

—There is no cream separator.

—No, and it would be necessary?

—How much does it cost?

—About 60 dollars.

—What happens. How does it work?

—You take 3 litres of milk. You fill the cream separator and after ten minutes, the radionuclides are concentrated in the skimmed milk, which you discard. The cream is clean. You dilute the cream with boiled water. For example, if the milk contains 200 Bq/l, there will only be 30. Then the milk can be given to children.

—And they don't have this machine here?

—We would need 500–600 of them to give to the most contaminated villages, but unfortunately we haven't got any.

This precarious financial position lasts for years. The salaries that Nesterenko can pay his staff are very low and some of his qualified colleagues, with families to support, have had to leave, even though they have been convinced and committed since the start of this venture, to working alongside him. Like the Siberian peasants, who cross flowing rivers during the thaw by jumping from one piece of ice to the next, Nesterenko has ensured the survival of his institute, at the cost of chronic high blood pressure, by borrowing, asking for help and relying on the patience of his colleagues. Every month he has to come up with something to keep the institute going. He has a heart condition since flying over the reactor just after the accident, and because of the persecution he has endured, but his optimism is astonishing. He stands aside from the absurdity of it all, in order to save the children which is his sole preoccupation.

Nesterenko.—There are 1,100 villages in Belarus, like this one, where the milk is contaminated at dangerous levels. For children 37 Bq/l is the maximum admissible. But in 350 villages, the milk contains more than 100 Bq/l. Obviously the children should not be drinking this milk. Today, we found milk that contained 400 Bq. And unfortunately they are drinking it. Of course, to do it properly, each family needs to have a cream separator to decontaminate the milk. It would be better still if the cattle fodder was treated with adsorbents, and then the milk would be cleaner. But since 1997, the government no longer does this.

1. LIONIA OF KOSHARA

This boy of 14 does not have any "health problems", as he describes it, but he cannot speak normally; he has difficulty pronouncing certain letters and syllables, he stammers and he finds it hard to know how many brothers and sisters he has in his family. The dosimetrist Pasha has just measured him. About fifteen girls, a little older than him, are waiting their turn.

Nesterenko.—536! Why is he so contaminated? This is the highest dose I've recorded today 536 Bq/kg! Can you bring us some milk? Half a litre. It will be tested, the milk will be kept and you can take it back home again, no problem. Can you do that? Maybe after school?

Lionia.—I live at Koshara.

Nesterenko.—Ah, Koshara, that's quite far! Tomorrow we'll be working here. Bring it to us straight away tomorrow morning.

Q.—How far is it to Koshara?

Lionia.—Five kilometres.

Q.—Five! We can just go there in the car.

Nesterenko.—But you have classes until 1 o'clock.

Lionia.—Till midday.

Nesterenko.—Alright, maybe just after. Have you got some milk in the house?

Lionia.—Yes.

Q.—Where do your cows graze?

Lionia.—On the marsh. *(This makes the girls laugh)*

Nesterenko.—That means that they take the hay from the marsh, so the hay is contaminated. *(In a low voice)*. The girls are laughing now, but we will be measuring them too… Then we'll do the calculations.

Q.—Will you bring some milk?

Lionia.—Yes.

At 12.20 we arrive with half a litre of milk and we give it to Pasha.

Nesterenko.—You have the highest levels of contamination of anyone I have measured today. No-one has a higher level. Have the doctors given you an examination? How did that go?

Lionia.—I'm OK.

Nesterenko.—No problems?

Lionia.—No.

Nesterenko.—Well done!

Pasha.—Here we are; the radiometer has beeped. Look at the result: 1042 Bq/l! (*The norm is 100. In Russia, it's 50…*).

Nesterenko.—We need to test the hay. Where is it from?

Pasha.—You get it near to the Vis, don't you? Is that it?

Lionia.—No.

Pasha.—Where do you get it?

Lionia.—Near to the house.

Pasha.—Where the marsh is? Near Morlitzy?

Lionia.—Yes.

Nesterenko.—Have you got any brothers and sisters?

Lionia.—Yes.

Nesterenko.—How many?

Lionia.—…

Q.—How many have you got?

Lionia.—My little sister, she's not walking…

Nesterenko.—That's one.

Lionia.—My brother…

Nesterenko.—That's two.

Lionia.—One brother who has done his military service.

Q.—Three. You're the fourth…

Nesterenko.—Where does your father work?

Lionia.—At the moment he…In winter, he worked as a driver.

Nesterenko.—And in summer?

Lionia.—In summer, before, he worked as a fireman… he doesn't do it anymore.

2. THE DOSIMETRIST AT OLMANY IS A LIVING LABORATORY

Pasha.—We write the date on the form. Here we write the measurement. Here the result: 207 Bq. And here is the norm. It doesn't appear on the print-out; we write it in ourselves. The norm for milk is 100 Bq/l. And we sign it. They keep it and when they come again they can measure and compare. Because they can't remember it. They look at the form again and read it: this month it was so much. They can find different hay, feed it to the cow for two weeks and do another measurement. So they can judge for themselves the quality of the hay from certain marshes and how much radioactivity there is in the milk. That's why I give them the forms.

Q.—In general, are they disciplined? Do they monitor it the way you would like them to?

—Well, not all of them. There are some families who take care of themselves, who bring the hay on their own initiative, who monitor it. There are some who can't be bothered. "I don't want to know about the radioactivity, I want to live in peace". And there are people who know and are prepared to do something about it but who are simply in a difficult financial situation. They can't buy clean hay, they can't buy food from the shop and they just have to eat what they can grow. The majority are like that. That is the largest category. Because they are poor, they cannot afford clean food and they just have to eat what's there. Salaries are so low, and they are never paid on time. You know what a disastrous situation we face here in our Republic. Everyone is poor, and miserable, about our shame and misfortune, I don't know what to say. We have to live, to survive. It's more about surviving. I'm concerned about these things because I live here and it really matters to me what I eat. I came to work here voluntarily, I asked to do it.

—Ten years ago, then.

—Yes, it was in 1991.

—What conclusions did you come to?

—I would say this. When I first started to work here, I was monitoring everything myself. I went on walks; I asked a lot of questions. I was curious to know how much radioactivity there was in this or that place. Blueberries for example, I would measure them in one place and then in another, and I asked people to bring me them. I would measure them and I always marked down which woods they came from. The name. I would go there myself, and in this one there was more, and in another there was less and in another, it was perfectly normal. So I would know that if I wanted to gather blueberries myself, I would go to this wood. And I explained all this to people. I could already draw my own conclusions, choose the least bad.. Even if nowhere was absolutely clean, at least I knew which was the least contaminated.

—What is the general state of health in people in the village? Have you noticed any changes over these last ten years?

—All the doctors say that it's getting worse. All I can do is repeat what they say... And the inhabitants…I sense that I am less well myself. I have insomnia. I don't sleep any more. I feel apathetic, in a torpor. I think I want to sleep, I lie down but I can't sleep. I can see in my own case that it is getting worse. I don't know what has caused it. I'm getting old, of course, but I'm not that old yet, I've a long way to go before I'm eighty. For myself, I think it's the radioactivity. Everything hurts. Sometimes, it's my heart, but I have a congenital malformation, so it could be explained by that but also by the other, but it's getting worse, I'm certain of that… I don't really want to talk about it. There's rarely anything good to say, and I don't want to remember the bad stuff. It's depressing. If I forget about it, everything seems OK but when I want to go down to the river, in summer, or into the woods, I remember that I'm not supposed to. Even if I go into the woods, I don't enjoy it the way I used to, there is no longer the same atmosphere… Nature is beautiful, but when you know that you aren't allowed there… it's unbearable. You can't experience the same unadulterated joy. You can't see anything, everything appears normal, but you are always aware…

After the disaster, as my son was six, I wanted to have another child. I tried twice and it didn't work either time. I had miscarriages. I don't know what they were due to, but I think it was the radioactivity. When I had my first miscarriage, there were many pregnant women who had abortions and miscarriages. I thought it was probably because of the radiation. Then I tried once more, and the same thing happened, I couldn't get pregnant. I didn't really want to try again. My health has been badly affected. I'm 40. I would like to be a mother again but I'm frightened now. I don't feel very well, and then I lose heart. I am on edge all the time, I try to control it but it isn't always possible. The insomnia is a torment. Many of the villagers complain of the same fatigue, apathy and irritability. In general, health is deteriorating. People complain, of course, but they all have the same symptoms, they are all very similar. When I hear other people talking, they could be talking about me.

3. THE DOSIMETRIST AT WORK

There remains the problem of the ash. The ash is incredibly radioactive. According to the regulations, this ash should be buried. It is considered as radioactive waste. The people throw their ash on the vegetable garden and then the radioactivity is transmitted to the vegetables they grow.

Q.—But everyone knows that the ash should not be used on their vegetable gardens. But they still do it?

Pasha.—So where should they put it?

—Underground, but not on the vegetable garden.

—That's right, people have been told, but not everyone does it. Take me for example, I collect the ash in plastic bags and when they're full, I get rid of them. Outside the village there is a waste dump. We brought this question up once and representatives from the Ministry came from Minsk. They looked at our "grave" as we called it, and they said it was not appropriate, and that the ash needed to be taken to proper "burial grounds", like the ones they have in the Gomel region. They left saying that they would ask their superiors to sort it out. It was a question of organising the collection of the ash in large containers and of taking it away but so far nothing has been decided.

—When was this?

—This year, about a month ago. Our rubbish tip is not equipped for that. It needs special treatment so that the contamination does not spread, either into the soil or into the surrounding area. If it rains, or if there is a flood, it must not be allowed to leak away. We don't have the appropriate rubbish tip.

—Is it close to the village?

—About 4 kilometres away.

—And is the contamination level there very high? Is the ash very radioactive ?

—It's terrible, terrible. I have tested it myself. Once I measured the level in my ash and sent the result to Nesterenko's institute. They told me that such a high level was not possible. Their instruments could not register it.

—How many becquerels?

—It was 116,000 Bq/kg. But their reply did not reassure me. I wanted to know the truth. When we were called there again on a training course, I said to myself. "I really want a clear understanding. Did I measure it correctly or did I make a mistake? Who is right?" I took the ash with me to Minsk. I brought it with me to the training course and they measured it during their work. Again, the level was too high to register on their machines: they weren't capable of measuring such high levels. They mixed the ash up and finally got a measurement. It was 200,000 Bq/kg.

—Where does your wood come from?

—Around the village, from the forest…The forest can't be changed and we're not going to bring wood from Minsk. It's not realistic.

—Do many people use the ash as a fertiliser?

—Lots. You can go and ask in the village. Not everyone follows the recommendations, not everyone buries the ash. They throw it behind their vegetable gardens, in abandoned fields. It spreads from there, the grass grows and animals graze there. They need to bring a gas supply to the village and instal central heating. That would be the only way to get rid of the ash. But that's not going to happen any time soon. They've said that we'll get a gas supply around 2005 maybe. So, we just carry on, living, stewing in our own juice.

Pasha's gestures are precise and professional. She puts on an immaculately clean white coat, and her blonde hair is tied back in a chignon on the top of her head with a piece of wool. On a table there are several bags of vegetables and hay, and half-litre bottles of milk waiting to be measured with the Belrad RUG-92 radiometer that is standing in a corner of the room. An elderly peasant woman gets out her vegetables. Pasha washes them in the sink, cuts them up into pieces before placing them in the radiometer container and closing it hermetically.

The peasant woman.—*(She is addressing us)*. Oh, you should come and see where I live and take measurements inside my house. I'm going to live in Japan, it's clean there. A lot cleaner than here anyway. *(To Pasha)* Do you think the carrots are clean?

Pasha.—They're clean.

Q.—And how much do you measure yourself?

The peasant woman.—632 Bq.

Pasha takes the carrots off the radiometer, puts on the potatoes and returns to her table opposite the window to record the results. The machine gives out a beep, the result comes up.

Pasha.—Look, the potatoes are showing only zero. Let's try once more… zero again. The vegetables are clean.

Q.—Where does the 632 Bq come from then?

Nesterenko.—From the milk. I wanted to have a milk separator installed in one family where it could be used by several households. Pasha, who was it that agreed to organise a group? You said it was one of your colleagues, didn't you?

Pasha.—Lena.

Nesterenko.—Do they live far from each other?

Pasha.—You live on the other side of the village, don't you? It's quite far.

The peasant woman.—Yes, we are at the far end, through there.

Nesterenko.—It's absolutely crucial that you measure the milk. Using the milk separator, you will end up with the cream which you then dilute with boiled water and you will have the same milk, but far less contaminated.

The peasant woman.—What milk am I going to separate. I barely get 3 litres from the cow!

Nesterenko.—You'll have 3 litres of clean milk.

Peasant woman.—Where should I bring it? Here?

Nesterenko.—No, no, we want to give a milk separator to one family that can be used by several families.

Q.—But she lives at the other end: the village needs two milk separators.

Nesterenko.—It's impossible for the moment.

A boy of about ten years old, who we previously saw splitting logs as skilfully as any adult, brings some hay to Pasha. She gives the peasant woman her produce back, but the peasant woman does not leave. She is listening to us, curious to know more about the hay.

Nesterenko.—The way he splits wood is incredible! I've never seen anyone do it so well. It's a real art. I'm serious. His axe is almost bigger than him. He lifts it up and down it goes, effortlessly, one log after another.

Q.—Vassili Borissovich, can you explain why the acceptable limit for hay is 1300 Bq/kg?

Nesterenko.—We know the coefficients for the passage of the contamination from the hay to the animal. If the government has fixed the limit in milk at 100 Bq/l, it means that the limit for the hay can be 1300 Bq/kg. Unfortunately, 100 Bq/l is too high. In Russia the norm is 50, half of that amount. It's the same in Ukraine. So, really, if the hay measured 600–700 Bq/kg, the milk would not be so dangerous, it would be less than 50.

Q.—On Paraskovia Pavlovna's register, the hay can measure up to 5,000.

Nesterenko.—I can see quite clearly here that the milk brought in by this woman measures 174 Bq/l, so her hay will measure more than 2,000 Bq/kg. I asked her where her cattle graze. There are some very marshy areas here. They put boots on and go and cut the hay. Normally this hay would be withdrawn, and, because the government is short of money, it would be fed to beef cattle, not to dairy cattle.

Of course, it would still have an effect on their health, it would slow their growth. But before being sent to the abattoir, they just need to be fed on clean grass for three months. They would purify themselves quite naturally down to acceptable levels. Three years ago, when we arrived, we organised some political action, and wrote to the president of the regional soviet to put pressure on them to bring in hay from uncontaminated areas. They took the contaminated hay from the peasant farmers and exchanged it for clean hay; that's how the system worked here. Unfortunately, they don't do that anymore. Last year, it was very difficult. To give you some idea, 15 to 20% of the districts in the North had to slaughter their cattle, because there was nothing to feed them on. It was the worst year since the war.

The peasant.woman.—*(she is angry)*—Yes, they brought us hay from the clean areas, to replace ours, which was radioactive. I went to the president of the collective farm and he said he would bring us some. They brought some to us and it was mouldy. The cow wouldn't eat it; it could only be used for bedding. That was the hay they brought us so that the milk would not be contaminated! It was so mouldy that the cow wouldn't even sniff it, let alone eat it!

4. FORCE OF HABIT

We leave Pasha's laboratory and take a walk around the village. For Romano, who comes from a peasant family in the Po valley (Italy), filming the villages in these areas is a joy. There are no pretensions towards any abstract architectural beauty. These wooden buildings are human, functional and blend in with the natural environment. The form of the houses, the fences and other objects, weathered and worn with the passing of time and constant usage, have an inexhaustible beauty.

In the courtyard of a khata *(an* isba*) smoke comes from the door of a small building. Two women, the grandmother and mother of a little girl that Nesterenko has measured, are at work at a strange sort of brick stove built around a cauldron, suspended in the centre. They are preparing food for the pigs, a sort of stew made from beetroot, potatoes, rye and bran. Under the cauldron, the wood crackles and flames releasing smoke from a bed of radioactive ashes.*

The mother.—We have five pigs. Our two cows give us a lot of milk. We keep some of it for us, and the rest is given to the collective farm, which pays us in cash. My children drink very little milk. I took it to Pasha to be measured, and it was 60–70 Bq. That's a lot for the children. But what can we do? They said we should buy milk or use a milk separator. If we had a milk separator in the house, I would use it. There are milk separators at the market, but you have to buy them. People from the local council came to measure the radioactivity here and they brought milk separators. We take our milk to the school canteen and put it through the milk separator. They said they would bring several milk separators, that they would be kept at the clinic and that Pasha would deal with it. Nothing more came of it.

Q.—They didn't do it?

The mother.—They didn't do it. They never brought the milk separators. Just one, to show us how to do it. They separated the milk, diluted it with water, and got clean milk. That's all.

Q.—If they had kept their promise, would you have gone?

The grandmother.—Of course!

The mother.—We would have done it for the children.

Q.—You need to acquire new habits.

The mother.—It's true, but you know, Ivan finds it hard to change his habits. Some French people came, the organisation ETHOS. They made some measurements and they said: "You need to put the ash in bags and take it away!" It's true. Other people would have done it. But we're used to putting it on the vegetable garden, and that's what we're still doing. Tell us or don't tell us, it will end up the same. I do it for a month and then I forget, we're always so busy. If I spent all the time in the house, if that was all I had to do and if I had any money, OK, I would use gas for cooking; I would buy clean food from the shop. But I haven't got any money. I get up at 5 in the morning, I need to milk the cow, do the housework... In summer, we have blueberries. We go into the forest to collect them and then we sell them. The wholesaler comes to the shop and buys them.

Q.—And you eat them too?

The mother.—We eat them (she laughs and looks a bit ashamed) We eat them, because we think of them as our own produce. The same way that we eat our cherries and our apples. When we bring blueberries from the forest, we try, in general, not to give them to the children, but it's the same thing; when we're not looking, they eat them. Or else, we make jam... (She stops, looking a little guilty). We do that sometimes.

Q.—But that's how the becquerels accumulate.

The mother.—Yes, the becquerels accumulate. They accumulate inside the body.

Q.—You need to take it more seriously.

The mother.—I know...

Q.—For the sake of your health.

The mother.—I should. But if we don't eat the food we've grown... which is radioactive, then we would simply go hungry. It's too expensive at the shop.

Q.—Yes OK, but you could be a bit less Russian and eat less of them. *(They laugh)* Be a bit more sensible.

The mother.—Yes, "Russian"... That's what we're used to.

We measure the ash from the oven in the house. Very carefully, so it does not get blown about, the grandmother uses a dustpan to remove it from the oven and pours it into the bucket. I lower the sensor of the Geiger counter into the ash. The ash measures 2.2 μSv/h. Two microsieverts per hour, that is twenty times higher than natural radioactivity.

The grandmother.—Yes, radiation. That's how we live.

Q.—You need to bury it. Do you put it on the vegetable garden?

The mother.—We throw it under the fence in the vegetable garden.

The grandmother.—We use it as fertiliser. We grow cucumbers under glass. You need peat, soil and ash... a mixture of all three.

When the grandmother empties the bucket under the fence in the vegetable garden, a cloud of ash rises, blown about by a current of air coming across the neighbouring field that has just been ploughed. We go to see the greenhouse where these beautiful cucumbers that they are so proud of are grown. There is nothing growing there at the moment.

The mother.—We cover it with plastic sheets. There, you can see the peat. We are going to plant our cucumber seedlings here, which have been grown in little bags. We make little holes and sprinkle the ash on top because it kills all the microbes and insects. Then we plant our cucumbers.

Q.—With this ash?

The mother.—With this ash ... with radioactive ash.

5. NO-ONE TAKES ANY NOTICE

At another khata. *The little boy who was splitting wood with such skill is hurrying across the courtyard. In one hand he holds a plastic bag of hay and in the other the piece of paper Pasha gave him. His father waits for him by a pile of wood that needs cutting. The man looks at the paper in silence for some moments.*

Q.—Were you expecting this result?

The father.—No... no.

—It's very high.

—Aha! (*Among Russian peasants this means yes*)

—The time before, did you have the hay measured?

—They measured it. There may have been some radioactivity, but no-one takes any notice.

—But you yourself, have you ever had your hay measured before today?

—No.

—This is the first time?

—Yes.

—Do you know what the acceptable level is?

—They said it was… (He looks at his son) How much is it, the norm?

—About 1,500, 1,300.

—It's 1,300.

—But this measures10,000!

—There you are…It seems the radioactivity is very high, but no-one takes any notice.

We film the two men working together, splitting the silver birch logs that are piled up in the yard.

The smoke and the ash from these logs are a permanent source of radioactivity in the households of the village, inhaled by the inhabitants. We will be reminded of them later at the WHO conference in Kiev, in June 2001, when we will hear Professor Mikhail Savkin, vice-principal of the scientific centre of the Institute of Biophysics in Moscow and member of the International Commission for Radiological Protection, the ICRP, that sets the "norms", asking conference members to stop spinning yarns about radioactive wood.

6. NOBODY TELLS US THESE THINGS

Standing on the beaten earth of the street, a peasant is looking over a fence at two villagers in the yard who are cutting up a pig that they have just killed.

Q.—You know that food products are tested for radioactivity at the clinic?

The peasant.—There's a laboratory at Stolin, where they test the food.

—No, here… at Olmany.

—They don't come here to measure.

—No, it's you who has to take a kilo of meat to the clinic and they will measure it for you there, to find out if it is clean or not.

—They don't bring it here?

—It's you that takes it there. You need to take it to Pasha.

—Alright, I see. But what difference will it make?

—You can then treat the meat to eliminate the radioactivity. In salt water.

—We haven't got the equipment to kill the radiation. We can't afford it.

—I understand that, but you can decontaminate the meat, to a certain extent.

—When the shop gets it, then yes, it is clean. But otherwise…

—You can put it in salt water. It reduces the amount of radioactivity.

—In salt water?

—Salt water eliminates the radioactivity.

—Really? Nobody told us that. Who knows… Maybe it's true. We don't take those precautions here… Maybe it will kill them , but… We don't take these precautions.

—No, it's you that needs to prepare the food so that it is less contaminated, and it will be better for you… and better for the children too.

A woman calls him.

The peasant.—Huh?

The wife.—Come to the house!

The peasant.—I'll be right there.

Q.—Is that your wife?

The peasant.—My niece. We need to plant the potatoes.

Q.—Is she calling you to come home?

The peasant.—Aha! (*means yes*) We need to plant the potatoes. Can I go?

Q.—Of course, off you go. Keep well.

The peasant.—Aha! (*And he waves goodbye to me and goes off*)

7. LIONIA'S MOTHER

There are twelve people in Lionia Yavorsky's family, three generations living in a three-roomed wooden khata, *on the edge of a marsh that is used in summer as a pasture for cows. The household is run by the mother, an energetic peasant woman aged about 50 and by Lionia, the Cinderella of the house. While the mother oversees the property with hurried step, occupied morning till night with the cows, the pigs, the hay, the backyard, the vegetable garden and all the unforeseen problems that arise, Lionia looks after the grandchildren and feeds them and cooks for everyone. The father is unemployed.*

Q.—We measured your milk because your son's level of radioactivity was high.

The mother.—It is high but we can't sell the cows. We are a big family. The children are young. We can't. I get 6 million roubles and my husband is unemployed. We have to live. My grandson is still very young; look at him. We don't know what to do to survive.

Q.—By the way, does no-one give you any clean milk?

The mother.—They bring it to us, but you need money to buy it. We haven't got any money.

Q.—The collective farm doesn't give you any?

The mother.—Oh the collective farm doesn't bother about us. One peasant went to ask for some fertiliser, and they said "No!" She began to cry so they gave her some.

Q.—Do the people working on the collective farm get clean milk?

The father.—Their cattle graze in the fields. They are less radioactive.

The mother.—We have three cows altogether. We were told "Sell them, Sell them!" But we're not selling them, we can't live without them.

Q.—Which fields are they pastured in?

The mother.—Over there… But they shouldn't be: there is too much radioactivity. Here is the hay. It was in the stack. We've put some of it in the barn. But there's a lot of radioactivity there too… What can we do? Before, the collective farm gave us hay.

Nesterenko.—We've just been talking to the president of the collective.

The mother.—Huh? Don't bother!

Nesterenko.—Listen to me. Just listen for a minute: they are offering you a clean field, at Kochtovassi.

The mother.—He said that?

Nesterenko.—He said it just now.

The mother.—Then give it to us, my husband would go anywhere.

Nesterenko.—You must go there now. The vice-president of the district committee, the president of the collective farm, and ourselves, were present. It was all recorded on camera. He can't refuse now. Take advantage. Cut the hay, Pasha will measure it (*To Pasha*) That way you will be able to see if it is clean. If it isn't, it's not worth the trouble of preparing it. If it's clean, then prepare it for winter. Secondly, before haymaking, he is offering you some pasture for your cows near the school. It's better to go there, because it's obviously more contaminated here. And graze your cows there because 80% of the contamination in your children is coming from the milk. The rest is from mushrooms, berries and fish.

Q.—And how do you feel yourself?

The mother.—Oooh! I don't feel well. I am losing weight, I'm not getting any better, I'm not in good health.

Q.—Where do you feel ill?

The mother.—My heart…stabbing pains…And when I start working, it will suddenly hit me…I start shaking. It vibrates, I have to sit down and then it goes away. Sometimes my blood pressure goes up and I have to lie down, like a log.

She goes off. How can I describe the natural elegance of this 50 year old Slavic woman, with slightly Asiatic features? (The Tartars reached as far as this area in the 13th century). In spite of the lines on her face, caused by exhaustion, worry and the poison that is eating away at her body, she has incredible energy; there is something of the Romany in her, her silhouette, and her nervous energy as she goes about her work busily, confers on her a real beauty. The rapid pace of her walk through the tall grass to fetch her cows from some distance away is like a rhythmic dance.

Nesterenko.—Did you see? She is thin. You can see how hard she works. A really brave woman. Everything in the garden is tied up, looked after. The soil is acid here, it needs lime. If they gave her potassium, in large quantities, the food she produces would be clean. Even here. You can see, the help that is needed for the people living here isn't enormous. We have analysed their needs. If each family just got this practical help, everything would be alright.

Q.—Are they contaminated?

Nesterenko.—Of course they are! We will measure the children and her also, and you'll see, they're all contaminated. The hay there is contaminated. What is needed is potassium, calcium and phosphorus. With these three fertilisers everything would be clean. And she could protect her children. Because 90% of their radioactive load comes from what they eat: milk, vegetables and the produce that is gathered in the forest. The food that she collects from the forest, she must learn to soak in salt water. It would be less contaminated. Same with the fish…but she says they don't eat much fish. So for them it's the milk, the vegetables and the mushrooms.

Q.—Can it be done?

Nesterenko.—Did I tell you about Karin Reese from Burgwedel, near Hanover in Germany? She took charge of the village of Obidovichi in the Bykhov district. Five years ago, each family was examined to see what minerals they were lacking. She found the funding, we bought them and used them. She has already bought eight milk separators. And we brought everyone to the institute: the head teacher at the school, the director of the collective farm, teachers, young mothers running busy households. We held a three day seminar. There you are, we started with levels of 700–800 Bq/kg, and today we have 40–50 Bq/kg of contamination. The aid given by Burgwedel to the village of Obidovichi is an example of concrete aid.

Q.—This woman might never have had heart problems?

V.Nesterenko.—Yes, of course. It's the caesium…These are all the symptoms that Yury Bandazhevsky talks about. We will measure her today. If she has caesium, it will be blocking the energy transmission at a cellular level. Her heart is incapable of any effort. You can see how she races here and there, because she needs to work, and that is why she has heart pain. She gets up at 4 in the morning, and does all the work here. If she had another piece of land, she would cultivate that too.

8. THE FATHER AND THE BLUEBERRIES

Q.—Can I taste?

The father.—Why?... Of course!

—And the caesium?

—So what (*He laughs*) I'm used to it all.

—Do you eat them?

—Yes…Take some, taste them. The juice won't do you any harm. Taste…They're not very good. A bit acid. They'll be sweeter later on.

—How much caesium do they contain?

—None.

—You don't know. Have you had them measured?

—No.

—Never?

—No.

—But everyone knows the forest is contaminated?

—Yes, of course!

—And the blueberries are contaminated…

—We have to live, don't we? So, you've seen, the people go there, pick them, bring them back and sell them.

—Where?

—In the towns. Everywhere, even in Russia. (*He laughs*)

—The caesium is dispersed everywhere. (*He eats in silence, looking at the forest*) You should get them measured. (*He carries on eating*) How much do you sell them for?

—3000 roubles a bucket.

—For a bucket. How many kilos is that?

—It's not full. About 6 kilos.

—They are bought by the bucket?

—They are sold in buckets or in bottles. If you buy them in bottles, they cost more.

—How about if I buy a kilo from you, can I do that? I will buy a kilo and I will get them measured straight away at Olmany…For me, just so I know. How much for one kilo in a bottle? I will buy one, maybe more.

—You will have to ask her. I didn't collect them; it's not up to me to decide. (*He laughs*).

—It isn't your business?

—No…Can you measure this wood that comes from the forest and this that comes from the barn?

—Yes, we'll measure them.

Contamination of the Yavorskys:
– the father, born in 1945: 419.52 Bq/kg;

— the mother, born in 1950: 444.39 Bq/kg;

—the son Lionia born in 1985: 630.92 Bq/kg;

— The grandson (Tolik), born in 1998: 1,538.27 Bq/kg. Inside the body of this baby, that weighs 8 kg, there are 12,300 artificial radioactive atomic disintegrations per second.

We measure Yavorsky's blueberries.

Pasha.—They contain 573 Bq/kg. Normally, blueberries with this level of radioactivity would not be sold on the market. *(The limit is 185 Bq).* But at the market no-one asks if it's been checked, the people don't care. If it was me, I would ask the vendor to show me a certificate of conformity for the blueberries. If he couldn't show me one, I wouldn't buy them. Those who buy them and don't ask for verification are at fault.

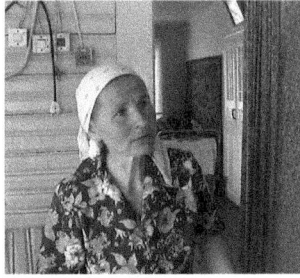

Lionia's mother

Chapter V

SLOBODKA

1.5 mSv/y, 5–15 Ci/km²; 70 km from Chernobyl

2000

Early in the morning on the 4th April, we arrive at the home of Lena Dubenchuk, the dosimetrist who runs one of the 83 LRMCs still under Vassili Nesterenko's control. The Belrad team, which has arrived before us, is measuring the radioactivity of the children and their mothers with the mobile spectrometer provided by Ireland. By the end of the day, everyone in the village will have come through. Lena wants to introduce us to the Skidan family so she takes us to their house. Romano Cavazzoni is filming as we walk along the beaten track that forms the main street of the village. The region of Polessie, spared the industrial development that has filled our landscape with "high tech" artefacts, remains a vast architectural and human museum from another era.

In the slanting morning light, a teenage boy reads a letter to an elderly peasant woman, who cannot read, from her grandchildren who have emigrated to the town. Two other elderly peasant women are talking in front of the wooden fence of their khata. The larger woman adjusts the shawl that envelops her face. She is both intimidated and exhilarated by the camera.

The larger woman.—There is radiation everywhere.
Q.—Where, everywhere?
The other peasant.—There's none left, it all blew away.
The first.—It went underground and off into the sky.
Q.—But, has it gone, or has it stayed?
The two peasants.—We don't know. We think it's all gone.

The Skidan family is poor. There are five children and the mother works at the hospital in the neighbouring town. The father is unemployed. All of them have health problems. What strikes us as we listen to the conversation between the two women and Nesterenko, is the knowledge of the subject of radionuclides displayed by these two country women, exasperated by the local bureaucracy. They have acquired this knowledge through necessity, faced with the changing nature of their environment, whose secrets have previously been transmitted from generation to generation. Today, they are disoriented, especially the mothers, who want to know and understand, for the sake of their children.

It is they in any case who run the centre for the monitoring of incorporated radioactivity in the human organism. The government gives them no help and they have to battle with the local officials, who, as at Poliske, cannot be bothered with anything unless it is in their own interests. The normal level of caesium-137 should be zero becquerels. Mrs Skidan knows this and can see the effect it has had on her own children whose development has been compromised.

Nesterenko.—When were your children last measured?

Mrs Skidan.—The last time your team was here. We get them measured each time.

Nesterenko.—And how much did they measure?

Lena.—90 and 70,

Nesterenko.—That's a lot.

Lena.—The youngest one had very little, didn't he?

Mrs Skidan.—A little over 30.

Nesterenko.—What did you do then to reduce it?

Mrs Skidan.—We bought vitamins. We economised on other items. And we take the milk to Lena to be separated. In any case, in summer, the children absorb more radioactivity. They play outside, in the sand… But the problem is the food. Lena, you remember, we couldn't find any cheap meat. We had a little veal calf, last year. Lena measured it… the meat was not clean. The meat we were buying from Kalinkovichi was four times the limit.

Nesterenko.—Before it is slaughtered, the animal should be fed for two or three months on clean fodder. While it is living it will purify itself biologically. If you want to sell your meat, prepare clean fodder, have it measured by Lena and give it this hay for two months. You will have clean meat without any caesium in it.

Mrs Skidan.—Yes, we have to sell our meat. There is always a problem with my salary which never arrives. The children want to eat something good, but we can't buy anything and we have to eat our own radionuclides. We do everything we can, we try to understand what we need to do, but the children are not in good health…

Nesterenko.—We have brought you some Yablopect. Give it to the children and we will come back in a month to monitor their levels of radioactivity.

Gradually Mrs Skidan begins to talk, more and more quickly, jumping from one argument to another. She gives vent to the feelings of humiliation and anxiety that have overwhelmed her.

Mrs Skidan.—I fell out with the district committee over some compensation claims for convalescence. I told them I did not want charity. I gave birth to these children; I knew what I was doing when I put them on this earth. I'm not asking for any extras. Why can't they give me what I'm entitled too? Three of my children are at school. They're all ill. This one has a goitre. The little girl's thyroid is enlarged, this one has problems with his sight. ..We barely make ends meet, we economise on everything just to buy food.... The radioactivity is having an effect. The little one here can't speak. We took him to see a psychiatrist at Kalinkovichi. He examined him and said that his mental state was normal, that we should take him to Gomel…

Nesterenko.—Have him examined by a speech therapist because your child is not the only case like this. In this area, many of the children have language development problems.

Mrs Skidan.—You need money to do it. If I ask for a car, I have to pay for the petrol. And you need to follow the treatment three of four times a week…They either think I'm not right in the head to have had so many children, or that I'm alcoholic and that I had them just to get benefits.

Nesterenko.—The children are here and you are responsible for their health. You are the only person who can help them. So don't feel ashamed if people offer you money. Take no notice of what stupid people say.

Mrs Skidan.—Members of the executive committee came to see me and said some very insulting things. One of them, a man called Obuzenko, addressed me with the familiar "you" form. Even if I am younger than he is, I am a woman, and he should show me some respect. He attacked us: "Why do you keep making complaints? You're telling lies…" They haven't brought us a gas supply. There is an arrangement whereby gas should be provided in the Chernobyl zone. Because our wood stoves are basically nuclear reactors. And they won't give it to us. Where should we get our wood? From the zone where they have organised a reserve for contaminated wood. There isn't any here, it all comes from there. If this is the arrangement, why have they not supplied us with gas? At least for cooking for the children. And he just shouted at me: "No, we haven't got gas, you can't have it!" Then he said the opposite. "No, you mustn't use the wood stove!" So where am I supposed to cook, if they aren't supplying us with gas? I had an argument with him one day about the milk separators. The villages are dying. How many families with young children have stayed in the village? If someone has stayed, the least they could do is to help them. Don't reject them! They had enough money to give 100 million roubles to the Kalinkovichi football team for an anniversary evening. But there was none for milk separators. Someone managed to get us one, on the sly, for two months, to share between three families!

Nesterenko.—How many becquerels were there in the milk? You said it was 200, didn't you? And after it was separated, how much?

Lena .—About 70.

Mrs Skidan.—Yes, it reduces the radioactivity by about a third or a quarter.

Nesterenko.—That's good. It means that the children are taking in three or four times less radioactivity.

Mrs Skidan.—What hurts, for example, is when they say "Chernovshchina is in the zone, Obukhovshchina is in the zone, but you, Slobodka, you are not in the zone". It's 4 km from here! We need clean fodder for our pigs, fertilisers for the seeds that we sow ourselves… For example, I have a hectare of land, because I can't buy everything from the shop. I grow half grain and half potatoes. They give fertilisers to Obukhovchchina and to Chernovshchina because they are in the zone. But the people in Slobodka, we're not in the zone! We called the executive committee. We asked them why we're not included in the zone. They came and they shouted at us. "Don't call us again!"

Nesterenko.—I've got a suggestion: there is a man called Ivan Albinovich Kenik. He is the president of the committee that deals with all the problems concerning Chernobyl (ComChernobyl). It's worth writing a letter to him about all your problems, including the fact that you have been talked to in an insulting way. Don't be afraid. Write to him. I will go and see him and I will tell him all about it. I have a meeting with him next week. Because I am in no doubt that you should get all the help you need. This is one of the 350 villages in Belarus where the milk is regularly contaminated.

Lena.—Here, it's not regularly, it's all the time.

Mrs Skidan.—Look at my children: Alexander was born in 1990. Hyperthyroidism. His blood sugar was too high, he is borderline diabetic. When these spectrometers arrived, they measured his levels of radioactivity and found that he already had an enlarged thyroid, aged 3. This is Nina. She was born in 1993. She has a congenital dysplasia of the hip joint. She was registered with the hospital in Gomel, and they suggested operating to straighten her legs. I didn't take her because I was frightened. Because in some cases, they still remain lame. So she is still like that. Her sight is also very poor. In all of them their eyesight is starting to deteriorate. Aliona was born in 1997. At 3 months she had chronic bronchitis. They tested her for allergies at Gomel when she was 3 years old. Lots of tests. I consulted our doctor. He is a good doctor, and he said it was too soon, it was better to wait until she was 7, otherwise the allergy could be aggravated and it would be worse. This is Vitali, who was born in 1994. He doesn't speak. His speech is affected. I worry about all of them: one is ill, the other needs something. It's never ending.

Nesterenko.—While they are young, you need to look after them; you have to.

Mrs Skidan.—That's not all. I had a little boy before Aliona. He fell ill and he died because of the flu. As soon as Aliona gets ill, I get frantic with worry. I immediately think it's something serious. Apart from that… I think my children are lovely. They work hard at school, they get good marks.

We are back in Lena's house where she continues with her measurements. The spectrometer has revealed a very high contamination, 67 Bq/kg, in a baby who is still being breastfed.

Nesterenko.—It's too high in such a tiny baby, much too high. Let's see how contaminated the mother is. Press your back hard against the back of the chair. What is your name?

The woman.—Poleschiuk, Liubov Konstantinovna.

Q.—When were you born Liuba?

Liuba.—In 1974

Nesterenko.—I can already see on the screen that the mother is very contaminated. We need to understand why?

Q.—Do such young babies need to be fed with breast milk?

Nesterenko.—The mother needs to be eating correctly so that her milk is clean. But in this case, when her levels of contamination are so high, the child needs, very quickly, to be given extra food that is clean. Because this is no good at all. Of course, nothing replaces mother's milk. It provides the baby with natural immunity. Babies do not develop allergies when they are breastfed. But if her milk is contaminated, the whole immune system will be destroyed. We need to do something quickly... We will be able to see the figures soon; we need to just wait three minutes. Can you try very hard, Liuba, while you are sitting there, not to move your back while we are measuring—which food do you think is to blame? Is it the milk?

Liuba.—Yes, I think so.

Nesterenko.—Did you bring it to Lena to be monitored?

Liuba.—No.

Nesterenko.—Why not? It's your baby we're talking about. Your little girl is the most at risk from the radioactivity.

Q.—But is it her milk that is contaminated or the cow's milk? How does it work?

Liuba.—No. it's me who drinks contaminated milk.

Nesterenko.—A mother who is breastfeeding passes all her radioactivity on to the child. The radioactivity is transmitted very efficiently through mother's milk. It's dangerous, especially if it is very concentrated. *(The figure appears on the computer screen showing the levels of caesium-137).*

Nesterenko.—There we are, Liuba, the level is 173 Bq. For a breastfeeding mother, that's much too high. We will give you some pectin pills. Give them to the baby too, but only half a pill, not more. She may get a bit of diarrhoea but don't worry. If she does, just give her a little less... Lena will give you the tablets. Take them and your milk will purify itself.

Q.—Should she stop breastfeeding?

Nesterenko.—She needs to stop drinking contaminated milk, then she can breastfeed. But for a few weeks she will have to be careful. She should give the baby a mixture of breast and formula milk, because the baby is becoming very contaminated through her breast milk.

Q.—How long will it take for her breast milk to be purified?

Nesterenko.—She is 26 years old… So I would think she will eliminate about 50% of the radionuclides in a month, if she takes the pills and if she stops drinking contaminated milk. A month.

So, let's recap, Lena. You need to find out if everyone has been taking the Yablopect as they should. Meanwhile we will send you all the information about what people have accumulated in their bodies and the ways to improve the pastures. You should try to obtain fodder containing adsorbents, fertilisers…

Lena.—We've been told that they won't give us anything like that in our district.

Nesterenko.—You need to write a letter to Kenik as I explained. If you write to him, I will go and see him. If you keep quiet, I can't really do much. But it can't just be you on your own, it needs several signatures , so that it isn't just your name. You need to get other villagers to sign it… five or six people. Send this letter to Kenik and a copy to me. I will go and see him. I'm going to write to him myself to say that I've been here and that I have seen the field that is supposed to have been treated, and it hasn't been done yet. I will tell him that you are not being given any clean fodder. I could also write to the president of the Vitiutkov district executive committee. They will have to do something; we're going to pester them. They'll kick up a fuss but they'll have to do something anyway. This is what we want to achieve together. The bosses move on, but we have to carry on living here.

Lena.—You know, immediately after the authorities have monitored the situation, everyone takes notice, but afterwards, they go back to how it was before.

Nesterenko.—So we need to repeat the operation. In my opinion, your village should be measured at the beginning of spring, then in the summer when the cows are at pasture, in autumn during the mushroom season and in winter when you are eating preserved food, dried mushrooms, dried fish etc. At least four times a year. If we could do it, then people would always understand the situation they're living in and they wouldn't fall into bad habits. Because we don't have sensory organs that can detect radioactivity. *(Nesterenko addresses a woman who has come to be measured.)* 12 Bq. Well done! That's fantastic! How much did you measure before?

The woman.—It was 33.

Nesterenko.—Well done! What do you put it down to?

The woman.—The cow is not giving any milk.

Nesterenko.—Mushrooms?

The woman.—No

Nesterenko.—Game? Meat?

The woman.—We only eat our own food.

Nesterenko.—You see, if everyone was like you, I wouldn't be worried. You're doing it perfectly. May God help you to teach others to do the same.

The woman.—Thank you, that's what we try to do. (*She smiles and goes off*).

Nesterenko was right. Kenik received the letters and took action. The village was given a milk separator, but then Kenik was replaced as director of Comchernobyl by Tsalko.

That was April 2000. In January of 2001, a French team, the consortium ETHOS, financed by the European Commission and headed by a representative of the CEA (Commissariat à l'Energie Atomique—the French nuclear lobby) had Nesterenko removed from the LRMCs in five villages that he had managed for ten years. They took his place, while having no mandate to undertake work in health or prevention. In 2004, Comchernobyl would close all the remaining LRMCs, apart from about twenty, which were financed by charitable associations from the West.

June 2001

I am transcribing the following conversation for the historical record. The reality of the ecological and health situation described by the witnesses has not changed and will continue for decades to come. "The bosses move on…", but the medical situation is getting worse every year.

The house that the Skidan family lived in before has burned down since our last visit. They are renting another house. The son, Seriozha, rides his bicycle in front of our car to the new house.

Mrs Skidan.—Hello. You have come back to see us?

Q.—We have come to see if your children are in any better health. Have they been measured since last year?

Mrs Skidan.—Yes. They were measured in school. In any case, as you know, when the summer starts their radioactivity increases. When they were at the sanatorium, or when they're in school in winter, their health improves. But in the summer, they stay here.

Q.—Are they taking pectin?

Mrs Skidan.—No.

Q.—Why not?

Mrs Skidan.—They were given it once in school. Seriozha was at the sanatorium, weren't you?

Seriozha.—Yes.

Mrs Skidan.—Nina had some, that's all. But the others haven't had any. Nor have the little ones either.

Q.—Do you know what the difference was between the measurement when we were here and the one that was taken a month later?

Mrs Skidan.—The other time they took pectin and it had gone down…I think by about 20%. They took the pectin for about a month and their radiation dose had gone down.

Q.—Are you treating the food the way Nesterenko advised? Are you eating mushrooms for example?

Mrs Skidan.—No. The mushrooms… *(She smiles, a little bit ashamed)* The only thing was that they once asked for a little bit of girolles with their potato. Otherwise, we don't eat mushrooms. But the milk is very radioactive too…

Q.—You eat mushrooms? Do you like mushrooms?

Seriozha.—Yes!

Q.—And does Mummy soak them in salt water first?

Seriozha.—Yes.

Mrs Skidan.—We soak them.

Q.—And blueberries?

Seriozha *(categorically)*—No!

Q.—Do they grow here in the wild?

Mrs Skidan.—They grow here, but we don't gather them. There, on the reserve the radiation is very high.

Q.—You need to be very careful with fish and meat also.

Mrs Skidan.—Here, we know we have to take care with all of the food. But we eat everything even so. Meat is monitored. The radioactivity in the meat exceeds the norm also. So does the milk and the potatoes, with strontium, doesn't it? All the food, and we eat it.

Seriozha.—So, what can we eat?

Q.—What did you say?

Mrs Skidan.—What can we eat? We don't get paid enough.

Q.—How old are you?

Seriozha.—10 years old.

Q.—And you already know about radiation and about the economy. You know that you can't eat anything else! How many are there in the family?

Seriozha.—Five.

Q.—And where are the others?

Seriozha.—At Grandma's house.

Q.—And how is their health?

Mrs Skidan.—What can I tell you? They are ill… two of them have enlarged thyroids… with Sergei, it's his eyes. He needs new glasses but we can't get them here, we'd have to go to Minsk. The little one has a congenital dysplasia. You see how many complications she has! Her joints here, her hips, they are not joined together properly. It's a malformation. When she started to walk—it wasn't diagnosed early enough—the tibias were not growing straight… She has a congenital malformation of the hips. I did not take her to be operated on because I didn't understand. I thought it was her hips that needed to be operated on if it was a malformation. But they wanted to break the tibias in order to straighten her legs. To break the bones and then straighten them. We didn't take her and so she stayed like that. And she also suffers from high blood pressure.

Q.—Is all this due to the contamination from Chernobyl?

Mrs Skidan.—Yes. When we took our children to the health commission, they already knew that they were in the critical group. All of our children. All of them.

Q.—What level of contamination is there in the village?

Mrs Skidan.—Well, before, they told us it was between 10 and 15 curies. But this year, they said it was 5 to 10 curies.

Q.—Why? They suddenly noticed that it had gone down?

Mrs Skidan.—I don't know. They just handed us the form: 10 to 15. And now it says 5–10.

Q.—And what does that mean? Do you have fewer rights?

Mrs Skidan.—Yes, when it was between 10 and 15, we got more benefits for the children. Now we get half of that.

Q.—Is this decision by the authorities based on any scientific data?

Elena Dubenchuk.—It's possible that they have made the decision on the basis of the years that have gone by since the accident, or according to their own measurements, I don't know. No-one has measured us here. Apart from Nesterenko, no-one has come to measure the inhabitants here; no-one has explained…

Q.—You haven't seen any scientists?

Mrs Skidan and Elena Dubenchuk.—No.

Mrs Skidan.—Our house burnt down. We are renting here. We lost everything. We weren't able to save much. I went to see the president of the district. My husband is unemployed, and we have a large family. I'm not saying they should buy us a house, but they could at least lend us something. We do actually live in the evacuation zone. They told us it isn't an evacuation zone any longer.

Q.—It was before?

Mrs Skidpan.—It was a voluntary evacuation zone. Some people evacuated from this zone. They gave them some money for their houses when they left. Now it's finished.

Elena Dubenchuk.—They just want to save money. You only need to look, for example, at the way the admissible level in milk is 100 Bq/l whereas in Russia it's 50 Bq/l. Think about it—here, 50 Bq in the milk means it's clean and in Russia it's contaminated.

Mrs Skidan.—They say categorically that this village is clean. But the cows have tuberculosis, and leucosis.

Q.—The cattle are ill?

Mrs Skidan.—Yes. They talk about improving the pasture, but the cows are still using the same fields as before. No-one treats the pastures. When you telephone they say they've been treated.

Q.—Have you noticed any changes in the health of people living here and in children over the last fifteen years? Have there been any changes?

Mrs Skidan.—Of course. First—I work at the hospital—there has been a big increase in tuberculosis. Secondly, there has been an increase in stomach cancers; that's very common. Liver cancer. All cancers. And there is even something new: in general, with cancer, even if it spreads rapidly, there are early symptoms, the person loses weight for example… but now there are hidden types of cancer. We have had people ill with a constant high temperature. They have been sent to town. Their lungs have been X-rayed, and everything appears normal. Then, they are examined: liver cancer. This hidden type of cancer is becoming more common. Really very common. The men who drive tractors, who work in close contact with the soil, if they have been working for twenty five years or more, they are given early retirement… But nearly all of them die. Cancer… and then there are many cases of acute cerebral disease, haemorrhages.

Q.—Are you talking about this village?

Mrs Skidan.—Everywhere, it's everywhere… Our hospital covers the whole district of Yurovichi. There have been many deaths. Tractor drivers dying at 55. There's Klempatch, cancer of the stomach.

Elena Dubenchuk.—Where's he from?

Mrs Skidan.—Chamochek.

Elena Dubentchouk.—Ah!

Mrs Skidan.—Then there's Morek, Nedozhiv… cancer. We have cancer and tuberculosis.

Elena Dubenchuk..—Because these tractor drivers work in the zone.

Q.—In the zone?

Elena Dubenchuk.—In the evacuation zone. All the fields are out of bounds, but, we don't know why, the collective farms ploughed, harvested, kept to the plan. All those workers are ill.

We follow her into the yard, towards the vegetable garden. A small elderly woman with a friendly open expression is working behind the fence. The fences do not hide the faces of people and you can talk or wave hello. Nina, the little blonde girl with deformed legs, is standing in front of the fence and gives Romano a cheeky smile.

As we are leaving, an old shepherd with a sad face—the very image of humility, like a character from a Dostoyevsky novel—goes by with his dog, and exchanges a few words with the peasant woman. In the West, he'd be seen as a tramp. The dog barks. The shepherd puts his hand on its head, strokes it and in a lilting voice, sings "She helps me, she helps me, Anna Ivanovna". The little girl strokes the dog too.

The shepherd. (*to Romano, who is filming him*)—Thank you! Thank you!

The peasant woman.—When you go, who do you leave with the herd?

The shepherd.—No-one.—I leave by myself. Alone, I am alone… little dog. (*The peasant woman teases him good-humouredly, repeating "little dog"*) Just us two, us two, little dog. Come to me, come here. There we are, my darling. She looks after me. Always. Come on… Thank you! Thank you! (*He takes off his hat before Romano in the classic submissive gesture of the poor in Russia.*

Chapter VI

VERY EARLY MORNING
IN VALAVSK

1.3 mSv/y, 5–15 Ci/km²; 96 km from Chernobyl

In the distance are some sheep, and a cow on its own on the road. Mrs Kokhan, who put us up for the night, gets her cow out and calls to her neighbour—"Vasilievna! Vasilievna!" and she goes off towards the herd, that is gathering at the end of the village. Vasilievna gets her cow out, two young men pass, carrying scythes on their shoulders. "Hello!" Behind the fence a small dog yaps and goes away satisfied. In the distance we can see fields and a group of houses.

A young woman puts two full buckets of milk down before the fence and covers them with a plank of wood. The milk collector lifts them up, puts them on the handlebars of her bicycle and rides off.

The milk collector—I collect the milk and take it to the cart and empty it into the milk cans…Yes, yes, I collect it and take it there. I am helping my husband. My husband works and I help him. Then we take it to the stable…At the state farm… We give it to the young calves to drink.

We go with her. We meet two peasant women returning to the village. One of them gives the collector a milk can and she empties it. Another appears with a full bucket.

Q.—Do you measure the radioactivity in the milk? Do you know what the levels are?

The peasant.—We know what they are. There is a family over there with high levels of radioactivity. Above the norm. That's what it's like there. The cow grazes in the wood. There was more fallout there.

Q.—The cows graze there?

The peasant.—Probably.

Q.—And this milk here, do you know if it is clean?

The peasant.—We collect it to give to the calves.

The milk collector.—This herd is more or less alright, but the other herd over there, the milk is bad. Our herd, it's alright.

Q.—What do they do with the milk that is bad?

The milk collector.—The same thing. They give it to the cattle.

The sun rises a little above the horizon. The village is getting busier. The milk collector goes off with her milk, passing women who are coming out of a side road, leading their cows towards the herd. Others arrive from the opposite direction.

Three milk cans in front of a house, a wayward cow passes in front of us. A man on his own greets us. A peasant woman with a stick returns to her house. An artisan, his bag on his shoulder, crosses the path of two women leading three cows towards the herd.

Q.—The milk you are collecting, is it clean, no radioactivity?

The milk collector.—Who knows?

Mrs Kokhan's father, an elderly man, passes on his horse and says hello. He is wearing a rather eccentric papakha[143] on his head to protect him from the morning cold, one ear flap up and the other down. He smiles at us, with a proud and slightly mocking regard. A group of women return to the house. Laughter and bantering from the men and women who are walking in front of us. They walk away to the fields. From the end of the road comes the milk collector on her bicycle with full buckets. She empties them into the milk can and puts them down in the garden. A man leads a horse out and leaves. Other houses. Two women who were standing with the herd approach the village. Others come back including the smiling woman and elderly woman.

The elderly woman.—Hello. What are you doing?

Q.—We are watching how you...how you...

The smiling woman.—How we work! (*She smiles.*)

Q.—What is the radioactivity like here?

The old woman.—Oh we're full of radioactivity.

The smiling woman.—My cow measured 890 Bq/l.

Q.—Who do you give the milk to afterwards?

The smiling woman.—We put it in the milk can.

Q.—Yes, of course, you empty it into the milk can but who drinks it afterwards?

The old woman.—It's taken to Elsk...They use it to make butter.

---

[143] Called an Astrakhan hat in English. Worn by men, made of wool, fur or sheepskin, sometimes with ear flaps.

The smiling woman.—Here, before the explosion at Chernobyl, it was so clean! We were so strong, so healthy! And now we walk... Just look at me... If they measured me I bet they'd find 300 of these things! I can feel it, when I walk; I feel it in my eyes. I can't look up and then down, everything starts to turn. I feel very ill. The heart, it thumps, it hurts. I can't sleep on this side. My heart beats. I had my blood pressure taken. It was more than 200. That was about three years ago. I can't go on much longer.

Life has changed. Oh! It's changed so much, oh! ... Before, relatives would come from Leningrad. Now it's St Petersburg, but before it was Leningrad. They came here, they took deep breaths. It was so pure, especially after the rain. And now when they come, they get ulcers round the mouth. Around the mouth, ulcers. Our lives have been contaminated. Before, we used kerosene. The lamps did us no harm. And now, this electricity, we could have done without it... . What do you think? We were used to the light how it was, it suited us. We didn't know anything else. We had good eyesight. But now, I can't see anything any more, I can't read any more. Nothing... The light is too faint unless I'm wearing glasses. It's not like it was for our ancestors...

Q.—Your ancestors have always lived here, have they?

The smiling woman.—Yes...Are you surprised by my face?

Q.—No.

The smiling woman.—My hair has gone grey. I had dark hair but I've already gone grey... Someone who didn't know me said I was from... Asia.

Q.—Maybe. The Tartars came as far as here didn't they? [144] Take care of yourself.

The smiling woman.—Thank you! You too. Thank you!

She goes off. The husband of the milk collector loads the milk cans onto his cart which is drawn by a horse. He goes off to collect the next milk can.

Q.—Is it given only to animals, or do people drink it too?

The milk collector.—People have to drink it at home; it's impossible not to. We give away any that's left over.

Q.—Do you know how many becquerels per litre it contains?

The milk collector.—The radioactivity? ...Wait a minute... It depends on the farm. One farm, for example, had 106... Too much, yes too much. The admissible level is something like 96; there were already some with more. In another farm, for example, it was 70. But when you give fodder that contains adsorbents—I've experimented with it at home, when my cow calved and I gave her some—well, then it was about 30, 50; it was very little.

---

[144] The Golden Mongol Horde in the 12th century.

She goes off one way and her husband the other, with his horse. We start talking to the peasant in the house next door.

Q.—Has life changed since Chernobyl?

The peasant.—Of course! People are more ill… Some suffer very badly with their joints, their liver, heart problems. It affects children's eyes, their thyroid. Many of the children are ill; they weren't ill like that before. This radioactivity can't be detected; it has no smell, no taste. And we're paid a pittance in compensation for Chernobyl. It measures 15 curies round here.

—Here, where you live?

—Yes. As high as 15 Ci/km², it's considered a radioactive zone.

—And do you grow vegetables in your garden?

—We grow everything!

—And it accumulates?

—Probably. Who's going to check? The onions, the garlic. We grow cucumbers, cabbages. Who's going to check them? We get ourselves checked but that's all. Nothing else is checked. Look at our village, Valavsk. It's big; it has a lot of potential. Before, when I was younger, when I left technical college, I became manager straight away—there were five agricultural teams here: Shia, Glazki, Novy Khutor, these are the names of the big farms. Then there was the village of Dubrovka, of Korma, dead villages… Well, Shia no longer exists, nor does Glazki… There were so many fields! There were lots of people, lots of cows, there were beef cattle and now there are no cattle there any more. There are only the ones we've got here. There are even villages that have been buried, like Kusmichi. You would never know that there had ever been a village there. Completely buried.

It was in those fields belonging to that village, Kuzmichi, where contamination levels were between 15–40 Ci/km², that Ermakov and his friend Yury drove the combine harvester up until 1991.

Chapter VII
EAST OF THE RIVER SOZH

2002

In September 2002, following Nesterenko's suggestion, we went to an area east of the river Sozh, a natural paradise that has metamorphosed into a kind of cursed no man's land between the river and the border with Russia which runs parallel to it. Nesterenko had told us of his discovery of a completely abandoned population, living in villages cut off from the world because of the Belarusian government's policy, supported by United Nations agencies, of minimizing the effects of radiation.

To understand the extent to which the people of this territory have been severed from the rest of the country you need to look at the map showing the contamination of Belarus, published in 2001 by ComChernobyl. There are fifteen large areas marked in a dark colour on the map within a narrow band of land, about 30 kilometres wide, which from South to North—from Gomel up to the Mogilev region—stretches for 150 km between the river Sozh and the Russian frontier, bordering the Bryansk region. They say that the radioactive clouds were seeded chemically from helicopters precisely on these areas so as not to contaminate Moscow[145]. The darkly coloured areas are concentrated in the eastern part of a much larger contaminated territory, covering about 15,000 square kilometres, that extends west beyond the river; on the map, the territory is marked in differing shades of colour depending on the number of curies per square kilometre. Within the shaded areas, of which six lie either side of the Russian border, the dark grey indicates contamination at a level *higher than 40 Ci/km²,* similar to the levels found in the immediate vicinity of the exploded reactor. It is extremely dangerous to cross the area or to visit it on a regular basis. Also as radioactive strontium and caesium, plutonium (half-life of 24,000 years) has been detected. The government, which does not have the money to evacuate all the inhabitants from this stretch of

---

[145] Urban myth? People who live there talk about it. Alla Yaroshinskaya quotes it in *Tchernobyl, verité interdite*, (Chernobyl, the forbidden truth), Arte-Editions de l'Aube, 1993.

territory, simply ignores them, to the extent of cutting off all aid and of dismantling the state and collective farms that used to provide them with food, transferring those farms to areas west of the river, that are scarcely less contaminated. In reality it is an incitement to flee individually or to disappear in silence, to become extinct in a kind of accelerated "natural" selection process. As Yury Bandazhevsky described in his report, it is "everyone for himself". The recordings we made in this highly contaminated area, 200 km from Chernobyl, reveal a people who have been left entirely without means, protection or information, utterly abandoned.

The joint communiqué from WHO/IAEA/UNDP, published on 5th September 2005 for the twentieth anniversary of Chernobyl, is more than a little optimistic.

> As for environmental impact, the scientific reports are equally reassuring, as their evaluations show that, with the exception of the highly contaminated zone in a 30 km radius of the reactor, to which access is still forbidden, and certain enclosed lakes and forests with limited access, radiation has mostly returned to acceptable levels. "In most areas the problems are economic and psychological, not health or environmental", claims Balonov, the scientific secretary of the Chernobyl Forum who has been involved with initiatives to restore normality since the disaster occurred.

The inaccuracies, the falsifications and the omissions within this text, that manages in seven lines to brush aside questions of such gravity, are astonishing in view of the supposed expertise of the authors. The first falsification concerns the highly contaminated 30 km zone. These circles and half circles with a radius of 5, 30, 60, 100 km...around the reactor, have no meaning, because there are areas equally contaminated with caesium-137, as far away as 420 km from Chernobyl, northwest of Minsk, with 1–5 Ci/km². The "forbidden" zone around Chernobyl itself extends far beyond the area defined by the 30 km radius. There is a "finger" of contaminated land that extends 60 km northwest of the exploded reactor, and, as a result, is forgotten and not included in the so called "forbidden access" zone.

Secondly, the fifteen "leopard spots" with similar levels of contamination referred to above, situated 150 to 260 km north east of Chernobyl, are totally ignored by the United Nations. It is not just a question of "forests with limited access". There are hundreds of villages in this area, with contamination levels that vary between 5 and 40 Ci/km².

Finally, knowing that hundreds of thousands of peasants are emprisoned in this radioactive territory, disinformed, without health protection and condemned to eat highly contaminated food, the inadequacy of the words they choose to minimise the danger of contamination—*"radiation has mostly returned to acceptable levels"*— would be scandalously superficial from the point of view of scientific precision, if it were not wilfully criminal, given the authority and influence that the report's

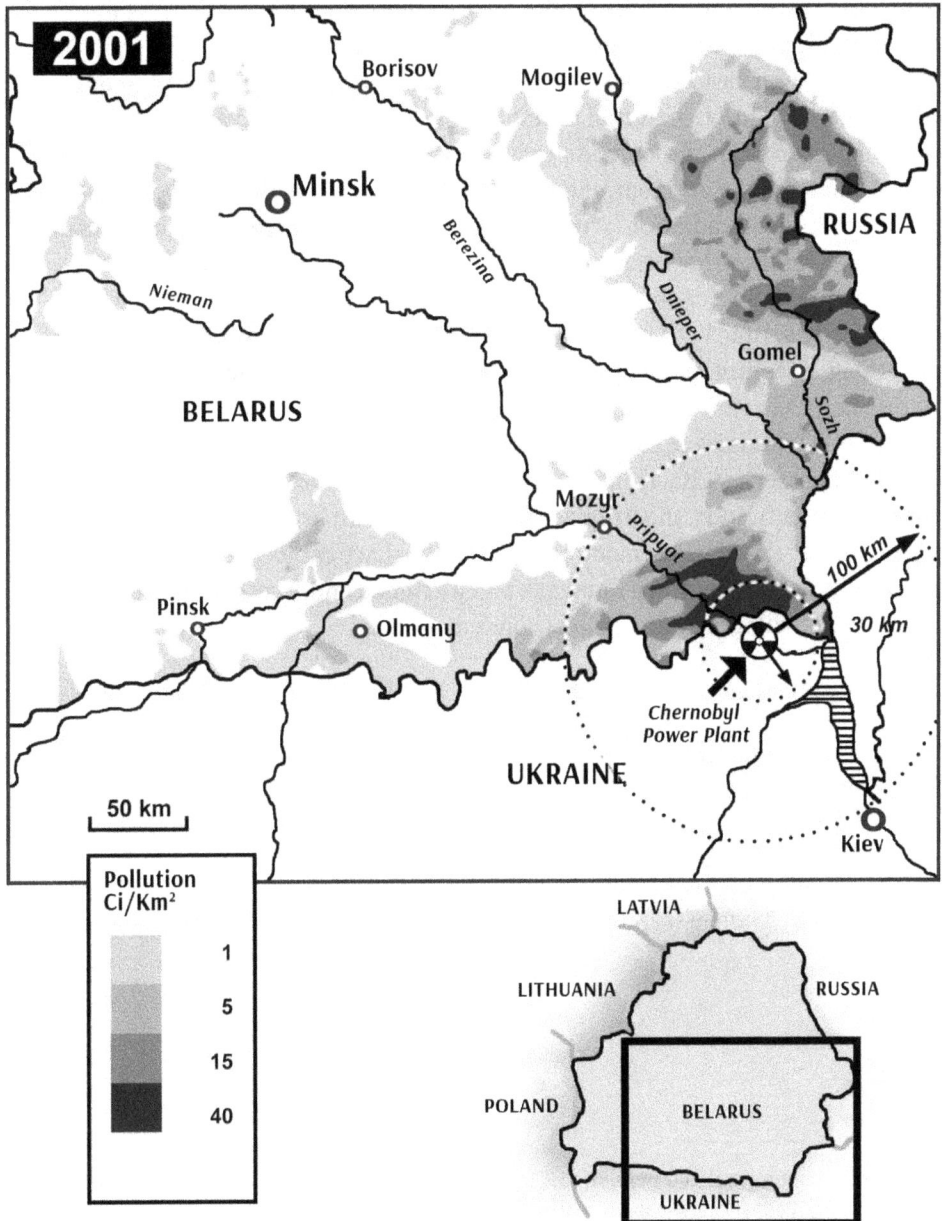

2001

Borisov
Mogilev
RUSSIA
Minsk
Berezina
Nieman
Dnieper
Gomel
BELARUS
Sozh
Mozyr
Pripyat
100 km
Pinsk
Olmany
30 km
Chernobyl
Power Plant
UKRAINE
50 km
Kiev

Pollution
Ci/Km²
1
5
15
40

LATVIA
LITHUANIA
RUSSIA
POLAND
BELARUS
UKRAINE

authors have over governments[146]. What does *"acceptable"* mean when we are talking about caesium, strontium or even plutonium lodged in the internal organs of a person who ingests these radionuclides on a daily basis? The most serious lacuna, as always with these authors, is their ignorance about low level internal contamination by radionuclides incorporated into the body through food. They continue to describe Chernobyl using the parameters defined by Hiroshima. They talk about exposure to external *radiation,* that has *returned to* acceptable levels, when, apart from in the immediate vicinity of the exploded reactor in the first weeks following the accident, the radiation has always been "low level". None of the three words—"radiation", "returned", "acceptable"- correspond to the reality of the phenomenon that occurred at Chernobyl: external radiation has never been the main threat. So it cannot *"return"* to acceptable levels. As for the term *"mostly"*, it is perfectly in tune with their style: it means everything and nothing. This text is a soothing, deceitful, public relations exercise, containing nothing scientific. The ignorance, "duly sanctioned" by these people, who preach the need for "precise information", finds its equal in cynicism only in the policy they support.

Accompanied by Lisa, from Moscow, who translates our urgent exchanges with our Belarusian friends, the irreplaceable Romano with his camera, and Emanuela who edited all the films we made about Chernobyl, we camped for eight days in an abandoned house in Staraya Kamenka and we filmed and recorded the inhabitants of four of the villages in this strip of condemned land.

> Gaishin, 225 km from Chernobyl, 5-15 Ci/km²; 12 km from an area where contamination exceeds 40 Ci/km²;
> Staraya Kamenka, 225 km from Chernobyl, 5-15 Ci/km²; 7 km from an area where contamination exceeds 40 Ci/km²;
> Volyntsy, 210 km from Chernobyl, 5-15 Ci/km²; 5 km from an area where contamination exceeds 40 Ci/km²;
> Kliapin, 205 km from Chernobyl, 5-15 Ci/km²; 7 km from an area where contamination exceeds 40 Ci/km².

---

[146] What is being formulated here, in this injunction to obey in the name of the "international community", constitutes scarcely veiled economic blackmail: "The governments of the three countries most affected realised that they must define the route to be followed clearly and go ahead only on the basis of a firm consensus as to the environmental, health and economic consequences, taking advantage of judicious advice and support from the international community". Judicious advice! But more to the point, who is the *"international community"*, which promises the carrot... and the stick?

Chapter VIII
STARAYA KAMENKA

225 km from Chernobyl, 5–15 Ci/km²; 7 km from an area an area with contamination levels above 40 Ci/km².

1. THE ABANDONED SACRED SPRING

Nesterenko and Lisa chat as they lead us towards the sacred spring. He tells her the history of Kliny, the buried village.

Nesterenko.—This stream, with its wonderful water, ran alongside the village of Kliny and all the vegetable gardens backed on to it. The people here put their milk and meat in a barrel and put it in the stream, to preserve it. With a temperature of 7 degrees C, the running water acted as a refrigerator. In this area, it measures more than 15 Ci/km². Everything they were growing exceeded the admissible levels. There was no choice, the decision to evacuate everyone was right. It was dangerous because all the houses were contaminated. To prevent anyone coming back to live there, they brought bulldozers, dug huge trenches, and tractors pushed the houses in and buried them. You can see, all that is left are the electricity poles, because the village had electricity. Now, there is no life here. There were thirty houses and about seventy to eighty people. It was monitored continually: not a single product grown here could be eaten. Up till then, what was known about this village was that its inhabitants were never ill, because drinking this water, bathing in it, guaranteed that they would live to one hundred.

Lisa.—Heaven and hell.

Nesterenko.—Yes, heaven and hell met here: the sacred spring and radioactivity. This water is amazing. It never becomes stagnant, in other words, it does not contain any bacteria, it keeps for between three to five years, it is aseptic, rich in trace elements, and chemical analysis shows that it contains a lot of silver. I'm told that the water comes from a depth of 2,800 metres.

Lisa.—How many villages like this should have been evacuated?

Nesterenko.—I think they would have needed to evacuate more than 140,000 people. They needed to eliminate a few hundred villages, where no matter what agricultural measures were taken, clean food could not be produced. If the inhabitants here could not produce clean food, given that 80 to 90% of the contamination comes from food, there is no sense in living here even though the surroundings are magnificent. The earth needs to purify itself. Within a number of decades the contamination will have sunk down deeper and it might be possible to live here again. If it was possible to evacuate Staraya Kamenka, it could only be of benefit to the inhabitants…

Lisa.—But caesium only disappears after three hundred years!

Nesterenko.—For its complete disintegration. But fortunately it sinks down in the soil. I would say in this village that is very contaminated, in sixty to eighty years the caesium would have sunk down through the soil to a level below which the roots grow. The soluble caesium will have left the superficial layers of soil, and migrated down deeper, and food grown in the soil will become clean again.

2. THE AUTHORITIES AT STARAYA KAMENKA

There are two villages on either side of the river Sozh. Gaishin is on a hill overlooking the river to the west. To the east, lower down towards Russia, Staraya Kamenka suffers agony during the seven or eight months of the year when it is cut off from the rest of the world.

Conversation between Tamara Golubina, president of the local council of Staraya Kamenka, Valeri Drosdov, president of the regional executive committee, Nesterenko and Lisa.

Golubina.—Now, during this season, we can get to the district administrative centre , but in winter we can't get there. They bring us bread from Krasnopolie once a week. They bring food to the shop once a week. Last year, the roads remained blocked. Only the main road that goes to Krasnopolie was cleared. Our road that links up all the villages was not passable. So what can we do? The people who live in other villages come here on horseback to buy whatever it is they need. If they haven't got a horse, they come on foot, and then there are people who can't get here at all. We stock up for the week. That's what I do in winter, I buy bread for the week.

Nesterenko.—There are five or six villages here, is that right?

Golubina.—There are nine villages. Under our administration. We need to get out of here…

Drosdov.—The sooner the better.

Golubina.—The sooner the better because we don't live here, we just survive. When we lost the collective farm, life became very difficult.

Drosdov.—We lost it because of Chernobyl, because the main village Starinka, where the collective farm was situated, the village council, the whole infrastructure, the social centre, everything had been entirely evacuated. Here, at Staraya Kamenka, only a sub division of the collective farm was left.

Golubina.—And the sub division had to be evacuated with all the rest.

Nesterenko.—The collective farm was managed from where?

Drosdov.—From Starinka, in the village of Starinka. It was evacuated and buried.

Neseterenko.—How many people lived there?

Golubina.—About 300 people… The farms were doing really well.

Nesterenko.—Strange… Why not here?.. How many curies were there?

Goboulina.—20…at Starinka. At Dobrianka there were 28.

Nesterenko.—And here?

Gobulina.—Here? 8.

Q.—But why are the inhabitants more contaminated here than at Gaishin?

Gobulina.—At Gaishin, people have jobs, they get a salary, so they have money to buy food. Our people have nothing to buy food with. They go into the woods, they pick mushrooms, berries. They hunt. They fish.

And they work in the woods cutting logs. Where else can they work? There is no collective farm any more, so they work in the woods. Fifteen, twenty people have work in Gaishin, on the other side of the river. But they only started work in June, because the river was flooded. Our fields get flooded and you have to cross the fields to get to Gaishin. They take a long time to dry out. There are ditches and everything. Besides it takes time to put the bridges back, for the road to be passable. So our people could only start work in June. They worked in June, July and August. Perhaps they will be able to work in September too. Then it's finished. The rains come, and the fields can't be crossed, and the floating bridge will be taken off and everything will stop.

Nesterenko.—OK, in winter the pontoons are lifted. But can't you get there through Slavgorod?

Golubina.—Last year, we went through Slavgorod. But it's just more expense for the collective: taking the bus there and back. Fifteen kilometres on our side, fifteen on the other, that's 30 kms to Gaishin: 60 kilometres there and back for ten to fifteen people.

Q.—Have you got a school? How many children?

Golubina.—Twenty nine and about forty pre-school children.

Q.—So, children are still being born here?

Golubina.—Yes, they are being born.

Q.—Are the people here aware of the radioactivity? Are they afraid?

Golubina.—No-one notices it. We live with it.

Q.—Aren't they worried about having children?

Golubina.—I'm sure they worry, yes.

Q.—But they still have them.

Golubina.—What else can they do?

Nesterenko.—Life isn't going to come to a stop.

Golubina.—Whether you're worried or not, you have babies. We were afraid too when our little one was born, we had to check to see everything was in order... For the moment, everything seems to be normal.

Drosdov.—Last year, in our ten villages, the contamination in milk was above the admissible level.

Nesterenko.—More than 100...

Drosdov.—More than 100 Bq/l. At Dubno there were two samples of milk that were contaminated, 106 and 120 Bq.

Q.—In an area that is contaminated like this one, isn't the Ministry of Health supposed to bring a spectrometer and measure? How many times?

Nesterenko.—Once a year minimum. And on the same subject, there is a Council of Ministers decree that says the president of each collective farm should be taking measurements.

Drosdov.—There is a spectrometer at the local hospital, which worked up until 2000. The technicians said it couldn't be repaired. So at the moment we have no spectrometer in our region.

Q.—In principle, the spectrometer should be brought to the villages?

Nesterenko.—No. In principle, the president of the collective farm should accompany the inhabitants in a bus and get them measured once a year. That is an official decree. Most seriously, the Ministry of Health states—I read it in the minutes of Parliament in April—that 98% of inhabitants in this area have been measured.

Gobulina.—Our people? No-one has taken them anywhere. When we had a machine that worked at the local hospital, people went. They were simply invited to get measured. If anyone was going to town on business, they would be told: "Go and get yourself measured".

Q.—Did all the inhabitants get themselves measured once a year, or not?

Goboulina.—No, they didn't all go.

Q.—So, the Ministry of Health is not telling the truth.

Nesterenko.—Of course not!

Q.—And if the measurement exceeds admissible levels, what happens then?

Nesterenko.—They have to write a report to the regional executive committee, contact the family, talk to them, find out what caused the accumulation, and give advice. We made sure that the CORE programme included the necessity of buying spectrometers for the villages of Slavgorod and Bragin. We made a proposal to the European Commission to give us funding to buy them, because they are absolutely indispensable. But it would be better of course if they were not fixed but mobile like ours. I am convinced that we need to approach the inhabitants rather than the other way round. Say to them: "OK, we measured you a month ago, today we're going to measure you again. And we will tell you the result straight away". The president of the collective farm can send the results to the public health department. They are obliged, on the basis of these results to take charge of anyone whose level is above 200 Bq/kg in body weight.

Q.—Have food products been measured here?

Gobulina.—The health authorities in Slavgorod measure food products. The milk almost every month. Other products, when required. People take their potatoes and their beetroot, anything they have grown.

Q.—Is it done here in the village?

Gobulina.—They take it there. Then they send us the information.

Q.—But on a day to day basis, families don't know exactly what they're eating?

Gobulina.—No, of course not.

Nesterenko.—To do it properly there would need to be a radiometer here at the clinic. So that everyone could come and measure.

Gobulina.—Before, a long time ago, it was like that here.

Nesterenko.—Of course, at Staraya Kamenka, even at Gaishin.

Gobulina.—We used to send someone from here to Minsk. They would receive training at your Belrad Institute and we had a dosimetrist at the clinic. Then everything disappeared, I don't know why...

Nesterenko.—Then Rolevich[147] arrived, in 1994. At the beginning, we had 370 centres like that.

Gobulina.—I know that afterwards they came and put all the machines under lock and key...

Q.—Vassili Borissovich, you had a local centre here...

Nesterenko.—We had centres here and in Gaishin as well. That's how we did it... At the time the government introduced as a general rule, a dosimetrist in these villages. They had to monitor the milk and the meat...Afterwards they announced it was finished, there was no danger any more. This is what they claimed in 1993[148], and by 1994, there were only 180 centres left out of the 370. By 1995 there were only 90, and now there are just 56, financed by ComChernobyl.

Q.—Did Rolevich close them because the government has no money or for another reason?

Nesterenko.—I think in this situation, the main reason is money of course. Afterwards, they started to look for reasons to justify it. Instead of saying: "There

---

[147] I.V. Rolevich was the vice-president of ComChernobyl during the years when 285 out of a total of 370 of Nesterenko's centres were closed. It was I.V. Rolevich who took the decision personally. Rolevich contributed to CEPN publications during the years 1996–1999.

[148] In 1993, Julich published its reassuring results, which omitted these villages. See p. 160.

is no money, let's find a way to keep these centres going", they said: "The danger has passed, we don't need them any more". It's Rolevich that made that mistake. When the government closed my centres, I made appeals abroad. Today, we have 19 centres that are maintained through partnerships between German and Belarusian schools. The German children make cakes, knit, and sell what they make, and collect the money. Here, we needed to do the same thing: find a partner. For example the German association Burgwedel, near Hanover, helps the village of Polessie in the Chechersk district (5–15 Ci/km²). They came last year, twenty eight people: "We want to judge the situation for ourselves". They talked to the teachers, the town council, the president of the collective farm and they made a five-year agreement. They decided to invite children to go and stay in Germany. They bought and installed a greenhouse near the school to grow clean vegetables. We invited doctors, agronomists, forestry experts, all our local specialists, and we passed on our experience about the best way to help. It's an example of the kind of project that shows how even in these contaminated areas, we achieved a six-fold reduction in the radioactive load in people living here. (*He is talking to Golubina*). We will try to find you a partner. In my opinion, the dosimetrist and the doctor should be the last to leave a village. Unfortunately that's not the adopted policy. They closed the centre and took everything away, all the equipment and the skills.

3. ADULTS COME TO BE MEASURED

The manager of the shop and an elegant young mother come to be measured at the school, where Nesterenko's technicians have set up the spectrometer.

Q.—How much did you measure?

The elegant mother.—A lot. 736.

Q.—And you?

The manager.—580.

The elegant mother.—My son, the youngest, had even more than 200. Has he got less now?

The dosimetrist.—Yes, half as much, but still too much.

The elegant mother.—It's odd, so quick…Let's see how much I measure, now.

Q.—Have you changed your diet since the last time?

The elegant mother.—No. Everything we measured was at admissible levels. We were told that it was practically uncontaminated and that we could eat it.

The dosimetrist.—I can see on the screen that your measurement is very high… It will be less than last time but still a lot.

(Nesterenko comes in)

Nesterenko.—What's new here? Ah, it's the Zhuravliov family. How much has she accumulated?

The dosimetrist.—385.

Nesterenko.—Is that all? ... So tell me, what have you eaten in the last month?

The elegant mother.—The same as before.

Nesterenko.—That's impossible, if the level is lower.

The elegant mother.—No really, nothing's changed.

Nesterenko.—There have been a lot of apples recently. Maybe you have been drinking a lot of apple juice? ...

The elegant mother.—No, we ate some apples, but we haven't drunk any fruit juice. We can't afford to buy fruit juice.

Nesterenko.—But you must have made fruit puree?

The elegant mother.—We eat fruit puree sometimes, yes.

Nesterenko.—The fact that the level has gone down can only be explained by having ingested pectin. It could be in beetroot, or fruit...

The elegant mother.—I don't know. Our food is practically the same as before.

Nesterenko.—The machine registers what you have.

The elegant mother.—I don't believe in those machines.

Nesterenko. (he laughs)—Where do you work? What do you do?

The elegant mother.—I am a chief accountant.

Nesterenko.—So, you've had an education. Well, if your body is giving off gamma rays...

The elegant mother.—If it's true that the level really has been halved...

Nesterenko.—It really has. It would be important to know how your diet has changed. The machine records what you have eaten, let's say, over the last two months. Something in your diet has changed since the beginning of the summer, since the spring.

The elegant mother.—Absolutely nothing.

Nesterenko.—Game?

The elegant mother.—No, why? ... I don't know.

The manager of the shop is measured using the HRS (Human Radiation Spectrometer)

Nesterenko.—The peak on the right shows that it is very high. They have accumulated a lot. Unfortunately, they eat a lot of contaminated food.

The elegant mother.—We are very contaminated.

Nesterenko.—You see it and you say "I'm not eating any contaminated food". It's not the Holy Spirit that's responsible...

The elegant mother.—Everything was monitored.

The manager.—All our food products were monitored and it was all clean. It's all admissible.

Nesterenko.—But was it contaminated?

The elegant mother.—No, it was all admissible.

Nesterenko.—Listen, if you bring meat and it measures 480 Bq, according to the norms, it will be "clean". But it's contaminated!

Q.—So, the norms are wrong?

Nesterenko.—They need to be stricter. They are wrong.

The elegant mother.—We are told lies about everything.

Nesterenko.—Yes, unfortunately.

The elegant mother.—There's no truth anywhere.

A conversation between the manager and Nesterenko.

The manager.—We brought berries and had them monitored. They were collected from a place that I was a bit suspicious of. They told us they were within admissible limits.

Q.—"Admissible", what does that mean?

Nesterenko.—If the berries measure 80 Bq, they will be considered "clean".

The manager.—They measured 60 Bq.

Nesterenko.—But the berries need to be below 30 Bq.

A mother, Natalia, interrupts.—Can I bring the children?

Nesterenko.—We're here for the children. The adults, that's one thing, but we are here for the children...

4. MAIA SAVCHENKO, COW HERDER ONE DAY A MONTH

At dawn, we go with Maia Savchenko, who takes the cows far from the village, beyond a ravine. Clouds of mist float above the fields and the ponds, and then dissolve as the sun rises.

A gentle strong peasant woman, Maia talks in a straightforward manner, apparently unsurprised at the arrival of these foreigners and their interest in her life. Her quiet distress contrasts with the beauty of the poisoned landscape.

Q.—You are from Staraya Kamenka?

Savchenko.—Yes.

—Were you born here?

—Yes.

—And your ancestors too?

—We have always lived here. Now there's the radioactivity. You managed to get here, luckily because the bridge is open, you can get across easily. But at the end of autumn, they take the bridge away for the winter so we can't get out any more. It's like living on an island, no-one comes to se us and we don't go anywhere.

Especially since they demolished the collective farm, and we no longer have the machines. We can't even call an ambulance if someone is ill.

—How do you survive?

—It's difficult, of course. At the school, the teachers are mostly self-taught. In a way, they're quite ignorant, you could put it like that. *(She laughs)* They haven't closed the school yet. They've said they're going to close it next year. There aren't many children at the school: twenty-seven children. We're in a difficult situation.

—Is there a nursery school?

—No.

—Are there any young children?

—Yes, there are some very young children, but no nursery.

—How do you see the future, for you and for your children? Do you have any?

—I have three children. All at school. The oldest is 14 and the youngest is 8. We live…you know… another day gone, everything is alright. We've slept, we've got up, we're OK. We have signed petitions. The authorities came from Minsk. We asked about leaving.

—To be evacuated?

—According to the register at the village council about 350 families stayed. So we are asking, we are writing to them to ask them to move us somewhere else… down there, beyond the river, to the West of Sozh. So that we can get to the hospital more easily. So that the children can go to a normal school.

—Do you have problems with the hospital?

—In winter, it's very difficult. We telephone another district, like Krasnopolie, if someone's ill. They either say they don't have any petrol,—sometimes they use that as an excuse—or they say "You're not from our district". That's what they say[149].

—They refuse to help you?

—It's difficult living here. Of course it's difficult, but we've got used to it. When I think about going somewhere else, I say to myself "Stay here, where you belong. Do everything slowly, it will be alright".

—How do you manage in winter?

—How? We're cold. We freeze in our houses. I live in a flat, in one of the collective farm houses that has heating. We have to find our own fuel. It's very expensive. With three children and a very low salary, it's difficult. I have to choose whether to spend money on food or fuel for winter.

—You have to buy it? You can't provide it for yourselves?

—No, we're not allowed. You have to buy it. You have to buy the wood. One cartload of wood costs 9,500 roubles. That's how it is.

---

[149] Krasnopolie is about forty kilometres from Staraya Kamenka, where the territory of Belarus widens a little towards Russia. The roads there are quite inaccessible in winter.

—How many days does a cartload last in winter?

—If I economise…I don't know. We use three cartloads of wood each winter and if it isn't enough…in secret in the forest, we pick up a branch, keeping hidden, put it on the horse or on the cart and get back to the house quickly so that no-one sees us. Otherwise they tell you off. You get fined. You have to give them money. In fact, no-one goes and gets wood here, because the fine is exorbitant. If you've got your pay, it's better not to cause trouble. People with parents who are retired are in the best position. It's something to fall back on.

—Are your parents retired?

—Yes, they are. My mother, not my father. She's retired. If I need money, she'll always help me out.

—For how long has your family lived here in the village?

—I was born here and I live here. I knew my grandmother. Not my grandfather. He never came back from the war. My grandmother lived with us for some time before she died.

—Did they talk about their grandparents?

—I don't remember. They may have talked to me about them, I don't remember.

—What is your name?

—Maia Savchenko.

—And your children's health?

—Tatiana, my eldest daughter, is invited to go abroad, to Luxembourg. This is the third year that she has gone. She is losing her sight. Probably because of the… The doctor, the hospital said. "What are you doing on your island? You need to leave". She said that as soon as she heard the name Staraya Kamenka. But where can we move to?

—What is the problem with her eyes? Did her sight deteriorate suddenly?

—I don't know. I can't tell you. She became shortsighted at school. The children complain about headaches, they're often feverish, probably because of the cold, the children are often ill. My eldest has started complaining about her back. It often hurts. I don't know, is it because of adolescence? I don't know. I took her to the doctor. She told me it was radiculitis. [150] They prescribed some pills. At the moment she is not saying anything. But at other times she complains of pain constantly. Her bones hurt…All that is due to the radiation. It's probably true that the radioactivity is having an effect on us. We live here. We were here when it happened and we stayed.

—And what about you, do you feel anything?

—I've got heart problems.

—What sort of problems? What do you feel?

---

[150] Inflammation of the nerve root.

—Pain, fatigue, my heart feels constricted. I have that from time to time. It will suddenly come on. The pain goes right through me. Sometimes I wake up in the night, I sit down and it feels like needles. Pain in my heart, here. I sit for a while then I lie down again. I don't take any pills. The pain goes away by itself.

—When did that start?

—About two years ago, maybe three.

—Since Chernobyl, have the authorities come here to find out how you are, to help you? Do you feel the government is helping you?

—I don't know if I feel that, but I feel as if we're talking to a brick wall. They listen to us and then they go off. That's it…And we stay here and nothing changes. Unless we find a place to move to ourselves, I think we'll end up staying here, and no-one will care.

—You live off your vegetable gardens?

—We plant our own gardens and that's what we live off. Our pay is minimal, 60,000 roubles. It's not enough for a month. A loaf of bread costs 416 or 460 roubles. I have five mouths to feed. We manage but only because my daughter started to make these visits abroad, and she sends us parcels. We do alright for clothes.

—Does your husband work?

—Yes, he works.

—What does he do?

—He's in charge of the heating at the clinic.

—Where's that?

—Here, near the shop, the whole building is the clinic. He's in charge of the heating. It's seasonal work, there's nothing else.

—And your salary, what do you do?

—I'm a nurse at the clinic at Kamenka

—What sort of nursing exactly?

—Cleaning, putting things back in order after the doctor has been. The doctor does his work, goes home; I come and put everything back where it should be, like in a hospital.

—Does the doctor live in the village?

—No, she lives in the next village, at Dubno. It's three kilometres away.

—The people of this village can come to this clinic, and get help? There are medicines here?

—Yes, we have medicines, only you have to pay for them. Everything has to be paid for. You need to pay.

—Who supplies the medicines?

—The doctor herself. She buys them and brings them here, only you have to pay. It's probably since March that you have to pay. Before March it was all free. You just needed a prescription. Then they introduced the paying system and now it's very difficult. Medicine is very expensive. Life is hard.

—It's the government that is forcing you into this misery?

—When it was free, more people came to get treatment. Whereas now that you have to pay, they don't come any more.

—Do you know what the levels of contamination are here, in this area? In the food?

—The food and health services came here the day before yesterday from Slavgorod. They took samples of food: beetroot, carrots, potatoes, tomatoes— depending on what people were growing on their gardens—they took them to measure the radioactivity. We haven't yet received the…results…They'll probably send them to the village soviet. We'll know later…

—Have you ever been told the level of contamination?

—Yes. For example, with the milk, they said: "The milk is very contaminated. You can't drink it". What can you do?…What about the children! They have to drink something.

—Is the milk from your cows contaminated?

—Yes. But this is our daily life. We only have one field for pasture. There are no clean areas here. The cows are contaminated. The cows were monitored recently. Nearly half of the herd has leukaemia. Nearly all of them. You can't drink the milk. I have a cow with leukaemia myself. I replaced her last year, and now it appears she has leukaemia. We exchanged them at the collective farm, one animal for another.

—And you give this milk to the children?

—Yes. Where am I going to get milk? If I buy it at the shop, a litre of milk, you know yourself, costs 400 roubles. Where am I going to get that? I boil it, or drink it as it is. We drink it and the children drink it, what else can we do?

—Has anyone told you how many becquerels there are in each litre of milk?

—This time, I didn't get the milk measured. I don't know. Because the cow is pregnant, there is nothing to measure. I took potatoes, beetroot, carrots along to be measured. I haven't had the results yet.

—Does anyone know the average level of contamination here in the milk, how many becquerels?

—I don't know. I haven't heard. They measure, the results arrive, someone telephones to see if we can drink the milk: "The radiation is very high, above admissible levels, you mustn't drink it". But how has it happened, whether it's the becquerels or whatever, I don't know. We live here on a kind of island. At the moment, there is the road, but when they lift the bridge off, you can't go anywhere. It's only if you have your own horse, and it's a cold winter and the river is frozen, then you can cross it to get to the other side. Then, some people can get to Slavgorod on horseback, if necessary. Otherwise, we're cut off.

—Are people still getting married in your village, having children?

—Yes, people are having children. This year, we've already had two new babies, and another two are expected. One at New Year, one after New Year. So people are gradually starting to have children here. I say that either you should leave or you should get married and have children. Especially as they're not going to evacuate us in spite of the radioactivity. What else can we do? In any case people say that the world is coming to an end.

—Today you are looking after the cows.

—Yes.

—Is that every day, or does everyone take their turn?

—Everyone takes their turn. We have thirty-two cows in the village: every thirty-second day, it's my turn.

5. MARUSIA AND HER SON

Solitude has had the effect of making this elderly mother and her son a little strange. We begin our conversation next to the fence in front of their khata. *Over the fence, in the yard we can see an apparently motley collection of wooden and metal objects, but all presumably useful in their own way. Marusia is very likeable. She talks very loudly, almost shouting abrupt phrases, with a roguish smile, that masks a gruff tenderness, and shyness towards these visitors who honour her with their presence in front of her house. She interrupts her son often and he looks on affectionately while she finishes her tirade.*

Q.—You've been cut off from the world?

Marusia.—We've been completely forgotten, yes! We are no longer human beings. They have made the village uninhabitable.

Q.—Forgotten?

Kolia.—We've been forgotten—oh yes!

Q.—Has your son got a family? Is he married?

Marusia.—He was married, to Kalita. But it didn't work out. Life didn't bring them together…

Q.—Don't you want to leave the village?

Marusia.—Leave the countryside for the town? The food is here.

Q.—You haven't got a pension?

Kolia.—Nothing, not a penny. We've only got one income: berries and mushrooms.

Marusia.—From the forest.

Q.—Are they clean?

Kolia.—For us, now…How can I put it…they won't do us any more harm now.

Marusia.—We've already eaten so many, we've filled our bellies.

Q.—But, don't you have headaches?

Kolia.—No-one takes any notice of that any more.

Q.—Yes alright, but don't you feel anything?

Kolia.—Yes, we feel it… We feel in pain when it is very bad. That's it. In any case, no-one goes to the hospital here. You won't get anyone here to go to the hospital. Anyway the forest is very radioactive.

Q.—Yes, the mushrooms and the berries…

Kolia.—Whatever happens, now that it's rained, everyone will be going into the forest to get mushrooms.

Marusia.—We live off the forest. My son collects the berries and the mushrooms, and that's what we live off. And we have pumpkins too.

Kolia.—The pumpkins…they're not worth much… If there was some work here, as they promised… Those with young children were evacuated. The others were told: "We're leaving an annexe with the machines for the workers." But you've seen for yourself the state of the cattle farm, they demolished everything.

Marusia.—They destroyed everything, demolished everything.

Kolia.—They want to make it all disappear, get rid of it, completely.

Marusia.—Take this pumpkin.

Q.—No! No!

Marusia.—Don't you want it?

Q.—No. How do you prepare it, this pumpkin? Do you make pumpkin soup?

Kolia.—With millet, milk, sugar and this pumpkin, steamed in the oven. What a feast!

Q.—Good?

Kolia.—Very good.—You will never have eaten anything so delicious. Millet, milk, sugar.

Q.—In pieces…

Kolia.—Yes, yes, little pieces. And salt, for taste. That's all you do. It's a feast! Or you can boil the pieces, roll them in flour and fry them. We like them like that too.

Q.—Do you measure the pumpkins? Are they radioactive?

Kolia.—No-one measures them here. Three or four years ago, they came and measured the potatoes and the milk… They measured and they told us "There's a little radioactvity but you can eat it". Alright. We're used to it. But not the children. The children, of course, they should be evacuated from here as quickly as possible. The sooner the better.

Marusia.—(*She calls to Emanuela in the yard*) Miss! Miss! Come in…See how I live.

Kolia.—Show her the apples. There's a whole lot over there.

Q.—Can we have a quick look in your house?

Kolia.—Yes, you can. (*We enter the yard. He opens a door into an outhouse where they keep all their preserves.*) Half is a workshop and half…there!

Marusia.—Look at the pumpkins!

Kolia.—This variety only grows in Russia.

Marusia.—And if we have to cook some *chachliks (kebabs)*

Kolia.—I can make them in half an hour.

Q.—You can feed yourselves in winter with that?

Marusia and Kolia together.—No!

Kolia.—That's for two months.

Q.—For two months, all that?

Kolia.—Yes! Yes!

Q.—It's for you two.

Kolia.—It's not for us. It's for the animals. It's for the cow and for the pigs. We don't eat them.

Q.—I thought it was for you.

Kolia.—No! We pick a pumpkin every month or two just to taste…No, it's for the business. Because at the moment there's no other animal feed.

Q.—What's that?

Kolia.—Cranberries.

Q.—The cranberries are contaminated too. It's dangerous to eat them.

Marusia.—So what?

Kolia.—And the kvas *(a beer made in Russia and Ukraine)*? … Look how big the cranberries are here. Have you ever seen them like that? You've never seen them? Mum…bring them here. I came here four days a go and took some. A little. To taste. But anyway in the winter, I ferment the kvas using cowberries and cranberries.

Q.—You collect them from the forest?

Kolia.—Yes, I'm not lazy. I collect them. Mum…give her some so she can make a compote. No, no, with the bag. You can make yourself a compote. Look at those berries! That's for you.

Q.—Can I take them?

Marusia.—Of course you can take them.

Kolia.—The cabbages, when we salt the cabbages…Pour them in, Mum. Here you are. When we ferment cabbage we always put cranberries in. We've always done that, from the dawn of time.

Q.—Thank you.

Kolia.—Eat them, they're good for your health.

Marusia.—For your health.

6. TWO FRIENDS IN THEIR SEVENTIES

The widow.—The milk is taken away every day. It's been taken before the Sozh has got out of bed. After the bridge is taken up, it stops. The Sozh freezes so sometimes you can get across on the ice. People have even drowned trying to cross it.

Q.—Your river, it's your destiny. The river brings you good and evil.

The widow.—It gives us fish, and it's good for washing in and it's a good trough for the cows in the pastures. But in winter…

The retired woman.—Fortunately we are retired and we receive a little money at least. Every month we get just enough for our needs. But the young people who are working, if they can't go to Gaishin or if the tractor doesn't come to fetch them? Go all the way round through Krasnopolie, it's very far, so they have no work. It's hard. Now in summer, I live here with my son-in-law and my daughter, but I'm from Msteslavl'. But if I wasn't here they wouldn't have any money to buy bread. It's only me that feeds them. I buy them food to eat. But in winter? They have no work, and they have three children: "Live, if you can". The winter is long, and how will they buy shoes for the children? There is no money for shoes. That's how we live *(She tries not to cry)*.

The widow.—There were lots of young people here before. The village rang with people's voices, in the evening, the young girls would came out, sit on the benches and sing. The singing filled the village. Singing, music, balalaikas, we were all so happy! Now in the evening it's silent. Nothing any more... Chernobyl did us a lot of harm. The young people have gone, they didn't have any more children, only one or two for each couple... Whereas before, there were seven children in every household, three or four at least, girls and boys. There was the war, seventy men from this village alone, who didn't come back, young men...Today, there's nothing left, everything is sad, and desolate.

The retired woman.—But if you look at this place, everything is blossoming. The forests are magnificent, they are so beautiful! And the mushrooms and berries, they're everywhere. And then...everything ruined, and the power station is to blame.

Q.—Isn't it dangerous to be eating all that?

The widow.—Dangerous to eat it? Well, we eat it any way.

The retired woman.—Immediately after Chernobyl they gave us imported food. Everything was imported...There were no cows or pigs, everything had been destroyed. Then they said the danger from Chernobyl had lessened. They allowed us to keep cows again, otherwise we would not have had any.

Q.   —Do the cows have leukaemia?

The retired woman.—Look at mine. Last year we exchanged the cow at the collective farm because she had leukaemia. We were given another, a younger cow. They examined her, and there you are, she has leukaemia too. I think it's because of the contami... something... from Chernobyl. And the children are so skinny, they aren't growing properly, look at that one.

I won't live much longer, I'm ill; I had a stroke two years ago. I am 78 years old. Old age, illness...If I die tomorrow, what will my children live off? This is the reality of our lives. Things are really bad here. I don't know what the government is doing. The old people like us, we can die. We're reaching the end. But the young people? How are they going to live, the young people?

Q.—For anyone who comes from outside the area and sees how beautiful it is, it seems like a paradise.

The retired woman.—It's like that at Msteslavl'. That's where my other daughter lives. She writes to me and telephones: "Mum, come, we need help". She has not even got the money to plant potatoes, nothing to plant in her vegetable garden. It's three years now since they harvested anything. And this year everything is dead. The people here aren't lazy, they work as hard as they can but they just can't do it. It's a very hard life. How will they carry on? Lukashenko came to a meeting in Slavgorod. He spoke and then he left. A commission came. They looked. Then what? They had a discussion and then they left. *(She goes off)*.

7. THE PEASANT WOMAN WEARING THE BLUE SCARF

The peasant in the blue scarf.—We live on a kind of island. Around Korma it's the same, everyone was evacuated. To get to Volyntsi you need a guide. There are three roads. One goes through... Bachovna, we go there to make hay... Before, there were so many people! So many young people! So many workers! They all left ... And on this side, this is the back road to Pezakovka. It's covered in bushes... You wouldn't get through. You'd have to take a saw to cut through all the branches... to clear the road. It's impossible to get through. There's no-one there any more... The people from Volyntsi come through. So, you have to go to Korma along the other bank, beyond the Sozh, it's big, like a town, like Slavgorod. And there, there is a pontoon bridge, like the one at Slavgorod. And from there it isn't far. But from here, I don't know. First, you have to cross a stream, and the bridge has been demolished. Can you see how our streams are overgrown with bushes? There, it's the same but worse. There's nothing left around there. No-one lives there. All the houses have been burned down. They buried the houses, it's like Khatyn, it's a real massacre. They burned everything and... buried it all. Buried! They dug ditches and demolished the houses. All they said to us was that we should leave... No-one wanted to go. They told us the radioactivity wasn't all that high, so we stayed. But five families did leave, even so. They were young...

Staraya Kamenka

From Staraya Kamenka to Gaishin, on the other side of river Sozh.

Chapter IX

GAISHIN

225 km from Chernobyl, 5–15 Ci/km²; about 12 km
from an area with contamination levels above 40 Ci/km².

*Before plunging into the wilderness surrounding the deserted villages to reach Kliapin
and Volyntsy, we accompany a group of men and women, squashed in with the milk
containers on the back of a trailer wobbling to and fro as the tractor pulls it along.
They are going to pick apples at the collective farm at Gaishin. There, in the orchard,
thanks to one woman's sincerity, we witnessed the discomfort felt by inhabitants who
had benefited, unintentionally, from the plundering of their neighbours' possessions.*

Q.—You're from Gaishin?

The woman.—I was born in Gaishin and I've lived here for fifty years. *(She
laughs a little at having inadvertently revealed her age).* And you know, as far back
as I remember, there were always lots of people here before Chernobyl. We built
little houses, and we were all quite happy here. But since Chernobyl it's all finished.

—How was it that the collective farm of Staraya Kamenka was transferred
over to you?

—For the same reason. *(She lowers her voice).* Beyond the Sozh, at Staraya
Kamenka, the radioactivity was very high. They decided to sack all the people at
the collective farm, and bring them here. We took them into our collective.

—But they should have been completely evacuated?

—It wasn't possible to evacuate all of them. Most of them were evacuated.
Those who stayed came here.

—Why wasn't there a complete evacuation?

—I don't know. Lack of money, probably. They needed housing. And anyway,
some people wanted to go, but others didn't. Most of them left Gaishin, especially
the young people. But I was born here and I can't leave. I can't! Whatever the
radiation. Most of it, we ingested long ago.

—Do you have children?

—Three sons.

—Already grown up?

—Nearly.

—Grandchildren?

—A grandson.

—How is his health?

—It seems to be normal.

—Do they live here?

—Yes, they live here. Apparently things are OK. They are still young. But later, of course, all that will make itself felt. It's certain... Of my three sons, one is doing his military service, the other is still at school, and the third works for the collective farm. We live...

—Do you remember when the accident happened?

—No-one even talked about it, at the time. Later, some time after, they began to tell us that there had been an accident and that a cloud had passed overhead. So then we knew. But what could we do?

—Three, four years later?

—Yes, about that. Two or three years later.

—At the time, you weren't given sufficient protection?

—But how could we have been protected, from a cloud passing over us? *(She points to the sky)*

—They give out iodine tablets for the thyroid, they should have done it as soon as possible.

—It took a couple of years... But now the children, the schoolchildren go abroad each year for cures in sanatoriums, in holiday camps, abroad. But what will that do, if they have already been irradiated?

—It's possible to purify the body.

—Is it?

—The body purifies itself, as always, of anything that is not natural to it. But you can help to purify it more thoroughly and more rapidly. By taking pectin, for example, and by decontamination in a clean environment.

—The countryside is still beautiful. Have you been down to the banks of the Sozh? It's only now that it's all covered in bushes. But before! The students from Minsk used to come on holiday every summer, to the banks of the river. They came from all the universities. It's incredible how it used to be here! And now it's all overgrown. It's finished. All because of Chernobyl.

Before visiting two other villages to the east of the Sozh, we visited the sanatorium, "Serebrennye Kliuchi" (Silver springs), at Svetlogorsk, where Galina Bandazhevskaya was examining some children from these villages.

Chapter X

A FIRST FOR GALINA BANDAZHEVSKAYA

In September 2002, Galina Bandazhevskaya undertook a clinical study at the sanatorium, "Serebrennya Kliuchi" (Silver springs), with Professor Nesterenko and his team. The study investigated two hypotheses: on the one hand, the correlation between caesium-137 load in the body and symptoms of cardio-vascular disease in children, taking into account the origin and the nature of their diet; on the other hand, the therapeutic effects of an apple pectin cure, which could correct the cardiovascular disorders caused by incorporated caesium-137. The results were published in the journal *Swiss Medical Weekly*. [151]

Important significant data were obtained following the standard double-blind protocol. It was established that children who consume food grown at home or gathered in the forest in the villages in contaminated regions accumulated a lot of caesium-137 in the organism. They present with functional alterations in the cardio-vascular system, manifesting as complaints, heart murmur at the apex of the heart, high blood pressure and alterations to the electrocardiogram. The study showed that blood pressure and alterations to the electrocardiogram were correlated to the levels of accumulated caesium-137 in the organism. One important fact—the pathologies observed are reversible: a pectin cure administered as a food additive leads to a reduction in the concentration of radionuclides in the organism—this had already been shown before by Professsor Nesterenko's work—and a new discovery—the electrocardiogram could be normalised. No-one had yet demonstrated the reversibility of heart problems in children contaminated by caesium-137 with a pectin-based cure.

This study, which was completed at very little cost, contradicts the WHO and IAEA theses, that attributes the somatic illnesses experienced by people living in the contaminated territories of Chernobyl to stress or radiophobia, and it poses an unavoidable moral and political problem: national and international institutions

---

[151] Swiss Medical Weekly, 2004, No 134, p.725–729; www.smw.ch.

responsible for the health of thousands of children contaminated by the accident at Chernobyl, can no longer justify their lack of medical intervention. The experiment proves that children affected by internal low dose radiological poisoning can be protected, and, in certain cases, cured by a simple and inexpensive cure. Continuing to refuse to finance these cures is inadmissible from the point of view of medical ethics and of institutional responsibility. Their negligence in this matter has already gone on for twenty-nine years, and during this time, children have suffered and died. Those who opposed the pectin cure all these years, without giving any valid scientific reasons carry a heavy responsibility that could lead to criminal prosecution.

We posed the following questions to Galina Bandazhevskaya before continuing with our investigation with the families of some of these children.

—Do you lack equipment?

—Yes, of course. Using an ECG is a crude method: it only shows electro-physiological alterations in the heart at a particular moment, and this could be influenced by different factors. For example, a child who is being examined may be nervous: a pulse of 120 per minute, and their blood pressure has increased. We could examine them at lunch, when the child is calm, and their blood pressure would have stabilised. With only one ECG measurement... we don't get very accurate data.

—What do you need?

—We would need a Holter monitor that can be attached to the patient to monitor them over a twenty four hour period; a good ergonomic bicycle and an echocardiogram. All that is expensive.

—Did you have those machines at the institute, at Gomel?

—No, no. We worked with the same equipment as we have now.

—Not very accurate, as you say?

—Yes. But we are dealing with a large group of children. And we also duplicated all these clinical examinations with laboratory tests and autopsy reports. So the illness we were diagnosing in the clinic could be confirmed by the pathologist. Bandazhevsky confirmed our observations in his experimental work. What we have done is only the first step towards a more in-depth study.

—You have had to work in the conditions you describe through a lack of funds?

—Of course. We have no money.

—Where did the money come from to do all that?

To start with, we were doing routine clinical work. We began with simple things, like electrocardiograms of children who were in good health. We knew these children were apparently in good health. They weren't ill, they were in nursery. We had no idea what we were going to find... But when we found a large percentage of alterations in the electrocardiograms, we began to think: "Could this be linked to what they are eating, to caesium, etc?" That was when we began to compare the results obtained from the spectrometer and on the ECG, and we found a correlation. That is what led us to test our hypothesis through experiments and from the results of anatomopathological examination. But the institute could not afford the machines that we needed. We were only able to do this research through the support of our fellow professionals in the region, who worked voluntarily. This research was done without any funding…

—Despite the lack of equipment, you still feel that you have shown a correlation between illness and Cs-137 in a sufficiently scientific manner?

—It is very well founded. Solid. Because all this data is not the fruit of our imagination. We have shown the correlation through matching up our data according to three research methods. And we have shown that this problem exists, particularly in cardiology, and that this study needs to be followed up. In our work as clinical cardiologists, we are coming across cases of death from acute heart failure in young men of 32, 42, 45 years old: a man comes home from work, he lies down and he does not wake up again. Or else he goes into the kitchen to drink a cup of coffee, suddenly he collapses, he is dead—we had this case recently. It is a problem that really demands investigation. And certainly, the connection with Cs-137 needs to be studied. Why is this happening to young men?

—Do you conduct autopsies?

—Yes. Anatomopathologists, in particular Bandazhevsky, have described the alterations in cardiac muscle and the correlation with radioactive caesium. So what we are saying is that material of this sort, for a scientific study, is there. All we are waiting for is the order to study it in more depth.

—You say the material exists?

—Yes.

—It could disappear?

—It could be forgotten, not disappear. It has already been published, it's in print. The slides that confirm the study are there.

—Who has the slides?

—These are Bandazhevsky's scientific results. The evidence exists. It needs to be confirmed or disproved. In other words, undertake the in-depth research ourselves, we must do that. Our cardiologists, in collaboration with pathologists, morphologists, etc.

—When they stopped your work, this material was not confiscated?

—No, no. We have all that. The histological material. We have the slides, we have photographs of the histological preparations. If anyone is interested, we can make them available. These studies have been published in monographs, they can be consulted. But unfortunately no-one wants to do it. Because they are closing their eyes to it all. I know our scientific research institute. They say there is no problem with low-dose radiation: that there is the problem of stress, "The psychological state of the inhabitants, the evacuations, the stress, all these are having an indirect effect on people's health, in particular on the cardiovascular system. But low-level radiation has no effect". They're not allowed to have an effect.

Using two cameras, we filmed the first phase of the work being done at the "Silver springs" sanatorium: on the one hand, the technicians from Belrad who were measuring, using an armchair spectrometer, the amount of Cs-137 accumulated in the children's bodies, and in another room, Galina Bandazhevskaya, with the help of a nurse, listens to their heart, notes down their complaints, what pain they have and records their ECG. Galina had no knowledge of the data from the spectrometer, which was recorded on the technicians' computer. The children did not know what levels of caesium they had accumulated. On their side, the technicians did not know the health status of the children, when they presented themselves to be measured. At the end of sixteen days, during which the children ate uncontaminated food and had a daily dose of pectin, the procedure was repeated in order to record any changes that had occurred either in the Cs-137 load in the organism, and any alteration in cardiac symptoms. This "double blind" process, in which neither side of the operation knows the results of the other, is essential to establish, without subjective interference, a quantitative correlation between levels of contamination and illness.

At the moment that we left to research the families of some of these children in the last two villages that we visited, the Belrad technicians gave us a copy of the measures they had recorded. So we knew the contamination levels at the start of the experiment but we did not know yet if the pectin had reduced, and if so by how much, the Cs-137 load, nor whether there had been any benefit to their health. At the end of the experiment, the Belrad institute sent us the results that had been validated by the ethics committee. I reproduce in summary the data along with Galina Bandazhevskaya's conclusions, at the end of each meeting with the families[152]. It should be noted that in certain children who presented with two symptoms on the ECG, one of the two symptoms disappeared after only sixteen days of treatment with pectin (In its place I have put a line of dots). [153]

I need also to specify that our journalistic enquiry is independent of the educational work provided by the teams from Belrad for the families of these children. In any case, the villages of Kliapin and Volyntsy that we were to visit are not monitored by Nesterenko. Due to lack of financial means, Belrad is only able to monitor a small proportion of the 1,100 villages in Belarus, where the milk is contaminated above 50 Bq/l according to the official figures published in 2000. Some of these villages are monitored by the German rehabilitation institute Julich.

---

[152] The names have been replaced with pseudonyms in order to respect the confidential character of the diagnoses.

[153] This study was published in the Swiss Medical Weekly, 2004, No 134, p.725–729; www.smw.ch.

Chapter XI

KLIAPIN

205 km from Chernobyl, 5–15 Ci/km²;
7 km from an area with contamination levels above 40 Ci/km².

1. THE BORISSOV FAMILY

Q.—Does your son complain about anything?

The mother.—He often has headaches. He needs to take pills.

Q.—Since when?

The mother.—I don't know. Since he was very young. They said it was his blood pressure. They said he has intracranial hypertension and there were problems with the heart. Later on they told me that it was due to his growth, that he was growing up, and that because of his heart…he would never be able…what was it…?

Q.—He wasn't developing properly?

The mother.—Yes, that's it. He complains of headaches.

Q.—And what about you, anything in particular?

The mother.—I'm alright…Except my head, because of low blood pressure.

Q.—We were at the sanatorium where all the children from Kliapin and Volyntsy were measured. They are on this list…

The mother.—How much did Stassik measure? Is Stassik on the list of levels of contamination?

Q.—182.

The mother—180!

The father.—He had less.

The mother.—Why is it so high?

In the spring, his levels of radioactivity were normal.

Q.—Who measured him?

The mother.—The Germans.

Q.—The difference in the contamination depends on what he has been eating since.

The father.—In general, we collect berries in the wood, cowberries, blueberries, girolle mushrooms…

Q.—Did you have the mushrooms tested?

The father.—The girolles were good but the ceps and the white mushrooms were very high in radioactivity…We do collect them from time to time, but in general we don't eat them. Rarely. Otherwise, it's the vegetable garden. Our vegetable garden is there, just next to us: potatoes, gherkins, tomatoes, onions, it's all ours.

Q.—The milk?

The father.—The milk is ours. We had it tested.

Q.—Your cow?

The father.—Yes.

The mother.—The Germans tested it and said there was no radioactivity at all.

Q.—Absolutely none?

The mother.—Zero…

Galina Bandazhevskaya's conclusions for Stanislav, born in 1987: aged 15.
- contamination before taking pectin: 182.7 Bq/kg;
- contamination after 16 day pectin treatment; 123 Bq/kg;
- heart: complains of headache, shooting pains in the heart;
- ECG (before taking pectin) Partial right bundle branch block, Arrhythmia;
- ECG (after pectin treatment)… …………. Arrhythmia.

2. THE LEBEDENKO FAMILY

Q.—How is the milk?

The mother.—It's our milk. The milk from our cow. We test it. It's clean.

Q.—Do you know how many becquerels per litre it contains? Do they tell you?

The mother.—They don't tell us, but they say it can be given even to young children.

The father.—We used to have a collective farm here. They closed it and annexed it to another, on the other side of the river. Now, we aren't useful to them, they're not interested in us.

Q.—But you say that you take more care here with what you eat than the families at Volyntsy. Why is that? Are they poorer than the people here?

The mother.—There are as many poor people in Volynsty as Kliapin.

Q.—According to our list, there many more contaminated children in Volyntsy than in Kliapin.

The mother.—That's because they eat more game there.

Q.—Why is that?

The mother.—We don't know.

Q.—Are there more hunters?

The mother.—Maybe. The forest surrounds us as much at Volyntsy as at Kliapin.

Q.—Have you measured the food products?

The mother.—We eat the produce of our garden. We haven't measured the potatoes or the beetroot, nothing. We don't test it, we just eat it.

The father.—Who knows?...The workers receive at least one meal at work. But, at home, people haven't got anything to eat. Because, as I've told you, we've been forgotten and abandoned...No money. Our son is going to university, he has enrolled at university, but there's no money for his lodgings. And then he needs money to feed himself. I don't know where to get it from. So he'll have to come home. But what will he do here? It's a vicious circle. All that's left is the game in the forest. .. And anyway, we don't know. If we moved to Korma, beyond the Sozh, we could get jobs, but everything hangs on the question of housing. Before, the State Committee for Chernobyl allocated housing, people rented houses. Some people left for the towns. But now, that's all finished. Now, if you want a house, you have to buy it.

Q.—Is that expensive?

The father.—Of course: about 1000 dollars. Where would we get that?

> Galina Bandazhevskaya's conclusions for Yulia, born in 1988: aged 14.
> - contamination before pectin treatment: 67.7 Bq/kg;
> - contamination after 16 days of pectin treatment: 39.9 Bq/kg;
> - heart: complains of headaches;
> - ECG (before): Partial right bundle branch block;
> - ECG (after): Partial right bundle branch block.

Chapter XII

VOLYNTSY

210 km from Chernobyl, 5–15 Ci/km²;
5 km from an area with contamination above 40 Ci/km².

1. SOBOLEV-KRESTINSKY FAMILY

Zinaida Sobolev holds her new born baby on her knee while we talk.

Q.—What can you tell us about Natasha's health. Does she complain?

Sobolev.—Yes, she has something…she has something here *(She touches her throat)*. Her glands are a little swollen…she has colds all the time. And her heart is a bit…

—The heart?

—Yes, I don't know what it is, pain…

—When she runs, does she get out of breath?

—I wouldn't say so. Our son Rouslan on the other hand, when he does gymnastics, if he runs a lot, his heart hurts. He has a problem with his heart. The oldest one, Victoria, when she was still quite young, they said she had a heart murmur. Something like that. But now I don't know. It's she who complains the most. She says: "Mummy, my heart hurts, My heart hurts".… Why? Have you tested them also at the sanatorium?

—It's the heart that they measure at the sanatorium. That's why we wanted to ask what you have observed. For how long have they been complaining of these symptoms?

—Every time they run or walk fast… But, sit down…

—Has anyone explained to you how to prepare food in order to take in fewer radionuclides? Mushrooms for example…

—They explain, but we have to live, we have to eat something. Our salaries are so low nowadays. We have no money.

—Do you eat mushrooms?

—Of course we do.

—Do you know what to do to reduce the radionuclides in the mushrooms?

—Boil them several times and then throw out the water.

—Not boil, but soak them in salt water. Didn't they explain?

—No, just soak them in salt water?

—Two spoons of salt per litre of water. Soak them for two hours, throw out the water, do it again, and throw out the water again. That way, the radioactivity is reduced by 80%, they are almost clean and you can cook them. If not, they are very contaminated. Do they come here often to measure you?

—Twice a year. In May and in September. It's the Germans from the Julich institute.

—Do they examine you?

—Well, they measure the radioactivity, and that's all…

Leaflet distributed by Julich

What you need to know about reducing the risk of incorporating radioactive elements in food.

The following products can be considered safe:
—grains and cereal products;
—vegetable and fruit from fields and gardens;
—milk and other dairy products;
—meat produced at home (for example: beef, pork, lamb, mutton, goat)
—non-carnivorous fish.
The consumption of the following products should be limited:
—mushrooms and berries from the forest (if they are not contaminated, it is not dangerous to eat them, if they come from the forbidden zone, they must not be eaten at all);
—game (in particular boar) and carnivorous fish (pike, perch etc) can be eaten once or twice a month maximum.

Commentary.—Reading these instructions at the end of our investigation in nine villages under Nesterenko's supervision is devastating. Readers can analyse the text themselves, formulate their own hypotheses on the basis of the information we collected and ask themselves if they would entrust their own children to Julich, if they feel they would be protected by its judicious advice—certainly inspired by the precautionary principle (apparently someone is interested in knowing how people can live with these levels of contamination in order to manage the next accident). Julich[154] says that there is no danger in eating the fruit and vegetables grown in gardens or gathered from the fields, but we know that these food products must be contaminated, because the levels of radioactivity in the soil here are between

---

[154] In 1993 Julich was financed by the German government and instructed not to include villages. Now they are included; are they still publicly financed? See Part Two, chapter IX, 2, p. 160.

5–15 Ci/km2; likewise for the milk and for meat raised locally; as for the boar and the carnivorous fish, which, according to the advice given by Julich, can be eaten twice a month, these are extremely contaminated.

Contempt for the victims reaches its height with this simple statement of truth on the yellow card according to which *mushrooms and forest berries do not present any danger if they are not contaminated.* Let's see now! Five to seven kilometers from Kliapin, is an enormous area where levels of radioactivity are from 15 to 40 curies per square kilometer. The berries there are really succulent!

After filming this precious document, we continue our conversation.

Q.—Is there any government organisation from the Ministry of Health that comes here to make measurements?

Sobolev.—No, no. It's at Korma that they test the radioactivity. If people want to go there, they can. As often as they want.

You can just imagine a child coming out of the wood and racing to Korma, a little town 15 km from his village, to find out...if he can eat his berries...as often as he wants.

Q.—Do they measure food products regularly?

Sobolev.—It is not a requirement to measure the food. When the Germans come, it's us who bring our milk, or mushrooms, our jam. We measure them for ourselves.

—And what do they usually say? That all of it's admissible?

—They say it's normal. They test it and say *"Gut! Gut!"*

—Gut, gut?

—It's good, and that's it. It's good.

—How many years have the Germans been taking care of things?

—Goodness me, I'd have to think, certainly five years...maybe four...three at least.

—And they think the situation here is normal? That it's alright to eat this food?

—Normal. They never said anything negative.

Galina Bandazhevskaya's conclusions for Sobolev Rouslan, born in 1995, aged 7:
- contamination before pectin treatment: 110.2 Bq/kg;
- contamination after 16 days pectin treatment: 56.9 Bq/kg;
- heart: complains of headaches;
- ECG (before) partial right bundle branch block;
- ECG (after) partial right bundle branch block.

Sobolev Natalia, born in 1989, aged 13:
- contamination before pectin treatment: 178.6 Bq/kg;
- contamination after pectin treatment: 111.1 Bq/kg;

- heart: complains of shooting pains in area of heart;
- ECG (before) partial right bundle branch block; myocardial metabolic problems[155];
- ECG (after) partial right bundle branch block; myocardial metabolic problems.

Krestinskaya Victoria, born in 1986, aged 16.:
- contamination before pectin treatment: 121.3 Bq/kg;
- contamination after 16 day pectin treatment: 106.6 Bq/kg;
- heart: sharp pain in region of heart, headaches;
- ECG (before) partial right bundle branch block. myocardial metabolic problems;
- ECG (after): partial right bundle branch block…

2. THE GRANDPARENTS OF NEKRITCH

This child was not included in the group that Galina Bandazhevskaya examined, because he had mumps and was in quarantine. We met his grandparents.

The grandfather.—Our little Yura is not afraid to dive into the stream which is very cold…

Q.—In this village, is the children's health normal?

The grandfather.—I wouldn't say that. They are ill too often. Before Chernobyl, it wasn't like that.

Q.—Are they more fragile now?

The grandfather.—Now, their blood pressure is too high.

The grandmother.—You know, many people are dying here at the moment. What a pity! One man, he was in good health, he was working, all that. He was rich, Pavlovich. He died nine days ago. He was fine, suddenly he said "Oh my head really hurts!" he said to his wife: "Go and get me some pills". He was outside, sitting on a bench. In the time it took to fetch some pills, it was over. He was already dead. Instantly. The heart.

Q.—Is that happening a lot?

The grandmother.—At present many people are dying like this. We say it's the easiest death. Otherwise, you could be paralysed, or unable to move for years. That's worse. A terrible thing. Better to die like that.

Q.—Yura's parents are not in the village at the moment?

The grandfather.—They left today…

The grandmother.—But the father is dead, our son, he is already dead.

The grandfather.—Also a heart attack. Within a day, gone. He was a forest warden.

Q.—How did he die?

The grandfather.—Well, how…

---

[155] Repolarisation problems, Wave T anomalies.

The grandmother.—He came home one night and he felt ill. They took him to the hospital in Gomel.

The grandfather.—They opened him up.

Q.—Why did they operate?

The grandfather.—The doctors didn't know. They said: "He's in pain, we're going to open up the stomach and we will see". What are they, these doctors! (*Silence*). And he died.

Q.—From what? They don't know?

The grandmother.—You say they don't know? They wrote lots about it! They wrote reams. But we don't know what he died of, or why! We don't know.

The grandfather.—They wrote on the paper: liver. Then, they wrote, burst spleen. They wrote about the heart. A piece of paper this long. (*He lifts his hand above his head*)

Q.—Did he eat a lot of contaminated food?

The grandfather.—No, why?

Q.—He was simply poisoned.

The grandfather.—Noooo! How could he have been poisoned?

Q.—By the radioactivity! It damages the liver, the heart, the kidneys, everything! If you have a high level of becquerels, it destroys all the vital organs.

The grandfather (*suddenly thoughtful*).—That's what they wrote... yes...

Yury, born in 1994, aged 6:
Contamination before pectin treatment: 348.4 Bq/kg;
Contamination after pectin treatment: 237.0 Bq/kg.

3. THE KONSTANTINOV FAMILY

We did not meet this family, whose three children were measured by the Belrad team and examined by Galina Bandazhevskaya.

Galina Bandazhevskaya's conclusions about:

Anastassia, born in 1991, aged 11:
- contamination before pectin treatment: 98.4 Bq/kg;
- contamination after 16 days of pectin treatment: 46.3 Bq/kg;
- heart: complains of headache, nose bleeds. Systolic heart murmur at apex;
- ECG (before): partial right bundle branch block. Repolarisation problems;
- ECG (after):... Repolarisation problems.

Valentina, born in 1988, aged 14:
- contamination before pectin treatment: 82.3 Bq/kg;
- contamination after 16 days of pectin treatment: 49.8 Bq/kg;
- heart: complains of headache. Muffled rhythmic murmurs, systolic murmur at the apex;

- ECG (before) rhythm problems;
- ECG (after) rhythm problems.

Sergei, born in 1987, aged 16:
- contamination before pectin treatment: 160.5 Bq/kg;
- contamination after 16 days of pectin treatment: 99.1 Bq/kg;
- heart: complains of headache, sharp pains in heart region, systolic heart murmur at the apex;
- ECG (before): partial right bundle branch block. Myocardial metabolism problems;
- ECG (after): partial right bundle branch block.... ....

4. THE ROMANENKO FAMILY

Yulia's father is busy in front of the house. We are greeted by the young blonde mother, holding little Ruslana, who is three years old, in her arms. Slim and sporty, she is dressed like someone who lives in the city. In this poverty stricken village, where the family faces a terrible and apparently inescapable destiny, the elegance of this beautiful woman appears to us as a tribute she pays to maintain her dignity as a person: an implacable determination to live despite the betrayal of the state that has abandoned them into the care of cynical foreign occupiers in the poisoned territories of the Sozh river.

The mother.—Ah well! Yulia's health is not brilliant. I took her for an electrocardiogram at Korma. She was complaining about headaches as well, they said it was because of her adenoids *(adenoid hypertrophy)* and that they would have to operate.

Q.—But even so, her radiation levels are quite high: 96 Bq. It seems she has accumulated it through her food. It could be the milk. Have you tested it, has it been measured?

The mother.—They told us it was alright.

Q.—Was it the Germans who came?

The father.—Yes.

Q.—They didn't have anything negative to report?

The mother.—No.

Q.—So where is the contamination coming from?

The mother.—Ruslana, the youngest, also had high levels of radiation last year, and they said that on no account should it be allowed to get any higher. Absolutely not!

Q.—But what can you do? Did they tell you?

The father.—Only eat clean food.

The mother.—But where can we get clean food when we live off what we grow?

Q.—Did they explain what you need to avoid to stop it getting any higher? What you should not eat?

The mother.—Berries, mushrooms, game. It's written on their leaflets.

The father *(looking at the leaflet from Julich)*.—Ruslana had 7 kBq in the whole of her body. *(7000 Bq per 15 kg= 466Bq/kg)*

The mother.—Yes.

The father.—You had 30 kBq! *(30,000 per 50–60 kg= 600–500 Bq/kg.)*

The mother.—So did you…

The father.—about 210 kBq *(210,000 for 90 kg= 2333Bq/kg)*

Q.—You?

The father.—Yes, all you can say is that at night I don't need a light. I walk around in the dark and can see quite well. I really shine!

Q.—Do you ever eat game?

The father.—Yes, of course. We live surrounded by forest.

Q.—So you eat it?

The father.—We have to eat something. This year, we haven't seen the Germans, yet.

The mother.—They told us it was the last year.

The father.—"Everything's fine. Get back to normal life! Everything will be fine. Absolutely. You will live!"

The mother.—The government says the same. They said; "We have got rid of the radioactivity". They "got rid" of our pensions too. For example, we were getting 2000 roubles per person per month. We're not getting that money any more.

Q.—So what you're saying is that they have taken those allowances away, that financial help?

The mother.—Yes. They took it away from pensioners too.

Q.—But it isn't because there is less radioactivity?

The mother.—I don't know why. No-one explained anything to us, no-one came here…We live in wonderland. Where does that money go? Where does it come from? It's obvious that it's contaminated here, very contaminated. Look, Yasen is 2 km away from here, and they were evacuated. Zhavunitsa, completely evacuated, Koliudy, half of the village. Over there *(she points to the east)* everyone. What about us? They say we don't have the right to be evacuated. We're all "reported missing". We don't exist any more. And it's even worse because, on this side of the river, we're cut off. In winter, the bridge is taken away and we're cut off.

Galina Bandazhevskaya's conclusions for:
Yulia, born in 1990, aged 12:
- contamination before pectin treatment: 96 Bq/kg;
- contamination after 16 days of pectin treatment: 73.2 Bq/kg;
- heart: no complaints. Systolic heart murmur at the apex;
- ECG (before): Sinus rhythm, regular;
- ECG (after): Sinus rhythm, regular.

5. THE TITOV FAMILY

The mother.—When she comes home from school, she does not start her homework straight away because she is tired. You can see it. After school she has a headache. That never happened, before. Because we weren't living here before, we've moved house. We have been here for two years. We lived in Russia before. It wasn't contaminated where we lived before.

Q.—What do you eat?

—We have our own vegetable garden. We have goats that give us milk.

—Do you go fishing?

—Absolutely! What would we do without fish?! Live near the Sozh and not go fishing!

—That wouldn't make sense, would it?

—Of course not! *(she laughs)*

—Berries?

—Absolutely! Berries, mushrooms, all summer.

—Game? Do you ever eat game?

—Oh no, we don't go hunting.

—Because your daughter has accumulated a lot of radioactivity?

—A lot?

—She has 149 Bq/kg.

—Oh dear oh dear!

—She should really be below 50Bq.

—Yes, it's a lot.

—And that could come from the fish?

—Definitely, and the berries, and mushrooms.

—Do you know how to treat food products to eliminate the caesium?

—Well… we cook them… we just cook them, we grill them, but as to eliminating…

—All you have to do is soak them in salt water for two hours, discard the water and do it again.

—We've never done that! We didn't know. No-one told us.

—The Germans come here to measure. Don't they explain it to you?

—No, they take measurements. They ask us and we give them our milk, fish and, if we have any, mushrooms.

—They ask questions and that's all they do?

—Yes. They ask questions but they didn't explain anything, about soaking the food or anything. You are the first person to explain that to me.

—So they are not really taking care of your health?

—No. They just take notes…

—You, yourself, how do you feel?

—I get dizzy and I get headaches. Sometimes I get a pain in my heart, but we don't run to the doctors. With my husband it happens a lot. He works in the forest, at a sawmill. He often has pains in his heart. Yes, it happens often with him. The truth is that a lot of the people around here have got high levels of radioactivity, even the adults. I myself have got more than 300.

—Have you got the yellow leaflet given out by the Germans?

—Oh! No, the children must have lost it somewhere.

—Do you remember how much you measured?

—300 or more.

—It's much too high. I wish you…

—Thank you. Goodbye.

It is not at all clear what this figure of 300 means. On the yellow cards from Julich, that we had seen and filmed, the contamination is always recorded in kilobecquerels. If this is the case for Mrs Titov, who relishes all nature's generous gifts, her contamination has reached astronomical levels. She must weigh between 65 and 70 kilos.: 300,000 for 70 kg = 4,286 Bq/kg! "Gut! Gut!" But even 300 Bq/kg is far too much and that was obvious from the symptoms she described.

Galina Bandazhevskaya's conclusions on:

Cristina, born in 1990, aged 12:
- contamination before pectin treatment: 149.1 Bq/kg;
- contamination after 16 days of pectin treatment: 141.8 Bq/kg[156];
- heart: no complaints. Muffled heart murmur at the apex;
- ECG (before): partial right bundle branch block. Myocardial metabolism problems;
- ECG (after): partial right bundle branch block. Myocardial metabolism problems.

6. THE SAZONOV FAMILY

The first questions concern Vassili, who was not able to take part in the experiment.

Q.—Has Vassili any health complaints?

The mother.—Dizzy spells. He wakes up in the morning: "Mummy, I don't feel well". Either he feels sick or he feels dizzy. He complains of headaches. The older ones complain often as well. They come home from school and straight away: "Mummy, I don't feel well" and they go to bed. They complain mainly about headaches, and feeling dizzy. And often about the heart as well.

—Do they get out of breath when they run or do sport?

---

[156] Practically no effect after 16 days of pectin. It is possible that in this case the pectin was not taken by the child for reasons we were not told.

—Yes, after gymnastics:"Mummy, when I run it hurts here and it's difficult to breathe. I can't breathe". I even spoke to my eldest daughter's PE teacher. I asked her not to ask her to run, because it makes her feel ill... she shouldn't do it.

—And what about you? And your husband? Do you have the same symptoms? The head, the heart?

—Yes, my husband, mostly it's his head where he feels very ill, and he has heart problems as well. He is registered at the hospital at Korma. As for me, before everything was alright, but now...I have rheumatism, my joints are beginning to hurt, especially my shoulders.

—Have you been measured by the German team?

—I went last year. My levels weren't that high: about 13 kilobecquerels... *(13,000 for 65 kg = 200 Bq/kg)*

—What do you eat at home?

—We have our chickens, our eggs. We raise pigs, our milk... yes. We have a cow, everything is ours, and then there's the vegetable garden.

—Have you measured the milk?

—The people from Korma tested it. It was alright apparently. Within admissible limits:—40–50 Bq/l. They said it was alright to drink.

—The mushrooms, for example? *(she seems to be defensive about her mushrooms)*

—Well, you know, we don't really even collect them. Very rarely.

—Absolutely never?

—Well, sometimes.

—You don't have dried mushroom soup?

—Well yes, but only the girolle mushrooms...

—"Only!" It's precisely those that are full of radioactivity. Mushrooms contain enormous amounts. And it accumulates and has an effect on the heart, the eyes... Your two daughters complain of the same things?

—Both of them complain about pains in their joints as well. Their legs hurt. The joints.

Galina Bandazhevskaya's conclusions on:

Tatiana, born in 1988, aged 14:
- contamination before pectin treatment: 81.9 Bq/kg;
- contamination after 16 days of pectin treatment: 55.9 Bq/kg;
- heart: complains of shooting pains in region of heart.
  Systolic heart murmur at the apex;
- ECG (before): partial right bundle branch block.
- ECG (after): partial right bundle branch block.

Elena, born in 1987, aged 15:
- contamination before pectin treatment: 59.6 Bq/kg;
- contamination after 16 days of pectin treatment: 44.2 Bq/kg;
- heart: complains of shooting pains in region of heart. Systolic heart murmur at the top;
- ECG (before): partial right bundle branch block. Myocardial metabolism problems;
- ECG (after): partial right bundle branch block.

Vassili, born in 1992, aged 10[157]:
- contamination before pectin treatment: 112.27 Bq/kg.

---

[157] This child was not included in the group that Galina Bandazhevskaya examined because he had mumps and was in quarantine in a sanatorium.

Chapter XIII

THE CHILDREN
IN THE HOSPITAL AT GOMEL

June 2001

1. CHILDREN WITH HEART PROBLEMS

The hospital at Gomel. We follow Galina Bandazhevskaya, who takes us into a room where her colleague is working with a group of about twenty children who are ill. They agree to talk to us in a break in the conversation they are all having with the doctor. The doctor addresses a little girl who looks about 7 or 8 years old.

—Olechka, tell us, how do you feel?

The little girl smiles, happy to be talking to these foreign journalists. She is sitting down, her feet together, her hands on her knees, with a white ribbon in her hair. She has large eyes.

The little girl (proudly)—I have been ill since I was 7.
Q.—How old are you now?
—I'm already 14... *(Her smile disappears. I did not realise I had touched such a sensitive spot.)*
—What illness do you have?
—Systemic collagenosis.

An illness that attacks the fibrous protein of the intercellular substance of connective tissue and cartilage. There can be retarded growth. The immune system that should defend the body from viruses, bacteria and cancerous cells, "mistakes" its target and attacks the cells and tissues of the child. These auto-immune diseases including serious cases of Type 1 diabetes in children, have increased in the contaminated territories.

—Where does it hurt? *(The little girl looks away. She is ashamed to be so small for her age.)*

—It hurt... my heart hurt... *(She cries silently. The doctor comes over and strokes her head.)*

—Calm down and tell us all about it, don't be sad...all that's over now, it doesn't hurt any more... Don't cry. *(She strokes her.)* Olechka, come on now. Wipe your eyes, your eyes are beautiful. You're a big girl. You were ill, but now you feel better. Tell us again. What was it that hurt you?

—I had headaches, then my knees hurt, so I came back here... Now I'm better... That's it. *(She sighs and begins to cry again, in silence. She hides her face in her hands, pressing her knuckles into her eyes. The little girl next to her, with blonde hair, looks at her furtively, on the verge of tears.)*

—And why are you here at the hospital?

—This is the second time I've been here. The first time I was eight.

—And how old are you now?

—I'm 9. Before I often had stomach pains... and then I came to the hospital. They told me I have gastritis.

—Do you have any pain in the heart?

—Yes. *(She is crying too. Little Olechka steals a glance at the other little girl.[158] A little boy with fragile shoulders is sitting next to her.)*

The little boy.—I have a congenital defect of the heart.

—How does that make you feel?

—How do I feel? Normal.

—Does your heart hurt?

—I get shooting pains.

An adolescent boy.—When I run, it hurts, it stings in my heart.

—Where exactly?

—Here. *(He puts his hand on his chest)* Then, my head hurts...

An adolescent girl.—I can't play any more, or run...

—Why?

—I get out of breath. It's difficult to run... It makes me go dizzy... It goes black before my eyes...

I turn to the doctor who explains:

—There are a lot of children here who need operations.

—From a statistical point of view, is this normal in childhood?

—No, no. First, I'd have to say that there aren't really any norms for congenital malformation, for anomalies. This illness has no known incidence or frequency.

---

[158] Olechka died a few months after this meeting.

But from a quantitative point of view, it's true that the percentage is increasing among children.

—Is it still increasing?

—Yes, it's increasing. In the last few years, children are being born with malformations.

—Is there a link to the area in which they live?

—There are many factors, the causes are multifactorial—heredity, the health of the parents—but ecology plays a role.

—Ecology meaning Chernobyl?

—Of course.

—Before and after Chernobyl—it's not the same thing?

—No, of course not.

—Has it got worse?

—Yes. It hasn't improved the health of the children nor of the rest of the population. *(She turns towards a little boy who is standing in the doorway.)* Come here little one, come! *(The little boy comes over to her.)* Here, this little boy lives in Gomel. You are nearly 9, aren't you?

—I was born in 1992.

—Sit down, please, don't be frightened. *(The boy looks round anxiously)*. He is very emotional. When he was 9, his blood pressure increased significantly, and he began to have seizures. A stroke, like adults have. He was in a very serious condition. He is beginning to get better, but unfortunately, the underlying cause of the initial rise in blood pressure is still the same.

The boy latches on to the last words spoken by the doctor. He addresses us, the visitors.

—I have high blood pressure. I get headaches when my blood pressure is high. In 1998, I had a stroke and hemiplegia and I was in hospital for three months…

—What happened after you had the stroke?

—I had a cyst in the front of my head. It's still there. My left leg, my left arm and the left part of my face was paralysed. I had cramps in my left leg.

—And how about you? *(the doctor addresses his neighbour, a boy with dark rings round his eyes)*

—I came to the hospital with polyarthritis in the right hand. Three weeks ago. My hand was swollen. Then my foot started to hurt, I couldn't press down on my foot. I have pains in my heart as well. I have already had three heart attacks…

—Three?

—Yes, and I have gastritis as well.

—When did you have these three heart attacks?

—When I was little, I was still at primary school.

—How old are you?

—I'm 12.

A timid adolescent boy, sitting near the window.—When I was three months old they discovered I had a heart murmur. I've been coming here ever since. I come here all the time to be examined.

My heart hurts but I can play football.

—Do you get out of breath?

—No, it just hurts my heart if I run a lot.

A little girl, wearing a pretty red dress, is sitting among the other children.

—And what about you?

The little girl in red.—I've got a heart murmur too. I didn't have any pain before... *(The doctor looks at her kindly.)*

—This little girl has alterations to her heart rhythm. She has to have treatment.

—Is it normal that they have the same illnesses as adults?

—No, certainly not. All we can say is that our ideas about the age at which these illnesses appear has changed substantially. Many illnesses that only appeared in adults before are appearing in children. *(She beckons to a boy to come forward.)* This boy has a serious congenital malformation of the heart. There's no doubt that he will need serious treatment, and surgery. *(The boy looks at her with his blue eyes and a little frown, without understanding the seriousness of what she is saying.)* Where are you from?

—I live at Stradubka, in Loev province.

—That's far from Gomel? *(His brother, who is sitting next to him and also has a heart anomaly, answers for him).*

The brother.—About 50 km from Gomel.

—Part of that area is contaminated. They've got their problems there as well...

—Do you use spectrometers to measure the children to see if their organisms are contaminated?

—We can't do that at this hospital. In areas that are strictly monitored, there are mobile laboratories and the children who live there are measured. But not all our regions fall into this category.

The children go out. The doctor approaches a boy who is sitting alone in a corner, looks at him and gently takes his hand and accompanies him outside.

In another room, near the neonatal resuscitation unit, four young mothers are looking after their babies.

2. THE YOUNG MOTHERS

The first mother.—It's a neurological diagnosis. He has spasms in his arms. They say it's congenital... But what is it really?

Q.—Which area are you from?

—We're from here, in Gomel.

—How do they explain it?

—They say there was insufficient oxygen during the pregnancy, or its hereditary... or else the ecology.

—Chernobyl?

—Maybe that as well, because...Maybe that as well.

—Have you been measured with a Human Radiation Spectrometer? Do you know what levels you had?

—Ah, yes, radioactive accumulation. Oh that was a long time ago. I was 10. That was the only time I was measured. No-one does that any more. I had more than 10,000 becquerels (*about 330 Bq/kg body weight*)

—How long have you been here?

—This is the fourth week. He looks a bit better. They seem to have got him back to normal. We're leaving tomorrow.

The second mother.—With us, it's the kidneys. They said it's something congenital. I had problems with my kidneys too.

— How old were you at the time of the Chernobyl accident?

— I was little. I was born in 1976, so I was 10. We went out walking, we went to the beach, everywhere. Sunbathing, swimming.

—Is it a girl?

—It's a boy. Vladik. He's a bit better now... He's eating, he's growing. For the moment...Tfu! Tfu! Tfu! (*She makes a gesture to dispel the evil spirits.*)

The third mother.—Mine is a little girl, Svetlana. She has infantile hepatitis. Now it seems that everything is normal, we're going to be sent home soon.

—Where were you when Chernobyl exploded?

—Here.

—How old were you?

—I was 9.

—Did they evacuate the children en masse from Gomel? Were you evacuated?

—No, we went to my mother's parents.

—Was that decision made by the family?

—Yes, when we knew more about what had happened.

—Did you learn about it late on?

—Of course. We finished school in May.

—Did they measure you?

—I'm trying to remember... They came to our area to measure, they dug up the soil. There was definitely something, but I can't remember now. They certainly made measurements. I know that I often had bronchial pneumonia. But was it linked?... Yes, I was ill very often, during that year. I was on the hospital list because it was so frequent. I don't know what caused it.

—When was she born, your baby?

—She will be a month old tomorrow.

—Well, best of luck!

—Thank you.

The fourth mother. *(she is rocking her baby in her arms)*.—I was 15 at the time of the accident.

—Do you remember it?

—Of course! At the beginning, when it first happened we were playing outside, no-one knew anything. We went on the May Day march, because it was kept secret from us.

—Were you at Gomel?

—No I'm from Mosyr. It's very close to the zone. It's 90 km from Chernobyl, as the crow flies.

—Did anyone measure your levels of contamination at the time?

—They may have taken our measurements but no-one said anything to us.

—Is it a boy or a girl?

—A boy.

—What is wrong with him?

—According to the doctors, it's because I was infected. My immune system has been weakened. During the pregnancy there were no complications, but the baby was born with a malformation...He has septicaemia and purulent meningitis. It's congenital.

—And you think there's a connection to Chernobyl.

—They say that the immune system has been weakened. It's congenital. 80% of women are affected currently. They told me that at the maternity hospital in Mosyr, 30% of new born babies have to be resuscitated...

—We know that a weakened immune system is one of the effects of Chernobyl.

—The doctors have more up to date information about it... The women who are infected don't even know and don't know that their child will be ill. Whatever the analyses show.

—All these mothers were young girls at the time of the accident?

—Yes, yes. At Mosyr the director of the maternity hospital said that out of 600 new born babies, 230 had to be resuscitated. This is what our children are like. I am from there. *(She looks at her baby as she rocks him in her arms.)*

—How is he now?

—Thank God, for the moment… Thank God!

The Ministries of Health in Belarus, Ukraine and Russia refuse to make the link between the internal radioactivity of these children and the illnesses that are appearing in hospitals. If they were to do so, and if the correlation proved by Bandazhevsky between caesium-137 and its negative effects on vital organs were to be systematically verified, which would require measurement using a spectrometer, more effective prevention measures could be taken to avoid the worrying increase in new illnesses in children in the contaminated regions.

PART SIX

THE ESTABLISHMENT TURNS A DEAF EAR

Chapter one

TWO LETTERS
TO THE UNITED NATIONS

Each year, in May, the Ministers of Health from 191 Member States take part in the World Health Assembly of the WHO in Geneva. Every year, NGOs urge the Ministers of Health in their own countries to revise the agreement signed in 1959 between the IAEA and the WHO that prevents the latter from fulfilling its role in the area of radiation. And every year, the WHO assembly turns a deaf ear.

On 12th February 2001, for the umpteenth time, the ritual was repeated; two anti-nuclear organisations, Contratom from Switzerland and CRIIRAD from France, held a press conference in front of the Palais des Nations, in Geneva, at which a number of questions[159] were posed to the Secretary-General Kofi Annan, and the Director-General of WHO at the time, Mme Gro Harlem Brundtland. We filmed the official handover of the letters to two high ranking UN officials.

---

[159] The questions were:

—Why did it take five years following the disaster for the World Health Organisation to visit the Chernobyl area, begin studies and offer any help?

—Why did three WHO delegates approve the decision, in June 1989, to revise upwards by a factor of two or three the acceptable level of radioactive contamination in the Soviet Union thus condemning hundreds of thousands of people to live in contaminated territories?

—Why, following Chernobyl, did WHO not continue to prioritise research into "genomic effects", as its own study group had recommended (1957) instead of "*mouth dryness and dental caries*"?

—Why have the proceedings from the international WHO conference which brought together 700 doctors in Geneva (20th November 1995) on the theme "The health consequences of Chernobyl", that were promised for March 1996, not yet been published?

—Why has WHO not intervened in the European project authorising the presence of radioactive substances in our food?

—Why is WHO absent from the crucial debate about the safety thresholds for contaminated materials from the decommissioning of nuclear installations that could be recycled into our environment?

—Why has WHO never intervened to prevent industry and governments from using and releasing into the environment so-called "depleted" uranium, which is toxic both chemically and radiologically?

Here is the substance of the letter addressed to Kofi Annan:

> After reading the moving words that you wrote in the forward to the 2000 OCHA report on Chernobyl, we would like to respond to your appeal and make our contribution to awakening public awareness of the consequences of the disaster.
>
> [...] Today, we have seen the immeasurable damage that the nuclear industry can cause to humanity and to the environment—damage that will last for generation upon generation, bequeathing to them the job of cleaning up our waste for thousands of years into the future.
>
> The time has come for WHO to break free of the unnatural constraints that tie it to the IAEA, whose official role is to promote the use of commercial nuclear technology [...] On subjects as worrying for the future of our planet as Chernobyl, low dose radiation and the growing use of depleted uranium for civilian and military purposes, the whole world expects WHO to be able to express itself freely, on the basis of irreproachable scientific research, and to act in accordance with its constitution.
>
> The undersigned ask that you take steps immediately to ensure that the WHO regains total independence.

In demanding the amendment of the 1959 agreement with the IAEA, the letter to Dr Brundtland stated:

> The organisation that you have the honour to direct is no longer able to fulfil its Constitution, that requires it to assist in developing an informed public opinion on the effects of the nuclear industry on people's health and the health of future generations. The time has come, for WHO to break [...]

At the Place des Nations, a former associate of WHO delivered his own personal testimony to the journalists present.

Professor Michel Fernex.—My links with WHO are as a colleague, who worked for fifteen years in tropical medicine. I was a member of the Steering Committee on Tropical Diseases Research into malaria and filariasis and I hold the organisation in great esteem. But since 1986, I have felt extremely saddened by WHO's absence for five years from the Chernobyl territory, and by its premature departure from Chernobyl, after only a few years of research that began in 1991.

—Why has WHO waited until 1st February 2001 to "launch a four to five year research programme, costing 22 million dollars" into the effects of depleted uranium in Iraq and in the Balkans, while the Americans published a report in 1990, available to the press, proving that it is an extremely dangerous radioactive material which is "politically unacceptable"?

—Why has WHO said nothing about the use of depleted uranium weapons that condemns the populations in areas that have been bombed to live the rest of their lives in a contaminated environment?

WHO's creation in 1946, which aimed to provide health for all was a magnificent achievement. Its primary function, set out in its constitution, was *to act as the directing and co-ordinating authority on international health work.* It is doubtless for that reason that for five years WHO was absent from Chernobyl. WHO left the IAEA to do the research in its place. WHO should have initiated and guided research, provided information, given advice, offered assistance in the area of health and finally,—and it did not do this either—assisted in "developing an informed public opinion among all peoples on matters of health".

With regard to this matter, it would be interesting to know what went on in the building just behind us, on the other side of the square. WHO tried to inform public opinion in 1995. 50 years after Hiroshima and 10 years after Chernobyl, the Director-General at the time, Hiroshi Nakajima, brought together 700 doctors and scientists, in this building. It was very interesting conference because a great variety of opinions were expressed. For example, a UN representative told us, in the first session, that 9 million people had been exposed to radiation as a result of Chernobyl. This was a Mr Griffith, if I remember correctly. When I say "if I remember correctly", it's because I took notes. The proceedings of this conference, which had been promised for March 1996, have still not been published. They have not been published because WHO signed an agreement with the IAEA that no information should be published that might harm the nuclear industry. We have been asking for a long time, and more recently have asked in writing, for an explanation for the delay of five years. We have not received a reply. It's a real shame. The reason I am here is to ask WHO to amend Articles I, II and VII of the agreement[160], to put an end to this handicap and regain its freedom in the scientific and medical domain.

---

[160] According to Articles I § 2, "…it is recognized by the World Health Organization that the International Atomic Energy Agency has the primary responsibility for encouraging, assisting and co-ordinating research on, and development and practical application of, atomic energy for peaceful uses throughout the world…" Article 1 § 3 states that "Whenever either organization proposes to initiate a programme or activity on a subject in which the other organization has or may have a substantial interest, the first party shall consult the other with a view to adjusting the matter by mutual agreement".

According to Article III, "The International Atomic Energy Agency and the World Health Organization recognize that they may find it necessary to apply certain limitations for the safeguarding of confidential information furnished to them. They therefore agree that nothing in this agreement shall be construed as requiring either of them to furnish such information as would, in the judgement of the party possessing the information, constitute a violation of the confidence of any of its Members or anyone from whom it has received such information or otherwise interfere with the orderly conduct of its operations"..

According to Article VII, "…the International Atomic Energy Agency and the World Health Organization undertake… to avoid undesirable duplication between them with respect to the collection, compilation and publication of statistics, to consult with each other on the most efficient use of information, resources, and technical personnel in the field of statistics and in regard to all statistical projects dealing with matters of common interest".

Q.—What significance is there in the fact that Dr Hiroshi Nakajima, Dr Brundtland's predecessor, will be the honorary president at the next WHO conference in Kiev?

Fernex.—It allows me to hope that the censorship of the proceedings imposed after the WHO conference in Geneva in 1995, will perhaps—I say perhaps—be lifted at the next conference. I have confidence in Dr Brundtland, the new Director General, and believe that she will not allow censorship. I am putting all my hopes on it. But the IAEA will be there too, don't be under any illusions. UNSCEAR, the IAEA, they have a fantastic amount of money. Buying a scientist from a poor country costs them nothing. With $10,000 you can buy many people...And meanwhile they remove Nesterenko, from the centres that he runs. It's monstrous. I was speaking on the telephone to the people from ETHOS, who are chasing him out of the villages he works in, and I said that aid to poor countries should always be used to strengthen local structures. Do you know what they said? "Yes, that's what we think. When we leave there will be nothing left". I told them "When you leave, if you give your support to Nesterenko, there will be a stronger structure. If you get rid of him, there really will be nothing".

I have only met Kofi Annan on a few occasions and in general from afar, at meetings and conferences. But I am struck by the fact that he attends pacifist and anti-nuclear NGO conferences. He goes out into the field. So he is someone who lives among the people, and he respects the NGOs. There's no doubt that he is a man of great sensitivity. And for someone sensitive, the suffering of the hundreds of thousands of children around Chernobyl, is not just a news item, not just a statistic, but a reality that must be followed up. He showed real wisdom when he said that fifteen years after Chernobyl, we had not seen anything yet—that the real problems would begin in years to come and that we should be very circumspect today. Kofi Annan upset UNSCEAR, whose report basically said "Erase it all, forget it all". That is the IAEA's role. It pays people to cover up the truth. But the truth is beginning to appear. What is the latency period for a solid cancer? A minimum of ten years. So all the research needed to be stopped after ten years.

Q.—To talk about future generations requires expertise.

Fernex.—WHO has this expertise. There have been plenty of scientific experts who have warned them of this, provided them with information. The Nobel Prize winners in 1957, the Nobel Prize winners this year. Why has WHO not brought in Jeffris? Why have they not called on Dubrova from Moscow? It would not cost them very much money. Why do they not ask Goncharova from Belarus, Ellgren from Sweden? Why? Why have these people, who were brought together in 1956, and published in 1957, all very highly qualified scientists, not been consulted by WHO? In 1986 a committee of experts should have been brought together. That did not happen. It was five years before WHO appeared at Chernobyl! Do you know what was going on during those five years? Just take as an example, infant mortality, between the first and the twenty eighth day of life, which increased by 4.8% in Germany. In the contaminated zones, it was more than 8%. But in Russia, Belarus and Ukraine, it went unnoticed because WHO was absent.

Against the background of these disagreements within the UN about the consequences of the Chernobyl disaster, we prepared for the next stage of the confrontation at the heart of the scientific community. It would take place at the beginning of June in Kiev. The association Doctors for Chernobyl had succeeded in getting Dr Hiroshi Nakajima, who had failed in 1995, to be honorary president. Would there be a confrontation at Kiev? What might be the result?

Chapter II
THE KIEV CONFERENCE

The tragedy of Chernobyl is the result of an information process, in which the interpretation of the facts imposed by the authorities plays the main role [...] As long as the media refuses to cross swords with these prestigious institutions, the WHO/IAEA power base will continue to maintain its order.

YVES LENOIR,
Tchernobyl, la tragedie optimisée, Bulle Bleue.

The 3rd International Conference, "Health Effects of the Chernobyl Accident: Results of 15-Year Follow-Up Studies" took place between 4th and 8th June 2001 in Kiev, in Ukraine. The participants included researchers and scientists from Belarus, the Russian Federation and Ukraine, as well as representatives from WHO, from the Office for the Coordination of Humanitarian Affairs (OCHA), from the IAEA, from the United Nations Scientific Committee on the Effects of Atomic Radiation (UNSCEAR), the UN Chernobyl programme in Ukraine and the International Commission on Radiological Protection (ICRP).

Participants at the conference presented 88 reports and 316 summary presentations. The aim of the conference was to *"provide a scientific basis for future decisions to be taken by national and international organisations about the medical consequences of the Chernobyl disaster"*. Thus, every word of the final resolution of the conference would have a direct influence on the destiny of all those human lives affected by the disaster.

The idea of filming this conference came to us after Michel Fernex described in detail what had gone on at the Geneva Conference in 1995. "The programme put forward by Dr Hiroshi Nakajima had convinced the health authorities in the countries most affected and 700 doctors and scientists to take part. The IAEA, for its part, had mobilised staunch supporters of the nuclear industry. Contradictory opinions were expressed which produced a very lively exchange of views. The representatives of the nuclear lobby had tried to stifle debate and Professor S.

Yarmonenko from the Centre of Oncology in Moscow (directed by Professor Ilyin) had insisted that in future, the organisers exclude from scientific conference programmes any speaker who introduced the subject of low dose radiation on living organisms[161]. Dr Hiroshi Nakajima's attempt had ended in failure. Blocked by the IAEA, the proceedings of the conference, which were keenly awaited, were never published. The truth about the consequences of Chernobyl would have been disastrous for the promotion of the nuclear industry".

Having heard Fernex's description, I could imagine what might happen at Kiev. He had described the fury of the representatives of the pro-nuclear UN agencies. Their power threatened by revelations about the health consequences of the disaster, they denied the seriousness of the radioactive contamination. It was a unique opportunity to observe and to film, at close quarters, the faces of those in power, imposing on the world the Holy Commandments of the nuclear industry, in conflict with the independent scientists. I explained this, both in a letter to Thierry Garrel, director of documentaries at the TV channel Arte, and then when I met him and put forward the idea of filming the whole event, in cinéma vérité style.

He told me in a meeting that he could not see the interest in filming the conference. My project did not interest him. On the other hand, our colleague Aldo Sofia, the producer of the programme "Falo" at TSI *(Swiss TV in Italian language—translator's note)* gave us his help.

With Romano Cavazzoni as cameraman and Emanuela Andreoli as editor,—the same team who worked with me on the previous documentaries about Chernobyl—we made the film with our own limited means, and a small amount of money from the television company in Lugano. A short version of the film *Nuclear Controversies* was shown on three national television channels, (Swiss Italian, Swiss German and Canadian). It is an astonishingly revealing piece of theatre about human behaviour but it also provides incriminating evidence about the way officials from the nuclear lobby ended up duping a scientific conference in which new and crucial information was presented and ignored. This film created problems for WHO. Dr Brundtland set up a commission to find out why the proceedings from the 1995 conference had never been published. But some time after, she herself was replaced as the head of the organisation.

The director of documentaries at the television channel, Arte, was right. No-one films these conferences, so we were the only people filming over the five days. Swiss Television, whom I represented officially, was simply fulfilling its legitimate role of informing the public, when we filmed the conference at Kiev. But it was also a Trojan horse, given that the filming took place in one of the bastions of the nuclear establishment where, in the face of media indifference, one of the most dangerous

---

[161] Cf. Dr Michel Fernex, *op. cit.,* L'Harmattan, 2001.

scientific untruths of modern times is officially ratified: that chronic low levels of Caesium-137 incorporated into the body are harmless. As we had imagined, we were able to film the fury of UN officials and their Soviet accomplices on hearing the revelations of the independent researchers and doctors about the radiological causes of the health catastrophe in the Chernobyl territories. The data and the radioprotection measures recommended by the independent scientists to protect the contaminated populations were arrogantly and disdainfully dismissed. They refused even to discuss it.

1. DR HIROSHI NAKAJIMA'S ADMISSION OF POWERLESSNESS

It was Dr Hiroshi Nakajima who presided over the conference at Kiev. We could not miss the opportunity to interview this important figure. We approached him during a coffee break, while he was talking to Michel Fernex.

Fernex.—… the one in 1995 was the best. That was an excellent conference…

Nakajima.—Unfortunately, even the Japanese did not take up the recommendations of that conference.

—Why were the conference *proceedings* that were ordered not published?

—Because the conference was organised jointly with the IAEA. That was the problem.

—And here, at this conference, does WHO have more freedom than in Geneva? Here in Kiev?

—But I am not the Director General any more—I'm a private individual.

Would the former Director General dodge the issue that really interested us? I questioned him myself.

Q.—Don't you think that the link between the WHO and the IAEA is contradictory? If WHO is to act completely freely? … What do you think?

Nakajima (*without any embarrassment, he replies clearly and honestly*)—I drew attention to this at the time when I was Director-General, and therefore responsible, but it was my legal department above all, my legal advisor that raised the issue… You need to understand that the IAEA reports to the United Nations Security Council. And the rest of us, all the specialised agencies, report to the Economic and Social Council. It's not a question of hierarchy, we are all equal, but in nuclear matters, in the military and peaceful use of the atom, or in civil nuclear power, it is the organisation that reports to the Security Council that has authority.

Q.—They give the orders.

Dr Nakajima looks at me in silence: silence means consent.

No-one in such a position of authority had ever admitted that the defenders of our health were subordinate to the promoters of the atom. The whole power structure, as it is exercised both in practice and legally by the military over the civilian wing of the nuclear industry, was delivered to us in a few words by someone whose competence in the matter could not be doubted. Nakajima's allusion to the "legal department" at WHO, confirmed that what was at stake was a real conflict of interest, enshrined in the legal framework from the very beginning. The unnatural agreement of 1959, that imposed mutual prior consent in any initiative in the nuclear domain, tied the hands of the organisation responsible for health from the first day of a disaster. It had prevented it from helping the victims of Chernobyl. Eleven years after our visit to the town of Poliske, Hiroshi Nakajima had answered Alla Tipiakova's desperate plea with this admission of powerlessness, this prohibition to offer help to the children she saw suffering[162]. Today, twenty-nine years later, just as in the first days following the disaster, this scandal morally disqualifies the United Nations and the states that make up its Security Council. Perhaps before Chernobyl, their attitude could have been explained by a lack of foresight and a disproportionate confidence in technology, but since the accident, this pact of silence is a crime committed by the very people in whom we place our trust. Fortunately, there was a righteous man who had shed tears of powerlessness when faced with the deaf ears of the powerful.[163]

As we had expected, these contradictions and tensions between the same official protagonists from the 1995 conference, resurfaced at Kiev. They divided the representatives of agencies. They were openly expressed by the independent doctors and scientists, who made forceful presentations but to no avail: the representatives of the IAEA and UNSCEAR took the podium. They played a decisive role on the last evening, when the wording of the final resolution was determined behind closed doors, and distributed after the conference had closed.

It is impossible to recount all the presentations and debates. I will relate only those exchanges and passages that illustrate the gulf that separates the two versions of the medical consequences of Chernobyl. A dialogue of the deaf. The testimonies that follow are authentically reproduced from film and recordings.

---

[162] See p. 6–9. Part One. Alla Tipiakova's complaint "But it's as if the policies, that might actually resolve our problems, are being blocked somewhere and at the political level no-one's interested in us".

[163] . See p. 94–95, Nesterenko "talking to the wall" with his academic colleagues.

2. OFFICIALS AT THE CONFERENCE

The representative from the Office for the Coordination of Humanitarian Affairs (OCHA), Dusan Zupka, agrees with Kofi Annan's estimate that there will be about 9 million victims from Chernobyl and that the tragedy has only just begun.

Zupka. The consequences of Chernobyl do not fade away, but actually grow increasingly, uncertain and in many ways more intense. The United Nations Secretary-General, Kofi Annan put it very clearly when he said that "the legacy of Chernobyl will be with us, and with our descendants, for generations to come". Given the magnitude of the disaster, its long term consequences, its international implications, the international community has a humanitarian obligation to assist 9 million people affected by the Chernobyl accident. However, while millions of US dollars have been pledged for the construction of a new and safer sarcophagus, comparatively little has been done by the international community to provide direct assistance to the populations affected by the consequences of the accident. We, the international community, can and should do much more to provide tangible assistance to the populations of the three countries. In the current highly competitive atmosphere when natural disasters and other crisis situations around the world are competing for donor attention, Chernobyl remains the only catastrophe with an unpredictable future. We do not know exactly what new health manifestation from the radioactive contamination the next years will bring. The United Nations programme aims at strengthening international cooperation, the mobilisation of resources and the coordination of efforts to study, mitigate and minimize the consequences of an unprecedented disaster. In ignorance, we had hoped that the situation could be returned to normal in a matter of years. Now, in retrospect it is clear that it will take generations just to bandage the wounds.

Michael Repacholi (WHO, director of the Radiation Department)—We all have our ideas of exactly what happened with the accident and there is no doubt that it was a tragic event. One of the aspects that has been highlighted in particular is the stress and trauma caused by the psychological problems linked to being evacuated from the contaminated zones, and having to abandon one's home. I would like to talk to you briefly about the efforts that were undertaken by WHO. The information available was very limited. I would like to add that it was three and a half years after the accident that the Soviet Government asked the IAEA for assistance with the catastrophe. Now most member countries have signed the first agreement on notification. This agreement specifies that in the case of an accident, the signatories inform the International Atomic Energy Agency, and that this agency, which has great expertise in determining contamination levels, and the possible contamination of different zones, is in a position to mobilise its teams of experts who, arriving in the field, can evaluate the situation and assess the significance of the impact. As I said, it was only in 1990 that WHO was officially requested to intervene and become involved with the health effects of the accident. The lesson we draw from all this, is that in the case of an accident of this scale, national governments need to be able to transmit information as quickly as possible so that the risk of losing information is reduced to a minimum. It was only in May 1991, so five years after the accident, that WHO was able to put a programme of health assistance in place for the Chernobyl accident. WHO was able to provide a great quantity of medical equipment, among the most up to date, that was able to detect thyroid cancer in children or other thyroid diseases and this allowed an enormous number to be identified. The three countries received modern medical equipment worth 16 million dollars that allowed them to bring their health system up to date. In 1996 OCHA participated in the development of brochures that were distributed to the population to help them understand the risks to health from exposure to ionising radiation. Additionally, the population were informed that they had nothing to fear from exposure in the long term to low level radiation.

I would like simply to say something about the lesson we have drawn from the Chernobyl accident: we absolutely must be ready for the next accident that could occur. A response is necessary and we must face the fact that this kind of disaster, like all major ecological disasters, earthquakes, flooding, has a traumatising effect on the population and there are psychosomatic effects that need to be confronted: psychological stress, social and economic upheaval in people's lives that need to be addressed in as humane a way as possible. We need to make sure that all personnel intervening to mitigate the effects of an accident in a nuclear reactor receive effective training so that they know exactly how to deal with the situation and to reduce to a minimum the risk of exposure of the population. Now that the Chernobyl reactor is quiet and not producing harmful radiation to anyone, we can reflect on what happened with this accident and what we need to do in the future to protect ourselves in the most rapid and appropriate way.

The IAEA representative supports the view that the Chernobyl disaster caused 31 deaths, a few hundred cases of acute radiation, and 2000 thyroid cancers in children. This UN agency only recognizes validated reports, in other words, reports confirmed by the laboratories at Los Alamos and at the Atomic Energy Commission (CEA) in France, manufacturers of the atomic bomb.

Gonzales (IAEA).—Our agency as you probably know is the only agency in the United Nations family with the specific charter obligations in radiation safety[164], radiation protection. We were already here with our Director-General in front of

---

[164] He uses the words "safety" and "the only...agency...with specific obligations..." when he himself is not a doctor, but a physicist promoting nuclear industry.

us a few days after the accident and since then, we have been fully committed to the Chernobyl tragedy. We were with the local people, the people who really suffered the consequences of this accident. And we shared not only their suffering; we shared also the tremendous change that happened since the accident until today in this area of the world. Capitalism brought good things, but capitalism brought a lot of problems as well. A lot of problems to people that were not prepared for the changes. Our challenge is to distinguish the health effects attributable to the radiation exposure, to the health situation created by the Chernobyl situation and by the political changes in the region.

What do we know today? Really not too much that is new.

Let me remind you that the significant radionuclides were two and only two: caesium and iodine, which cause tumours in the thyroid in children and which, unfortunately, were not blocked, and could have been blocked without any tablets, without any emergency intervention, simply by banning the drinking of milk[165]. We launched a large scale program to verify seriously how many people had been exposed to radiation. We made 16,000 measurements in the different villages. This data was monitored by the Los Alamos Laboratory and also by the French CEA, and were published by our agency. The conclusion that we drew was that the levels of contamination revealed by our measurements were very much inferior to those predicted from theoretical models[166].

Now, the question of the one million dollars is the following: are these predicted effects, which are not detected, are not detected but real? This is what people ask you permanently. My answer to that is the following. This is an epistemological, insoluble problem. There are not grounds for direct knowledge at this stage. We don't know.

In conclusion, our conclusion at least... either Chernobyl produced around 30 deaths at 200 sieverts, from injuries clinically, clinically attributable to radiation exposure. 2,000 avoidable thyroid cancers in children. Until now, no other confirmed, international confirmed evidence of the public health impact, directly attributable to Chernobyl exposure—and I underline—radiation exposure. If you need more information, this is my address. For your pleasure, the agency will be able to provide copies to you of all the reports that I have mentioned here today. Thank you for your attention.

The representative of the United Nations Scientific Committee on the Effects of Ionising Radiation (UNSCEAR) supports the view held by the IAEA that from

---

[165] In France, unlike in the neighbouring countries, the government did not make this recommendation to its citizens.

[166] See the results obtained by Professor Pellerin with his 8,000 dosimeter films, mentioned by B. Belbéoch: Part Two, Chapter VIII, p.155.

a radiological point of view, "generally positive prospects for the future health of most individuals *(in the areas affected by the Chernobyl accident)* should prevail". *This agency's reports provide the scientific basis on which governments establish safety and radioprotection norms.*

Gentner (UNSCEAR).—We said there, that the risk of leukaemia does not appear to be elevated even among recovery operation workers. And that there is no scientific evidence for increases either in overall cancer incidence, or in other non-malignant disorders that could be related to the accident. Dr Gonzales has said that we will never be able to show there is no effect, and that it is difficult, because of a lack of statistics, to show that an effect exists. The great majority of the population, and you saw from Dr. Gonzales' presentation the doses, need not fear serious health consequences as a result of the Chernobyl accident having occurred.

This closing slide is one I use for almost any topic. It says: "For those who believe, no explanation is necessary. For those who do not believe, no explanation is possible". That can apply to every situation each way but we hope to be in the committee on a scientific basis, not using emotion, but using [...] the most rigorous possible data so that the people and the decision-makers can get the right information.

Gentner's joke on his last slide was a cause of great hilarity to Professor Yarmonenko from Moscow. He was the one at the 1995 conference in Geneva who demanded that any speaker who tackled the subject of low level radiation on living organisms should be banned.

DURING THE BREAK
Alexei Yablokov, President of the Centre for Ecology Policy in Russia at the Academy of Sciences, opposes Putin's policy of accepting onto Russian territory radioactive waste from the nuclear industry worldwide.
We film him against the noisy background of the hall as he comments on the presentations made by the UN representatives.

Alexei Yablokov.—It was a shameless presentation lacking any objective facts. All that is understandable from the point of view of the governments. They don't want to know the truth. They want to avoid paying a lot of money to deal with the consequences. That is why any research that shows that the consequences are worse than they thought are rejected. They say the research is invalid. What worries me is that this is being said openly, that it is being presented as if their conclusions are scientific. In reality, they are making unscientific claims and conclusions. They don't put forward any arguments. They are slickly presented, but they have no foundation.

Q.—Who supported this point of view today?

—The worst was the IAEA of course, and UNSCEAR. The most reasonable, of course, is the Office for the Coordination of Humanitarian Affairs (OCHA). It's the only United Nations programme that tries to understand the truth.

—What is the name of its representative?

—Zupka.

Among the crowds of participants, we spot Solange Fernex who is introducing Vassili Nesterenko to Dusan Zupka. We push our way through.

Nesterenko.—... I brought along this presentation that I gave to the European Parliament, on 4th October. I wanted to speak to you, because we have started to monitor the contamination in children and the first map we made was, not of the territories, nor of the food, but just of the children, indicating their level of contamination...

Zupka.—Can you tell me how many children have thyroid disease following Chernobyl?

—More than 1,500 have already been operated on.

—Why don't you take the floor and tell them? That's what this conference is for. No-one's saying anything.

—There is no discussion here.

—I will speak to them. When they quote their figures, you need to interrupt and say: "No, that's not true. In Belarus alone, there are more than 1,500".

—Among adults, in Belarus, there are 7,700 people. Professor Demidchik has stated it officially...

—You know, the real problem is that at the UN we don't have access to these statistics. No-one in your government will send them to us. That's the problem.

—In that case, we will send them to you.

—We need official statistics. If we don't receive official statistics, we can't do anything at the UN.

—Maybe we can give them to you anyway. I would also like to give you Professor Bandazhevsky's presentation. You know that he was rector...

—Of course I know.

(I join in their conversation)

Q.—Are you surprised that you don't receive this information?

Zupka.—Exactly.

Q.—He has been giving out this information for years already, but he is being persecuted for it.

Zupka.—I have to tell you that we don't receive information from either Ukraine, or Belarus or Russia...

Nesterenko.—I'm going to give you Bandazhevsky's submission, showing the effects of radioactivity on the kidneys, on the eyes and on the heart

Zupka.—I know…it's very important. Have you told the IAEA about it?…

Nesterenko.—You know how it is. For the IAEA, the more nuclear power stations there are, the more money it has. They are not the place to go for help.

Zupka.—What you need to do is make your arguments and your data known, the information that you have, the academics, the scientists who live there.

Q.—They are in a very difficult situation there, the independent scientists who live there.

Zupka.—We have no data, no good data, from any of the three countries. The official data that we get, that's the only statistical material we have.

Q.—Is the material false, or is it insufficient?

M. Zupka.—Insufficient… insufficient.

Q.—They don't do any research.

In order to film simultaneously different situations and the interactions going on in the hall and on the podium, we had brought in a second film crew. On the first day, we found ourselves at a press conference where representatives of the UN agencies were responding to questions from journalists in a separate room. At the same time, Romano was filming in the main hall where Alexei Yablokov was speaking to the whole assembly.

3. THE PRESS CONFERENCE

Yury Dranchkevich (*Tovarichtch* newspaper).—I have a question for Mr Gonzales. I am a journalist from Chernobyl, I have worked for more than ten years in the exclusion zone and I have many acquaintances that I have known for a long time in the Chernobyl area. I am absolutely convinced that the consequences of the Chernobyl disaster have had a catastrophic effect on health and on mortality among the liquidators and other victims. So I have personal motives for what I am saying as well. And so this is my question. What percentage of Chernobyl's population and the liquidators have to die before the effects of the Chernobyl disaster on health are recognised by the world?

Gonzales (IAEA).—Chernobyl is a tragedy and our duty is to discover exactly what you have asked. But it is not simple because a lot of information has been lost. We have observed that the general state of health of the population in the Chernobyl territories is not good. But all the international studies have also said that this poor state of health is independent of the contaminated territories. The fundamental question is the following: are the health problems attributable to Chernobyl? In spite of all the information that we have at the international level, we have been unable to establish a correlation between radiation exposure due to Chernobyl, except in the case thyroid cancer in children. I repeat, up to now, at an international level, no correlation has been established between illness and radiation exposure, except for thyroid tumours in children.

426

Dranchkevich.—The economic situation in the whole of Ukraine has deteriorated, but it is only among the liquidators and the inhabitants of the contaminated territories that there is so much illness and death.

Gentner (UNSCEAR).—All disasters have a negative effect on health. For example, unemployment, losing a job, problems caused by alcoholism, hospitalisation, divorce, all increase the incidence of poor health indicators. Unemployment influences people's health. So an event like the Chernobyl accident and the evacuation, which is another event that can cause negative health effects, has certainly had an impact on health.

Natalia Preobrazhenskaya (*Le Monde Vert* newspaper).—I want to remind everyone that Ukraine suffered terribly in the war. In 1945 the economic situation was much harder than it is today, but there were no thyroid cancers, no leukaemia and young people did not die. So, stop repeating all this rubbish about an economic crisis…By the way, what is your area of expertise? I'm a biochemist. What is your specialism?

Gonzales.—I have worked in radioprotection since I got my degree.

Preobrazhenskaya.—What is your area of expertise, Sir? (*She is addressing Pr I. Likhtarev, who is chairing the press conference*)

Likhtarev (ICRP Ukraine).—Radioprotection…

Preobrazhenskaya.—Yes, OK. We're all philosophers here, but what is your area of expertise?

Likhtarev.—Radioprotection.

Preobrazhenskaya.—Me too, I work in radioprotection. What are you? A physicist? A biologist? An engineer? A doctor? What are you?

Gonzales.—Yes, I am a physicist.

Preobrazhenskaya.—Ah. You are a physicist. OK. So now, before I ask my question, I want to say this: you are a physicist working for the IAEA; you need to merit the money that you receive to help sustain and develop nuclear energy. Quite simply, you can't say anything different. And so, you ignore the real effects of radioactive isotopes and the consequences of the global nuclear catastrophe…

Gonzales.—Because you're talking so much, let me tell you what I wanted to say before that you prevented me from saying. I assure you every hour of every day over the last fifteen years spent on the project, I have devoted myself to the good of the people of Ukraine, Belarus and Russia. You can disagree with me, I can disagree with you, but I will not accept you casting doubt on my integrity, otherwise I will question yours. I think that you, and what you are doing, could do more harm to the people you think you're protecting, than the harm that you're supposedly protecting them from. You must realise aggression is a double edged sword: if you attack me, I'll attack you. If you want to talk to me to understand the situation better, I'm prepared to talk to you till the end of my days. I am willing to stay and talk to anyone who wants to, at the end of this press conference.

Likhtarev *(He is trying to bring the press conference to an end as it is not going well)*.—We can carry on with the press conference informally. We must talk about the significant issues. Otherwise the journalists will go away thinking that we're "still covering things up".

Fernex *(ignoring I. Likhtarev's request)*.—I also have observations to make about the IAEA, I would like to say that Chernobyl has been a catastrophe for your agency and that the construction of new nuclear power plants has been blocked by it. You are under a professional obligation to reduce the impact of the radioactivity on health to a minimum, or even less. There is conflict of interest so fundamental and total between medical research and your institution that all studies on the health aspects of Chernobyl, sponsored or presented by you in whatever form, should be banned from this conference. When there are such conflicts of interest, the IAEA's contribution should be reduced to zero... To summarise, I think that the presence of the IAEA at medical scientific meetings should not be accepted.

The whole "conversation" has been very lively, and the hapless Ukrainian interpreter has barely been able to string three words together. Likhtarev finally manages to close the debate, and invites those who wish to continue the discussion individually to do so in the corridors or café. The two sides go their separate ways— the lobby relieved to be able to stop reciting its litany and those in the opposite camp, tired of listening to it.

Meanwhile, in the main hall things are hotting up. Yablokov, his white beard moving vigorously up and down, emphasising every point he makes, is talking at a fast pace, using every second of his allotted time, in a precise and clearly aimed counter-attack. Instinctively, Romano, who does not speak Russian but has grasped the emotion behind Yablokov's words, is filming in close up, without a tripod, his camera on his shoulder, capturing every gesture of his indignant tirade.

4. CONFERENCE PRESENTATIONS

Alexey Yablokov (ecologist).—We have heard today that from the first days following the accident, the IAEA has been actively involved in the work at Chernobyl. I met the director of the IAEA, Hans Blix. This is what he had to say in the first days after the accident: "Even if there were an accident of this type every year, I would still regard nuclear power as a valid source of energy". That is the ideology with which the IAEA deals with Chernobyl.

Hans Blix is an excellent director, I know him and have nothing against him personally. But he was expressing the view of the IAEA! The apotheosis of all that was expressed just now by the scientist from UNSCEAR when he said "generally positive perspectives exist for the future health of most persons in the Chernobyl region". We heard it here, it wasn't a lapse. They present this as the conclusion of their work of the last ten years. We can't agree with this. It is a political conclusion. It is the conclusion of government representatives, who do not want to see the obvious consequences of Chernobyl.

How can UNSCEAR present these conclusions when they are based on clearly falsified government data? Professor Korblein and I tried to collect figures on infant mortality in the Russian federation. They are ridiculous. They should be thrown in the bin! The real figures, the monthly figures about infant deaths don't exist. Everything has been falsified. From top to bottom. As for the absence of up to date medical figures, let's be honest. Surely you know that the directors of the Government Committee on statistics were arrested two years ago, for falsification of data? And this is not something peculiar to this country. Look at what happened in Great Britain. This is the testimony of a British scientist: he was prohibited from publishing certain information under threat of dismissal. This is a British ecological service. Someone who was trying to tell the truth was prohibited, obstructed and dismissed. It's still going on. Half of the presentations here are about cancer. We have been shown some very beautiful images. I am amazed at the quality of the images. I would simply remind you that all the cancers in the world are monitored by the International Agency for Research on Cancer. And what does this agency say? "The data from the Republics of the former Soviet Union may underestimate the number (official) of patients who are ill". Please note that I found that quote in an UNSCEAR document. UNSCEAR knows. UNSCEAR knows it is using falsified data! And it carries on using them, in order to be able to say that the consequences of Chernobyl are not that serious!

As for perinatal mortality of newborns; it has indeed increased. We have the figures to prove it. These figures are presented in the conference materials. Why aren't we talking about it? Perinatal mortality has increased everywhere! What could have caused this increase? There is no other explanation than Chernobyl, of course.

Here is another interesting example. Yury Bandazhevsky—he is on trial as we speak; the accusation against him could land him in prison for nine years; I think this conference should send a message on this subject—Yury Bandazhevsky has shown that unexpected deaths, sudden deaths, are directly associated with the incorporation of radionuclides. If this is confirmed, it is an enormous source of data to include in the consequences of the disaster.

To say there are no genetic effects following Chernobyl is incredible: the genetic effects will be the most serious. Dozens of scientific papers, published in serious scientific journals, show that the genetic effects are serious with mutation rates that will be transmitted exponentially from generation to generation. How can one refuse to see that these congenital malformations are a consequence of Chernobyl? Of course they are! I am only quoting from a few studies, but there are dozens of others in existence. All you need to do is refute them. They exist!

Loganovsky's studies are very significant. He is a Ukrainian psychiatrist. I don't know him personally but I am very impressed by his work. It is a very important study. Schizophrenia cannot be confused with anything else. This illness has been found in people living in the Chernobyl area. Among the liquidators! You should read this study. It is illuminating! He has calculated very precisely how much of the psychiatric illnesses is due to stress and how much to the radioactivity. He says: 50 to 70% of the neuropsychological illness is due to radiation and not stress...

And here's another new syndrome. Bandazhevsky, the same Professor Yury Bandazhevsky: "syndrome of incorporated long lived radionuclides". A completely new syndrome. But we can't do anything: there it is, it exists. How can you possibly deny it? It's science. It seems to me that to refuse to talk about it is simply wrong.

I'm going to give you a summary of what is really happening to people's health in the Chernobyl territories. Increase in miscarriages. Increase in infant mortality. Increase in the number of babies born weak and ill. Increase in the number of genetic alterations and in congenital malformations. Increase in cancer. Retarded mental development, increase in psychiatric illness, changes in the immune system and hormonal imbalance, diseases of the circulatory system, etc. Abnormally slow growth in children, abnormal levels of exhaustion, delayed recovery from illness and accelerated ageing. Surely this list of illnesses has to be recognised at least!

I began by quoting Blix, the director of the IAEA, who said that we could have an accident like Chernobyl "every year". Now think about the particular breed of "specialist" who defines radiation safety norms. "Effects such as the temporary weakening of haematopoiesis, minor burns to the skin, a temporary decrease in libido, are not *too* serious, given that these effects do not last long and have no further consequences. Cataracts that cloud the normally transparent crystalline lens of the eye do not affect visual acuity". This was written by Professor Kiril Markus, one of the nuclear ministry's ideologues and one of those responsible for the material published by the Ministry. Well, I don't want my grandchildren's libido to decrease. I don't want my great grand children to lose their sight, even if it is only temporarily... We need to talk to these people. The people who work at the IAEA and the WHO are intelligent, talented people. Let's start a dialogue. They are stewing in their own juices. They don't understand what's going on, they are under pressure from the government, which... Of course, I understand quite well, if I was Minister of Health in Belarus, I would certainly be worried. Ukraine has run out of money, it's already spending 8% of its budget on the consequences of Chernobyl. Belarus spends 20%. It's clear that no-one, no government has any interest in spending any more money. It's obvious. But the rest of us, the scientists among us, need to say: "You must spend more!" If we don't say it, they will always spend less and we will all be worse off. Thank you!

Adopting an expression of deep outrage, Professor Yarmonenko moves onto the stage and takes the microphone.

Samuil Yarmonenko.—On behalf of the radiological community in Russia, I would like to apologise to the international organisations whose contributions towards helping us to mitigate the health effects of the disaster at Chernobyl have already been described in many previous presentations. I would like to respond to the completely unprofessional statements made by Yablokov, given that he is neither a radiologist nor a radiobiologist and has no expertise in the matter. We are very appreciative of the work…

Preobrazhenskaya (*from the hall*).—Who has authorised you to say this?

Yarmonenko.—… we appreciate the work being done by the international organisations and hope that it will continue with the same success in the future.

Yablokov.—I can't let that go without responding. I am not against the international organisations.
I want them to work. I know that they can do good work. There are some excellent people working there, I know many of them. I hope that they will take notice of what I have said here, and that they won't close their eyes to the real consequences of the Chernobyl disaster.

Author of a manual on human and animal radiobiology, Professor Samuil Yarmonenko is part of Professor Ilyin's team, that prevented the evacuation of a large part of the contaminated territories with his famous theory of 35 rem as an acceptable level of radiation for all: children, pregnant women, the old and the ill. In reality, the level accepted by the international authorities on radioprotection for members of the public is five times lower.

S. Yarmonenko.—I have to say that I support the objective view about the development of atomic energy, to which there are no alternatives in particular for Russia, because we cannot squander our mineral deposits forever. Can I remind you that Dimitri Mendeleev, in the notes he wrote for posterity, said that using oil for heating is like burning bank notes.
As for coal, our second most important energy resource, you must know that, leaving aside the terrible work of a miner, whose life differs little from that of a serf in bygone days, in the fifteen years that have passed since the accident at Chernobyl, there have been five times as many deaths in mining than in the nuclear industry.
The following diagram presents the estimate of the medical consequences of the Chernobyl accident, by experts from the *UNSCEAR 2000 report,* a document of the

highest international competence. These estimates make a distinction, naturally, between, on the one hand, those somatic health effects[167] that are independent of radiation, and caused by a number of harmful factors such as socioeconomic and psychological conditions, and stress, and on the other hand, truly radiological medical effects, caused by the radiation itself. What are these effects, according to UNSCEAR's data? Above all, there are no deterministic effects[168] at low levels of radiation. The only effect is radiation sickness in 134 people who took part in liquidation work in the first few days after the accident. The prognosis for all other conditions categorized according to deterministic effects is very favourable. As regards long term stochastic effects[169], we have, unfortunately, cases of thyroid cancer here. We are no longer talking about 1800 cases, but many more. According to this prognosis, the possibility of an increase in thyroid cancer is very small, but the probability of other cancers exists, and it is even smaller for hereditary illnesses.

Our society is not only waiting anxiously to see what serious consequences arise from the accident at Chernobyl, but finds itself, still today, fearing the supposedly extremely damaging health effects of low-level radiation. This is the counterpoint to my communication. There are three ways in which society's objective understanding of the effects of radiation on human health has been distorted. What are these three ways? First: we are forgetting our classical heritage, intangible truths about quantitative radiobiology including human radiobiology, in which our fellow

---

[167] Somatic: concerning the body (rather than the mind). Used here to designate physical illnesses, other than cancer. All the illnesses described by Bandazhevsky resulting from low level internal contamination by Cs-137 are somatic illnesses. The official doctrine, based on the Hiroshima experience, where the victims studied had been exposed to very high levels of external radiation, does not recognise radiation among the causes of these somatic illnesses, that have appeared in large numbers in the areas contaminated by the Chernobyl disaster.

[168] Deterministic effects: early damaging effects of radiation on living tissue (the death of an organism, damage to organs or tissue, cataracts...) that appear in general above a certain dose threshold and whose severity depends on the level of absorbed dose (definition of the IRSN report DRPH/2005–20). The correlations established by Y. Bandazhevsky between observed somatic effects and the concentration of the radionuclide caesium 137 incorporated into the body show deterministic effects of low level radioactivity. The prognosis for all categories, according to these effects, is extremely unfavourable.

[169] Stochastic effects: long-term or delayed damage from radiation (leukaemia, tumours, for example) whose severity is independent of dose and where the probability of its appearance is proportional to the dose received. It is assumed that there is no threshold below which stochastic effects appear. Stochastic effects therefore appear at lower dose levels than the levels that produce deterministic effects and can appear after a long delay (years, decades) following exposure to radiation. *(Definition IRSN)* If Bandazhevsky's research constitutes an advance in scientific knowledge and establishes a law concerning the effects of low dose radiation on vital organs and systems, the final proposition of the IRSN where it says "stochastic effects therefore appear at lower dose levels than the levels that produce deterministic effects" is no longer accurate.

countrymen were the pioneers, and ignoring the experience from all over the world in radiological medicine. Secondly: the dissemination in society, by journalists seeking to cause a sensation, of rumours that make people anxious, and are based on pseudo-scientific "theories" and "discoveries"—I use quotation marks, because neither one nor the other exist in reality. Not to mention the various propaganda techniques used by populists and politicians, with pseudo-humanitarian aims. Finally, thirdly: incorrect information communicated between the scientific world, government administration and society about the medical effects of the radiological accident, and consequently—this is very important—the way to alleviate them.

There are four errors in the way low dose radiation is estimated.

The first is the *overestimation of different events at the molecular, biochemical and even cellular level*[170]. I would like it to be noted straight away that I have never wanted nor want now, in any way, to diminish the significance of this research. It is very important. It may be useful at some time in the future. But today, there are insufficient grounds on which to work and they cannot be used in the field of human health.

The second: *when the consequences of hypothetical and unfounded deductions are substituted for facts.*

The third: *these are deductions made on the basis of research undertaken in uncontrolled or inadequately controlled conditions that support results that contradict the findings elsewhere in the world.* We need to understand quite simply that radiobiology is an experimental science. If there is no data, nothing can be affirmed. If the data obtained does not correspond with classical data, there is an obligation to explain it. If you find the cause, then this constitutes data. Tangible data. This corresponds perfectly with Pavlov's testament that "Facts are the air of the scientists".

And finally, the fourth: *information that is notoriously false or from unqualified sources.* All information supplied to governments, to the media and to the public, should come from scientists, professionals. If instead of scientists, we have pseudo-scientists giving out whatever information, wherever and however, it will give rise to rumours that create panic. This even applies to governments, who, understanding nothing of the subject, then take the wrong decision.

Here is an example of this kind of absurd tittle-tattle: "In the years that have passed since the Chernobyl accident, more than 16,000 liquidators have died". This is what Sergei Chaigou said, standing in a cemetery. He knows nothing, of course, someone just handed him these figures, but even so, he is the Minister for Emergency Situations.

One can understand, in this situation, people's terror. They believe neither the government, the scientists, the doctors nor any other experts so they fall prey to

---

[170] See "2. Effects of proximity", Part One, Chapter V, p. 54.

rumours about the genocide of hundreds of thousands of people through radiation sickness, and widespread epidemics of cancer and hereditary diseases.

To ask scientists to demonstrate the absence of these effects is not justified, because fundamentally science cannot prove a negative, whatever it might be, even if it can be postulated. The problem is the reverse. It is a question exclusively of demonstrating the presence of effects. Until it has been demonstrated, the effect cannot be recognised.

It has been demonstrated[171]. The demonstration exists, but the national and international bodies with responsibility for people's health are doing everything they can to prevent the demonstration being presented. And when it is presented— Bandazhevsky's research was presented at the conference in Kiev—they refuse to discuss it. Barricaded behind this thick wall of deceit by omission and censorship, Professor Yarmonenko allows himself the luxury of haughty pronouncements like the truism in the sentence above.

Buzunov.—(Scientific Centre of Radiological Medicine, Kiev) (*He addresses Professor Yarmonenko*)—Samuil Petrovitch, can I tell you what I think of your observations? In 1988, mortality among the liquidators was 0.95‰. Please note 0.95‰, one in a thousand. In 1998, 10.5‰...more than 10‰. In other words an enormous rate of increase over that short period. Here, the comparison with the rest of the population is not justified; 80% of the liquidators in 1986 were young lads in good health! How many of them are in good health and able to work now? You tell me. People in perfect health were dead within ten years. That's what we should be investigating.

Ivanov (Chief Medical Officer of the Russian Federation) .—More than 200,000 Russians were involved in the liquidation work following the accident, and most of them received doses of radiation from 50 mSv up to 250 mSv. Today, this register brings together information about more than 570,000 people, of whom 184,175 are liquidators and 336,309 are inhabitants of Bryansk, Kaluga, Tula and Orel, living in territories that are contaminated at levels above 5 Ci/Km2. What does an analysis of the basic health indicators of the liquidators reveal? The proportion of liquidators in good health has changed over a period of fifteen years from 95% in 1986 to 4% in 1998–1999. Almost 75% of the liquidators suffer from several chronic illnesses. The total number of invalids among the cohort of

---

[171] Y.I. Bandazhevsky, *Swiss Medical Weekly*, 2003, No 133, p.488–490; G.S. Bandazhevskaya *et al.*, *Swiss Medical Weekly*, 2003, No 134, p.725–729; Y.I. Bandazhevsky and G.S. Bandazhevskaya, *Cardiomyopathy of caesium-137, Cardinale,* Tome XV, No 8, October 2003.

184,175 registered Russian liquidators is over 50,000 people: nearly 30% of the total. The total number of deaths among the liquidators is 15,000 people, nearly 10% of the total number of liquidators registered in Russia. Among the liquidators, we note a predominance of respiratory tract diseases, diseases of the nervous system and sense organs, diseases of the circulatory system, digestive system, the muscles, the bones, conjunctive tissue, diseases of the endocrine system and metabolic alterations.

Babadzhanov (Tashkent, Uzbekistan).—Here are the changes in rates of illness among the liquidators ten years after they received the excess radiation in comparison with five years after. Firstly, there is a statistically significant increase in morbidity between the fifth year and the tenth. Secondly, the clinical progression of illnesses becomes more intensive. Functional illness diagnosed in the fifth year, has become organic disease during the course of the second decade. Neurocirculatory dystonia has become hypertonic illness, cardiac ischemia, circulatory encephalopathy: duodenal and chronic gastritis have transformed into duodenal and gastric ulcers. As our research shows, the liquidators suffer a series of different illnesses, or polymorbidity, long after the excess radiation received. Thus 69% of the liquidators suffer from four or five illnesses. Naturally, this leads to the loss of the ability to work and their invalidity.

All in all, ten years after radiological exposure, nearly three quarters of the 10,000 liquidators living in Uzbekistan were invalids and 500 had died. We can predict with a high level of probability that the same progression towards increasing ill health and that the same evolution towards a nosological spectrum in their illnesses will continue into the future.

Alarm bells about Chernobyl have been ringing for fifteen years across the planet. It is not only a requiem for the victims of the disaster who have already lost their lives there. It is a reminder to learn the lessons from Chernobyl, the sign that Chernobyl is continuing. We must not forget it, if we want to protect people from the nuclear nightmare. The most important lesson, in our opinion, is that the effects of low dose radiation represent a much more real and present threat than the larger atomic cataclysms.

It is essential that we take new measures to prevent the progression of the consequences to the additional radiation exposure to which people have already been subjected.

G. Rumiantseva (Centre for Social and Judicial Psychiatry, in Serbsky, Moscow).—I would like to move away a little from the problem that we have been discussing here for two days and focus on two postulates.

The accident at Chernobyl has been called a catastrophe. A radiological catastrophe in terms of its biological content, but also, a "catastrophe" with the same factors as the war in Chechnya, or in the Afghanistan syndrome, the Vietnam war, or the inhabitants of areas that have been victim to an enormous ecological catastrophe like Bhopal. All these catastrophes have one thing in common: the catastrophe itself. Human beings are not only enduring the fact of being subjected to the effects of the radiation, but they endure the whole catastrophe in its entirety. It is simply impossible to divide the effects into different categories, psychological, chemical and physical. We need to study the consequences of the catastrophe as a whole. The second factor, which is fundamental, is that man is a being equipped with a cerebral cortex. He cannot be considered simply as an amalgamation of radiosensitive organs. He does not simply suffer the effects of radiation, he assimilates it into his conscious mind and thinks about it. Without the conscious mind, the consequences appearing today would not be there.

As for the liquidators, they can exhibit a range of symptoms with varying degrees of severity. The fundamental characteristic appears to be an underlying asthenia.[172] Changes in the behaviour of the liquidators often manifest themselves in the so-called "autistic asthenia".

In the end, we cannot absolutely deny the emergence among a majority of the liquidators of overestimation. It is even perhaps a collective symptom. Overestimation as a form of hypochondria, in fighting for one's rights. You must be aware of the number of court cases, of deviant behaviour in the form of demonstrations, hunger strikes… This sort of behaviour is widespread among the liquidators. Also the accident itself has become mythologised which equally has an effect on the psyche.

G. Sushkevich (WHO, Geneva, Vice-President of the Conference Organisation Committee).—These questions are linked to the psychological rehabilitation of the liquidators. It is just one aspect of the work undertaken by psychologists with the liquidators. I think we need to work with society as well and change people's perceptions a little, so that they begin to see the liquidators not as victims but as heroes. So that, the children of liquidators will feel, right from childhood, that they are the children of heroes and not of victims. In this psychological work on society and in the healing of society and of the liquidators, psychologists and the media could play an important role. Fifteen years have gone by since the accident. The liquidators should be regarded as heroes, heroes of a patriotic war. The children of war heroes were always taught to feel proud and to have a sense of dignity. We need to remember the words of Pirogov, when he said that the wounds of the victors heal quicker than the wounds of the vanquished. This principle could be adapted to the psychological rehabilitation of the liquidators.

---

[172] Asthenia: lack of strength, of physical and mental vitality; a state of depression, of weakness.

Sushkevich duty, as a highly placed WHO official, was to encourage research and medical aid for the liquidators rather than to give a funeral eulogy for those whose heroic lives have been destroyed.

Angelina Guskova (Institute of Biophysics, Moscow).—I would like to say something on this subject. The specific nature of radioactivity is that we have no sense receptors that can feel it or measure it inside our bodies. It is very important to us that we can distinguish between useful heat and dangerous heat, useful cold and dangerous cold. This internal measure does not exist in our bodies for radiation. And so the main way we perceive radiation is through our information systems. And these information systems create conditions for the mythologizing of the degree of danger. This has been described very well by the social psychologists today. I thoroughly approve of their work and I believe today that for the situation to return to normal we do not need to lower the dose of radioactivity which is already very low but rather to rebuild social structures, employment, medical aid for those, whose illness is not caused by radiation but in whom illness has appeared. I think that this is the most important and useful approach for them.

In a chemical accident, the effects are easier to diagnose clinically. Bhopal not only caused a colossal number of victims but also a great many cases of abnormalities and serious illness. In other words, the gravity of the accident was demonstrated by the concrete existence of serious illnesses. With radioactivity, we are talking about the potential effects in future generations. We have not found this to be the case at Hiroshima, where no genetic effects have been found. We need to monitor the situation, but for the moment we do not have many reasons to think there are immediate effects, and even fewer reasons to think there may be effects in the next twenty five generations.

An anonymous woman from the back of the hall asks to speak.

Woman.—The doses recorded in the liquidators' documents do not reflect the real doses that they received. When the dosimeter is calibrated to measure a maximum of 2 roentgens, we wrote 2 roentgens on their papers. Often the fire chief had a collective dosimeter. In those cases, they would average out the doses received by the whole brigade. When the maximum admissible dose was set at 25, they marked 24.8 or 24.9...if it was set at 15, they marked 14.8 or 14.9...

The methods that we use when we study changes in health in these people are objective. I see these people from every corner of Russia. They could not have agreed amongst themselves, yet they all have the same complaints: headaches, pain in their bones. A neurosis cannot appear in so similar a fashion in everyone, not a mosaic, but a precisely drawn syndrome. We are not talking here about a direct radiological lesion to the brain, Angelina Guskova. We respect, read and honour the work you do, but we are dealing here with completely new data. It requires further investigation. Yes, we have our own experience, particular dogmas from the past but we should not rely on these alone. We are talking here about long term effects that do not result from "radiation sickness". Long term effects preceded moreover by a latency period. They do not appear straight away but one, two, three, five years later. There is a kind of mechanism that is set in motion. Today we are finding changes in the immune system. We are even finding changes in the liquidators' children. They have psychogenetic alterations. These children are developing a type of immunodeficiency. It is our genetic heritage that we are concerned with here, and we absolutely have to protect it and restore it.

The president asks if one of the psychiatrists can give their opinion.

Rumiantseva.—The men involved in this accident who received a dose…(*she corrects herself*) We will leave aside the question of your dose, we are looking at the perception of this dose. A perceived dose can often produce the same effect as the biological dose. And it is not by chance that we end up at this point because prolonged stress and prolonged biological action have the same effect.

In the same year, 1986, when I was working at Chernobyl, there was the *Nakhimov* boat accident, do you remember? The psychological trauma there was terrible. And the victims of that traumatic event are still suffering today. Nobody could deny that what those people are suffering today is due to the shock they received fifteen years ago. Why should we deny that what is happening with our liquidators also is due to serious psychological trauma, *of which the radiation dose they received is certainly not the principal element?* I am not going to discuss it, I am not a radiobiologist, but I can confirm that if I have a liquidator in front of me, and their clinical record does not mention this psychological trauma and at the end of fifteen years, it has become a severe psycho-physical syndrome, the logical conclusion for me would be that this man's illness was linked to this trauma.

Anatoli Saragovets was able to crack jokes with his dog while his living body decomposed. Piotr Shashkov decided to be optimistic, even though the flesh was coming away from the bone on his leg (the one that had been pressed against the side of the tank he was driving). Victor Kulikovsky, crippled with illness, put his own troubles aside and thought only of his son, who had been born with a birth defect, caused by his father's psychological trauma, according to Rumiantseva's theory… These healthy young men did not realise that, in 1986, the real catastrophe was not the smoking ruins of the reactor, vomiting out its tons of radionuclides, but the fragility of their minds. The dose they received was certainly not the principal element that caused their illness, claims Rumiantseva!

Preobrazhenskaya.—Can I make a brief comment?

Zupka (*he is chairing the meeting*).—Briefly, if you don't mind, we haven't got much time.

Preobrazhenskaya.—I am extremely concerned and alarmed that an eminent scientist such as Mme Guskova keeps repeating that it was "fifteen years ago and they were very low doses" But she doesn't mention the fact that these doses accumulate. You do not say that you are worried about future generations. You don't want to know. But we already know the future generations. Because the children who were irradiated have become mothers and fathers, and we already have their descendants. But the most important thing, Dr Guskova, is that we cannot go on comparing the tragedy, the terrible tragedy of Hiroshima and Nagasaki, to Chernobyl. Because there, the chain reaction has come to a halt, whereas here it is still going on: plutonium lasts 24,000 years, americium has appeared here and the radiation continues forever, as does the internal radiation. Strontium dissolves in water and we are irrigating our fields all along the Dnieper river—don't interrupt me!—with this water. Radionuclides are taken up by the plants, the cows eat them. This explains precisely the slow accumulation of low doses… Think about it and don't go along with the IAEA when they cover up the tragedy. We don't want demonstrations, we don't want to frighten people, we want a scrupulous approach, international, communal.

BREAK

Standing near the podium, Professor Yarmonenko is talking to a member of congress not far from our microphone.

Yarmonenko.—You have eaten, and you evacuate. Everything you have eaten today, has been eliminated, including nuclides… All the rest of it, it's just rubbish. You need to understand that!

Member of congress.—No, you're right, but you need to look at the balance that is established from ingestion on a daily basis and what is evacuated.

Yarmonenko.—What balance? The balance is in one direction only: it's e-va-cua-ted! It doesn't increase, it decreases. This is how the world works!

The organisers announce a round table discussion after the break. I go up onto the stage to ask how the discussion will take place in the hall so that I can organise filming around it. Yarmonenko is up on the stage also talking to Madame Nyagu, the president of the organising committee, because he is worried about the round table discussion. Romano sees us and joins us quickly on the stage. As he passes by he films their conversation. We owe this little masterpiece to the instant reflex, presence of mind and skill of the cameraman. Capturing a unique moment such as this is only possible thanks to the creative equality and mutual understanding between the members of the crew.

Yarmonenko.—It's a disaster, there's no doubt about it. But not from a radiological point of view.

Nyagu.—Samuil Petrovich, calm down! You are like Angelina Konstantinovna, an unquiet spirit.

—How can I calm down? I'm going to have to rewrite the next edition of my book, my manual, and it's a shambles.

—You need to understand that there is going to be a great deal of agitation now with the final resolution. We will work on it again with you tonight.

—Willingly.

—Even the IAEA recognises that we need to work around the difficult issues but you, you shoot on sight.

—That's true. It is good to be diplomatic, but ignorance has to be excluded. We can only accept expert opinions.

Q.—So, according to you, it wasn't a radiological accident?

—Radioactivity is one factor among many. But it is the least significant.

—Really!

—Who is this?

—A television journalist...

—What? Television?

—Swiss television.

—Ah, that's nothing. Swiss television is like ours.

A little later, we come across Yarmonenko under siege in the centre of the hall from a group of women doctors, to whom he is giving a science lesson about radiation. Other conference members are curious and have gathered round to observe the scene.

A woman doctor.—We do not know what effect these low doses of radiation will have, that constantly...

Yarmonenko.—(*cutting her off in a professorial manner*) Why do you say "We don't know"? How can you say "We don't know"? We know more about radioactivity than about any other environmental factor. Why? Because we have been studying radioactivity for a hundred years already and it's just one factor. With other factors, like toxic substances for instance, there are hundreds. To say we don't know when we know everything about radioactivity! Chernobyl has not taught us anything new. It may have taught something to people who don't know anything about it... and who are starting to learn now.

—This is a unique situation, different from all others.

—Absolutely not.

—Chronic and permanent.

—Absolutely not! ... Absolutely not!!

Second woman doctor.—No, listen, we have presented our reports in Moscow. We are conducting long term experiments, we are introducing low doses of caesium...

—On who?

—On animals, rats...

—So what? I've been doing these experiments all my life! Rats, alright!

—600 becquerels, is that a high or a low dose?

—What becquerels? Give me the incorporated dose...Have you calculated that?

—Incorporated dose: 0.3 centigrey[173].

—That's a low dose. For rats, yes, it's tiny.

—A tiny dose.

—Yes, tiny. Not acute, not chronic...

—Are alterations possible?

—No, there shouldn't be! ... And if you have found any, you need to verify it ten times. Because if this is what you are claiming, it contradicts the existing scientific conventions. You need to understand this once and for all.

—What must we "understand", if this is what we are "seeing"?

—Look at it again ten times.

—We have looked at it ten times.

—No, you haven't. Look me in the eyes.

—I'm looking at you.

—I'm telling you, you haven't looked at it ten times.

—This is research done with an electronic microscope. We have published; our research has been published...

—That has no importance.

—...and we ask that someone refutes it.

—Why refute it?

—Why has no scientist come forward to refute it, I can't understand it!

—No. Listen. Listen to me. Okkam's principle—listen to me!—is the only scientific principle here. This is what it says: we can never reject anything, at whatever level of knowledge. It isn't possible.

—No, of course not.

—But, if we claim that something is so, we must be able to demonstrate that it exists.—The value of that demonstration comes down to two things. The first is that your findings can be reproduced by any other person, and initially by you, and the second is that they should not contradict a large body of previous findings. If they do, you need to try and understand yourself why they contradict it.

---

[173] See glossary, page 599.

—We are talking about internal radiation.

—It's not important whether it's internal or external.

—I have analysed an enormous amount of material published on the effects of low dose radiation. Chronic low dose radiation is something else altogether.

—What difference does that make? There is less effect.

—I think that if these low level doses are incorporated chronically, they represent a danger.

—But have you studied my course book or not? I'm sorry but you can just go to hell! It is written on every page that because there are molecular mechanisms that repair DNA, it repairs much more easily with a low dose of radiation than with an acute dose. We have established norms about the effects of acute radiation and the effects of chronic radiation: chronic radiation has been shown to be three or four times less dangerous, according to different of indices. But you, you're saying the exact opposite. It's appalling! There is a whole repair mechanism that has evolved…All this talk about greater harm at low doses—that Burlakova is explaining—she's not explaining, she's rambling—it's absurd.

The third woman doctor.—How do you explain the thyroid disease then? How? It's on the increase.

—But it's not increasing at all. We all understand that the dosimetry at Chernobyl was faulty. We should not forget that. All that is true. But there is the clinical evidence.

—That's why we shouldn't be talking about dose, but simply about illness.

—There you are, you have been shown today in a series of talks with excellent data that an elementary cure in good conditions, at St Petersburg, gets rid of these symptoms within a month. What are we talking about?

—It's temporary. It is a temporary improvement. After a year, it starts again…

—Have you been there? Where have you heard this?

—I have been working for fifteen years on this. After a year, it starts again.

The first woman doctor.—You talk about repair mechanisms. But the repair processes are not infinite.

—What do you mean, not infinite! You can't say that! Can't you read? You said you've read my material. Go and read the chapter about repair mechanisms again and come and see me tomorrow, we will talk about it again. What you are saying at the moment has no basis in reality. We are going to adopt the resolution, I insist categorically on educating people. This incredible ignorance and lack of training in radiological matters is inadmissible in doctors. If, God help us, something else happened tomorrow, we would need to know what to do. There is no guarantee that it won't happen. The issue of radioprotection is very important.

Preobrazhenskaya.—We are going to hear from Professor Grodzinsky in a minute, he's going to talk about low dose radiation too. Alice Stewart, a well known American scientist, also talks about it, an excellent scientist, she has also studied it...

Yarmonenko.—It's not important, Stewart or no Stewart...

Preobrazhenskaya.—Yes, Yarmonenko or no Yarmonenko, it's not important... What is important is that children are dying because of low dose radiation.

THE ROUND TABLE

Mikhail Savkin (ICRP vice-director of the Scientific Centre at the Institute of Biophysics in Moscow).—This is a scientific conference essentially. To this effect, we would like specialists, if they are specialists, to talk about their own field of expertise, rather than recounting half baked theories about radioactive wood that we have heard here. We have known for a very long time, it has been shown—and the member of the Ukrainian Centre for Radioprotection must know this—that the part played by internal radiation is insignificant in comparison to external. People have been terrorised for fifteen years...

Preobrazhenskaya.—The smoke, does that mean anything to you? What is smoke?

Savkin.—I've finished.

Solange Fernex (Honorary Deputy of the European Parliament, Solange Fernex was at the time of the conference, president of the Women's International League for Peace and Freedom, French section).—I totally agree with the professor from Moscow who thinks the experts must speak about their area of expertise, and that health is really a matter for doctors. This is the motive for our great determination to amend the agreement between the World Health Organization and the International Agency for Atomic Energy. If we want an informed public opinion and a good psychological response to accidents, we must have transparency. We must eliminate conflicts of interest and make sure that the people who talk about genetics are the geneticists, nuclear physicists talk about the nuclear industry, about radiation and not about the effects on the cell, on the pancreas, on mitochondria, on DNA. That is the task of doctors and we must listen to them.

Fifteen years after Chernobyl, the scientific world, as well as the United Nations, are profoundly divided on the health consequences of the Chernobyl disaster.

According to UNSCEAR, there are "so far, no increase in congenital malformations, stillbirths, or premature births that can be attributed to radiation exposures caused by the accident; no overall increase in the incidence of cancer that can be attributed to radiation has been observed; the risk of leukaemia, one of the most sensitive indicators of radiation exposure, has not increased, even in the workers responsible for the clean up (liquidators) or in children...

From the radiological point of view and based on the assessments of this Annex, generally positive prospects for the future health of most individuals should prevail…"[174]

In contrast, according to the OCHA report on Chernobyl, a report described by UNSCEAR as "lacking any scientific basis" and liable to provoke "unnecessary panic", the situation is extremely serious and requires immediate help from the international community. In his preface, Kofi Annan, Secretary-General of the United Nations, writes: "[…] The exact number of victims may never be known. But three million children require physical treatment, and not until 2016, at the earliest, will we know the true number of those likely to develop serious medical conditions. The most vulnerable victims were, in fact, young children and babies, unborn at the moment when the reactor exploded. Their adulthood—now fast approaching—is likely to be blighted by that moment, as their childhood has been. Many will die prematurely. Are we to let them live and die, believing the world indifferent to their plight?"

The UNSCEAR report was severely criticised by the delegates from Ukraine and Belarus, during the United Nations General Assembly. Nevertheless, it was adopted without a vote on 8th December 2000.

The UNSCEAR delegate, Dr Gentner, has explained that his organisation only employs reputable, *"peer reviewed"* studies, which means "validated by their peers" All other research is dismissed. According to its own terms, these "peers" are scientists from the Los Alamos laboratories and from the French Atomic Energy Commission.

This shows clearly that the scientific research is dominated by conflicts of interest. No-one is allowed to question the dogma according to which ionising radiation is harmless. It is pronuclear groups of researchers, those who develop the bomb or those with commercial interests in nuclear energy, who decide on behalf of the international community, and against all the evidence to the contrary, that Chernobyl offers "a generally positive outcome from a radiological point of view".

This has serious socio-political consequences for decision makers, for doctors, for the future of patients, and for researchers.

How can we stop the "donor fatigue", deplored by the UN representative, as long as UNSCEAR continues to insist that the only consequences have been 1,800 thyroid cancers in children and adolescents, and nothing more? If the future should be considered as "generally positive", why spend contributors' money mitigating non-existent illness and risk? How dare Belarus, Ukraine and Russia continue to solicit funds from donor countries? This was the question posed by Dr L.E. Holm

---

[174] 2000 *Report to the General Assembly*, annexe J, "Exposures and Effects of the Chernobyl Accident", paragraphs 383, 413, 421, United Nations, New York, 2000.

(UNSCEAR) to Kofi Annan, in a letter in which he said that their complaints had no scientific basis and were motivated by economic considerations. This reversal of the facts and blaming of the victims, for claiming unnecessary financial aid, is at the very least, surprising.

Thankfully, there are devoted scientists and doctors who ignore UNSCEAR's claims and work for the victims' radioprotection in the three worst affected countries, and elsewhere in the world. These researchers devote their lives to mitigating the painful consequences of radiation. However, they find it impossible to publish material about what they see on a daily basis, and inform public opinion about it.

Professor Bandazhevsky, who has presented five *abstracts* to this conference, has undertaken hundreds of autopsies of children and adults who had lived in the contaminated territories. Using direct radiometric measurements, he has shown that different organs of the body have different concentrations of caesium-137. He has demonstrated a correlation between tissue concentration and organic damage. He has described, following experiments in animals and humans, anatomic and histological lesions in the organs where caesium-137 accumulates.

In 1997, when I visited his Institute of Pathology, he had prepared, for a Parliamentary Commission from Minsk, a collection of foetal malformations and still births, collected over the two preceding weeks. The number corresponded to what one would normally have observed over a whole year.

Quite obviously, these results were completely incompatible with the dogma propagated by UNSCEAR: "So far, no increase in birth defects, congenital malformations, stillbirths, or premature births could be associated with radiation exposure following the accident. (paragraph 383) "

Bandazhevsky's research needed urgently to be stopped. Like Galileo, he needed to recant or be silenced. Given his refusal to conform to UNSCEARS's dogma, his imprisonment was the next logical step.

As Bandazhevsky had proved, no organ or system in the body is exempt from caesium-137. That means that the nervous, digestive, immune, hormonal, reproductive, excretory, cardiovascular systems all need to be studied, in relation to concentration of caesium-137, and the results published.

This will inevitably change not only our view of the dominant risk model based on Hiroshima, which has no relevance for Chernobyl, but also the international radioprotection norms. On 26th April 2001, the European Parliament adopted a resolution on Chernobyl, "calling on the international organisations to re-examine the risk model".

We need to provide clean food and a clean environment to the populations concerned, in particular the children, and give them adsorbent food additives to reduce the concentration of caesium-137 in their organisms.

Theoretical calculations based on reconstructions, incorrect dose and inadequate risk models (Hiroshima) are no longer acceptable.

The era in which the Holy Inquisition persecuted Galileo for his scientific discoveries is in the past. The truth about the health consequences of Chernobyl can no longer be covered up, as was the fact that the earth revolved round the sun during the Renaissance. Science should no longer be muzzled by the commercial interests of the promoters of nuclear electricity. As citizens of the 21st century, it is our responsibility.

We owe it to the victims of the disaster to prevent another Chernobyl.

Artificial sources of radiation need to be isolated and the whole nuclear cycle, including nuclear power stations should be stopped if we want life on earth to continue.

Chris Busby, the next speaker, is a member of the British government committee researching the risks of incorporated radionuclides. Scientific secretary of the European Committee on Radiation Risk (ECRR), he has coordinated the publication of recommendations on "The Health Effects of Exposure to Low Doses of Ionising Radiation"[175]

Chris Busby.—I must say that I am a bit confused by what I have learned so far at this conference, because it seems to me that the outcome of the Chernobyl accident in Belarus and the Ukraine is actually very little, and seems to me that there has been a much worse situation in my own country Wales and Scotland and also in other parts of the world where the doses were much less.

If you look at the increase in cases of leukaemia and cancers near nuclear sites, this is called inductive philosophy, and what this suggests is that you try to see what is common between all of these situations and what is common is low level radiation exposure, from internal radiation, not external, internal radiation. In order to try to tease out what the real answer is, we decided to look at some results that were produced in a number of countries relating to infant leukaemia, these were leukaemias of children who were in the womb at the time of the Chernobyl fallout. You can see that in Scotland, in Greece, in America, in Germany and also our own result from the United Kingdom that there was a sharp increase, a statistically significant increase in infant leukaemia of those children who were in the womb [...].

We can say without a shadow of a doubt that this is a real effect, it could not have occurred by chance. Because it occurred in a category of people, a group of people who were in the womb at the time of Chernobyl, we also believe that it could not have occurred by any other reason, because you cannot have population mixing in the womb and chemical effects are unlikely to have occurred in that period

---

[175] ECRR 2010, Regulators' Edition at *http://www.euradcom.org/2011/ecrr2010.pdf*. See page 60

that would have been different from periods before and afterward it [...] I myself presented a paper with Molly Scott relating to an increase in infant leukaemia in 5 different countries of Europe and also the United States and this was certainly and unambiguously an effect of Chernobyl[...]

Professor Rose Goncharova, member of the Institute of Genetics at the Academy of Sciences of Belarus, has studied genetic anomalies, in fish and rodents, that are worsening from generation to generation in areas with relatively low levels of caesium-137 contamination, 200 km from Chernobyl.

Rose Goncharova.—Nobody will contest what our esteemed Professor Yarmonenko says about the intangibility of the postulates and foundations of a certain scientific discipline. I am talking about the field of radiobiology. But there is another side of the coin, and that is the acquisition of new knowledge. As we know, there have not been any major problems in genetics since the central tenet of genetics, the transmission of DNA information in protein, was refuted. Put simply, new knowledge, completely new knowledge, was acquired and genetics continued to develop. In this context, radiobiology, radiological genetics, is no different from other sciences. Completely new data, published today by scientists from the Research Foundation on the Effects of Radiation in Japan, are now available[176]. It is a first publication, in 1999, about the existence of somatic morbidity caused by radiation.

So, one of the previously uncontested paradigms of biological radiology has been refuted. Of course, you need to understand...

Yarmonenko.—What paradigm has been refuted?

Goncharova.—That there is no radiation-induced somatic morbidity.

The chairperson.—Please, Professor Goncharova, if you don't mind, we are not discussing the paradigms of molecular biology of genetics, we are discussing socio-psychological consequences.

Goncharova.—I'm talking precisely about the socio-psychological aspects of the acquisition of new knowledge to find out to what extent it is understood by scientists. It is clear that if it is hard to understand and if the new knowledge is not taken into consideration when analysing this or that effect of Chernobyl, this will have social consequences. It will also have very particular consequences for scientific researchers. And then, I need only mention that the last publication of Peers and Preston[177]—anyone who works in radioprotection well be completely

---

[176] Y. Shimitsu, D.A. Pierce, D.I. Preston, K. Mabuchi, "Studies of the mortality of atomic bomb survivors, report 12, part II. Non cancer mortality: 1950–1990", Radiat. Res., 1999, 152 (4), p.374–389.

[177] *Ibid.*

familiar with these two names—shows that there are statistically significant effects within the 0–100 mSv range.

Yarmonenko.—It's appalling!

Chairperson.—Would you be so kind…

Goncharova.—I've nearly finished, please allow me. Don't interrupt me!

Voice from the hall.—Let her carry on! Let her speak!

Goncharova.—Dear Roxana Garnets, I will finish my last sentence. Thus I think that Yarmonenko and his supporters sooner or later we will accept this new information that we have, and then we will be able to draw new conclusions and disseminate this new teaching material. Thank you for your attention.

Rose Goncharova passes in front of Yarmonenko who looks at her and says furiously:

Yarmonenko.—It isn't new information. Aren't you ashamed? This is rubbish. (*He demands the microphone*) Today, fifteen years after the accident, it is perfectly clear that the factor of radiation that occurred and still persists is nevertheless not the predominant factor. Whatever caused the acute effects is over. It still plays a part, not by itself, but in combination with a large number of other social factors. In this connection, the socio-psychologists have suggested that we replace this ridiculous term "radiophobia" with "radiological anxiety", which is excellent. The existence of radiation in itself causes anxiety to humanity and to the victims. But more important, obviously, is its combination with this enormous quantity of other social factors. This is the correct starting point for rehabilitation measures. There is no need today, when everything is over, to be thinking about eliminating this or that isotope, battling against this or that illness, caused by radiation. No!

Preobrazhenskaya (addressing A. Guskova).—You are a highly qualified specialist and you know the effects of radiation, don't you think that society, the planet, all countries, all of us doctors, should say no to nuclear energy, because it will be the end of life on earth? We won't have anywhere to bury radioactive waste.

Guskova.—I feel sorry for you. I am very worried about you. New reactors should have been built here, where the environment has already deteriorated ecologically. You have such an urgent need for energy in Ukraine, your lives are so difficult, I worry enormously about you. This should not prevent this technology from being managed with enormous caution, with appropriately trained staff, responsible, prepared, because they hold in their hands not only their own destiny but also that of the people. I firmly believe that humanity will not turn its back on atomic energy. It will simply make it safer.

Preobrazhenskaya.—It's against nature.

The chair of the session, Roxana Garnetz (UN Chernobyl programme) announces a fifteen minute break and makes a promise.

"The discussion has been recorded. We will include the material from this round table in the conference materials and we will publish them".

This promise was not kept: neither the round table discussion nor the proceedings of the conference were published. The same scenario as 1995.

BREAK

I ask Norman Gentner from UNSCEAR if we can film his conversation with Professor Michel Fernex. He agrees.

Fernex.—We have followed the conference and we see that according to your organisation there were 31 deaths, of which 28 were due to radiation, 200 irradiated people and 1,800 thyroid tumours attributable to Chernobyl. Is this still your opinion?

Gentner.—That's what comes out of the data. I am upset that more thyroid cancers are predicted; fortunately the percentage who survive is high. We expect some increase in the incidence of cancer, among certain categories who received the highest dose, and that could be statistically demonstrable.

—You continue to confirm only the figures provided for acute radiation. You have not been influenced for example, by what has been said today?

—If you look at the symptoms and the internationally established progression of the illnesses, these are the figures on which all the international organisations are in agreement and that are supported by the data. Claims have been made by people who have suffered ill effects or believe they have suffered ill effects, but the sources to which we refer are the health services of the Member States. We try to collect the most complete and the most validated data possible. Whatever the result of this data indicates, if it is scientifically supported, we will make sure it is given the widest possible distribution.

—We are very concerned about the fact that nobody in your group is working on the effects of other radionuclides.

—They are always a concern. We are convinced that the unifying principle is the dose, so we look for what the dose is and where it comes from. The majority of the dose comes from radiocaesium. Other particularly poisonous radionuclides, or acting on particular people, have in certain circumstances, the potential to cause serious irradiation.

—In your opinion, where, in which pathology, is caesium concentrated at the highest levels?

—The major problem with caesium is with external radiation. Caesium is also absorbed by vegetables and Annex J of the UNSCEAR report contains some discussions on this subject.

—Don't you think that during the last 15 years the majority of caesium is internal?

—No, it's not internal, are you talking about the external exposure that people receive?

—I am speaking of the internal one, which they receive through food in very large populations. In regions following the absorption of food they are cultivating or even through products from the forest.

—The models that were used at that time to measure the exposure, in that early period, suggested that internal and external exposures were more or less equal in importance.

—Don't you think that today it's 90% internal and 10% external?

—I am not a specialist in dosimetry.

—Don't you have whole body measurements of people?

—We have these measurements and they suggest that levels of exposure for the majority of people are very low. But in the case of caesium, the way the exposure has taken place has no importance.

—In the Gomel region, it is still very low in children, isn't it?

—Do you want to see the validated data? You can look at it.

—Who has the validated data? In whose possession is this validated data? Where are the validated data?

—I can take you to Dr. Kenigsberg, deputy director of the Belarus government register, for example. You can talk to him, he's an expert. We can go there right now.

—Yes, I know Dr Kenigsberg and I have some idea of his measurements. There are other measurements, they have been done, and are still being done, and it's shocking to see how high the values are today among children.

—We have to see what the data says, but I refuse to think that whether a radiation dose is internal or external... what counts is the dose received by whatever mechanism and to prey on people's consciousness to say that somehow because it's internal it's worse does a disservice to the people.

—There are cardiac diseases found in humans which may be lethal, found in children which may lead to sudden death.

—OK but we know the complications, these things are arising but to simply say that these things have occurred following the accident and to infer from that a blind acceptance that they are radiation related doesn't allow the public health authorities of these countries to serve their people.

—They are dose related in children.

—I have not seen any information on that.

—There are universities in Belarus having worked on this topic for nine years, and you have never had any interest to what they did.

—They may be working on it, but what counts is whether this information is published in the peer-review literature. Give us the information, don't come and tell me that the information exists out there. We have official links with Belarus, with the Russian Federation and the Ukraine. There are people whose responsibility it is to pass on information and to submit any information that we may not yet have received. These are conscientious people, devoted to improving health. They are scientists, experts, and we use them as sources of information.

When I asked Michel Fernex what he thought about this interview, he answered in three words: "It was useless" I was worried because I did not think we would get anything out of it. In fact, this interview is interesting. Gentner's replies are a clear demonstration of the strategy of ignorance in the two senses of the term: they reveal the dogmatic desire not to know the truth, "I refuse to make a distinction…" and in so doing, probably reveal a real ignorance about the situation (unless someone at UNSCEAR is conducting research in secret). His replies belong to the rhetoric of the nuclear lobby as a government system. Officials from the nuclear lobby can only send bureaucratic responses, devoid of scientific meaning, and are unable to understand simple questions based on real facts. They are living on another planet. They emit a string of automatic slogans, from within an extremely narrow intellectual bandwidth. Replying to the question about the percentage of internal to external contamination, Gentner even admits he "is not a specialist in dosimetry". What is his area of expertise then? Replying to the crucial question of which illnesses are related to the highest concentrations of caesium in the body, he cuts the discussion short, claiming that the major problem is with external radiation. Yet he concedes that vegetables are contaminated. He is obviously unaware of something the IRSN has only just discovered, and that Bandazhevsky has demonstrated[178], that the accumulation of caesium concentrates differentially in the organs that are most active. Struggling to answer these concrete questions, Gentner ends up hiding behind Dr Kenigsberg, who, in his turn hides behind the IAEA, while he acts as their official purveyor of information, information validated *by their peers…* at the CEA and at Los Alamos. A chain of solid endorsements with no head or tail other than the bomb, power and money. The final speaker is Professor Vassili Nesterenko.

Nesterenko.—I would like to talk to you about the radiological situation in Belarus, remind you of certain figures. As we know, following the Chernobyl accident, 23% of the territory of Belarus was contaminated. More than 2 million people, including 500,000 children live there. Over the last fifteen years,

---

[178] See Part Three, Chapter 2. "2. The French Context", p. 179.

135,000 people have been evacuated, 1,700,000 hectares of cultivable land have been contaminated and as much again of forest. About 260,000 hectares have been taken out of agriculture and today constitute an ecologically radioactive reserve. The economic damage to Belarus is estimated at around 235 billion dollars.

After the release of long-lived radionuclides on the territory of Belarus, the long-lived isotopes accumulated in the topsoil and this determines the radioactive charge on the people today. I was astonished to hear people claiming that we have already assimilated 70% of the radioactive charge and that from now on, we can concentrate on something else. I would like to remind you that over the last fifteen year, new human beings have been born, and they, thank God, did not experience the first radioactive shock. But for the last fifteen years they have been eating contaminated food. The inhabitants of these areas receive between 80% and 90% of their radioactive charge from consuming locally produced food. I would draw your attention particularly to milk, because the inhabitants and in particular the children in rural areas, receive between 60% and 80% of their annual radioactive dose from milk. In second place is the contamination from products gathered from nature and from the forest. These are the dietary habits of the inhabitants of Polessie, and they will not change.

At Olmany, in a village in the Stolin district, in the Brest region, 220 km from Chernobyl, the milk contained 2,600 Bq/l when we measured it last January. This sort of information puts us under an obligation to be vigilant because, I repeat, the main radioactive charge comes from milk. Recently we have discovered something particularly alarming. In certain areas of Belarus, 80% of the milk and cereals contain strontium in excess of the admissible levels. In this case, radical measures are needed: grain that is already stored in silos needs to be disposed of.

Food products spread across the whole republic like grasshoppers, and it is not surprising that in Minsk, a "clean" town, we found children with a radioactive charge of 700–900 Bq/kg. The children receive higher doses because the dose coefficient in a child is 3 to 5 times higher than in an adult. The safety standards in food are the same for adults and children, and since all members of the family eat the same food, children receive a higher dose.

I would like to draw your attention to the work of Professor Bandazhevsky. We worked together. He concluded that, in children, 50 Bq/kg is the threshold at which pathologies of the vital organs such as the kidneys, the liver, the heart, etc appear.

The state of children's health today is such that, if we do not take urgent measures, I do not know what the future holds for them. Suffice to say that according to the estimates presented at parliamentary hearings that I have attended over the last five years, their health is getting worse each year. In April our Ministry of Health reported that, before Chernobyl, 85% of our children were "practically healthy"—this is the expression used by our doctors—but this figure has fallen today to less than 20% and less than 5% in the Gomel region.

Savkin (ICRP, vice-director of the Scientific Centre at the Institute of Biophysics in Moscow).—You have put forward the figure of 50 Bq/kg per body weight, which is less than 1mSv per year. You claim that at this precise level of internal radiation, changes in vital organs, as described by Bandazhevsky, are possible.

Nesterenko.—I think what we are forgetting here is the considerable inequality in the accumulation of radionuclides in the organism. With an average of 50 Bq/kg in the body, there will be a concentration of 1000 Bq/kg in the kidneys and more than 2500 Bq/kg in the cardiac muscles. These figures are taken from examinations by Dr Bandazhevsky on corpses. Illnesses arise at these levels, these are established facts.

Savkin.—But caesium has always been regularly distributed, within a margin of about 10%, throughout the soft tissue. So we are talking about a new scientific phenomenon which needs to be analysed and evaluated very carefully. I have another question: after conducting your radiological examinations, how do you explain the results to people and what advice do you give them? Who does this? The dosimetrist, the teacher, or the doctor? To what extent is there a distribution of roles in a task requiring such delicacy, linked as it is to psychological effects and to the effort to mitigate those effects.

Nesterenko.—Medical advice,—doctors participate in our programmes—is given by doctors naturally. Teachers are given recommendations which they then communicate to parents. These recommendations are written in collaboration with the doctors. They are given out along with the gamma spectrometry measurements. And then, as I have said, we have centres where teachers and doctors work together. They help to make the public aware.

Savkin.—You said that it is the children who are the most irradiated population compared to adults. The data using a spectrometer that has been presented by the Ministry of Health, the official data, shows that the critical group are workers: adults. But in your area, it seems to be the children. How can you resolve this fundamental contradiction in Belarus: children have now been found to be the critical group for internal radiation by caesium?

V. Nesterenko.—We measure the whole population, and then we isolate the critical group. This is standard practice. Whether you look at radioprotection norms in Russia or Belarus, they are the same. We isolate ten to fifteen people, those who have the greatest accumulation. In general, this group—I'm talking now about dose not accumulation—eight will be children. These are our figures. We have 110, 000 measurements[179]. Our spectrometers are certified by the government homologation centre. You are welcome to visit the institute and see for yourself.

---

[179] Gonzales talked about 16,000 measurements in total, undertaken by the IAEA.

Sushkevich (WHO Geneva).—The questions posed by Dr Savkin are really very important. As a member of the dosimetry committee of the International Commission on Radiological Protection (ICRP), he is trying to elucidate the elements that could help us to adopt, or, on the other hand, to critically evaluate the data that you have presented. And so, these questions in themselves and the responses to these questions compel us to consider the need for complementary verification.

Nesterenko.—We are open to cooperation with anyone. When I started this work—because I had already worked with spectrometry before Chernobyl at the institute—we had always thought that children, with their very rapid metabolic rate, would not accumulate much. But, unfortunately, I repeat, it is in children that we find the highest accumulation.

DISCUSSION OF THE FINAL RESOLUTION

The importance of this session, led by Dr Hiroshi Nakajima, lay in the adoption of the final resolution, whose recommendations would provide the basis for governments in the area of radioprotection. Every word had been weighed and debated. The destiny of millions of people depended on it. The reputation of the nuclear industry would undoubtedly depend on it also.

Some time after, I requested the definitive text of the resolution and I compared it with recordings of the discussion. In the minutes written up following the arguments that had caused the biggest debates, I checked whether the proposed amendments, sometimes accepted on the spot by whoever was presiding, had been included in this crucial text. Gennadi Sushkevich, the WHO representative in Geneva, vice-president of the organising committee, had presided over the session. The draft text submitted to the assembly was discussed chapter by chapter.

• Introduction

The first point to be debated was about the extent of the consequences of the disaster. The draft text was geographically restricted: "Significant exposure has been found in large parts of the population of Belarus, Russia and Ukraine".

After a wide ranging discussion with contributions from Professors Yablokov, Nyagu, Guskova, Sushkevich, Busby, Solange Fernex and Bruno Chareyron from CRIIRAD, the proposition put forward by the latter was accepted and appeared in the text of the final resolution: "Significant exposure has been found in large parts of the population of Belarus, Russia and Ukraine and in parts of other European countries[180]".

---

[180] See *Contaminations radioactives: atlas France et Europe*, CRIIRAD and André Paris, Editions Yves Michel, 2002.

• Medical lessons from Chernobyl

1. Statistics

Yablokov immediately raises the question of statistics. "We need to write: "lack of reliable medical statistics". This is a lesson that is costing us very dear now. The absence of good medical statistics has cruelly affected all our conclusions. We need medical statistics that are credible at an international level. This is a lesson that we must learn from Chernobyl".

Not accepted: "Lack of reliable medical statistics" does not appear in the text. Statistics are not mentioned at all. We know that at the UN, it is the IAEA that deals with the matter, in order to reduce costs[181].

2. Aetiology

Draft text: "The worsening of people's state of health could be caused by radiation, or by non radiological effects, as well as by the deterioration in the socio-economic situation".

Preobrazhenskaya.—We need to change the words "could be" to "are". And get rid of "socio-economic situation". I agree, a man who is starving is not going to be in a good state of health. But we are talking now about the damage from radiation, in combination with other factors. Can I remind you that our countries, in particular, experienced a much worse economic situation after the war, but there were no cases of thyroid cancer or leukaemia.

Sushkevich.—But we need to take into account all the factors that could have had an influence.

Nyagu.—Factors common to all catastrophes.

Preobrazhenskaya.—Listen to me, you must know that cancer, brain tumours and leukaemia are not caused by starvation. Quite the opposite, it stimulates the immune system.

Souchkevitch.—But stress can cause cancer.

Preobrazhenskaya.—I've put forward a proposal. Decide for yourselves. Thank you.

Buzunov (Scientific Centre for Radiobiological Medicine, Kiev) Gennady Nikolayevich, I think she is right because no study has shown an influence on health from the deterioration in socio-economic situations.

---

[181] Agreement WHO-IAEA, Article VII: "…the International Atomic Energy Agency and the World Health Organization undertake, bearing in mind the general arrangements for statistical co-operation made by the United Nations, to avoid undesirable duplication between them with respect to the collection, compilation and publication of statistics, to consult with each other on the most efficient use of information, resources, and technical personnel in the field of statistics and in regard to all statistical projects dealing with matters of common interest".

Preobrazhenskaya.—We've been living here for fifteen years. The people who come here and think we are happy, and are coping, are behaving like Marie-Antoinette. When she was told "The people are demanding bread", she replied "Let them eat cake!" Well, the people who live on patisseries need to understand… we're ill because of radiation not because we haven't got bread. Thank you (Applause)

Yarmonenko.—This applauding, it's inadmissible. It's absolutely inadmissible. For heaven's sake, we're not at the opera, this is a scientific conference! As far as I'm concerned, it is completely incorrect to take out the words "socio-economic factors"…

Preobrazhenskaya.—Only "economic"!

S. Yarmonenko.—Removing socio-economic factors from the list of causes is quite simply scandalous. We have tried for four days—I don't know how well we've succeeded—to show that it is only the combination of influences that has an effect. I was very impressed by the comment made yesterday that we are dealing with radio anxiety rather than radiophobia. These somatic illnesses are not caused by radiation. It is a combination of causes. We need to leave the text as it is: it is precisely the combination of all these factors that has led to these disturbances.

Not accepted. The words "could be" and "economic" remain in the final text: "The worsening state of health could be the result of a combination of radiological effects, non radiological factors, as well as the deterioration in the socio-economic situation".

Dimitry Mikhailovich Grodzinski.—Given that the aim, written in the first part, is that those who make the decisions will be guided by this resolution, I think that the "lessons" constitute the most important chapter, because the decisions will be taken on the basis of these lessons. That is why I am proposing to reinforce them and add two others.

First: Dose monitoring after the accident did not function at all well. As a consequence, governments that have nuclear reactors should take into consideration this experience of how limited the possibilities are of evaluating dose over such a large area. It seems to me that it is a very important lesson to be drawn from this experience.

Second lesson: I am worried here about raising a complicated issue, but I would introduce a chapter on "insufficient knowledge of the effects of chronic low dose radiation" This is how it seems to me. And I have to tell you that the director of the ICRP thought it was possible that we would reach an impasse, if we only use standard epidemiology as our base. We absolutely have to study the mechanisms of effects at the cellular level. Then, and only then, will we have a result that allows us to predict the situation in which human beings may find themselves in the future.

This is why, I am asking you now, to be indulgent towards a man of my respectable age and introduce this point.

G. Souchkevitch.—Thank you Dimitry Mikhailovich. These questions are entirely in line with what we discussed about chronic irradiation. There are indeed many points that need clarification and we will have to return to the subject again. It really is an important lesson from Chernobyl. Thank you.

Not accepted[182]

• **Medical consequences of the accident.**
1. Stochastic radiation effects[183]

Buzunov.—On the first point about "stochastic effects", instead of saying "we predict", it should say *"we are seeing* an increase in the number of thyroid cancers among the liquidators of 1986, *among those who were evacuated and among the inhabitants of the areas contaminated by radioactivity"*. These materials were presented, they were sufficiently convincing, and no-one put up any objection. I do not understand why it has been cut.

Not accepted. The text of the resolution reads: "We predict an increase in thyroid cancer among the liquidators of 1986".

2. Deterministic radiation effects

Buzunov—I would like to propose something on this point: *"There are indications that cardiovascular and cerebrovascular diseases increased"*. But the fact is that other non-tumoural diseases have increased also.

Sushkevich.—What is your proposition?

Buzunov.—That's why we need to add: *"that cardiovascular diseases, cerebrovascular diseases and other non-tumoural diseases have increased"*.

---

[182] The obvious sign that it is *the fundamental lesson* from Chernobyl is that it is treated with respect, reverence, and then dismissed. It is a lesson that causes real embarrassment for upholders of the Hiroshima dogma. Up until now, the correlation established by Bandazhevsky is the only serious scientific explanation for the aetiology of the somatic illnesses found at Chernobyl. The Hiroshima model cannot explain these illnesses. Never in the modern era has the world of official science sunk so low or opened itself up to ridicule as in the invention of concepts like "radiophobia" and "radio-anxiety". When Solange Fernex said "The era in which the Holy Inquisition could persecute Galileo for his scientific discoveries is in the past. The truth about the health consequences of Chernobyl can no longer be covered up, as was the fact that the earth revolved round the sun during the Renaissance". Alas, history has periods of regression. The dissimulation has lasted decades.

[183] See notes 167 and 168 p.432

Accepted: "There are indications that cardiovascular diseases, cerebrovascular diseases and other non-tumoural[184] diseases have increased".

3. Other health effects

An anonymous woman sitting next to Chris Busby speaks:

The anonymous woman.—I noticed at the end of the chapter on "other health effects" there is a list of the factors that could be correlated with a certain number of health effects, but curiously radiation is not included among these factors. It seems to me that one of the things that happened at Chernobyl was exactly this, there was an enormous amount of radiation. It would seem therefore unscientific not to include it as a possible cause of health effects, given that there is no proof that radiation was not the cause. Also, I would like to propose that we include "radiation" at the beginning of the list of possible factors causing these other health effects.

(There is a barely audible exchange away from the microphone between the woman and the chair. She asks to speak again, but this is refused)

Yablokov.—I am completely in agreement with what has just been said by our colleague delegate, who noticed the absence of radiation among the possible factors. I would like to propose the following formulation for the last paragraph. It reads "They are probably linked to a certain number of factors. I suggest: *"Apart from radioactivity, they could be linked to other factors as well"*. In other words, the word "radiation" should be mentioned here. Moreover, we have not discussed the matter of food. There has been no report on food in the control groups etc. There have been no reports on vitamins, no reports presented on the evacuations either. I would ask you to include without fail the word radiation in this final paragraph: *"to radiation as well as to other factors",* it could be written like that. Thank you.

I. Likhtarev (ICRP, Ukraine).—*(this portly scientist is standing against the wall near to the rostrum. He is wearing a headset and has a hand held microphone that refuses to work. This adds to his visible embarrassment, that he overcomes heroically in the end)*—Firstly... firstly... I would object to the fact that... simply... . is it working? *(People in the hall tell him to plug the microphone in. Someone helps him.)* Is it plugged in? *(The people in the hall can hear him but not the interpreters in their cabin. They tell him to give it a tap. He starts shaking it and hitting it with his fist.)* Is it working? Yes? Firstly, we cannot include abstract... *(The murmuring in the hall interrupts him. Perplexed, he regards the microphone as if it were a living creature)* Abstractly, we cannot inc... *(crackling)* Have we got another one? ... *(He is looking*

---

[184] Illnesses described as *somatic*. There is an epidemic of these illnesses in the contaminated territories today.

at the camera man's microphone) Is this one working?... (*Finally someone passes him another microphone. Two coughs to check for sound then off we go)* Firstly, in general, I don't agree that we should put the word "radiation" first... By itself, the presence of radiation does not in any way mean that there are medical consequences. That's why like here... there are a huge number of problems associated with storage, in the 30 km zone, in the soil, it's simply absurd. Secondly, if we had to include something, it should be radionuclides... Radiation is a common notion.

Not accepted. The word "radioactivity" *does not appear in the list of factors causing illness apart from oncological diseases (in other words, somatic):* "A number of factors linked to the accident at Chernobyl, including the deterioration in the socio-economic situation, continuing to live in the contaminated territories, insufficient food, lack of vitamins, the evacuation and psychological stress can have an influence on these effects".

Yablokov.—I would like to comment more generally on Point 3 "Other health effects". We seem to have a rather curious construction here in the resolution. We have stochastic effects, deterministic and then "others". This is not possible. They are either stochastic, or they are deterministic, a third category does not exist. I propose that we take out the heading "other health effects" and include them in deterministic. Without changing anything else...They belong quite clearly in "deterministic".

Likhtarev.—I am completely in agreement with Dr Yablokov about the fact that there are no effects at the moment that are not either stochastic or non-stoch... but others. But we do need to keep that category perhaps...in the sense that... in this chapter... because they are perhaps linked with the CA-TAS-TRO-PHE (he enunciates each syllable) of Chernobyl, and not with the radioactive component. Only in this sense.

Sushkevich.—I think perhaps we could adopt this; there is the proposal to call it the "catastrophe", precisely because these are the effects that are linked to the notion of catastrophe.

Likhtarev.—Absolutely. That's right. Effects linked to the CATASTROPHE but not necessarily with a radioactive component.

Not accepted: The word "catastrophe" *does not appear in the text of the final resolution. It was superfluous, because the word* "radiation" *does not appear in the list of factors either. So, according to this resolution, neither the catastrophe nor the radiation explains the health problems that have hit the contaminated territories of Chernobyl. Quite simply, a list is made up and no-one does anything about it, neither in 2001 nor in 2006... nor in 2015 and...*

Adopted text:

"3. Other health effects

Fifteen years after the accident other health effects have appeared. In the main these are neuropsychiatric and cardiovascular illnesses, but other illnesses have also appeared:
- deterioration of health among the liquidators;
- increase in invalidity among the liquidators;
- a decrease in birth rate;
- deterioration in the health of new born babies;
- increase in complications during pregnancy;
- illness in young children".

A blonde woman who appears upset.—Mr President, information has been presented to this conference that indicates a deterioration in children's health linked, among other things, to the radiation factor. We have shown that there are alterations in health following irradiation to the thyroid gland above the threshold, and alterations following irradiation in utero of children... (*Dr Nakajima—who is presiding at the end of this session interrupts her*). Excuse me. I put my hand up at least twenty minutes ago...

Solange Fernex.—At the top of Page 4, we have "chronic low dose exposure to long lived radionuclides". I would like to add the word "*incorporated radionuclides*". I would like to emphasise that we must always insist on the fact that it is incorporated through food, as we heard in Professor Nesterenko's presentation.

Not accepted.

Chris Busby.—I would make the point that the European Parliament recently passed a resolution asking for a reassessment of the dose risk model because of low dose effects which emerged after the Chernobyl accident in the countries that I mentioned earlier and I mentioned in my paper and I feel that this conference should call for re-examination of the dose risk models [...] now there is a very great deal of evidence in the last ten years and certainly evidence that has been presented at this conference and by us and by other workers that very low doses of radiation have a significantly high effect, a proportionately higher effect and this calls into question the whole radiation risk model, and I do not think attention should be focused only on the groups that were significantly exposed to radiation because if such is the case then many radiation effects from Chernobyl will be entirely missed.

Nakajima.—Yes, you will remember that in 1996 or 1997, WHO and the IAEA held a meeting in Seville, in Spain, and we determined definition of low dose ionising radiation... the minimal level of risk from a radiation at low doses. We already have a clear recommendation. So I think that your statement is correct and it should be included in a more precise way.

Busby.—Alright, well if we can recommend as a conference that more research should go into the evaluation of the effects of very low doses.

Nakajima.—So, this is an additional sentence.

Not accepted. On page 137 of the "Recommendations" of the European Committee on Radiation Risk (ECRR)[185], coordinated by Chris Busby—among the fifteen papers and publications about Chernobyl used as a basis to study the effects of the accident— there is this reference: "WHO 2001, Kiev Conference. "The final resolution of the conference recommends a re-evaluation of the risk models". Professor Busby and the ECRR trusted the word and honour of the honorary president of the conference, Hiroshi Nakajima, who gave his assurance at the plenary session that this formulation would be included in the final resolution. It was recorded on film.

Adhering to the line maintained by the three UN agencies, the resolution does not mention the new scientific data presented at the conference about the effect of caesium 137 incorporated at low doses in the human organism.

The IAEA, UNSCEAR and WHO, who do not study the effects of internal contamination by incorporated radionuclides, cannot explain the unexpected increase in somatic illness from which the contaminated population are suffering.

Contrary to the promise made at the assembly, the proceedings from the Kiev conference have never been published. There are many words in our language to describe the different ways in which people were duped at Kiev: mystification, pretence, obfuscation, deception, fraud, faking, faithlessness, massaging of the facts, lying by omission, trickery, betrayal, arrogance, contempt...

Finally, I must return to Alla Tipiakova's appeal[186]. "I am calling on the whole world, I am speaking to all kind people. You must help us…We no longer believe that our children will be safe here. We think no-one is interested in us".

---

[185] ECRR, Ed. Green Audit, 2003. ISBN:1 897761 24 4, page 110

[186] See Part One, Chapter 1, p. 6.

Chapter III

THREE MASTERS OF THE SUBJECT

The following text is the response of three academics from the three countries most badly affected by the policy of trivialising the Chernobyl disaster. It is a rational analysis of the methodological errors and the inadequacy of the international assistance given to the contaminated population. It is also, and most importantly, a programme of possible concrete action to tackle the real problems that the international scientific community continues, obstinately, to ignore. Two of the three authors, Nesterenko and Yablokov, have already been introduced. D.M. Grodzinski, a biologist, and member of the Academy of Sciences in Kiev, is president of the National Commission on Radiological Protection in Ukraine. He opposes the policy of concealing the health consequences of the Chernobyl disaster.

D.M. GRODZINSKI (UKRAINE), V.B. NESTERENKO (BELARUS),
A.V. YABLOKOV (RUSSIA)
(NOTES IN THE MARGIN OF THE 2002 UN REPORT)

In July-August 2001, six experts from Russia, Belarus and Ukraine were asked by a number of different UN organisations to collect "useful and reliable information about the humanitarian consequences of the accident at the Chernobyl nuclear power station". Based on "a rigorous scientific analysis of the factual data", obtained through observation and from material provided by the local authorities in the contaminated territories, these representatives of Goscomhydromet and from the Chernobyl Committee from Belarus, the Ministry for Emergency Situations, Centre for Radiological Medicine, the Institute of Sociology and the High Chamber of Ukraine, as well as the NGO Taifun and two scientific establishments of the Russian Academy of Sciences, that do not appear on the official list of academic establishments (a so-called "laboratory of ecological and medical dosimetry" and the Institute of Nuclear Safety of the Russian Academy of Sciences), produced a series of recommendations for the international community and the governments of Belarus, Ukraine and Russia concerning problems linked to the consequences of the disaster at Chernobyl. They made use of the reports from national programmes on Chernobyl (the materials already quoted of Goscom-tchernobyl and from the Ministry of Education in Belarus, from the Ministry for Emergency Situations of Ukraine and the Russian report on water quality in the Bryansk region) and from the analysis of "scientific articles, legislative measures, and other publications". The UN experts produced a report, published in book form in English and in Russian, and entitled *Humanitarian Consequences of the Accident at the Chernobyl Nuclear Power Station,* which was solemnly presented to the media and the public in February 2002 in New York, Minsk, Kiev and Moscow.

There are two diametrically opposed views about Chernobyl. These two positions are reflected within the structure of the UN; on the one hand, UNSCEAR, the IAEA and WHO claim with one voice that apart from 1,800 thyroid cancers caused by irradiation during childhood and the death of a few dozen liquidators, there have been no other consequences due to radiation from Chernobyl that can be established with any certainty. On the other hand, the UN Secretary-General, Kofi Annan, wrote in 2000 in the preface of a publication from the UN Office for the Coordination of Humanitarian Affairs OCHA that: "The exact number of (victims) may never be known. But three million children require physical treatment, and not until 2016, at the earliest, will we know the true number of those likely to develop serious medical conditions [...] Their adulthood—now fast approaching—is likely to be blighted [...] Many will die prematurely. Are we to let them live and die [...]?"

The 2002 UN report is an attempt to reconcile these two points of view: thus we find among the organisations that financed it, WHO, known for its sympathies with the nuclear industry, but also the UN children's foundation (UNICEF), the UN development programme UNDP, and the Office for the Coordination of Humanitarian Affairs (OCHA).

The Chernobyl tragedy has affected millions of people. It is an event of global significance that testifies to the adventurism of the promoters of the nuclear industry, but also to cowardice and heroism, suffering and solidarity among the earth's inhabitants. Chernobyl continues to pose a number of problems to humanity but this is the most pressing: what can be done to reduce the human suffering and to normalise life in the contaminated territories? The economic scale of the problem is proportional to the human disaster: tens of billions of dollars have already been spent but future expenses will be even larger.

The UN report quite rightly talks of the importance of effective international aid; there is no doubt that the measures proposed will help to consolidate the efforts of the international community. We completely agree with the authors of the report when they say that it is "necessary to have complete, truthful and precise information about the consequences of the accident" and that all arguments must undergo scrutiny by "detailed and honest expertise". But careful reading of the report leads us to conclude that the UN report itself is lacking in truth, and that the information contained within it is neither complete nor objective.

The report claims for example that the radioactive fallout "will continue to have an effect on the rural population for decades". This is not true. Contamination by caesium and strontium, even though it will be become weaker over time, will continue to have an effect over hundreds of years (ten half lives), and as for those territories contaminated by plutonium and americium, they will remain a danger for ever, for several millennia. Remember also that even after the reduction in radioactivity due to the natural transformation of radionuclides, the contamination of people may not decrease and, as experience has shown, may even increase: this is exactly what we are seeing everywhere today in the areas contaminated by Chernobyl.

It is also incorrect to claim that the risks linked to the initial impact of the radioactivity "have already manifested themselves". We know that radiation causes a transformation of the genetic material (mutations) and that these genetic changes are hereditary. For this reason alone, the radiological shock of Chernobyl will unfortunately make itself felt for many generations to come. Also, we know that radio-induced cancers do not appear straight away: breast and lung cancer appear after twenty years, cancer of the colon after thirty years. Therefore, the risks for those who received the first radiological shock in 1986 will not manifest themselves until after 2016.

Unfounded phobias are no help at all, of course. However, refusing to take the effects of radiation seriously is equally dangerous. When we are told that it is possible "to create a favourable environment" in the contaminated territories, we are being told lies. The environment there will always be unfavourable. On the other hand, even in such an unfavourable environment, it is possible to organise life so as to avoid the danger to a certain extent by adopting a whole series of rules and respecting a whole series of prohibitions (see below). But it is clear that over the centuries to come, if people are to continue living in these areas, a good number of precautionary measures will need to be taken.

Continuing to insist that it is possible to live in the contaminated territories, the authors claim that "there are various agricultural crops that can be cultivated safely on radioactively contaminated soils". Again, we are being told a half truth. It is true that certain plant species accumulate smaller amounts of radionuclides than others. For example, there is five times less radioactive strontium in wheat cultivated on contaminated soil than in barley or peas cultivated in the same soil; half as much accumulated radionuclides in potatoes than in beetroot, etc. It is even possible to distinguish different species of tree by the amount of radionuclides that they have accumulated. However, there is no plant species that does not absorb radionuclides from the soil. This means that radiological monitoring of food products will have to continue for a long time.

The central thesis of the chapter in the report on ecology is equally false from a scientific point of view. It is proposed that we make use of "the potential" of the contaminated territories to "fulfil the international obligations of the three countries concerned in the protection of biodiversity", and to use the ecosystems of the forest and marshes "to conserve biodiversity". In the area contaminated by Chernobyl, it might seem that the wildlife, freed from human influence, is particularly rich but this area can in no way be considered as a normal healthy reserve for diverse life forms. Studies undertaken not only in the areas contaminated by Chernobyl but in the area to the East of the Urals and in other contaminated areas (for example the area of Orenburg following the Totsk explosion) have shown that ten generations later, living organisms that suffered radiation exhibit genetic instability. Moreover, there are health problems in animals and plants born in these areas. The fact that most of the birds that overwinter in these areas do not return, for example, shows that they are dying in huge numbers during the first winter. Alterations can be seen in the genetic apparatus of animal and plant species that have been studied in the contaminated areas. Biodiversity in these areas is a superficial illusion. In reality, the health of this environment has been seriously disturbed. These areas do not constitute a living reserve; they are, on the contrary, a sort of malignant tumour growing within the living body of nature. Scientifically speaking, the problem is not how to use this radioactive biodiversity but how to protect people from it.

But it is probably in their study of the morbidity among the population of the contaminated territories that the authors of the report distort the truth most seriously.

They claim that the increase in congenital malformation linked to excess radiation is not corroborated by statistical data. This is untrue: the data exists. Between 1986 and 1995 across the whole of Belarus there has been a 40% increase (it rose from 12% to 17% in new born babies) in significant congenital malformation (harelip, cleft palate, anomalies in limb formation, alteration in the development of the central nervous system and the circulatory system, closure of the oesophagus or the anus etc) and if we take into account the number of foetuses aborted because of malformation, it has gone up 80% (up to 22 cases in 1000). Other statistical data shows that from 1988 to 1999 the frequency of congenital malformation in Belarus has more than doubled. Statistical data confirms that there has been the same increase in congenital malformation in the contaminated territories of Ukraine and Russia.

Number of cases of congenital malformation per 100,000 newborn babies
in the administrative regions of Bryansk and of Kaluga
(Balaïeva et al, 2001)

Region	1990	1998	Augmentation
Kaluga	104.7	352.6	x 3.4
Bryansk	32.3	404.2	x 12.5

The usual argument against statistics of this kind is that it is the result of improved screening, meaning that it has appeared through more attention being paid to these cases. But this effect could not come about when the study concerns the same area and has been undertaken by the same people using the same methods. Yet it is precisely in the contaminated areas that this significant increase in cases of congenital malformation has been observed. These figures exist for numerous areas of Belarus (Gomel, Mogilev), and Ukraine (Zhitomir), and Russia (Bryansk). Thanks to its advanced statistical services, Germany has produced similar figures. Following minutely detailed analysis of medical statistics, it has been discovered recently that in Bavaria, the region of Southern Germany that was most affected by fallout from Chernobyl, the number of cases of congenital malformation reached a maximum in November and December 1987, seven months after the peak of concentration of caesium-137 in the mother's body. A developing foetus is at its most sensitive to the teratogenic action of radionuclides in the second month of pregnancy. In the contaminated territories in Belarus, an increase in the percentage of deaths of new born babies, following defects in the development of the nervous system, and an increase in stillbirths has been observed...

All of this shows that even relatively low doses of radiation, cause alterations in foetal development that are incompatible with life. The catastrophic deterioration in the health status of vulnerable young children with regard to all illnesses in the contaminated territories leaves no room for doubt: in 1985 more than 80% of children living in these areas enjoyed good health, but in 2000 this figure was only 20%. In the South of the country, the areas most affected in the Gomel region, there are practically no healthy children.

The claim made by the authors of this report that the increase in mortality "cannot be as a result of Chernobyl", on the grounds that this phenomenon has been observed over the whole of the former USSR, lacks scientific rigour. It is true that mortality has risen all over the former Soviet Union but, firstly, this phenomenon appeared after 1986 and it cannot be excluded that one of the reasons could be fallout from Chernobyl, which fell on the area in which more than half of the population of the former Soviet Union lives. Secondly, the increase in mortality is particularly significant in those areas which were badly contaminated.

Mortality (per thousand) in the Bryansk region 1998–1999
(Komogortseva, 2001)

Mortality	In the whole region	In the 3 districts most heavily contaminated by radionuclides
Infants	10.2	17.2
General population	16.3	20.1–22.7

The report claims that "the structure of morbidity in the contaminated territories is similar to that of other regions of the former Soviet Union" but this is a lie. Where there are reliable statistics, it is apparent that in addition to increased mortality, after the disaster there was a significant increase in miscarriages and in stillbirths. As regards other changes in the structure of morbidity among people living in the contaminated territories (in comparison with neighbouring areas where social and economic conditions are the same) it is apparent that there is:

- an increase in the number of babies born weak and ill;
- an increase in the number of genetic alterations and in congenital malformation;
- an increase in the number of cancers (and not only thyroid cancer);
- an increase in disorders of mental development (delayed development), neurological and psychological problems;
- an increase in psychiatric illness (including schizophrenia);
- changes in the immunity and hormonal status (endocrine);
- an increase in the number of illnesses of the organs of the circulatory, lymphatic, respiratory, urinary system, of the skin, internal secretion glands and sight organs;
- growth problems in children and abnormal weight loss;
- abnormally long recovery time following illness;
- premature ageing.

There are dozens of illnesses caused by the fallout from Chernobyl. They cannot be explained either by the effect of screening or by socio-economic factors, because the areas that have been compared differ only in their level of contamination. In the UN report, certain illnesses are indeed mentioned with an accompanying phrase such as "which is not absolutely certain" or "it is possible, it is not without foundation", "it is not confirmed by the statistical data". These remarks obviously cast doubt on data that is absolutely reliable. Here is a concrete example: among the consequences caused by the disaster, the authors of the report mention the appearance of cataracts among the liquidators who received high doses of radiation, but they do not say that this phenomenon has also appeared among the people living in the contaminated territories. Among those people who were evacuated from the strictly monitored zone (more than 40Ci/km^2) cataracts are even more pronounced than among the liquidators.

Frequency in appearance of cataracts (per thousand) in Belarus from 1993 to 1994. (Goncharova, 2000)	1993	1994
Average frequency in the country as a whole	136.2	146.1
In areas contaminated between 1–15 Ci/km²	189.6	196.0
In areas contaminated above 15 Ci/km²	225.8	365.9
In people evacuated from areas contaminated above 40 Ci/km²	354.9	425.0
Among the liquidators	281.4	420.0

In the analysis of the data on health and biological consequences of the disaster, the authors of the report allow two methodological errors to slip into their reasoning. The first concerns the logic of their argument. To justify their refusal to take into account existing data, in several passages of the report, they claim that it is necessary to " undertake rigorous scientific research, that is recognised by international community"[187], to extract "scientifically reliable consequences", to initiate "the unbiased scientific research that is objective and methodologically sound", to be "faithful to internationally recognised scientific protocols", to obtain "authoritative proof", to undertake "high quality scientific research", to have "an internationally recognised programme of scientific research", to obtain "reliable and objective results", all of which implies that much of the collected data does not correspond to these criteria. One can only ignore existing studies if the comparison between data that do not conform to "international scientific protocols" and data from studies that do conform to these protocols shows us that the studies done in the same areas according to different methods give different results: in which case it would be normal to demand verification. As long as no such comparison has been made, it is methodologically incorrect (and morally inadmissible) to ignore the results of scientific research that has been done previously. Those who ignore this research are proving their bias against the results of research that show that the people's health in the contaminated territories is deteriorating. Even if we accept the point of view expressed by the authors of this report and if we put aside the

---

[187] The authors apparently only consider as "recognised by the international community" those journals whose editors and readers' committees include people from the nuclear industry. Doctors conducting research in Belarus, Ukraine and Russia, working with patients from Chernobyl, have not got the time, in general, and are not in the habit of publishing their results in journals abroad that are "recognised by the international community". At best, they present the results of their research at regional conferences and publish articles in national newspapers. As of today, there are thousands of published articles from Belarus, Ukraine and Russia about the consequences of Chernobyl, of which no more than 1% or 2% appear in Western journals. Nevertheless, a great deal of high level research has been undertaken, thanks to modern methods of dealing with statistical data.

thousands of studies which, according to them, were undertaken using incorrect methods, no-one has the right to conclude that there are no health consequences purely on the grounds that the data is unavailable.

The second methodological error made by the authors of this report is that they ignore the precautionary principle. The history of humanity shows that where we cannot claim with certainty that our acts will not cause harm, we must suppose that they could have dangerous consequences. The authors of the report recognise that there are many aspects of the Chernobyl disaster that are unclear, and have uncertain consequences: we do not know everything about the dose that people received in the first few days after the disaster, nor the particular geographical distribution of the radionuclides that fell on the soil, the future irradiation of the people living in the contaminated areas, the health and genetic consequences of the radiological impact. What is more, the report talks about the necessity of studying "the possible link between breast cancer in young women and women who were breastfeeding at the time of the accident, and radiation" "between radiation and breast cancer, between thyroid cancer in adults and the health of those who took part in the liquidation of the consequences of the accident"; "the distribution of caesium in biological tissues and the risks of specific alterations", "the possible impacts of radiation on intrauterine development". How can the authors of the report claim that "the health consequences of the radiation are exaggerated" when they recognise themselves that our scientific knowledge in these areas is still very incomplete? While admitting that we do not know all the possible risks, how can they claim that there is no danger!

It is quite astonishing that while they refer to aspects that need to be studied scientifically in the future, they do not even mention a whole series of questions about the impact of radiation that need studying just as urgently. For example, among many other things, the effect of radiation on:

- the genetic health of the population;
- the central nervous system and the sensory organs;
- lesions of the endothelium (the vessel walls);
- the lowering of immunity;
- damage to the hormonal system;
- the premature ageing of the body;
- the increase in mortality in certain age groups.

In their recommendations the authors of the report ignore a number of very important questions of principle, and this casts doubt on the whole strategy that they are proposing, including their "new approach" (page 158–159) which consists of resolving the problem of Chernobyl in three stages: the stage in which "the most urgent problems are resolved", 1986–2001; the "rehabilitation" stage, 2002–2012; and the "management" stage after 2012. It is worrying to see that the authors propose leaving till 2012 any "… in-depth analysis of the current state of affairs

[...] to identify current needs in areas like public health, ecology and scientific research". We believe that all this needs to be done now and continuously rather than waiting till the end of the next decade.

The authors of the report declare their objectivity (the essential aim of the report) but exhibit quite openly their favourable opinion of the nuclear industry, the primary cause of the disaster. Those who work in the nuclear industry have continued over a long period to say that Chernobyl was simply a technological accident that caused the death of about thirty people, less than 2000 thyroid cancers (easily treatable), that its consequences have been exaggerated and are mainly the result of the stress caused by radiophobia and from too rapid an evacuation of the population, in short, that it's time "to forget Chernobyl". Even though the authors declare from the first few lines of the report—one has to wonder why—that they undertook their study "without any pressure from interested organisations or persons", they claim, in unison with the defenders of nuclear power, that people's fears about radioactive contamination and its consequences are "unfounded" and even that they are due to "provocation".

The authors of the report do not hide their anxiety to see nuclear power develop and say that "the fate of the population in the contaminated towns and villages will remain at the centre of any new discussion on energy development in the years to come" and that "the energy industry worldwide is very concerned that these problems should be resolved and that the future of nuclear energy be examined, without emotion, but reasonably on the basis of arguments and facts". Is it not surprising that the authors of an analysis of this humanitarian disaster should be so concerned with the development of nuclear energy?

They support the nuclear lobby's thesis that the most clearly determined evidence about the consequences of Chernobyl are the deaths of 39 people from high levels of radiation but they say nothing about the data from the Chernobyl Union, an association bringing together those who participated in the liquidation of the consequences of the accident (the liquidators) according to which 70% of the liquidators are ill (endocrine problems ten times more frequent than the average in Russia, psychiatric problems five times more frequent, problems of the circulatory system and the digestive system four times more frequent than the average in Russia). The liquidators become ill four times as often as the rest of the population. In general the fate of the 600,000 liquidators is one of the most important humanitarian aspects of the accident deserving the UN's attention. We know that the alterations in genetic material to which the liquidators were subjected will be transmitted to the generations that follow.

The authors of the report agree, *de facto*, with the promoters of nuclear power who have said for many years that one of the most tragic consequences of Chernobyl is that it brought a halt to the development of nuclear power, and that "it is time to forget Chernobyl". Of course, the governments of the three countries most affected

470

by Chernobyl are interested in reducing the costs of mitigating the consequences of the accident. For all of them, the less they know about radiation induced illness, the better. This refusal to recognise the sad truth is manifested in the cancellation of government research programmes on Chernobyl, the diminished status of those organisations dealing with the social aspects of the problem of Chernobyl and in the exclusion of the most active and honest researchers, (as is the case with Professor Bandazhevsky in Belarus).

As for the analysis of the health consequences of the Chernobyl disaster, we are seeing the same phenomenon that occurred with the study of the health consequences of Hiroshima and Nagasaki at the beginning of August 1945. The occupation forces prohibited any research at that time into the effects of radiation. It was only authorised in 1950, four and a half years later, by which time the most significant information on the effects of radiation had been lost for ever[188]. It should be noted that it is precisely this truncated statistical data that formed the basis for all the radiological safety norms currently in force. These norms were developed without taking into account the increased mortality rates among the most vulnerable members of society, children, the old, the ill—and therefore cannot ensure our protection. One of the most renowned Russian specialists in radioprotection, the director of the Radon Complex in Moscow, has recently admitted: "It has been apparent from the start that radioprotection safety standards were developed in deference to the nuclear industry". This attitude is the primary cause of several million deaths in the twentieth century. These deaths are due to the development of the nuclear industry and, surely, above all, to atmospheric nuclear testing, but also to the use of medical X-rays, nuclear fuel processing and the ordinary functioning of nuclear power stations.

Data about Chernobyl is suffering the same fate, at the hands of the nuclear lobby, as the data about Hiroshima and Nagasaki. They are proposing that we discard all the data collected by numerous researchers from Belarus, Ukraine and Russia because it has no scientific value, and that we start again from zero to study the consequences of the disaster, while fifteen years have elapsed and an enormous amount of data is irrecoverable.

What should we do? What strategy should we adopt to take effective action? Let's outline some of the priorities.

While there is no question of rehabilitating the contaminated territories entirely, we could and should take a raft of measures to reduce the humanitarian consequences of the disaster. Above all, we need a reliable method for measuring. At present, the calculation of radioactive charge in the population is made by

---

[188] See note 15, p. 53.

measuring levels of contamination in the territories and this has been shown to be very inaccurate. It needs to be replaced by an objective and precise evaluation of the radioactive charge incorporated in each individual. Within the same village, the dose received by inhabitants can vary considerably from one individual to another. There are many reasons for this, including the phenomenon of "leopard spot" contamination and the effect of individual diet. An effective strategy has to be tailored as far as possible to the individual and should focus initially on those people who have suffered the most and who are most at risk. This kind of individual approach is entirely possible, the equipment exists: there are machines that can measure human radiation (Human Radiation Spectrometers or HRS), methods to analyse accumulated levels of radiation over a lifetime through an examination of dental enamel, and modifications occurring in protein molecules (FISH method). Other objective measures of individual dosimetry could almost certainly be developed if the problem was brought to the attention of scientists and they were given the necessary financial means.

Reconstituting the contamination of the first few days and weeks following the accident forms part of the objective study as we understand it. During the first few days, the amount of radioactive material was hundreds or thousands of times larger than now because of the short-lived radionuclides. These included not only iodine-131, but lanthane-140, tellurium-132, neptunium-239, xenon-133, barium-140 and many others. It is possible that the more obscure effects that we are observing today can be explained by the brief powerful impact of these rare radionuclides.

A United Nations aid fund needs to be set up for the victims of nuclear disasters. There are 430 nuclear reactors in operation in the world today; as they age, the risk of an accident increases. There is no doubt about this: we should expect there to be new disasters at nuclear power stations. This fund would be constituted from obligatory payments, representing a percentage of the revenues from the sale of electricity, from countries with nuclear power stations.

Because today the inhabitants of the contaminated territories receive 90% of their radioactive load from local food products contaminated by radionuclides, it will be necessary for several decades to monitor contamination levels in food and levels of radioactive charge incorporated by the inhabitants (using anthropogammametry). Maps need to be drawn up showing the contamination of the population by radionuclides (and first and foremost the children) and marking those regions that need particular surveillance.

Radioprotection of the population should be based on the annual body charge of the critical group, in other words, the most contaminated group in the population. HRS measurements should be made in each locality from a reliable sample made up of representatives of diverse social groups (20% of the population is sufficient). Thresholds (1mSv/year for adults) should be established taking into account the critical group of inhabitants in the village (more than 10).

A law on social protection for the victims of the Chernobyl disaster was passed in Belarus in 2001, and contains an extremely important stipulation that protection measures need to be maintained even if the annual charge drops from 1 mSv/year to 0.1 mSv/year.

More than 4 million people, including 1 million children, live in areas of the former Soviet Union that were contaminated by fallout from Chernobyl. Establishing strict limits on the concentration of radionuclides in food, and ensuring that they are properly respected, would have a very significant effect on radioprotection. The contamination of milk with radionuclides in a particular locality is a very good indicator of the level of danger posed to children's health from living there. According to 2001 figures from the Ministry of Health in Belarus, there are 1,100 villages where the levels of caesium-137 in milk exceed 50 Bq/kg and 350 villages where levels exceed 100 Bq/kg.

Even though children in these villages receive two or three meals a day at school and nursery, even though they have regular medical check ups and periods of convalescence, and even though the soil in these areas is treated with a dressing of extra minerals, we have not managed to bring down the levels of incorporated caesium-137 in children's bodies below 30–50 Bq/kg; we need, therefore, to make the admissible levels of radionuclides in food products even stricter. European limits currently in force for emergency situations (1 mSv/year radiation threshold, 1,000 Bq/litre, the permissible level of cesium-137 in milk for adults and 400 Bq/litre, for children) were established on the basis of calculated risk coefficients from data from Hiroshima and Nagasaki and are *completely inadmissible.* For the situation at Chernobyl, where people are receiving chronic radiation doses, the limits need to be ten to twenty times stricter (the annual admissible levels for internal radiation should be lowered to 0.1 mSv/year, which corresponds to 30–40 Bq/kg per body weight).

Medical research must include international projects to determine the correlation between illness and the concentration of radionuclides in the organism. This is the only way to establish the cause and effect relationship between illness and the consequences of the Chernobyl disaster. Before his arrest, Professor Bandazhevsky had established a cause and effect relationship between internal levels of radiation and alterations to the ECG, and diseases of the eye (cataracts). Physicists and doctors need to pursue joint studies in this area: examination using a HRS (Human Radiation Spectrometer) of children to determine the concentrations of radionuclides in the organism and complete medical examination.

It is equally urgent to undertake an information campaign to give people simple radioprotection training to reduce the amount of radionuclides incorporated into the organism through the food they eat. Soaking meat, mushrooms and fish in salt water (2 dessert spoons of salt in a litre of water), reduces the levels of caesium 137 by a factor of 3 or 4. Given that 60% of the annual charge comes from contaminated milk, people need to learn to separate it. The addition of chemical adsorbents

(Prussian blue) in fodder can reduce by 35% to 75% the level of caesium-137 in milk and meat.

In every district and locality, programmes need to be set up to put mineral dressing once every three years on cultivable soil (above all on people's vegetable gardens), the fields and the forests (where people collect berries and mushrooms, in other words in a radius of 10 km around localities). 3 tons of calcium and 100 kg of phosphate per hectare can reduce the transfer of radionuclides into the plants by 80% to 90%. Adding calcium or lignin to a forest ecosystem can reduce levels of caesium 137 in berries and mushrooms.

Taking natural adsorbents in the form of pectin-based food additives has proved very effective in eliminating radionuclides from the organism: pectin-based food additives should be taken for a month at least four times a year. The raw materials for producing pectin-based food additives (the waste material from making jam and fruit juice) is not lacking in either Russia, Belarus, or in Ukraine.

The catastrophic deterioration in health (particularly in children) sixteen years after the Chernobyl disaster is not due to stress or radiophobia, nor to mass evacuation (only 140,000 inhabitants out of 2 million people living in badly contaminated areas were evacuated from Belarus and the same applies to Ukraine and Russia) but to the chronic effect of low dose radiation.

Lack of money is not the only reason why the necessary protective measures have not been taken: ambiguous and contradictory government policy has had an equally serious effect (in order to spend less, governments have a tendency to hide the true scale of the tragedy).

Nevertheless, a small team of a few dozen people was able over a few years to undertake HRS examination of 140,000 children, distribute 45,000 pectin treatments to them and implement, in some contaminated areas, all the necessary radioprotection measures. It is therefore not an impossible task but it should be carried out on a different scale. All that is required is to use existing funds sensibly to substantially reduce the negative effects of the Chernobyl disaster on the countries concerned.

Our countries (and particularly Belarus) will never be able, at any rate in the next few decades, to overcome the consequences of the Chernobyl tragedy unless they receive extensive international aid. We need to find the financial means to set up all these international projects in all the contaminated territories. These programmes will need to be maintained for several decades until there are no longer any radionuclides in the topsoil and "clean" food can be produced for the entire population.

PART SEVEN

THE BANDAZHEVSKY AFFAIR.

AN INNOCENT MAN UNJUSTLY TREATED

Chapter I

A STALINIST TRIAL

The trial of Yury Bandazhevsky and Wladimir Ravkov started in Gomel on 19th February 2001 before the Military Chamber of the Belarusian Supreme Court. The trial ended on 18th June when the scientist Bandazhevsky was sentenced to eight years in a *maximum security* prison. No proof was ever presented. The main prosecution witness, Vladimir Ravkov, also sentenced to eight years in a maximum security prison, declined to testify. The Advisory and Monitoring Group of the Organization for Security and Co-operation in Europe (OSCE) in Belarus, noted eight different violations of the Belarusian Criminal Code and concluded that, from a legal point of view, the entire trial "testifies to the higher standing of 'expediency' rather than the rule of law". Amnesty International, that considered Yury Bandazhevsky a prisoner of conscience, detained solely for exercising his right to freedom of expression, launched an appeal for his immediate and unconditional release.

In his appeal for judicial review to the President of the Supreme Court, Bandazhevsky declared: "I have every reason to claim that these legal proceedings have been brought against me with the aim of getting rid of me, as rector of the institute, and as a scientist, who is known in our country and abroad, in the field of medical radiology, whose scientific research, discoveries, conclusions and recommendations conflict with the interests of a number of government officials who have a different point of view about the consequences of the fallout from the Chernobyl disaster on the territory of Belarus, and about the effects these could have on the health of the population, the flora and fauna of the country, and therefore conflict with the policies being pursued by those in charge in this domain."

Apart from the main prosecution witness, Vladimir Ravkov, a lieutenant-colonel in the medical service and vice-rector of the institute, who, under heavy pressure (classified as torture), made an accusation against Bandazhevsky of accepting bribes to help young applicants in their admission to the institute, the six other people accused of corruption were all freed after having delivered the testimony required of them. The conditions under which their statements against Bandazhevsky were made can be judged by their declarations in court, which were recorded by the OSCE observer. We were able to obtain the stenographer's notes from the observer and the essential parts are reproduced below.

1. THE UNDERWORLD OF THE BELARUSIAN JUDICIARY

SHAMYCHEK, ONE OF THE ACCUSED AND A WITNESS:
MINUTES OF THE HEARING ON 27TH FEBRUARY 2001
IN THE BANDAZHEVSKY-RAVKOV CASE
RECORDED BY THE OBSERVER FROM OSCE

The court considers the evidence in the case against the accused Shamychek.[189]

At 13:00 hrs, the court adjourns. In the morning the room was half empty.

During the break representatives from Amnesty International, Natalie Losekoot and Matthew Pringle arrive. They talk to Bandazhevsky about his research in radiological medicine and his attitude towards the trial.

After the adjournment, at 14:05 hrs, the clerk of the court warns the public that bringing cameras into the courtroom is prohibited. He will confiscate any cameras if he sees them. I ask him to introduce himself but he refuses.

Today, Ravkov has been given neither breakfast nor lunch.

During the cross-examination that follows, Shamychek states that the investigating authorities demanded that she make a statement saying that she had transferred a total of 11,000 dollars to Bandazhevsky.

The judge reads extracts from the testimony taken from the minutes of the investigation that contradict her statements in the witness box. Shamychek repudiates some of the testimony explaining that when she made them, on 17th July 1999, she was under particular stress. She then claims that she was threatened, that the sums mentioned were those that the examining magistrates demanded that she indicate, that she had been shown lists of people that she did not know and who had supposedly testified against her.

Then Shamychek wants to name the people who had threatened her, but her attempts are interrupted by the presiding judge.

When Bandazhevsky's lawyer demands that the contradictions be resolved after they have been examined in detail, the judge tells him he has no lessons to teach the court.

At 17:45 hrs the hearing is adjourned until 28th February 2001 at 10:00 hrs.

MINUTES OF THE HEARING ON 28TH FEBRUARY 2001.

The hearing begins at 10:00 hrs. Following cross-examination, 15 charges remain against the accused Shamychek, a 60 year old woman.

---

[189] The accused and "witness" N.I. Shamychek, was a teacher who gave revision classes to prepare candidates for the examinations. She was supposedly the intermediary who accepted the bribes and then handed the money over to Bandazhevsky.

According to the minutes of the cross-examination in the preliminary investigation of 17th July 1999, Shamychek asked Bandazhevsky to give her the questions for the chemistry examination.

The candidates would pay her for them.

She did not give him any money. Today Shamychek repudiates these statements: Bandazhevsky never gave her the exam questions.

10:30 hrs. She is being questioned by the defence. There are six people in the room (not participating in the trial), including myself.

Shamychek repeats once again that during the investigation she was pressured into stating that she gave 11,000 dollars to Bandazhevsky. At the time of her arrest, no-one explained her rights to her. This was only done three days later. Until then she had been questioned without a lawyer present.

She was questioned by different people, including Alexandrov who questioned her between eleven o'clock at night and two in the morning. Another interrogation had lasted until four in the morning.

After her arrest, she was put into a small cell with five other detainees. The window did not open.

There were no chairs.

She had to sit on the ground (she is 60 and has worked for forty three years as a teacher).

She asked to see a doctor continually. Morning, midday and evening she asked to be given Validol: she suffers from ischemia, arterial sclerosis and chronic gastritis. She was given no medical aid. She only had the medicines that her daughter was able to bring in.

She spent thirty days in detention. After she left, they discovered she had focal tuberculosis. Sometimes, she was interrogated by six people at a time who threatened her with their fists, and banged a metal ruler on the table: "They threatened to put me in a cell and burst my liver".

When Bandazhevsky asks: "Can you confirm that you were forced to state that you gave me 11,000 dollars?" she replies: "Yes, I was forced to say it but it wasn't true. I was in such a state that I would have said anything".

The presiding judge asks the lawyer Baranov to withdraw the question: "Do you clearly remember having given the money?" He also asks him to withdraw a question about the sum given. Baranov claims that he is being prevented from questioning the accused.

Bandazhevsky asks Shamychek if she had been given the psychotropic drug, Phenosipan. She replies in the affirmative.

Following a request from the prosecution, the minutes of the interrogation on 17th July 1999 are read out in which it is stated that Shamychek refused the services of a lawyer. Shamychek responds that they had started the interrogation, and that this item had been added in to the minutes afterwards. She never refused the services of a lawyer.

THE ACCUSED E.D. ZHELEZNIAKOVA AND HER DAUGHTER T.S. ZHE-
LEZNIAKOVA

At 15:25 hrs, the prosecution begins their cross-examination of 58 year old
E.D. Zhelezniakova.

She was held in detention for three months. She was accused of corruption:
it was alleged that she had offered a bribe of 3000 dollars so that her grandchild
could be accepted into the medical institute. She denies bribery. She was arrested
on 3rd January 2000 at the canteen where she works as director. Her daughter
was also arrested without being given any explanation. She was in police custody
for six hours.

She was interrogated in a small room by three police officers. They were
smoking and blew smoke into her face. There was no lawyer present. They placed
her in a cell with twelve other people. She spent three days and three nights on
the floor of the cell without any blankets. The pre-trial investigation had shown no
interest in her case for two months. "Two months later, the examining magistrate
Terekhovich came to see me and said that my detention pending trial could be
shortened if I confessed". She suffers from bladder stones, cardiac insufficiency
and a thyroid cyst. From 24th March to 18th May 2000, she received treatment in
hospital, where she was transferred after she lost consciousness. When she regained
consciousness in hospital she saw that she was handcuffed to the bed. After leaving
police custody she was declared an invalid. The case collapsed.

Next, it is the turn of T.S. Zhelezniakova (daughter of the last witness), also
accused of having offered a bribe to get her daughter admitted to medical school.
She denies the charge.

She went to the police station after receiving a telephone call. It was at the police
station that she was arrested. They used force and put her in handcuffs after she
refused to confess, as the investigating judge proposed in exchange for letting her
go. When she refused she was incarcerated with twelve other prisoners. For two
months, she slept on the floor in a corridor where there was not enough room to
put a mattress. People walked on top of her. Between 8th October and 3rd February
2000 she was not questioned, and she had no access to a lawyer. "They threatened
to leave me there for five years if I did not confess that I had offered a bribe, and
I did not receive any paper, nor a comb, nor any bedding throughout the time I was
there and not being questioned".

Afterwards, she sent five letters of complaint to the regional prosecutor. None
of her complaints reached their destination. In November 1999, the investigating
judge told her that she would soon be interrogated. "Soon" turned out to be 3rd
February 2000.

The hearing is adjourned at 17:15 hrs. It will resume on 1st March 2001 at
10:00 hrs.

THE ACCUSED RAVKOV

MINUTES FROM THE HEARING OF 1st MARCH.

10:10 hrs: the hearing begins. There are only six people in the room not participating in the trial.

Ravkov states: "On 12th July 1999 I was apprehended at 4:30 p.m. in the neurology department. Two policemen (they introduced themselves) grabbed me, twisted my arms, banged my head against their car, and handcuffed me before putting me in their vehicle.

They drove me to the regional prosecutor's office. I demanded they explain the reasons for their actions. They gave no explanation. I demanded to be allowed to see a lawyer, my garrison head and the rector. They refused. From the regional prosecutor I was taken to the Office for the Prevention of Organised Crime. There, I was threatened. They gave me Bolotov's statement to read, in which it was stated that I came to find him in his office on 23rd May and received the sum of 2,500 American dollars from him. The statement was dated 12th July 1999.

They demanded that I state that Bandazhevsky had set up a criminal organisation at the institute. They asked me to give marked banknotes to Bandazhevsky. If I did it, I would be freed. The proposal came from the investigating judge, Krugliakov.

I was interrogated by six men. They threw me into a cellar. It was hot. I asked them to give me some water, which they did. After I had drunk, I felt funny. Then I don't remember anything. Everything became blurred. I found that funny, I kept laughing uncontrollably.

During the night I was taken to the institute, where they searched my two offices. At 4 o'clock in the morning they took me to the IVS (military detention centre). I felt sick. I started vomiting. I had pain in my liver. In 1987, I had viral hepatitis. I asked them to get a doctor. Their response was to ask me to confess. At midday, I had written everything they wanted: I had received money, I had handed it over to Bandazhevsky... I don't remember everything I wrote. I have looked through six volumes from the case. My statement isn't there. I spent ten days at the IVS.

The investigating judges threatened to put my wife in prison. The next day, the lawyer that my wife had employed was not allowed to see me. During the interrogation, the investigating judge, Alexandrov, was drunk. They had drunk the bottles that they had found in my car. Alexandrov told me that they were searching my house. That they had found 20,000 dollars. He put me on the phone to my wife. I could hear noises and my wife crying. Afterwards I promised I'd say anything they wanted as long as they left my family alone. At the detention centre, the interrogation lasted twenty two hours until 2 o'clock in the morning.

I had pain in my liver and I demanded to see a doctor. I wrote to the regional prosecutor through the investigating judge Alexandrov. They never allowed the lawyer to see me.

At the detention centre, they put me in a cell where there were 16 other detainees with eight beds. I lay either on the floor or on the bed for several days asking for medical help.

At the beginning of August, they did a blood test. They discovered that my liver was inflamed. My skin and my eyes had turned yellow.

During the eighteen months that I spent in detention, I only saw the investigating judge three times: at the end of September 1999, when he asked me to confess in exchange for a shorter period in custody; in February 2000, when I was in a critical state, and could hardly walk. At the beginning of June 2000, he repeated his offer and said that I would stay in prison until I became an invalid".

At 13.00 hrs, the court is adjourned until 14.00 hrs.

After the break, there are only four people in the courtroom, including myself and Bandazhevsky's wife.

Ravkov continues: "During the interrogation they kept saying "You are just a pawn. Our main goal is to get Bandazhevsky; it's him we need".

Ravkov then says that during the year 2000, the death rate in Gomel and in the Gomel region exceeded the birth rate by 8,500. "Over the last five years the number of deaths caused by cardiovascular problems had doubled. Bandazhevsky had openly tried to convince (*of their error*) the ninety two scientists who had written to Gorbachev in 1989 to say that the tragedy at Chernobyl no longer posed any danger. The Institute of of Radiological Medicine and Endocrinology[190] had conducted an experimental simulation: rats were exposed to radiation by X-ray imagining that in this way, they had created a model of the situation in the contaminated territories. Bandazhevsky did things differently: he got hold of grain from the contaminated territories and fed it to the animals. We proved that food from the contaminated territories is unfit for human consumption. I think this is the real reason for the legal action taken against Bandazhevsky".

Ravkov points out that volume 11 of the dossier contains a "signal" (the document uses this word) from the vice-rector at the Institute of Medicine, Sokolovski, addressed to the secretary of the Security Council, Sheiman, informing him that Bandazhevsky was being bribed, that he had given out the codes of examination papers that the papers were coded by Ravkov and that everyone knew the exam questions. That Bandazhevsky had lost his head. That the institute was full of "enemies of the people" who, by publishing their papers and articles would bring about the downfall of the Belarusian people.

Sokolovski states a little later that the signature was not his. The document is not dated.

---

[190] This is the institute whose work was criticised by the commission made up of Nesterenko, Stozharov and Bandazhevsky. See Part Three, Chapter III, starting on p. 197.

At about 16:00 hrs, Ravkov asks that the hearing be suspended to 2nd March because of his ill health. The president asks the opinion of the doctor who is in the courtroom; he confirms that Ravkov's blood pressure is significantly higher than in the morning and that his request should be accepted.

The presiding judge announces an adjournment until 2nd March 2001 at 10.00 hrs.

MINUTES OF THE HEARING ON 2ND MARCH 2001

The prosecution's cross-examination of Ravkov's continues. To the question: "Did you ever give money to Bandazhevsky?" he replies "No, I never gave or received any sums of money".

The prosecution reads out the minutes of Ravkov's interrogation as a witness of 12th July 1999 in which he admits having received 2,500 and 1,500 dollars.

Ravkov retracts his confession explaining: "I was interrogated by the prosecutor Pavlova. During the interrogation they were in constant communication over the telephone with my apartment. I insist on the fact that there was a psychotropic drug in the water I was given to drink. In the state I was in, I would have confessed to anything. Over the last twelve years I have not had any problems with my hepatitis but after the interrogation, my health deteriorated rapidly".

The prosecution reads out Ravkov's statement to the regional prosecutor on 13th July 1999 (page 72–73, t. 38) where he voluntarily states that he received a bribe and expresses his regret. In the statement he says that in 1997 he had a conversation with Bandazhevsky in which "we decided to find parents who wanted their children to be given help with the admission exams and to ask them to pay". Ravkov gives the sums and the names for the years 1997–1999. "I gave the whole sum directly to Bandazhevsky in his office". There is a declaration in Ravkov's hand at the bottom of the page saying that only these witness statements should be taken into account and that the statements he made on 12th July are not accurate.

Ravkov objects that this text was dictated to him. The judge asks him "Who asked you to dictate it?", and he replies that he does not know the man, but he was short and had a moustache.

The prosecution reads out the minutes from the interrogation that took place on 13th July 1999, with the investigating judge Alexandrov. The time according to the minutes was: from 21.35 hrs to 02.00 hrs. The interrogation took place at the detention centre. (In fact Ravkov was only transferred to the detention centre ten days after his arrest on 12th July 1999). According to these minutes, Ravkov admits having been bribed. In 1998, three times with 4,900 dollars. He states that he took the money to help Bandazhevsky and he gave the programmes and the exam questions to people who wanted them. That the money was given to Bandazhevsky once the candidates had been admitted to the institute. He kept nothing for himself. In 1999, he allegedly gave Bandazhevsky 6,900 dollars, a total of 11,800 dollars: "From 1997, in total, I received 11,800 dollars". To a question posed by Alexandrov: "Was the sum that you were going to hand over to Bandazhevsky agreed on in advance?" The minutes record that Ravkov replied in the affirmative.

In the witness box, Ravkov states that the interrogation did not take place in the detention centre. He gave them the figure of 11,800 dollars after he had been put on the phone to his apartment and he heard his wife crying. He had been pressured. There had been no agreement with Bandazhevsky. The statement from the witness Bolotov that stated that Ravkov had received the money from him at work could not be true because 23rd May was a Sunday, a public holiday.

The prosecution reads out the minutes of the confrontation between Ravkov and Bandazhevsky that took place on 15th July 1999, at the military detention centre with the investigating judge Alexandrov. In the minutes it says that Ravkov and Bandazhevsky refused to have a lawyer. Ravkov states that he gave Bandazhevsky the list of names and the money, that Bandazhevsky knew he was being bribed and that he himself had given the sums and the dates.

Reply from Bandazhevsky: "That is not true".

In the witness box Ravkov explains that before the confrontation, they had spent three hours telling him what he needed to say. "I hadn't eaten, or slept for five days", he says. So of course, he capitulated to all their demands.

The prosecution reads the minutes from the interrogation that took place on 19th July 1999. The investigating judge: Alexandrov. From 10.00 a.m to 11.30 a.m. According to the minutes, Ravkov gave some of the money to Bandazhevsky, as they had agreed, and that these were bribes.

In the witness box Ravkov explains that he had not been allowed a lawyer, that he was shown the lists, that he was told that if he refused to testify they would carry on interrogating him all night, and that he gave these testimonies when he was extremely ill.

The prosecution reads out the minutes from the interrogation that took place on 21st July 1999. Investigating judge: Alexandrov. According to the minutes, Ravkov confesses his guilt on all counts. He was taking bribes, and this was at Bandazhevsky's instigation. He gives the sums and the dates.

In the witness box, Ravkov retracts these statements and says that he was in such a state that he did not care about anything.

At the request of the prosecution, there now follows a confrontation between the two accused, Ravkov and Khomchenko. The prosecution asks Khomchenko, for each of the episodes concerning her and Ravkov, to confirm the statements she made during the preliminary investigation and whether she still stands by them. Khomchenko replies in the affirmative for each episode. Ravkov himself denies having received or given any money.

The prosecution tries to pose questions on this subject to Bandazhevsky. But the presiding judge does not allow it, as Bandazhevsky was not part of the confrontation.

Baranov, Bandazhevsky's lawyer, asks Khomchenko (*he does not have the right but the judge allows it*) "The prosecution has asked you to confirm all these questions and you have done so. Have you made some kind of deal with the prosecution?"

She replies that she is not confirming her statements for all of the indictments and does not stand by all of them.

Therefore from a legal point of view, the result of the confrontation has almost no significance.

MINUTES OF THE HEARING ON MARCH 19TH 2001

The hearing begins: 10.15 hrs. Cross examination of V. Ravkov.

To the question from the defence "Did the people who arrested you know your military grade?" *(lieutenant colonel),* the reply is in the affirmative. Ravkov then tells them that a week before his arrest an attempt was made to corrupt him: he was offered a bribe. After his arrest, his home was searched without an official search warrant.

Ravkov comments on the lack of objectivity in Khomchenko's statements. He says that this woman had been treated in detention the same way as he had and knowing this, it should be obvious that she would say anything to avoid any repetition of that experience. Ravkov reiterates that he never took any money from Khomchenko.

Cross-examination of Ravkov is over. The presiding judge adjourns the proceedings until 14.00 hrs.

The hearing begins again at 14.10 hrs. The presiding judge reads Bandazhevsky his rights.

The prosecution: "Begin from the time when you became director of the institute". Bandazhevsky explains that in 1990 the Minister of Health of the Soviet Socialist Republic (SSR), of Belarus Kozakov, suggested that he organise and become director of a medical institute.

"Forty per cent of the staff at the institute were specialists from Gomel and the Gomel region. There were no other staff to be found elsewhere. At the time, the country was in a state of collapse. At the beginning, we invited staff from Grodno, Vitebsk and from Russia. We even considered closing the institute at one time because of staff shortages. So we set about training specialists from the staff at our disposal.

The institute concentrated on the effects of radioactive caesium on animals and humans. The institute was not subsidised at all. But within a short time we had trained six PhD students. Our institute was awarded the Lenin-Komsomol prize from the Soviet Socialist Republic of Belarus.

By 1999 there were already twenty-four PhDs at the institute. We could compete with the old established universities. We made a number of films about our research. We had developed a methodological approach to the problems of Chernobyl.

Of course we had to dismiss members of staff who could not or would not do their work. For example, the first vice rector, the deputy dean, the vice-rector in charge of administration (he has now been re-appointed to this post), the chief accountant.

All students were judged according to the same criteria. In 1997, 49 students were sent down because of poor results, in 1998, 36 and in 1999, 67. These were first and second year students".

The prosecution: "What was the atmosphere like at the institute?"

Bandazhevsky: "Some people weren't up to it. Not all the specialists had the necessary level of competence. There was some tension, some dissatisfaction. Some people wrote joint letters of protest. The institute was inspected. In 1999, I had to go to Warsaw, to receive an international prize. It was that very day that an important commission came to inspect the institute.

After he was sacked Sokolovski engineered a denunciation and sent it to the secretary of the security council. Two students, Miskova and Pachaiev, who had been sent down, also played a part in this story. Certain people wanted to have *"their own"* rector.

Khomchenko, Ravkov are pupils of mine. Within a few years of study, they had their doctorates. Fourteen doctorates awarded in 1997 and 1998 alone.

I only met the examiners in my role as president of the admission board the night before the exams, when I would give them their instructions. Lenkevich posed particular problems for me. He worked in sanatoriums examining children. The results of his examinations differed according to whether he was reporting to me or presenting a report at an official meeting. I had even considered appointing him as head of the department of nervous diseases. But then I learnt that at an official meeting, he claimed that caesium had no harmful effects on a child's body. This was 5th August 1998. He never attended meetings of the scientific council. He was malicious.

I never conspired with anyone, let alone for criminal reasons. My life is devoted totally to my work.

There is also a lot of malice towards me in the media. I am under enormous pressure because of the opinions I hold and that I have made known throughout the world. There is enormous opposition from the IAEA. A secret agreement was signed between the IAEA and the WHO in order to control information about the consequences of radioactivity.

In 1999 we showed our film *Сердце на ладони (The heart on the palm)* and then we brought up the question of transferring the Institute of Radiological Medicine of Minsk to Gomel before the Academy of Sciences committee. I took part in the commission charged with monitoring the activities of this institute. But I did not have the time. I was never present at the admission examinations but I never let anyone approach the board either. Tsepkalo, who monitored the work of the board, can confirm this.

No-one ever asked me to intervene on behalf of a candidate for admission. I never told anyone any information about the content of examination questions. I did not know that professors on the examination board were involved in cramming candidates. This is against the rules laid out in "Regulations for examination board professors on the selection of candidates for admission".

After my arrest, I lost 35 kg in weight. I suffered a gastric haemorrhage. They wouldn't send me to hospital".

At 17.15 hrs, the president adjourns the hearing until 20th March 2001 at 10.00 hrs.

2. JUDICIAL AND PROCEDURAL GEMS

During the investigation, the lawyer Baranov had asked for the case against Bandaz-hevsky-Ravkov to be dismissed but the main prosecutor, Judge V. K. Terekhovich rejected the case for dismissal in a decree dated 5th December 2000.

> According to the defence, the preliminary investigation has not presented evidence proving the guilt of the former rector of the medical institute at Gomel, and the accusation is based on supposition and contradictory statements from other defendants.[191]
>
> The lawyer A.P. Baranov has said that no "direct evidence" has been provided that Y.I Bandazhevsky received bribes for the admission of candidates to the institute. In a criminal trial, evidence can be either direct or indirect, and neither has more force or priority over the other. The evidence collected is evaluated in its entirety.
>
> Unlike the other defendants who are being tried, Y.I. Bandazhevsky had no direct contact with the parents of candidates, did not conduct interviews on their admission to the institute and did not receive any material recompense from them.
>
> However, having analysed all available information about this affair, the investigation has concluded that the quality of the evidence is sufficient to bring Y. I. Bandazhevsky to trial. [...]

---

[191] Y.D. Yankelevich claimed to have given Bandazhevsky 7,000 dollars. Even though he had been accused of corruption himself, found guilty, and then accused of perjury, he was called to give evidence in court as a "witness" against Bandazhevsky. The military prosecutor interviewed on Moscow television, said: "Look at Yankelevich's testimony! He declared "Yes, I gave Bandazhevsky 7,000 dollars to allow two students admission to the Institute. Well, doesn't that constitute proof?" This statement, in the absence of any direct material evidence, after a year and a half of investigation is nothing short of slander. (cf *Nuclear Controversies*, by W. Tchertkoff, Feldat Film).

Taking the above into account, the case for dismissal on the grounds of lack of evidence, presented by Y.I. Bandazhevsky and A.P. Baranov is rejected.

3. PROFESSOR BANDAZHEVSKY'S FINAL STATEMENT, 24th MAY 2001

"Judges, like doctors, have to deal with human destinies. They both base their conclusions on a number of objective indications. It is these indications taken as a whole that allow the judge to formulate a charge or the doctor to diagnose the illness. It is only the proof of the existence of several objective indications in a patient that allows the doctor to diagnose a particular illness correctly and to prescribe the right treatment. It is impossible to make the right diagnosis, to determine the illness precisely, on the basis of one indication, especially if it has not been confirmed by objective methods. I think the same principle applies in law.

However, in my case, in order to prove the crime that I am supposed to have committed, the Republic's state prosecution is based only on statements from people who have their own interests in the outcome of the case—other defendants. The case has been fabricated from start to finish: the charge is based on unfounded statements. These seem to suffice to accuse me of a serious crime. The investigation has no witnesses to the crime I am supposed to have committed; neither do they have any material proof, or incriminating evidence. I have the impression that the prosecution does not itself believe the charges against me are true. In accusing me of an odious crime, and corruption is certainly that, they produce as evidence, slanderous statements made by other defendants, Ravkov (for the prosecution), Shamychek and Yankelevich. All of these statements, without exception, are absurdly confused and illogical. It is probably the reason why the prosecution asked the court not to allow television and radio to record the case for the prosecution. Otherwise everyone would have understood instantly why Professor Bandazhevsky was on trial.

The prosecution knew, long before they began investigating the case, what conclusions they would reach in order to accuse me of having perpetrated a serious crime. One only has to read the provisional detention order. Ravkov's statement and of course, the signed denunciation by the vice-rector Sokolovski, were quite sufficient. The conditions under which these statements were obtained did not matter to them, as long as it achieved the desired result: get rid of Bandazhevsky, turn him into a criminal despised by everyone. So, all his achievements in science and in medicine, would be forgotten. I do not know the exact circumstances in which Ravkov made his statement, but it is likely that conditions were extremely harsh. How else can one explain the text recorded in the minutes of his interview on 13th July 1999 (t.1 LD 126–133): "At the end of 1997, Y. Bandazhevsky, the rector of the medical institute telephoned me and said I needed to honour my function...I made a list of the candidates in my own hand...I can confirm that throughout

1997, I also received 1,800 dollars from pupils' parents". How can one talk about receiving bribes on behalf of candidates for the entrance examinations, at the end of 1997, when the entrance examinations to the institute were already over? The need to dream up a criminal act quickly seems to have deprived the prosecution team of elementary logic. In the minutes of the first of Ravkov's statements, apart from whole sentences that were prepared in advance with the participation of the people mentioned above, there is absolutely no concrete information about how the money was transferred (neither dates, nor amounts).

If I had not known the content of Sokolovski's denunciation of me, I would have had more reasons to believe that Ravkov wanted to offload his responsibilities onto me, the rector. But knowing how they harassed me, I have every reason to believe that Ravkov was forced to vilify me.

I would like to avoid talking in clichés about conscience and honour but at present, there are very few people who have remained faithful to me and to my family. And I am deeply grateful to them.

I would like to express my gratitude to all those who have supported me and continue to support me today. I have discovered that I have a huge number of friends all over the world. And that they are not indifferent to my fate. They are convinced of my innocence. Their confidence gives me strength. I have always tried to work professionally whether as a scientist or as a doctor or teacher. I am proud of the results of my work: the discovery of new facts, that were unknown before now, the research into problems that are of vital interest to the whole of humanity; I am proud of my medical institute, even if today it is on its knees and humiliated; I trained more than a thousand doctors and most of them will make excellent consultants.

To the prosecution I say, you cannot prohibit me from thinking. To do that you would have to take away my life. Thinking is my professional activity. To the great regret of those people who wish me ill, the crime of corruption does not carry the death penalty. That would probably have made them very happy".

Chapter II
THE SHOCK OF PRISON

After sentence was passed, Yury Bandazhevsky began his own Journey to the Cross for which he had not prepared himself. Right up to the last minute he believed he would be found innocent. As free men and women, our task had been to support him and his family, morally and materially and to publicise his case as widely as possible. We were going to discover also that he would need our help now to avoid the traps, the provocations and the manipulations of people who wanted to sever the links between himself and his wife Galina Bandazhevskaya, and through her, to the growing support of the international community. Only Galina had the right to visit him, periodically, in prison. Without her, no-one would have known anything about his years in detention.

Bella Belbéoch reacted straight away. On the very day he received his sentence, she wrote to the ambassador from Belarus in Paris, Vladimir Senko. In vain. But the wheels of history turn, the truth remains. Here is an extract from her letter:

> None of the corruption charges against him were proven, which confirms the fact that it was for his scientific work that Professor Bandazhevsky has been condemned. This is the first time in the history of science, at least in the West, that a scientist has been tried and found guilty by a military court for his scientific research. It takes us back to the Stalinist era, which we believed had been left behind when your country gained independence.
>
> You are, Ambassador, a Permanent Representative to UNESCO. I have heard you express yourself with feeling on the subject of the health problems experienced by the people of Belarus following the Chernobyl disaster. I find it incomprehensible that, in this situation, through your indifference, you join in the condemnation of this scientist, Yury Bandazhevsky, whose principle motivation is the same as yours: to help his fellow countrymen.
>
> As secretary of the Scientists' Group for Information on Nuclear Energy, I am asking you to transmit my words of protest about the judicial procedures used against Professor Bandazhevsky to the authorities in your country.
>
> I hope that you will also pass on to His Excellency the President, Mr Lukashenko, my indignation at the condemnation of this man of science whose only concern is to determine the real impact of Chernobyl on the deteriorating health of people, in particular children, living in the contaminated territories and to come to their aid;

but also to undertake this pioneering work on the previously unsuspected effects of internal contamination on the human body which could lead to the re-evaluation of international norms of radioprotection.

We were in Minsk when Russian and Belarusian television announced the verdict. We had come from Kiev where we were filming the WHO conference. Bandazhevsky's work had been presented by Vassili Nesterenko, Alexei Yablokov, and by Michel and Solange Fernex. But his work had been greeted by stony faces among the UN officials, and by visible irritation from Professor Ilyin's scientific team.[192]

Two days later, Nesterenko, at the wheel of his car, thought out loud as we drove to Galina Bandazhevskaya's house. She had come from Gomel to meet us.

V. Nesterenko—You told me I was wrong to worry, that I was exaggerating, that he would hold up psychologically. But I went through all this in 1986 when I was betrayed by most of my colleague at Sosny. What's happening to Yury is exactly what happened to me. Recently, he has been telephoning frequently and saying: "Excuse me for telephoning you every day, but I have no-one else to talk to. Everyone has turned their back on me". Eight years in prison, that's very hard…

I think Galina has been allowed to make a first visit to the prison…She wanted my advice. What is to be done? She is asking herself "How will I carry on with my life?" The situation is difficult. Those who really did take bribes have been freed… In his case, nothing has been proved. Given that he is the only one to be convicted, with his vice-rector, I think it is yet more proof that it is politically motivated, and specifically to do with Chernobyl. One thing is certain; they have produced no proof, absolutely none. It reminds me of 1937 with the famous purges. They're using the same methods.

What is truly regrettable is that his research has been stopped… Yury has obtained unique results. It was a revelation for everyone, both at the Minsk conference, in March, and at the conference in Kiev. Professor Savkin claimed at Kiev that "according to classic radiobiology, caesium is distributed evenly throughout the soft tissue". But Bandazhevsky made direct measurements of organs from bodies he was dissecting, and gave caesium at different doses to rats and guinea pigs, and he discovered inequalities of accumulation up to a factor of 10. The very high level of caesium in vital organs such as the heart, the kidneys, the liver, in comparison with the average over the whole body, explains precisely the dramatic rise in illness. It seems to be of great significance. Official science appears

---

[192] See "Nuclear Controversies" by W. Tchertkoff at *https://www.youtube.com/watch?v=MZR_Fvp3RrQ* .

to know nothing about it. It's a shame…I had just arranged, with Professor Tsyb in Kiev, for a pathologist to come from Moscow to see Yury and look at his histological sections of tissue that has been altered by caesium. I showed him the correlation curve established by Yury between alterations in electrocardiogram results and the accumulation of radionuclides. It was our collaboration that made these results possible. I provided him with my data on the incorporation of radionuclides in children and he compared them with the results of his clinical examination. It seems to me that this is the key to the whole question. We complemented each other's work. I am not a doctor, I am a physicist. I need medical expertise to know the limit at which there is a risk of danger. As a radioprotection specialist, I know that we need to reduce to 30% below this limit, because this is the threshold above which interventions are needed. But they stopped our research…

Galina Bandazhevskaya was waiting, overwhelmed by the enormous weight of responsibility that had now fallen on her shoulders. She knew that she would have to bear this responsibility alone for many years to come.

G. Bandazhevskaya—I have to do everything I can to get him out of there. Maybe the European Parliament could do something, make a request…Because if we do nothing…eight years… he won't make it. And to think that all those teachers that did accept bribes have been freed! By the way, Yury told me that in the police van that took them from the court to the prison, Ravkov asked his forgiveness for having slandered him and he told him that he had not been able to resist the pressure…They are all dressed in black there. Everything is black. I was able to bring him the regulation 30 kg parcel, of clothes, medicines and food. It's absolutely essential because the food is terrible. He told me that in the morning they get a ladle full of a sort of gruel, without butter, made with water. At midday, a soup made of who knows what, with water and a few bits of macaroni and beetroot floating in it. And in the evening, the same gruel again. That's all. Bread by itself, no butter, no cheese…That's why they allow you to bring these big parcels.

V. Nesterenko—Thank goodness they did not confiscate your flat in Gomel. As soon as they have finished making an inventory, you must try to exchange it for a flat in Minsk so that you can come and live here. I would like to invite you to come and work at the Belrad Institute, I would like to make you director of research in the medical field. The main thing is to set up some worthwhile projects because you need to earn your living. In fact, I think they would make life very difficult for you if you carried on working in Gomel.

G. Bandazhevskaya—Yury Bandazhevsky always lived in a dream. During the eighteen months while he was temporarily freed, instead of writing books, he could at least have given me power of attorney for the sale of the house…The night before the trial, I thought that Yury had almost regained the health he had before he was first arrested. In a way he had adapted to his new situation, no longer directing an institute. He wasn't worried about the investigating judges anymore. He got on with his work, which was what he loves more than anything else. He seemed to have gained a certain level of serenity. And it was in that generally positive spirit that he approached the trial.

Q—But did he try to intervene during the investigation; to know what the investigating judge was doing, to make requests to official organisations, etc?

G. Bandazhevskaya—He seemed indifferent to these questions. For example, the lawyer, it was me that chose him. He was completely convinced of his innocence and he was sure that the court would exonerate him. He approached the trial with 100% optimism. He was convinced right up to the very last day that he would prove his innocence. The lawyer and I were present for every session of the hearing. And I thought that he would come and talk to me after each one of these sessions about the way he should construct and organise his defence…During the trial, he was very alert. He asked lots of questions, he noted everything. The rational attitude of a scientist, quite distant in his attitude. He noted all the contradictory statements made by witnesses, but afterwards, he didn't talk about any of it. He didn't have total confidence in the lawyer. He was convinced that he should defend himself, and this is what he was doing, as an experiment, as a scientist. The lawyer made a lot of use of the analyses Yury had made during the investigation and the trial, and used them later when he was making his case for the defence.

Yury was really shaken for the first time when the public prosecutor asked the court for a nine year prison sentence. Suddenly he felt completely undermined. I still remember it now. He was panic-stricken. He had been sure right up to the last moment that the judge would dismiss the case for lack of evidence.

Chapter III
THE DESCENT INTO HELL

Extracts from *News from Prison* that I disseminated to supporters of Yury Bandazhevsky in the West, while he was in prison.

1. INITIATION INTO THE PRISONER'S WORLD

20th JULY 2001 (ONE MONTH AFTER THE VERDICT)
CONDITIONS IN PRISON

On the 18th July, after fifteen days in quarantine, during which he had been registered, photographed and submitted to various medical tests, Professor Bandazhevsky was assigned to "Detachment 21" in Minsk prison, at 36 Kalvariyskaya Street.

A large cell with 80 detainees in bunk beds. He was assigned a top bunk, which meant that he could not sit up to read, as the ceiling was too low: "I write my letters on the toilet". There are no tables or cupboards of course. He keeps his papers and his books in a bag. Contrary to information received initially, he can receive newspapers and books, by registered mail, with a maximum weight of 2 kg per parcel sent. He can correspond freely, but the process takes a long time because of prison monitoring. A month after his incarceration, Bandazhevsky received his first packet of letters from Belarus, which had all been opened and read. Galina Bandazhevskaya had asked as many people as possible to write to her husband, even if it was just a short note. He reads French and English. All personal messages from the outside world brings him great joy, a balm to a prisoner's heart.

Professor Bandazhevsky had only been in prison one day when his wife came to see him. He would be living there for eight years (minus the five and a half months when he was remanded in custody). What do the other prisoners do during the day? Nothing "They fool around", he said. He thinks that some of them go to work in a furniture factory. These are uneducated people, a bit lacking intellectually, no-one with whom he could have a sustained conversation. But all the detainees, irrespective of their level of intellect, have a lot of common sense and understand the situation perfectly. Nobody is taken in. They understand the reality of the regime. They are freer than the "free", and relationships with them are excellent. It is the humanity of prison.

Bandazhevsky continues to devise scientific projects in spite of everything. He is working on an article about a subject that has haunted him ever since he was deprived of his research facilities. He has dictated a list of books to his wife. All the same, she found him disoriented. "I want to concentrate, I still have things to say", he told her.

Galina Bandazhevskaya has asked me to pass on to those European Members of Parliament who supported her husband by awarding him a Passport For Freedom[193], her request that they write to the prison director asking him to allow Bandazhevsky to continue with his scientific work. Minimal material conditions that would allow him to write, arrange his papers and his books. Galina Bandazhevskaya has also asked that scientists and doctors who supported her husband to write to the Chief Medical Officer of the prison asking him to ensure that he gets the right food because he has an ulcer that bleeds, and to authorise the delivery of several food parcels each year.

2nd AND 4th OCTOBER 2001

The mother and the wife of Yury Bandazhevsky were allowed to spend three days with the prisoner, from 28th to the 30th September, in a "hotel room" made available within the prison walls for "extended visits". They brought him a 30 kg bag of food, which all three had to use up during their three day stay, since no-one is allowed out of the prison to go shopping. The next extended visit will take place in six months time in April; and a normal visit, a conversation by telephone through a glass partition, at the end of November (four months after the preceding one in July).

Lost among the mass of common criminals, Yury Bandazhevsky feels isolated because he is not like the others. He is finding it hard to adapt, and to find a place within which to draw breath. He veers between, on the one hand, depression and a disoriented state of panic and on the other, a euphoric creativity, when he is overtaken by scientific reflections. But these moments of intense activity, where he finds himself again, are interrupted and come to nothing because of a lack of means: minimal material conditions and a total lack of scientific literature.

Like all the other prisoners, he wears an all-in-one uniform made of black material with his name on it, his head is shaved and he wears a round black cap. The final insult—it was his wife who had to provide these clothes (the prison didn't have any). She went round all the shops in Minsk to find them, and ended up getting them from a workman in the building in exchange for half a bottle of vodka. There

---

[193] See Part Three, Chapter I, p. 169.

is a tiny television in the communal cell but it is impossible to get anywhere near it with the mass of other people crowding round.

Yury Bandazhevsky is a scientist through and through, a sensitive man, polite, and respectful of others. He cannot defend himself physically, stake out his corner. Disoriented, humiliated, clumsy, a fish out of water, "but however humiliated I feel, I would be alright if I could work" he says. He is trying to hold on, but against a background of anguish and depression. His wife has noticed he has a persistent cough, a nervous tic. He has told her, several times, with tears in his eyes: "I won't be able to hold out here. My nervous system will never adapt". She is convinced of it. So is the prison officer in charge of Yury who advised her to bring him vitamins and better food, "otherwise, with this length of prison sentence, his health will begin to deteriorate, first his teeth will fall out, and then the rest".

While he was under house arrest in between the two periods of imprisonment, the Bandazhevskys had taken refuge in the hope that the growing support from the West would spare them this ordeal.

Since Professor Bandazhevsky was sentenced, without appeal, to eight years in prison, no-one talks about the charge against him of extortion or corruption in Belarus. If a journal in official scientific and medical circles mentions him, it is simply to denigrate his competence as a scientist and to cast doubt on his mental health: "I am a dangerous man" he concludes. The fact is that the people who put him through this shameful trial wanted to break him and to discredit him, and they will succeed in this if he is not released very quickly.

Concretely

The governors at the prison have not granted him any special privileges. He is treated exactly the same as all the other prisoners in this type of camp (or penal colony as they are called today). He can go to the library to work, but it is devoid of scientific literature. His wife wanted to bring him a typewriter, but it is not allowed. He has asked his wife to bring him a set of cassettes so that he can learn English. But cassettes are not allowed. The camp administration that receives letters from abroad, keeps saying: "Professor or no professor, in the "zone", it's the same for everyone. It is unkind to encourage his illusions. He needs to get used to the idea of eight years. If not he won't make it".

Request for a pardon

Bandazhevsky's wife, mother and daughter have all written to the President of Belarus, Lukashenko, to ask for a pardon. They received a reply saying that the presidential administration would consider their demand when Bandazhevsky himself made the request. Yury Bandazhevsky does not intend to repent for a crime he has not committed; he has prepared a letter to Lukashenko in which he claims his innocence and asks the president to exercise his right of pardon to allow him to take up his scientific work again as a free man.

Professor Nesterenko has learned from the French ambassador that on 16th August 2001, European ambassadors in Minsk approached the Minister of Foreign Affairs on Bandazhevsky's behalf. They told him that they considered the severity of the sentence to be politically motivated and asked for a presidential pardon, on the grounds of Professor Bandazhevsky's scientific merit and the lack of any proof against him. They also asked, in the meantime, that the conditions under which he was being held be improved. The Minister said he would refer the question higher, but that no such request had been received by the government from Bandazhevsky himself. The management at the penal colony say that he has not fulfilled the necessary conditions for his request to be transmitted.

Demand for a retrial.

The lawyer has prepared a case for the Commission on Human Rights in Geneva and will have it translated into English. The report will be sent as soon as the translation is ready. The same report will be delivered to the Supreme Court in Belarus, in the form of a complaint denouncing the irregularities during the trial and demanding a retrial. The ambassadors' intervention has, actually, cast doubt on the verdict ("political case") and constitutes the new fact, the appropriate grounds, for which the lawyer was waiting, to file his complaint. His case is due to be heard at a consultation at the Supreme Court on the 5th November.

Request for support

There are two things that are saving Bandazhevsky and protecting: his scientific work and the many letters of friendship and solidarity that he receives, although these have decreased in number recently.

Professor Bandazhevsky has asked for the following:

—That Mrs Adi Roche, who received the Frantzisk Skarina medal from Lukashenko for the humanitarian work that she undertakes on behalf of the victims of Chernobyl every year, write to the president of Belarus to ask for his pardon;

—That each of the signatories to the Passport for Liberty that was awarded to him by the European Parliament—Marie Anne Isler Béguin, Mário Soares, Jacques Santer, José María Gil-Robles Delgado, Michel Rocard, Elizabeth Schroedter, Paul Lannoye, Daniel Cohn-Bendit, Ari Vatanen, Fodé Sylla, Thierry Jean-Pierre, Ole Karupp—do the same thing, as well as IPPNW, who awarded him a medal at their 14th International Congress.

He will send his letter to Lukashenko next week.

1st DECEMBER 2001
THE PRESIDENT REFUSES TO PARDON BANDAZHEVSKY

Professor Bandazhevsky and his mother received a reply from the presidential administration of the president of the Republic of Belarus dated 20th November 2001:

> We are writing to you to say that your request for a pardon for Y.I. Bandazhevsky has been examined and found unsatisfactory due to the fact that he has served so little of his sentence, the seriousness of the crime he committed and the social danger that he represents.
>
> *Signed V.I. Samusev, head of service for citizenship and pardons.*

Bandazhevsky receives about thirty letters every day from all over the world, from strangers expressing their solidarity, their warmth and their friendship. "They keep me alive" he says with gratitude, hoping that they do not stop. He is holding on and prepared to resist during the years to come whatever the cost, knowing that if he gives in psychologically, it would mean the end.

Apart from his use of the library where he can read and write during the day, his conditions are the same as those of the other prisoners. His wife had asked that his diet be adapted because of his ulcer. After an examination by endoscopy, this was refused because the ulcer was "healing". In the cell, with its concrete floor, where between 80 and 100 prisoners sleep on three-tiered bunk beds, the rats fight over scraps of food. The tiny television set, which at least shed some light on the outside world, is broken.

Bandazhevsky suffers most from the lack of information and not being able to work as a doctor and researcher. He longs for intellectual activity, for research, for creative exchanges. His wife is not confident that after several years of this existence, Yury's spirit will remain so lively and clear. In the meantime, he had the idea of setting up a science experiment that had greatly interested some of those employed by the penitentiary (inmates selected by the management). They grow plants in pots in which the earthworms are multiplying. Earthworms, like humans, like tea. Bandazhevsky suggested they test the possible toxicity of various different teas (green, black...) by pouring the used tea leaves into the pots, and then observing the vitality of the worms. The young research apprentices of the gulag were enthusiastic about the idea. This is how the man that Professor Nesterenko considered deserving of a Nobel prize is passing his time. He would like to work in the prison's clinic but that was refused: he is a prisoner like everyone else, in a "maximum security" prison.

12th MARCH 2002

LETTER FROM GALINA BANDAZHEVSKAYA

"On 10th January I was allowed a brief visit to my husband in the maximum security penal colony. "A short visit" means a meeting with the prisoner that takes place on the telephone through a glass partition. It is true that this time the telephone was not working and we were forced to communicate with gestures and shouts. But this is not the worst of what we have endured recently, and these minor inconveniences are accepted as completely normal.

Yury is doing everything he can to cope. He continues with his work, despite the conditions in which he finds himself. He is allowed to visit the library where a place has been reserved for him to work. He works a lot of the time, and the lighting is not that good and it is affecting his eyesight. It is possible that his eyesight is worsening also because of the stress he has been under. I could be mistaken, of course; it could simply be a question of age.

To improve his work conditions, he absolutely must have access to a computer, even if this access is limited and subject to certain restrictions. We are in the 21st century and even in prison some exception should be made for a scientist. His research is not a whim. It is important. Asking for a computer is not like asking for a softer bed, a room of his own, or an improved diet. It is an indispensable tool for a scientist. In any case, work is not forbidden in the colony! And the court did not strip him of his title as a scientist. He must be allowed to work, to have access to a computer, even if it is under the watchful eye of a warden, on certain days, or at certain times. The lawyers say that we need to insist, to make demands, and then perhaps we will be heard.

The lawyer from the Helsinki Committee in Belarus, G.P. Pogonyailo, is still working on the letter of appeal to the UN (Commission on Human Rights). The work is complex and arduous. The most important thing is that the letter of appeal be presented in a way that meets the Commission's requirements.

A few words now about the research at our institute (At *this time, Galina was still working at the institute founded by her husband in Gomel. Author's note*) We received a visit from a delegation of scientists from the university of Nagasaki to conclude a contract of cooperation between the medical centres in our two cities, both victims of nuclear technology—the atomic bomb and the accident. The current rector at the Institute of Medicine, who is actually a professor, a doctor and a research scientist, *asked the Japanese what subjects the doctors and scientists should be studying at Gomel,* what work they should be undertaking to understand the impact of Chernobyl on people's health! At the moment, the researchers at the institute are studying anything as long as it is not the effect of caesium-137 on the human organism.

Two months have gone by since our last meeting with Yury. We received authorisation for an extended visit from 8th to 10th March. We are entitled to two of these visits each year. They last three days. They are of course essential for us to discuss things, to lift his morale, renew his strength and his faith in the future. This time, he arrived for the visit a bit ill. The flu epidemic spread to his dormitory and he caught it, but I think he only had a mild form. Within three days he felt a little better. He was quite sombre and depressed but the publication of his book, thanks to Nesterenko, cheered him up. The book is entitled *Pathological processes in the organism in the presence of incorporated radionuclides*. His faith in science and the feeling that he is helping people gives him strength. He very much wanted all the scientific data that he collected over his nine years of research in the Gomel region (and it is an enormous amount of material) to be used by doctors and scientists who care about the health of the people living in such dangerous ecological conditions, contaminated by radionuclides. That is why he is doing everything to publish his material even at the expense of his strength and health.

At the moment, he is continuing with his work, but with the passing of each hour, each month in prison, he feels more acutely the lack of scientific information. That really upsets him. He is afraid he will lose touch with current scientific ideas. He needs to exchange ideas with scientists all over the world.

He has asked me to express his enormous gratitude to all those who support him, who send him letters and books in the colony.

The evening before the visit, I received a parcel of food for him, from Amnesty International. The joy of knowing that people are thinking of him brought tears to his eyes. He says that he never imagined that one day he would have to be looked after by others.

On the train returning to Gomel from Minsk I was reading the newspaper (*Svobodnyié Novosti*, No 9, 7–14 March 2002) and found a statement from the Minister of the Interior, Vladimir Naumov, under the title: "There is no food to give to the prisoners". In the article, the country was informed that the Minister of the Interior would be presenting a bill to Parliament for a new amnesty. The last amnesty took place in 2000. But there is almost no possibility of an amnesty for the crime for which Yury has been convicted. The only possibility would be a reduction of his sentence by a year. In any case, up to now, that is how it has been, and we can't really expect any miracles, can we?

9th-12th MAY 2002
NEWS FROM PRISON, TELEPHONE CALL FROM GALINA BANDAZHEVSKAYA
Today I have received bad news from the prisoner in Minsk. Professor Bandazhevsky is not coping well.

During his first stay in prison, he was examined by the prison doctor and by the surgeon at the prison hospital, who made a complex diagnosis involving five illnesses, three of them chronic, two that are worsening and *a depressive state of mind due to his situation.* At the moment we do not know how serious these illnesses are, but his physical and mental energy is giving way. Prison and the passing of time are doing their work.

Bandazhevsky's lawyer has got hold of the UN Commission on Human Rights. The case was handed over to the UN representative in Minsk, on 22nd April 2002, to be translated into English and sent to Geneva, where he is at the moment. In the conclusions of the complaint it says:

> There is every reason to believe that the court case that led to Bandazhevsky being sentenced to a long period in prison was organised by Lukashenko's government in order to oust Bandazhevsky from his position as rector of the medical institute and as a renowned scientist, within his own country and abroad, for his research in the field of radiological medicine. It is obvious that his discoveries, his conclusions and his recommendations conflict with the interests of various officials in Belarus who view the ecological and health consequences of the Chernobyl disaster from a different angle and would like to pursue Belarusian government policy in this domain.

The shared aim of the nuclear lobby and its henchmen in Belarus to prevent this great mind from ever working again risks becoming a reality. Their desire to destroy Bandazhevsky the scientist was clear from the zeal and the brutality of his arrest, in the accusations made against him, and from his sentence, which was delivered without the right of appeal. If he is detained much longer, there is the fear that we may not see the same man come out of prison.

In the short anguished letters that he writes to his wife, Bandazhevsky says very little now about the scientific work with which he had hoped to continue, even though he was in prison. This would have helped him to resist better. But that is exactly what they do not want.

> I work as best I can when my brain lets me, recalling the knowledge that I had before... I try to keep going. It is very, very difficult for me to be here. I am losing contact with reality.

The famous library that the authorities vaunted to make people believe that the prisoner was enjoying certain privileges, is the end of a dark corridor, with no windows, a chair and a table against a blank wall, with his eyesight deteriorating... This little façade was organised in response to demands from all over the world, without substantially changing anything.

When the UN Commission on Human Rights gets hold of this story, there will need to be a surge of protest from international public opinion, from the academic world, and from those worthy of the name within the scientific community, so that democratic governments who are committed to Bandazhevsky's cause will demand and obtain the immediate and unconditional release of this outstanding scientist. He is the only scientist to have studied and revealed, during nine years of research, a proportional linear correlation between the quantity of radioactive caesium in the organism and the seriousness of pathologies in vital organs and systems among the abandoned populations of the contaminated territories of Chernobyl.

2. THE TRAP IS SET

12th JUNE 2002
NARRATIVE OF EVENTS, AS THE BANDAZHEVSKYS LIVED IT AND RECOUNTED OVER THE TELEPHONE TO ME BY GALINA.
A first glimmer of light at the end of the tunnel in the prison in Minsk

On *28th May*, Bandazhevsky's former secretary, who lives quite close to them in Gomel, receives a mysterious phone call, telling her that a highly placed person has visited Bandazhevsky in prison and has asked her to go to Galina Sergeievna's house, fetch two important documents and bring them, the next day, Tuesday, to the governor at the prison in Minsk: the documents in question were the famous report entitled *Monitoring the effectiveness of scientific research undertaken at the Scientific Institute of Clinical Research into Radiological Medicine and Endocrinology, on the basis of research in 1998* and the *Project for restructuring the public health system in Belarus*.

These two documents, denouncing the shortcomings of government health policy regarding the consequences of the Chernobyl disaster, had been written and sent by Bandazhevsky to Lukashenko, to the Ministry of Health and to the Security Council of Belarus, just before his arrest in July 1999. On 29th May in the morning, Yury's secretary brought the documents to the prison governor in Minsk. The chief prison officer told her that the prisoner was writing a memoir.

On *Tuesday 4th June*, Galina gets a phone call at work from Parliament, telling her to ready herself for a meeting in Minsk the next day, at a time which would be communicated to her. At 21.00 hrs she is asked to present herself the next day, early in the morning, at the prison governor's office, where a member of the government will be expecting her. She travels overnight, takes a shower at her brother's house in Minsk, and presents herself at the prison.

Galina knows this high ranking politician, who had visited the institute in Gomel several times. He got on well with Bandazhevsky, and often stayed late in the office with him chatting, particularly on the occasion when his daughter had come to register at the institute. It was this man that Galina had approached when Yury was arrested in July 1999, and it was while she was with him that she had begun to feel alone and let down: on that occasion, he had been quite guarded and evasive. I will refer to this person as *the Friend*, so as not to create more problems.

On *Wednesday 5th June*, *the Friend*'s tone and the whole atmosphere had changed. He was no longer afraid to see her: "It was me that asked you to bring the reports…I have visited the prison. I talked for an hour and a half with Yury Ivanovich and I have understood a lot of things…I want to write a report to the President. I saw Yury again yesterday to get the documents and I brought him two bags of food". The *Friend* showed her a folder with the two reports dating from three years ago and another manuscript several pages long that Yury had just written, explaining the motives for the persecution that he had suffered.

"Tell me, how are you for money?" he asked Galina.

"People help me. Yury has not had any income for three years, but I work and I get help from some kind people. Thank God".

"But you have two children. How are your daughters? I want to help you".

The *Friend* asks Galina to tell her what Yury needs in prison, because he is going to see him again in the morning and would like to take him some more food. But Galina refuses. He gives her a piece of paper to make a list.

"No, I don't want to write anything".

"But why?"

"I don't want to ask you for anything. I just want to know I can take fruit and vegetables to him when I visit in July, because there is nothing fresh to eat in the prison. Especially in summer".

Galina says that the man seemed a bit disconcerted that she was not tearful. She remained stony-faced in front of him.

"You can trust me, he assured her. I have talked to Sheiman, the prosecutor. He says that Yury Ivanovich is guilty. I've also talked to Soukalo, president of the Supreme Court… he says the only outcome is a pardon. In my opinion, Yury Ivanovich has been the victim of a plot at a lower level. The order did not come from above. It's a local matter. It isn't the President… I advise you not to politicise the issue. Ask for a pardon".

"But a pardon has already been refused".

"What do you mean?" (*Galina shows him the brief three lines written to Yury's mother from the director of the administration for pardons*) Ah, that's just a standard bureaucratic response".

"It's possible, but that's the response".

The *Friend* then turned to the assistant accompanying him, with the words: "We need to contact Soukalo and get his notes from the trial.

Galina had the impression that he was bluffing when he said this, because it seems that his assistant was about to make a gaffe. She had said "I know that he is in his office, we can call him now". "No, no" the *Friend* stopped her immediately. "That's not how we should do it" Then addressing Galina: "Listen, you shouldn't politicise this problem. I think we can sort it out ourselves, in Belarus. The delegation from the Parliamentary Assembly of the Council of Europe (PACE) are coming on 10th June and have asked for a meeting with you and Yury Ivanovich. Of course, we'll have to organise that…"

Galina Bandazhevskaya did not believe one word of the politician's good intentions any more than the people who sent him to seduce her. She is worried about Yury. She does not know how he will react to this manoeuvre. When she asks the *Friend* a question about freeing the prisoner, he replies:

"Of course, that will take time. How long has he been in prison?"

"A year and a half".

"OK, two years is a lot better than eight, at any rate".

Galina is convinced he is lying and that, if Yury believes in these illusions and abandons international support, he will not get out for another six years. And then he will no longer be the same man.

Comment—Professor Nesterenko, who knows this little fraternity, thinks that this is a very clever move on the part of the authorities. It is the only thing that might persuade Bandazhevsky to compromise and it could be fatal for him. He is at rock bottom and very isolated and those in power are dangling before him the hope that his ideas on public health will be taken into consideration at the highest governmental level, as part of Lukashenko's initiative to overhaul and revise policy at the Ministry of Health. That he could be brought on board on condition that he gives up his "politicising", in other words his support from the West. It is a possible scenario.

11th June 2002. An unexpected turn of events.

This information and the crucial documents were handed over to Wolfgang Behrendt, head of the delegation from the political Commission of the Parliamentary Assembly of the Council of Europe (35 countries), on an official visit to Minsk between 10th and 12th June.

On the morning of 11th June, Galina Bandazhevskaya had a long interview with the PACE delegation at the Planeta Hotel in Minsk. The situation was quite different from what one might have feared. Behrendt starts by telling her that when they first arrived they were not certain that they would be able to see Professor Bandazhevsky, even though they had asked in advance. But the prison visit did take place. They met him in a bedroom with three beds, with curtains at the windows, a table and a chair. Permission was given for him to have the use of a computer, and this would be at his disposal in two or three weeks. The night before, Bandazhevsky had been allowed to telephone his family. He had spoken for a few minutes to his youngest daughter, who was at home.

All the same, Behrendt warned Mrs Bandazhevskaya: "I wasn't born yesterday. This could all be a smokescreen. Be prepared for all eventualities. We asked about the possibility of releasing him. They told us: "two years", which we were not happy with. We insisted on his immediate release. That's our role". Galina had the impression she was talking to determined people who took the matter seriously, and had a clear mandate. She is impressed with the level of their preparation and the fact that they knew the Bandazhevsky case down to the last detail, and understood every nuance. Galina has a slightly more optimistic view of things, given the general political climate in Belarus at the moment, regarding its relations with Europe. The forthcoming visit from PACE is being talked about in the media—newspapers, radio and television—in optimistic and glowing terms. The Belarusian parliament has been invited, for the first time, to a meeting in Strasbourg and this has conferred on it a legitimacy that it has not previously enjoyed. Europe only recognised the Parliament of the third legislature, which was dissolved by Lukashenko when he first took power and set up a new parliament that would be at his beck and call. It is hard to imagine, in these circumstances, why anyone would be *bluffing* in the Bandazhevsky case, given that his release from prison (retrial) is one of the conditions set by the European Commission for Belarus' entry to the European Union (see the letters from Romano Prodi, José Maria Aznar, Hubert Védrine, Christopher Patten, Louis Michel, Peter Hain, EuroScience, French Academy of Sciences…) Moreover, the Belarusian media gave ample coverage to the meeting yesterday between Putin and Lukashenko, in which they talked of uniting the two countries and of the entry of Belarus into Europe and NATO.

What we are probably seeing are the political negotiations that we have been expecting since Bandazhevsky received his sentence. It cannot be excluded that Lukashenko is trying to raise the stakes. In any case, at the moment, it is Parliament— for the purposes of legitimacy—that is playing this role. Behrendt is certainly right to be cautious. The appalling violence and humiliation that have been inflicted on this man, who deserves a Nobel peace prize rather than exile in the gulag, requires total redress. It is in his own country that Bandazhevsky needs to be reinstated, honoured, returned to his Chair as professor, to his research materials, to serve the population who are threatened with veritable genocide[194], caused and covered up by the nuclear lobby.

This morning *12th June,* Galina received a letter from her husband showing her that what Bandazhevsky suffers from most is being unable to work. Scientific research is his destiny and how he makes sense of the world:

> Today I am in a new situation. The new prison block is at No 1, instead of No 21 (postal box 35-1 *at the same address*) Conditions are very good. You probably already know this. In these conditions I can even work. My position has not changed: I remain faithful to my science as before, and now I am able to continue with it. Don't worry

---

[194] The term "genocide" has been crassly trivialised by a sensationalist press in an era when the world seems to be unhinged. But I am using it here deliberately. The production of nuclear weapons and nuclear power, combined with secrecy and deceit, poses a limitless threat to the whole of humanity, particularly to the human genome as it is transmitted from one generation to another.

about me, I'm going to make it. I need to hold on. I understand all that. Everything is much better for me now and I can finally start doing some work. My view of life has not changed. An enormous thank you to all those who are helping me.

He tells her that he is starting work again on his book *Congenital pathology in humans,* in preparation for publication, a project for which he had asked for financial help from the European Parliament a few months ago.

Returning to Gomel on the train, Galina felt, for the first time in three years, a little of the weight lift from her shoulders.

3. THE FALL

In the middle of summer, the news we received came like a thunderbolt from a clear sky. Yury and his family appeared to be in serious crisis, almost breaking up. It seemed inexplicable and we would only understand the reasons a little later in the autumn.

4th OCTOBER 2002
I am writing this *News from prison* after spending twenty five days in Belarus (from 5th to 30th September) filming in some impoverished and very contaminated villages, to the East of the river Soje, cut off from the world. We have therefore been able to talk in more detail to Galina Bandazhevskaya, Professor Nesterenko, and to Bandazhevsky's lawyers: Garri Pogonyailo, who filed the complaint to the UN Commission on Human Rights, and Andrei Baranov, who defended Bandazhevsky during the trial and visits him in prison.

Yury Bandazhevsky has finally emerged from a period of psychological torment and mental confusion which has damaged his health, caused pain and incomprehension even within his own family, and made us afraid during the long summer months (July, August and the beginning of September) that we had lost him. Now we are aware of the facts and of his changed behaviour, things that were unclear at the start, we have had to conduct a patient analysis to understand what was going on, working in silence, questioning him in such a way as to discover how he has been manipulated and to disable the trap laid by the "KGB psychologists".

Since the "improvement" in the conditions in which he was being detained last June, Professor Bandazhevsky has undergone an intense and systematic process of brainwashing and disinformation. This distorted his perception of reality. The politician who made out to Bandazhevsky and to the delegation of the Parliamentary Assembly of the Council of Europe, directed by Behrendt, that he was receptive to the requests from the West, was using him as bait in the show of strength between the European Union and Belarus about the Bandazhevsky affair. We know now, from Behrendt, that releasing Bandazhevsky was indeed one of the conditions for admitting Belarus to the Council of Europe (the Belarus parliament wanted this to happen very much, but Lukashenko less so).

The cell with the three beds, the television and the computer, in the area of the prison hospital, where Bandazhevsky was transferred in June, functioned in reality, far from the view of the other prisoners, as a laboratory of neuro-psychological repression and manipulation, aimed at dragging a confession of guilt out of him and a request for pardon. People who know something about (or have experienced) the totalitarian Soviet system are convinced that psychotropic drugs were administered to the prisoner to break his spirit and to confuse him. This is the only plausible explanation for the state in which his wife found him after three months in these *"improved"* conditions. On 17th September, she wrote:

> I have only once seen him in the state he is in today: it was the day he came out of the Prosecutor's office, flanked by prison guards, when he had been informed of the accusations made against him. But then he had been alone, for a month, isolated from the world, with no contact, no lawyer. Under pressure from the "authorities" who wanted him to admit his guilt.
>
> I don't understand what is happening to him today. I ask myself over and over again what has caused this brutal change in his state of health, this sudden deterioration over such a short period of time. He is suffering from insomnia, he complains of pains in his heart, and in his stomach. He says that he is incapable of making an objective or professional judgement about anything he is doing. He says "Even I don't even believe what I am writing any more". He feels useless and disempowered, which makes him feel even more depressed. In the state he is in at present, I think it is unlikely that he will even last a year.

Thrown into confusion by the sudden improvement in his conditions at the beginning of June, and the renewed hope it inspired in him, by the promises that he might soon be freed if only he would take notice of the advice being given to him to suspend contact with his wife—because she "talks too much" and gives useless information to the West, by the harsh criticisms of Galina made by his own mother, who had herself been taken in by the *"Friend's"* promises, telling her son that she was "taking advantage of Western support", Yury kept his distance, adopted a cold attitude and refused contact with his family who did not understand him any more. This distance had been very painful for his daughters. He never wrote to them, did not invite them to visit (according to the regulations, it is up to the prisoner to invite those closest to them) even though their grandmother was permitted several unscheduled three day visits from the prison governor during which she criticised Galina harshly (Galina only learned of this by chance in July). Galina, already persecuted and humiliated at the Gomel Institute who had removed her from her post as head of department, and now unjustly slandered by her mother-in-law, somehow found the strength to rise above this manipulation. At the height of his confusion, Yury went so far as to say that he no longer trusted his wife, that her appeals for help were doing him more harm than good, that she and her friends in the West were perhaps even taking advantage of his imprisonment and of his scientific work.

How much of this confusion was real, and how much a calculated pretence to protect himself and his family? Yury knows that when he speaks "they" are listening, when he writes "they" are reading. He cannot speak or write freely. When he opens his mouth or puts pen to paper to address those closest to him he has to think about "them" at the same time. We have to decode it: ostensibly turning away from those he loves may be to protect them. When, eventually, Yury wrote to his daughters inviting them to visit, the letter never arrived. His daughters, who were still very young, had no awareness of the way the system cynically manipulates people's perceptions and feelings, could not understand his silence and were very hurt. Their immediate reaction was indignation. In any case, today, these machinations appear to have been thwarted. Galina and Yury have broken the vicious circle, regained their humanity and have found each other again.

Of course, Bandazhevsky is still frightened. One of the two policemen, who sleep in his cell, has killed three people: "When I go to sleep at night" he confided to his lawyer, "I'm not sure that I will wake up the next morning. I don't know what he's doing here". But he has pulled himself together and seems to have decided not to give in to his fear but to fight it. Galina, for her part, says that she would not be surprised if a car ran her over in the street. But she also is determined to keep us informed, despite her fear. She asked me: "Don't forget my daughters". She knows that her existence, her resolve, her freedom of speech is causing real problems for their enemies.

Bandazhevsky's mother has also understood that she was wrong to believe in the politician's promises and to follow his advice to sever the links between her son and Galina. I was present during one of the many conversations between Galina and her mother-in-law, when she apologised for the wrong she did to her son, and asked Galina to forgive her for having slandered her. She also recounted that on one of the last of the unscheduled visits she had been granted, she witnessed an attempt by his gaolers to wrest a confession of guilt from Yury and a request for pardon. Twice, the text that Yury had written himself was rejected by the prison governors and sent back to him on the grounds that it was insufficient. On the third occasion, Yury was called into the governor's office, and after a lengthy period, he came back, afraid he might have committed a folly, in signing a third option. The lawyer went to see him the day before yesterday and asked him if it was true. Yury confirmed the facts adding that he had rewritten the text in front of five (five!) uniformed colonels who demanded in a threatening manner that he sign a complete confession of guilt. Yury dictated to the lawyer the main part of the text that he had signed:

During my time at the centre for re-education through work I have reflected a lot about my destiny and about everything that has happened to me. Having analysed my situation and perceiving it now as a divine punishment for the sins that I committed in my life, I am using all my strength now in expiating them. I am asking that the act of remission be examined with humanity, if only in a minimal way, and to be relegated from the re-education centre to the penal colony. In pursuing my scientific work to protect the inhabitants in the Chernobyl region and in order to preserve the nation, I would be of more use than I am in the re-education colony where I am detained.

GOING BACK IN TIME TO RECONSTRUCT THE FACTS

On *16th July 2002,* Galina, who was visiting her parents in a village near Grodno, receives a message by telephone from the son of Yury's secretary: he says that she needs to take 500,000 roubles to Yury in prison, by the 18th July at the latest. She is told that it is very important. Galina does not have that amount of money. She does not understand but asks her brother Sasha who lives in Minsk, to find it. Sasha borrows, gets together 400,000 roubles and takes them to the prison at the required time. The young prison officer in charge of Yury has never handed over so much money to a prisoner and does not dare take it. Sasha asks him to go and ask Yury. The officer said he can't do that because it is the last day of the prisoner's extended visit from his mother. This was the occasion on which Galina learns that her mother-in-law has been given permission to make unscheduled visits without the knowledge of the family. Several weeks have gone by during which Galina has not received any letters from Yury whereas before she would get one every two or three days. Later she will learn from her repentant mother-in-law that in mid July (when the demand for money was made), the *Friend* had promised her that Yury was going to be got out of prison soon and extradited abroad. An absurd promise that never came to anything but that kept Yury in a state of impatience and exhausted stress, in expectation of favourable events. It was like banging your head against the wall, like a cat playing with a mouse.

To avoid despair in prison, one needs to give up any hope in the short term. It is only in these last few days that Yury has broken free of this psychological blackmail.

On *26th August,* Galina obtains an interview with the prison governor and reproaches him for the way he has organised visits to Yury without his wife or his children's knowledge. She demands an explanation. The governor, his face red with anger, denies vehemently that any meeting took place between 16th and 18th July and forbids her from asking him this sort of question in the future. The same day, he allows Galina a brief visit to Yury where she speaks to him by telephone through the glass screen. It was at this moment that Galina who has not seen Yury for three months notices the physical and psychological deterioration in her husband's health. (On 6th September she will write her appeal for help to the UN Commission on Human Rights. The letter has since been widely disseminated[195].) During this meeting she learns from her husband that the twice yearly three day visit with his family, planned for 5th September, will not now take place, because he has already "used up" his allotted visits with his mother.

Yury has lost a lot of weight since the last time: his eyes are staring and have dark rings around them. His teeth are crumbling and falling out. But it is his psychological and emotional state that she is most worried about. He is in a state of deep depression, indifferent to everything, even his scientific work. He just keeps saying during their meeting "I don't care, I don't care, I don't care... I can't explain the enormous pressure I'm under here. My brain is like a broken record that has got stuck in the same groove. I'll never write anything again. I am not going to work on Chernobyl again.

He tells her that his two cell mates are cops. One of them is a murderer. The promised telephone call to his family, granted to him during the visit of the delegation from the Parliamentary Assembly of the Council of Europe, has been cancelled. His daughter Olga has never received the letter he wrote inviting her to visit.

Galina is struck by his strange indifference to everything. By his leaps of logic— he contradicts himself several times during their meeting. He says "You're working for the KGB. I don't believe you want to help me". At the same time, he begged her on bended knee to bring his two daughters on 17th September, the day that had been fixed for a visit by telephone through the glass screen.

Galina had a long telephone conversation with Yury's mother to tell her the news and ask her why she had said such terrible things about Galina to her son. Yury's elderly mother was very shaken by Galina's description, recognised her mistake and asked Galina to forgive her. It is the sad old cliché of the difficult relationship between a mother and her daughter-in-law. Bandazhevsky's enemies made the most of the situation to incite him to break his links with the West.

---

[195] Request for urgent action to be taken at the UN. "[...] After 5th June (the date from which conditions improved), letters from my husband arrived less and less frequently, he did not want to talk about science, he was no longer interested in his children, in family affairs. When I saw my husband again after an interval of three months (during which time he had no visiting rights), I did not recognise him. The man in front of me was a stranger, someone who had been crushed, indifferent to his surroundings. His eyes were empty of all emotion and expressed only extreme suffering. His identity had been crushed, split in two. He asked me for a divorce but added that I should not believe anything he said or did at the moment. He begged me to remember the situation he was in and the plotting that was going on all around him. I could see his suffering and that he could not talk to me as openly as he wanted. But in any case he seemed incapable of expressing his thoughts clearly. He told me that his thoughts were jumbled up inside his head and that the same ideas would keep coming back like a record that has got stuck in a groove. "I don't know what's happening to me, I can't think clearly about myself" he told me... "

On *6th September*, we meet the lawyer, Baranov, coming out of the prison. He confirms what Galina has said about her husband's state of health and tells us we need to get him out quickly: "By next year, we could have lost Professor Bandazhevsky", he claims. Yury has given Baranov a short letter, written in his presence, to give to us.

> Dear friends!
>
> An enormous thank you for your support. I am aware of it all the time, it helps me to live. I am working as much as I can on the problems I was involved with before. The lawyer, Baranov, will tell you about my health. In spite of a series of difficult circumstances, including some of a family nature, I will never abandon my task. I thank you from the bottom of my heart for your understanding.
>
> With profound gratitude, I remain always respectfully yours.
>
> *On the other side of the paper in larger writing:*
>
> If you need any objective information about my case, please ask my lawyer, A.P. Baranov.
>
> *Bandazhevsky*

The lawyer Pogonyailo asked if he could visit. This was refused. Besides, Bandazhevsky is afraid of having contact with him: he was vice president of the Helsinki Group of Belarus…Lukashenko did not like him.

On *17th September*, a brief scheduled visit. Galina has decided to confront Yury with the underlying situation, to reason with her husband explaining how he has been duped, and manipulated. She was afraid that he would refuse to hear it, but he listens attentively and begins to share her analysis. "Carry on, carry on, I'm beginning to see more clearly, I agree with you, I think your analysis is right". After this meeting, Yury gradually frees himself from the psychological grip of his gaolers, cautiously accepting the risk of communicating more freely with his family, of showing his attachment to them.

On *3rd October.* Yesterday, on 2nd October, the lawyer, Baranov, met Galina at the station in Minsk, after he left the prison. Galina says he was very agitated, worried, even fearful: "They are putting pressure on him, seriously ill-treating him", he told her. "A few days ago, he was taken out of his cell while they searched it top to bottom. They are looking for something in his papers. They haven't found anything. Letters arrive rarely. Nothing from Galina. He has written to Olga, but the letters never arrive".

Galina had given a letter to the lawyer to give to Yury. Baranov took it reluctantly, because it could have been confiscated and its contents were controversial. Yury read it, wrote his reply on the spot and gave it back to the lawyer. Baranov was not searched but he is frightened. He is very concerned about how important the Bandazhevsky case is to the authorities. He warned Galina: "I will not be going next week; I need to get myself together".

Galina tells me that she has just had a conversation with the cultural attaché at the French embassy, Sylvie Lemasson, who told her: "Your letter of appeal to the UN has come at the right moment. The European ambassadors sent a telegram to the Ministry of Foreign Affairs in Belarus, , who replied by telegram saying they would look into the case and that they would allow one of the ambassadors to visit Professor Bandazhevsky".

What to do? Inundate Alexander Lukashenko with letters asking him to grant an individual amnesty? The President of Belarus has that prerogative. This humanitarian gesture (towards his own people as much as the scientist) would have a good effect on public opinion in Europe and all over the world.

4. HIS STATE OF HEALTH DETERIORATES

7th NOVEMBER 2002
Yesterday evening I talked to G. Bandazhevskaya.

On 4th November, Bandazhevsky had been allowed to see members of his family for three days, during the regulation extended visit. This time, a larger room had been provided with three small beds. Yury's mother had stayed for two days, his two daughters two days, and Galina stayed two hours, giving up her place so that the others could stay (the regulations stipulate a maximum of three visitors at a time).

In spite of the joy she felt at seeing how happy Yury was when he saw the three of them approaching him, Galina has not yet recovered from the feeling of anxiety that remained after the meeting. She sees a young and dynamic man fading away. "It's difficult to say at what point I noticed the change that has come over him in a very short space of time". she told me "He is ill. He is very weak; he has no energy any more. At the end of our two hours together, he was covered in sweat and needed to lie down because he had no more strength. He has a headache all the time; he has got used to the pain in his heart; he has no appetite any more and forces himself to eat but the reality is that he hardly eats anything. He is continually depressed. He is constantly afraid of the murderer who sleeps in his cell and keeps him under surveillance.

There is still a lot of pressure being exerted on him. He has confessed to his wife that his gaolers have made him sign a false declaration": a statement saying that he does not want to meet anyone outside his family, no human rights defenders, no representatives from NGOs, no politicians. On the other hand, the notorious *Friend* continues to visit frequently, with no prior notice, and has convinced him that he really does want to help him, but that his "powers are not limitless".

The one positive element of the meeting last Monday was that after long months of silence and incomprehension, Yury and his two daughters have been completely reunited. The two girls came home reassured that they have not lost his love. But the meeting left them with feelings of fear, anxiety and pain. Because the father they saw is unrecognisable. Galina still does not understand what it is that is wearing him down so badly. She asked him:

"What message would you like me to give to your friends abroad?"

"That they get an independent medical opinion about my health. I am a doctor, I know what our country is like, our own doctors will only say what the authorities tell them to say".

"But will they let someone come and examine you?"

"I don't know…"

Galina does not know what else she can do. She feels powerless, close to having a nervous breakdown. Her blood pressure has risen to 18. I reminded her of the situation she found herself in, in July 1999, when Yury had been imprisoned for twenty two days and she saw him just for an instant through the bars in the prison courtyard, having lost 20 kg, stumbling about before being taken to an unknown destination. At that time she had sent a telegram to Lukashenko expressing her fear and asking him to help her to find her husband. He was found, close to death, in a police cell at Mogilev (200 km from Minsk and from Gomel) , and was transferred from there to the Ministry of the Interior hospital in Minsk. I told Galina that she must remain active, that according to her description, Yury's state of health was at least as bad as it had been then, if not worse, because of this constant progressive decline brought about by "normal" prison life, and that this situation must be brought to an end by formally confronting the authorities with their responsibilities in writing. I suggested that she ask for an urgent meeting with the "nice" *Friend,* as the prisoner's wife and as a doctor, because if the situation was not to become irreversible, action must be taken immediately. The authorities were, in fact, in the process of destroying a scientist of great value to their country … I suggested that she also write to President Lukashenko. These letters could be transmitted to Sergei Kovalev, ex-Soviet dissident from Moscow, who was to preside over the meeting on November 24th of the sub-commission of the Council of Europe on the "disappeared" of Belarus. Kovalev's assistant Valentin Mikhailovich Gevtor, has met Galina. He is the director of the Committee for the Defence of the Rights of Scientists in Moscow and he has said that he will make sure that Bandazhevsky's case is heard. He suggests that his defence committee and the Human Rights Committee at PACE write to Lukashenko and ask him to modify Bandazhevsky's prison conditions, so that he can work and stay in good health.

I also asked the lawyer Garri Pogonyailo if asking Lukashenko for an individual amnesty for Bandazhevsky was a valid demand, given that it is a legislative measure.

He told me that in fact, in their country, the President also exerts legislative power and is quite open about it. He makes his own decrees and it is he in the end that decides cases of amnesty. He calls up the president of the Supreme Court of Belarus, Sukalo, tells him what to do… and boasts about it on television. He freed an Italian spy sentenced by the court to four years in prison (it was not a pardon, but an individual amnesty), which was considered a humanitarian gesture. President Chirac could ask his counterpart, Lukashenko, to show humanity towards Professor Bandazhevsky and get him to understand how much this gesture would be appreciated.

P.S. Galina took a walk around the showers close to the visiting room. There, sitting quietly, was a huge rat, weighing about 1kg. There are people with tuberculosis in this prison. She asked her husband if he was coughing. No, he wasn't coughing. Not for the moment…

5. MANIPULATION: SECRET PRESSURE EXERTED ON AN INDIVIDUAL

9th DECEMBER 2002
Thanks to my telephone conversations with G. Bandazhevskaya I am now in a position to imagine the sort of psychological torture to which her husband has been subjected. I am reminded of what the judge Terekhovich said after a two hour conversation with Yury. "If they wanted to get a confession out of him, they went about it the wrong way". This time their efforts had been more effective. But even so, they failed. [196]

Today, Yury Bandazhevsky, born 9th December 1957, is 45. For his birthday, his family was granted, exceptionally, a special four hour prison visit by the governing board of the prison. Galina and her daughter Olga were expecting the visit to be painful or even dramatic given the information they had been receiving shortly before. But, once again, the meeting this time was very joyful.

Since the prisoner had been transferred from the large dormitory to the room with three beds, the other detainees, who had been sympathetic, are now quite guarded towards him when they meet. Yury feels as if he is in quarantine.

This time the meeting between Yury and his family took place under heavy surveillance and was quite eventful. Before entering, they had to undergo a whole lot of formalities. Never before had Galina been searched so minutely, after waiting for an hour and a half, out in the cold in temperatures of minus 15 degrees C. The official in charge of surveillance emptied her handbag on the table and scrutinised

---

[196] See Part Three, Chapter IV, p. 210.

the piece of paper she had brought to make notes on, against the light of the lamp, for a lengthy period of time. He rummaged around in the bag containing food and presents.

THE VISIT

When Galina, her daughter Olga (Natasha, the youngest, had a cold and didn't come) and her mother-in-law enter the area into which Yury would be brought, the deputy governor of the prison, a man called Los, goes with them. He is a hard man, feared by the inmates, who only makes an appearance when physical threat or intimidation is needed. Yury finds himself outside at the other end of the corridor and hearing their voices, calls out to them, through the open door. Galina starts to answer and is brutally reprimanded by Los: "Shut up! You'll speak to him when he is here!" Once they are alone, Yury tells them immediately that the room is most probably bugged. He is impatient to see them and very happy. Galina has brought him some presents: his latest book, containing three monographs put together and printed by his brother, a present from colleagues at Belrad, not to be opened until New Year's Eve (some caviar, a chocolate Santa, a little tart and some candles) and a copy of CRIIRAD's *Trait d'Union,* (the bulletin produced by CRIIRAD three or four times a year) which she has hidden under her clothes. They have a good conversation. Yury talks non-stop. Having been alone for so long, his head is bursting with things he needs to tell them. After half an hour someone knocks at the door and in comes an official from the sentencing board: "I have come on orders from the Ministry of Foreign Affairs". Bandazhevsky went white.

"What's the matter? What do they want from me?"

"I have come to give you a health check".

Galina asked if he was a doctor.

"No"

"Then how can you give an assessment of his health? Even if, from the outside, his health might appear more or less normal for a prisoner, I can tell you, as a doctor, that he is not at all well. This man is ill".

Yury knows the man and starts to talk to him: "Why are you doing this to me, why are you hounding me? You know I'm innocent. Why destroy my brain, when it could be still useful to my country? Why are you punishing me so badly?" The man says, in a kindly way: "I understand what you're saying, but what can we do? We are not the judges, we are not the court... Is anyone bullying you here, insulting you? " "No". And off he goes to write his report.

Three quarters of an hour later, the door is opened again, and the guard shouts: "Bandazhevsky, out here!" Yury is very tense "Why so soon? We were given a four hour visit! " "You've been called to the governor's office. The women can wait for you here". Anxiously, Yury asks if Galina can come with him. "No" Once more in a state of distress, at the end of his tether, he goes to see the governor. He comes back after half an hour, looking amused: "I don't understand anything. It was the political *Friend's* deputy who interrogated me: "How is your family? Are there any disagreements between you and your wife? Are you thinking of separating?" "No. I love my family. I have grown up children. I love them. I would never do that..". "Good, we are pleased to hear it..." It seems clear now that the aim of these "protectors" had been to break the couple up. With Galina's help, we were able to piece together the little game they have been playing over the last few months.

THE NIGHTMARE OF THE LAST SIX MONTHS

Although we have no proof that they used psychotropic drugs, what we can be certain of, and can prove, is that Yury was subjected to a form of brainwashing and psychological suggestion techniques. Throughout the summer, they "worked" on Yury, destabilising him by exploiting his psychological vulnerability, his anxiety, the panic attacks that overwhelmed him, by giving him false hope, making promises that could never be fulfilled.

Already by the winter of 2001, among the many letters of support he was receiving, came a letter from someone called Orlova. She had begun by writing to Galina, saying how moved she had been by what had happened to Yury and pledging her support. She wanted to send parcels to Yury. She wrote to Yury saying that she predicted an improvement in his situation very soon, that she had the gift of clairvoyance and that she could help him from a distance. She started to send him mystic literature and doling out comforting advice and predictions about the future. If he behaved in the way she suggested, she could guarantee that he would bear up well in prison and that he would soon be freed. She copied out prayers and incantations that he must learn by heart... A mixture of religion and magic, which blended subtly into the state of unreality and mental confusion from which he was already suffering. "I am losing contact with reality", he had written.

For us, who are on the outside, fighting for his immediate release and moving from one initiative to the next, we cannot say with confidence that we will ever succeed! But Yury, with no bearings to guide him about the reality of any action, has no escape from the daily hopes and expectations, and it is inevitable that he will find himself in free fall. He was responding to Orlova, who was asking him questions and drawing him into her logic. Once she had his attention, Orlova began to suggest that Galina had a negative influence on his destiny, that she was ambitious and was profiting from his fame, and that she would end up losing him. He should not trust his wife, and her behaviour would damage him. If he had the courage to break up with her, he would get out soon. Galina learned all this, first through her mother-in-law and through Yury himself, then through Orlova's letters, when she began attacking Galina directly. Yury, while he rejected the more outrageous suggestions, was influenced by this woman: "You work for the KGB. I don't believe you are really helping me" he said to Galina in one of his confused moments. We can say, in fact, that Orlova had finally won his confidence. Then, in June, grafted on top of this, the promises and advice arrived from the political *Friend*, explaining to Yury in a rational way, how Galina, by informing the West, was not furthering their cause to get him released. From that moment on, Yury's behaviour was divided. On the one hand, he tried to keep up his links with the West in the hope that he would find support there, and on the other, he was cultivating his friendship with the *Friend* and being subjected to Orlova's insinuations. His mind was constantly being pulled in different directions and this led to a crisis of confusion, bordering on depersonalization: "[…] I am not able to make a sober assessment of myself. I no longer believe in what I write […]" He was and still is very aware of losing his mental equilibrium. Galina describes it as a split personality: two parallel personalities expressing themselves alternately, the friend and the enemy, one minute trusting, the next doubting, against a background of depression and exhaustion. This passionate, sensitive and demanding man, who has always sought obsessively to verify his intuitions as a research scientist, is now as deeply buried in his suffering as he was in his science. His thoughts spin round endlessly. He is simultaneously strong and weak.

In the pleading letter Galina writes in September to the UN, she says:

> It is clear that this man is ill, a victim of our system: they have managed to divide his personality, to make him doubt himself, to disorient him. He has become incapable of resistance, a sort of clay that can be modelled into whatever anyone pleases. In his letters, he says one thing to me, to my questions he replies another. I have before me another man, a broken man, indifferent to everything around him. His eyes, emptied of expression, reflect immense suffering. This is a man whose mind has been broken.

If Orlova is not a KGB collaborator, she works along curiously similar lines, increasing the pressure on him gradually, until she lost all sense of boundaries. In the winter of 2001, she had presented herself initially to Galina as an Estonian from Tallinn. She said she had learnt of Yury's case in an article published in *Argumenty I Fakty*. Galina went to meet her in St Petersburg to accept a parcel from her that had been prepared for Yury by doctors in Estonia, or so she said. Afterwards, Orlova sent food parcels directly to Yury, which the prison accepted quite happily, even though, according to the regulations, only the family had this right, and only at specified times. It is a question of security because who knows what could be smuggled in by unknown people within the food products? When Galina suggested over the telephone that she send Orlova one of Yury's books, there was now uncertainty about her address: she did not live there any more, there was a friend staying there at the moment, she had changed her passport… in the end, Galina let it go.

After her attempt to convince Galina that she was harming her husband, Orlova began to insult and curse her, writing that she would lead Yury to his doom if she continued to be involved with him. At the same time she continued to flatter and indoctrinate Yury, who said and wrote to Galina that Orlova was a good person. But ultimately, this manipulator made a fundamental error: she asked Yury to give up his scientific research on the problems of Chernobyl. When he wrote that he could not abandon his science, she blurted out: "You're an obstinate ass. Your science is shit (sic)". That was in September. She has not written to him since.

A little before the meeting on 9th December, an orthodox priest from Gomel, Father Andrei, visited Yury in prison. After the interview, the priest told Galina: "Life is very complicated. Yury is well, he is calm and lucid. He offered me tea with biscuits and French chocolate. He told me things about you: "Galina does not believe in God. Galina does not love me. Galina is using my name. She wants to steal my fame, my brain. She wants to be Bandazhevsky's Bonner (*the wife of Sakharov*)... I believe in the *friend*, he is my anchor and my salvation" During the visit on 9th December, Galina repeated these phrases to her husband. Yury told her he had never said these things, that the priest was lying, that it was monstrous what they were trying to do to them. He was being sincere. His daughter Olga, who came to see him, forewarned and angry (She said, "Don't play cat and mouse with us", and was prepared to break off relations with him) is also certain that he is telling the truth. As they left, they said to one another that the man who had stood before them was Yury, the loving husband and father he had always been. When Galina called his mother in the village near Grodno, she told her that Olga had recounted the meeting to her over the telephone, in happy and jubilant terms. This priest is probably another tragic case within the Soviet system, a soul-destroying machine.

Galina has needed incredible courage and inner calm to withstand these storms, to avoid falling into the contemptible traps that have been laid in her way for months. Her daughters, young and intransigent, as young people are, rebelled against their father, and then resumed the relationship, but Galina herself has the difficult role of providing a solid anchor. Very few people know what it has cost her, is costing her now, and will perhaps cost her in the future. Her attitude recalls the old myth of Ulysses, crossing the Messina Straits, attached to the mast of his boat, resisting the charm of the sirens. Galina is prepared for anything, so long as Yury remains in prison.

Galina has confirmed to us that everyone in Belarus, including Bandazhevsky's own prison guards, knows that an enormous injustice has been committed. But it does not change the situation. The decision is and will always be political and depends on Lukashenko.

As for Lukashenko, Bandazhevsky respects him. He thinks the President is badly advised by those close to him. In a way, he is right. This peasant, ex-director of a kolkhoze who has become head of state, knows nothing about nuclear physics, or medicine. Chirac, Putin, Bush, Schröder and Blair...do not necessarily know any more. All these statesmen, upon whom the future of humanity largely depends, get their information about the subject from the nuclear lobby that has infiltrated all the international institutions responsible.

17th AND 18th MARCH 2003
LETTER FROM GALINA BANDAZHEVSKAYA

You know that we were supposed to have had an extended visit from 17th to 19th March but it was cut short because of a flu epidemic. My youngest daughter and I spent two days at the prison. The visit was very painful for both of us. Especially for Natalia because children feel things differently from adults. She was so shaken and upset when she got home, that she vomited. She asked me not to take her to the prison again because she cannot bear to see her father suffering any more.

When I saw my husband, it almost made me ill. This man, who used to be full of energy, cheerful and strong, has been transformed into a creature turned in on himself, weak and humiliated. Yury has visibly deteriorated in the short period of time since my last visit (20th January). He has lost between 5 and 7 kilograms, he has aged, his hair is completely white. The problems with his teeth are worse. During the conversation, which was just small talk, our voices seemed too loud and our conversation too fast and too difficult for him to follow. The first day of the visit, he kept saying: "Speak more quietly and slowly, I beg you". I understood why when I learned that for the past two months he has been alone in his cell. His cellmate was transferred somewhere else. I think that in this situation, being alone has a negative effect. Yury has turned in on himself, closed up, he can stare fixedly at something for a long time murmuring something to himself, or laid out on the bed with his eyes closed. He says: "Pay no attention to me. Talk amongst yourselves. I'm listening". He has started to believe in God, almost fanatically. He prayed four times during the day. He says that his faith is a great help to him. That only God is just. He refused to change lawyer or to ask for a new trial. "All that is just vanity. In any case, there is neither truth, nor justice".

He doesn't want to know anything about the outside world, which he will not see in the near future, he says. "If I let myself go, it's very hard to get back to the reality of my cell. It's better not to know anything about the outside world".

Galina did not say everything in this letter. She has given her permission for me to transmit the essential content of our telephone conversation.

I talked to her for a long time after I had read it and, on the basis of what she told me in more detail, we were able to modify somewhat the very dark picture that she had painted about her husband's psychological state. In replying to my questions and remembering other moments during their meeting, she saw that the situation was perhaps not quite as catastrophic as she had first described.

6. THE TRANSFORMATION OF YURY BANDAZHEVSKY

Psychologically, Yury has suffered and continues to suffer, but his suffering could be purifying. His scientific mind is still functioning and this period in prison, his sudden fall that precipitated him from "fame" into humiliation, shattered his outer shell, the social side of his personality, forcing him into a painful examination of himself. He meditates on his life and radically questions everything about what he was before this traumatic event. The change is so far reaching that Galina, accustomed to the man he was before, was frightened. She is discovering a new man, a man she does not recognise and whose new reference points, like his religion, are strange to her. I told her not to be frightened because, given the situation he is in, it is a good development; it helps him to hold out without losing his sense of self.

One night, sitting opposite each other on stools, while his elderly mother and his youngest daughter slept, he spoke at length to Galina, who listened to him for a long time, in a low voice from the heart. He told her that he had made a decision to get hold of himself, to make amends, to rid himself of everything that he perceived now as intolerable in himself. He made harsh judgements about his egocentric, proud, authoritarian character, which had affected his family for all these years, subordinating them to his needs as a scientist and as rector of the institute. He repented to her: "I was full of pride, I always put *me me me* in first place". Yury wants to rely on all that is good in his character to develop an honest relationship with himself and with Galina. He told her that this nocturnal conversation had returned him to his real self.

Even though he is closed up and isolated from everyone, Yury's mind remains clear. He was very pleased to receive photocopies of certain scientific articles that had been sent from Paris recently.

The only real worry is his health, because his diet is poor and he eats very little of it. The fact that he lives completely alone, day and night, in a mental state that was unrecognisable to Galina, worried her. She fears a reckless, hasty gesture when there is no-one to witness it. Even though Yury has told her that killing oneself is a great sin and that he would never do it.

A recent piece of news that needs to be treated with a great deal of suspicion:

The political *Friend* informed Yury that the documents needed for his release were ready: "We are waiting for President Lukashenko's signature. If it fails, I will let you know". But, in the meantime, given that Yury could end up disappointed again, why have they left him alone in his cell?

31st MARCH 2003
VISIT FROM THE LAWYER BARANOV
Last Monday, 31st March, the lawyer, Baranov, spent an hour and a half with Bandazhevsky. Baranov confirms the observations made by Galina when she saw him two weeks before. He found the prisoner thinner, weak and completely changed.

He really seems to have emerged from the psychological fog of the last few months. According to Baranov, he is in complete control intellectually of the situation in which he finds himself and has a clear understanding of what is at stake. The lawyer, who himself often speaks in a veiled and reticent manner, told Galina that Yury knows and understands a great deal more now about the various plots going on around him than he did on previous visits. Yury asked him, and also asked his wife in a letter, to express his gratitude to those who are supporting him in the West, not just for their support, which helps him and protects him, but for "their accurate assessment of the errors that have been made since last summer, and the interpretation they have given of the reasons for the repression to which he is being subjected".

It seems that the regime, which is keeping Bandazhevsky in prison, is focusing even more attention on his qualifications, his research and his knowledge. Recently, he was given a language test to establish his level of competence in French and English. The correspondence he receives from his wife and from abroad (mostly in French, the language he knows best) takes a long time to arrive sometimes and is often crumpled and dirty as if it has been handled by several people. An official from the sentencing board was sent by the Ministry of Foreign Affairs asking him for a list of his latest work[197], published in the period between his two arrests.

People in positions of responsibility in the political and scientific domain, both in Belarus and in the West, take shelter behind the official *omerta,* and it is not in their interests for Bandazhevsky to take up his scientific research again too soon.

19th APRIL 2003

Yesterday, Galina Bandazhevskaya was invited to the French Embassy. The European Union must have exerted a lot of pressure on the Belarus authorities if the ambassadors of France and Germany were allowed to visit Bandazhevsky in prison (9th April). An ultimatum must have been given by the EU to the Minister for Foreign Affairs, Khvostov, stating that if they were refused, they would make public the intransigent attitude of the government in this affair. This had an effect, probably because of the softening of relations between Belarus and Europe. (Other official requests from Europeans to visit Yury were initially granted and then refused on the pretext that it would remove Galina's visiting rights, which were strictly regulated by law).

The meeting with Yury and the ambassadors lasted an hour and took place in the presence of the prison governor and the director of the sentencing board.

---

[197] The titles of his most recent work are as follows: *Medico-biological effects of radioactive caesium incorporated into the body,* Minsk, 2000. 70 p.; *Radioactive caesium and the Heart: Pathophysiological Aspects,* Minsk, 2001. 63p. ; *The role of radioactive caesium in the pathology of the thyroid gland,* PSR-IPPNW Switzerland, 2001; *Radiocaesium and the intrauterine development of the foetus,* Minsk, 2001, 59 p.; *Pathological processes in the organism in the presence of incorporated radionuclides,* Minsk, 2002, 130 p.

Yury was very happy and encouraged by the visit. He has already written to his wife saying that it is a long time since he has enjoyed eating or had such a good appetite as after this meeting. The interview inevitably was constrained by the presence of the governor and director. The ambassadors noticed that Yury seemed to be experiencing some problems with his speech, as if he had taken sleeping pills. Galina confirmed that he takes a lot of tranquillisers. They found him pale and thin—not looking very good.

The director of the sentencing board said that Yury could be relegated in June 2004, unless in the meantime there was a general amnesty that reduced his sentence by a further year, (in which case it would be January). Galina does not think that Europe will have much influence in this process. The fact that they gave an ultimatum that resulted in a formal visit was of great comfort to Yury, but it is hardly a victory for European law. Yesterday, or the day before, the UN Commission on Human Rights voted unanimously on a resolution noting that human rights are not being respected in Belarus.

15th JULY 2003

Yury is on his own again, having recently shared his cell with a young man serving a sentence for commercial malpractice, whose company was preferable to the murderer. The guards systematically separate and divide up groups as soon as any relationship is established, but Bandazhevsky, who has started work again, does not complain because, in any case, intellectual conversation is minimal and does not satisfy him. *"Bandazhevsky is a brilliant scientist whose only crime is to reveal the lesser known, but nonetheless fatal, effects of the Chernobyl nuclear disaster. His intellectual powers are mouldering away in a prison cell in Kalvarijskaia Street in Minsk. Had he been born in a free country, his life would have been celebrated as a gift to humanity"*. This article, with its slightly rose tinted view of the free world, appeared in the Italian newspaper *La Repubblica* on June 28th last year. Bandazhevsky is not complaining about being alone at present because he has started work again, after a period of months in a state of illusion and mystical hope. He is grateful to the French ambassador who made him a gift of some beautiful books; the lives of famous men and valuable scientific works. Yury is doing a lot of writing and reading. At the moment he is studying the pathogenesis of the illnesses caused by caesium-137.

7. PROVOCATION OF A PRESIDENTIAL PARDON

23rd AUGUST 2003

A number of things have come together that make us think that we might soon have news of Professor Bandazhevsky's release. On 8th August, the Chamber of Representatives at the National Assembly of the Republic of Belarus wrote to the Bandazhevsky Support Committee in Grenoble inviting the prisoner, to repeat to

President Lukashenko his previous request for pardon, that was refused in October 2001.

> In July 2003, a group of Deputies from the Chamber of Representatives at the National Assembly of the Republic of Belarus wrote to the President of the Republic of Belarus asking him to pardon Y.I. Bandazhevsky.
> The President is considering the possibility of examining the question of a pardon for Y.I. Bandazhevsky a second time, if the latter submits his request in the required form.
>
> Vice-President of the Commission, A.V. Svirid

Bandazhevsky does not recognise his guilt and has never asked for *his pardon*. While protesting his innocence of the crime of which he was accused he asked Lukashenko to free him *as an act of humanity* in order that he might pursue his scientific research.

On Friday, 22nd August, on Nesterenko's advice, Galina Bandazhevskaya went to see the director of the sentencing board. The commission and the prison governing board already knew about the response from the Chamber of Representatives, had told Yury about it and had advised him to write his letter of request. The director of the sentencing board told Galina that the day Bandazhevsky left them they would feel enormous relief. His case, which is now the subject of debate at governmental level within Europe, at the Parliament in Strasbourg, at Amnesty International, within various humanitarian organisations and now at the Belarusian parliament, was making them lose sleep and preventing the smooth running of the prison: "As far as we are concerned, we will do everything to ensure a positive outcome", they assured her.

In the visiting room, Yury told Galina that they had been nagging him for several days to get his letter written. The prison governing board is meeting next Thursday, on 28th August, to conduct a kind of internal court case to examine the behaviour, motives and content of the detainee's request, and their comments will be communicated to the sentencing board. And the sentencing board, in its turn, will write to the President's office. But they need Yury's request, and it still hasn't arrived! They are getting nervous. It's a curious reversal of roles.

Yury explained to Galina that the whole humiliating pantomime would get them nowhere. Against the background of injustice, insincerity and stupidity in which he has been immersed for the four years of his incarceration, his brain can no longer formulate the words to make this request. In the two hours they spent together, Galina provided him with a few acceptable phrases. "If you want to understand why I am innocent, please take the trouble to re-read the notes from the trial, in which there were at least eight violations of the penal code and of legal procedures, and was not able to bring any proof of my guilt. I am asking President Lukashenko to use his power of pardon to free the innocent scientist that I am, so that I can continue to be of use to my country". At the end Yury thanked Galina for having helped him to make the decision. "In any case, you have nothing to lose", she told him. He will write his letter at the weekend and attend his "trial" on Thursday.

Zinaida Gonchar, the wife of one of Lukashenko's "disappeared" opponents, advises Galina not to expect too much. In her opinion, the letter from the Chamber of Representatives is just a *standard* reply which will have no consequences. There have been precedents. Everyone speaks his lines in this theatre of the absurd, but in the end, it is Lukashenko who makes whatever decision he sees fit. Galina confides to me that she only half believes it.

Bandazhevsky has finally written his request. The delay means he has missed the commission which only meets once a month. But they hurried to his cell to fetch the letter and reconvened the committee outside the normal time frame. It is clear that this initiative has been greatly encouraged from above.

23rd SEPTEMBER 2003
THE PRISON DIRECTOR PUSHES BANDAZHEVSKY'S REQUEST FOR A PARDON
On 22nd September, Galina brought some medicines to the prison and had an interview with the director, who introduced her to the doctor who is looking after Yury. During the night of Monday 8th, Yury had a serious heart attack: severe pain with shooting pains in the arm and in the back, difficulty breathing, blue lips, grey complexion.

> It was a good meeting between colleagues, open and humane. The doctor reassured Galina. For the moment, nothing catastrophic has happened. In his opinion his state of health can be explained by the deep and continuing depression he suffers in prison and the excitement over the past few weeks about his case. I have copied the words of the doctor, as I noted them down when Galina told me about the meeting. "A man of his intelligence, with no social relationships other than with the other prisoners, is a man who lacks air to breathe. He is suffering from depression. And with depression anything is possible. He has also been given a psycho-neurological examination and we are helping him with this side of things too. The medicines that you bring him will help him. But the best medicine will be his release from prison. As soon as he leaves prison, these symptoms will disappear. And while we're on the subject, the papers have been sent off. I have to tell you that, during my time at this prison, this is the first time I have seen the governing body making efforts to help a prisoner's request and asking for his release. Try not to worry, we know the professor, we know all about it and we understand it all. We have his best interests at heart. Take care of your own state of mind, because everything depends on you. We will do everything we can to help him".

It is quite surprising. It is true that this was the first time she has ever spoken to this fellow doctor. Up till now, she has only spoken to the guards, who are part of the military and adhere strictly to the regulations.

Whatever the personality of the new prison governor, who seems more humane than the last, he is also a member of the military and it is difficult to believe that he could bend the rules let alone break them. He was pushing for Yury's release even though Yury has not admitted guilt. Galina is aware of a change, which can only be explained by a political decision from "above". This decision has allowed people lower down the hierarchy (the governor, the doctor…), to express their opinions about the case and their feelings towards the prisoner more freely. Let's see what happens next.

THURSDAY 2nd OCTOBER
YURY BANDAZHEVSKY UNDERGOES A DELICATE OPERATION IN PRISON
On Saturday 4th October, at 9 o'clock in the evening, Galina Bandazhevskaya receives two telephone calls from relatives of detainees saying that her husband has undergone a delicate operation, that he was in a very poor state and that he needed to be transferred to a good hospital urgently, "if not you might lose him". They could not tell her when the operation had taken place or the reason for it. Over the weekend, no contact was possible with the prison.

Galina was only able to talk to the prison governor and to the chief surgeon on *Monday 6th October* at midday. At one o'clock, she was allowed to see Yury face to face, not on the telephone and not through a glass partition, but only after she threatened to create a scandal at international level. As her visit was unexpected, there was no room available, and it was against regulations to enter the hospital or his cell. They allowed them to meet in the guards' locker room. Yury Bandazhevsky walked there, accompanied by a warden. He had drawn features but other than that he looked well, better than he had the time before when he had his heart attack. What had happened?

During the day on *Monday 29th September*, he had pains in his stomach that soon became unbearable. To begin with, he thought it was his ulcer that had reopened. On *Tuesday 30th September*, he spoke to the young doctor on duty.` This doctor told him he was exaggerating the problem. He gave him an enema and a painkiller by injection, which not only ran the risk of making the problems in his abdomen worse but also reduced the possibility of a clinical observation. Instead of feeling better, he felt much worse, his stomach was swollen, tight like a drum, he felt sick and he was in continual extreme pain. By chance, there had been another detainee in his cell for the past few days and he was able to alert the doctor. On *Wednesday 1st October*, he told him that Yury was asking to go to hospital. He was put on a mattress with no sheets, where he spent the next night trembling with pain and a high temperature. He asked for a blood test, thinking now that it could be appendicitis. He was told that it was impossible as the laboratory assistant was not there.

On the morning of *Thursday 2nd October*, the hospital realised that he was in a serious condition and it was all hands on deck. The Chief Medical Officer of the central hospital of the Ministry of the Interior, situated in the prison grounds, Doctor Tuchinsky, who was a surgeon, took matters in hand and averted a catastrophe. *"A few hours later, and you would have come here to bury me. Tuchinsky has eyes on the ends of his fingers"*, Yury told Galina.

The blood test revealed severe infection (18,000 leucocytes, "movement to the left of the differential leukocyte count"). On examination and during the operation, they confirmed that it was a neglected gangrenous appendix with abscess and peritonitis. The operation lasted an hour and a half. Now the results of his blood test were normal. Yury himself reported that he was feeling better.

The authorities directly responsible for the prisoner Bandazhevsky had been really frightened. On *Sunday 5th October*, representatives of the sentencing board came three times to ask Yury how he was feeling, what he thought about his state of health. He told them they needed to ask the doctors who were taking care of him.

Latest news: on Wednesday 8th October, in the afternoon, Yury and Galina's daughter, Olga, gave birth to a little girl, Ekaterina.

On *9th October,* Galina got a letter from her husband, dated March 7th: "I am feeling much better. Don't worry. Try to relax now. The infection has gone".

22nd NOVEMBER 2003

It is November 2003. In fact, since 18th June 2001, Yury Bandazhevsky finds himself in the same situation, with no objective certainty, other than that he is facing an *eight year* sentence.

As always, Galina tried to reason with him, reminding him of the positive signs that have appeared recently. "They were insistent that you re-submit your request to be released…"

"Who is to say that it isn't a charade?" The absence of any certainty grinds his spirit down, gives him a permanent knot in the pit of his stomach and prevents him from thinking calmly.

14th DECEMBER 2003

The amnesty: on *2nd December*, the director of the prison, Kovchur, was categorical: "Thanks to the legal amnesty which is undergoing its second reading in Parliament, he will be out at the very latest in April and if the law is voted for before the end of the year, in February".

The pardon: still nothing. Galina has asked Kovchur to telephone the director of the service dealing with pardons at the President's office to find out more. Kovchur was reluctant to do so. So Galina asked her mother-in-law to do it, given that in 2001 she had been granted an interview with the aforementioned director. He replied: "We still have many cases to consider". This reply confirms the information that was circulating among journalists in Minsk according to which Bandazhevsky would be included in a list of candidates for presidential pardon. They are in the process of deciding now.

Yury Bandazhevsky's scepticism: Yury Bandazhevsky is deeply convinced that they can do and say anything. In his eyes, it could all be just another deceitful charade. He told Galina that if he is released, he will need at least six months to recover any objective and normal perception of reality.

On *3rd December*, the current representative at OSCE in Minsk came to see him. Before the meeting took place, those in charge of the prison had appeared rather agitated. The director came to check on the state of his cell, and ordered some repair work to be done. The visitor and the prisoner were invited into the governor's office, where they were given a cup of tea. The governor and his assistant were present, discreetly absorbed in their work in one corner of the office, but obviously, the theme of the conversation could only be very general.

His state of health: Galina thought her husband had lost weight. The slightest violence—and violence is the norm in prison—causes him to lose control of his nerves. The parcel that Galina had brought for him was quite heavy. The scar from his appendectomy is still fresh, and he is not supposed to lift anything. The regulations demand that prisoners deal with their own parcels. He had come to an agreement with one of the guards to help him carry the parcel to his cell. But when he left the visiting room, it was a different guard. When Yury asked for his help, the guard shouted at him viciously. Galina could hear him behind the door: *"So, what's this then? You think you're something special here? In here, you're the same as everyone else! Do it yourself"*. Yury tried to explain that he could not and that if he was not helped, he would have nothing to eat, but the guard pushed him away roughly. When he came back in, he was white as a sheet, sweating, shaking uncontrollably, completely disoriented: "I just won't eat. In a month it will be Christmas (*7th January in the Orthodox calendar*) who knows when you will be able to come again". A few minutes later, another guard, who had seen what had happened and saw the state he was in, came in. "Calm down, we'll sort it out". In the end the parcel was carried by another prisoner. "They humiliate me all the time" he told him.

Scientific publications: Yury was literally transfigured when he saw his articles published in the scientific journals *Cardinale* and *Swiss Medical Weekly*, that Galina managed to bring to him discreetly. (These publications are the result of a symposium organised by Michel Fernex at the University of Basle). As always, he looked sad when Galina and his youngest daughter Natalia first arrived. But when he saw his own texts written in French and in English in these prestigious scientific reviews, his eyes shone, he was completely bowled over. *"This is truly my rehabilitation! A huge thank you to Professor Fernex for this magnificent present. Thank you"*. Galina says that he stayed up till three in the morning, leafing through the journals, reading and rereading his articles.

8. DERISION

2nd FEBRUARY 2004

THE PRESIDENTIAL PARDON IS REFUSED

On *12th January*, Galina Bandazhevskaya called President Lukashenko's office to find out if any decision had been made about a pardon for her husband. She was told:

"Last week, the commission for pardons, that considers each case, reached a positive conclusion. But it will all depend on the Head of State.

—When will I know?

—Telephone us in a week.

At the end of a week, the same response. On *29th January*, Galina learned from the sentencing board that a pardon had been refused and that Yury had signed the request for relegation to the Vetka district. The amnesty, that came into force on 14th January, reducing his sentence by a year, gave him this right. In the letter refusing him a pardon, Bandazhevsky was told that he could put in another request, on condition that he repent and pay back the thousands of dollars that he owed the government.

Lukashenko who had everyone believe that he was prepared to make the good-will gesture and had set in motion the procedure to request a pardon, fooled everybody, including the Commission for Pardons, who had given a favourable opinion. He mocked appeals from public opinion in the West, who hoped, in view of the insistence with which the prison administration had put pressure on Yury to write to the President, for a favourable outcome. The administration could never have taken this initiative by itself without an order from above. (The Director of the Committee on the Execution of Penalties said to Galina: "I ask myself, why did they put so much pressure on me to send them the case notes on his request for pardon?")

Lukashenko made a mockery of the diplomatic initiatives from rich countries, because he knew perfectly well that the control he exerted on the population that had been contaminated by the nuclear industry suited many of them. The 8 violations of the Belarus Penal Code, denounced by OSCE during the trial, were of no concern to him. He treated with disdain the request made by Russian scientists, who asked him to *"bring to an end the trial, which is not justified in our view, in which Y. Bandazhevsky is the victim, and to offer him the possibility of returning to his research, whose importance cannot be overestimated.*

Lukashenko had fooled everyone, but strangely, had not duped Bandazhevsky, who had never believed it. Their provocative behaviour had liberated him. In choosing relegation rather than waiting another year in prison for his sentence to be reduced, Yury had taken the easier path.

Lukashenko had even ignored the wishes of his own Parliament. In July 2003, a group of deputies from the Chamber of Representatives, members of the National Assembly of Belarus, had made an urgent appeal to him ending with these words:

"[...] we are writing to you, Alexander Grigorievich, to implore you to take the political and humanitarian decision to release Professor Bandazhevsky, in consideration of his scientific merit and of the services that he could still bring to our people, the victims of the disaster at Chernobyl.

[...] you alone can allow Yury Bandazhevsky to resume his research, that is vital if we are to receive the aid from the outside world that we urgently need for the victims of radioactive contamination. In the interests of our people, in the interests of all the victims of this disaster, and for the health of those who could become victims in a similar disaster, we beg you to bring to an end what has become known throughout the world as "the Bandazhevsky affair".

On *8th August*, the vice-president of the Commission on Human Rights at the Belarusian Parliament wrote to the Bandazhevsky Committee in Grenoble:

The President is considering the possibility of examining, for a second time, the question of granting Bandazhevsky a pardon, if he submits his request in the requisite form to the president of the Republic of Belarus.

The requisite form meant confessing to a crime he did not commit, after a year and a half of investigation that had failed to prove anything. Bandazhevsky's request did not fulfil these requirements.

9. THE INTERNAL LIBERATION OF YURY BANDAZHEVSKY

21st MARCH 2004
"BELIEVE NOTHING, FEAR NOTHING, ASK FOR NOTHING"
The cruel game played by the Belarusian authorities to destroy Bandazhevsky had the opposite effect from the one they had counted on. Going down on bended knee to ask to be pardoned once again, knowing full well it would never happen, and then the humiliation of Lukashenko's refusal, that he had predicted all along, seems to have cured Bandazhevsky from a long illness. He is still in prison, but this time, as a free man, determined to fight. He has lost all his fear. The false promises, the delusional tactics, the irrational hopes, that shook his nerves and paralysed his mind, have turned to dust and ashes. In pushing their hoax to the limit, and showing him clearly that he can expect nothing from them, Yury's enemies, who manifestly thought they could bring him down once and for all, have, in reality, lost their authority over him and allowed him to rediscover himself.

An old adage from the Soviet camps in the North identified the frame of mind needed to remain free in the Gulag, briefly, in three renunciations:

"Believe nothing, fear nothing, ask for nothing" This sums up Bandazhevsky's attitude today. The person that Galina listened to and observed during her recent three-day visit is a new man—calm and determined. "He has grown. He is a better man than he was before his arrest", she confided to me.

The few letters that Galina has received from Yury, following the authorities *bluff*, already have a different tone. She has the impression that he is much stronger. "Don't worry about me. I know my captors", he writes, careless of the fact that his letter might be censored. Whereas before, Yury's letters were full of complaints and lamentations, now he comforts his family and exhorts them to hold on. It is as if he is preparing to spend a long time in prison. Galina hardly dared hope but her visit confirmed this change.

Yury cannot forgive himself for having asked for a pardon a second time. He expresses himself without fear. "If they refuse me relegation to which I am entitled since the amnesty, we need to attack hard on all fronts: legally, diplomatically, in the national press and abroad. In reality, they are cowards. They are afraid of people who talk. Why has it taken all this time to understand".

Galina is struck by the psychological and physical change in Yury. He looks really well. He is doing gymnastics and takes long walks every day in the prison yard. He seems calm and well balanced. He sleeps as soon as night falls whereas Galina has difficulty getting to sleep. He no longer looks grey. He is in good health and is full of energy. "You wouldn't believe it! I've regained the energy I had in my twenties, I am writing a long paper, you'll be amazed…I will never give up my scientific work. If I've survived these last three years, I can get through to the end with dignity".

In preparation for relegation, Yury has filled nine large boxes with the letters he has received since being in prison. Galina suggested that he reduce the amount, keep a select few. "Absolutely not. I'm keeping them all".

10. THE PRESIDENT'S PERSONAL PRISONER

Yury Bandazhevsky has benefited from two general amnesties that reduced his sentence by two years. His sentence is now six years. Having served half of it, he has the right to one year's relegation.

EXTRACTS FROM "NEWS OF RELEGATION"[198] On *29th May 2004*, after a delay of five months, Yury Bandazhevsky has been relegated to an institution, 200 km from Minsk, in a "clean" zone, where conditions are far better than was expected at the beginning of the year,. The transfer of detainees from different prisons took

---

[198] Translator's note: In Belarus relegation usually entails hard labour in a remote part of the country. In Bandazhevsky's case, his conditions were improved and he did not actually live at the penal settlement but in an abandoned house in a village on the Neman river 30 km from the colony.

place during the night of Friday/Saturday, in typical Soviet style. The detainees were handcuffed and escorted by soldiers with dogs. During stops, their bags were thrown to the ground, kicked open by the soldiers, the contents scattered around, and then they were ordered to put it all back. This humiliating performance was repeated three times during the night. By contrast, their welcome at the new penal settlement signalled a return to a more civilised environment: the officer who met them was unarmed, there were no dogs and his attitude was humane.

DECEMBER 2004

FORGET BANDAZHEVSKY

As we approach the end of 2004, Bandazhevsky is alone, in poor health, deprived of dignity and of any means of existence. What his enemies had not achieved five years ago—the quiet disappearance of this troublesome scientist into the prison system—seems more achievable now, by erasing him, forgetting Bandazhevsky. International opinion had been reassured by him leaving prison, but the no man's land of relegation may, in the end, prove more dangerous than prison. Bandazhevsky is no longer protected by anyone and, since last summer, the regime has engaged in a more underhand and harsher repression.

What is most worrying about this period of isolation, is that after a period during which he appeared full of energy, encouraged by the euphoria of leaving prison, Yury Bandazhevsky's health has deteriorated again. He complains of stomach pain, has problems with his liver and his kidneys, inflammation of the Achilles tendon, a torn ligament in the biceps. Galina Bandazhevskaya describes it as a sudden and overall deterioration in health. Is it prison that has done this? Is it some unknown infection?

Having lost all his rights, and no longer having a home, Bandazhevsky has no entitlement to free health care other than that provided at the penal settlement or at a clinic. When he learnt of Bandazhevsky's state of health, the director at the penal settlement gave permission for him to seek medical help in Minsk but at his own expense. Nursing, hospitalisation, medicaments, operations, anaesthetics cost a lot of money…Three eminent consultants, his colleagues, refused to see the patient and did not dare to intervene to make sure he received hospital treatment free of charge. People avoid him like the plague.

In Minsk, Bandazhevsky consulted a specialist about his Achilles tendon and about the rupture of the ligament in the biceps of his left arm, and he was examined at the health clinic and after seven days, without waiting for the results of the biopsy of the liver and stomach, he was taken to hospital urgently. The results from the clinic showed, among other things, that he had chronic hepatitis of unknown origin. But Bandazhevsky did not have hepatitis before he was arrested. They operated on his shoulder. The operation went well, his arm was in plaster for four weeks, but he had to leave the hospital before the time prescribed by the doctors through lack of funds. No-one knows the cause of these torn ligaments that keep occurring. The illness that led to him being operated on has no name, no cause, no recognised aetiology. The doctors say that it is very rare. It is certain that prolonged incarceration has ruined his health.

On *6th January 2005*, Bandazhevsky should legally be on parole. But…

On *10th January*, problems are starting again… Galina Bandazhevskaya tells us:

> "Today, I have the firmest conviction that the "powers that be", whose job it is to enforce the law, are doing their utmost to delay the presentation of Yury's case to the court for a reduction in sentence, for as long as possible. It's obvious to me that I need help and that by myself I will never break through their defences. You know that the 6th of January this year is the date on which, according to the law, Yury has the right to parole. But they have started a new game of cat and mouse. The sentencing commission says that these questions are not in its remit, that everything depends on the governor of the re-education relegation colony no 26. As for the prison governor, he says he is waiting for orders from above. […] Now the relegation colony no 26 is justifying its view that it is impossible to consider parole because, according to the President's reply to Yury's request for pardon, Yury needs to admit his guilt and compensate the government for the damage he has caused. Now, all the President's subordinates are afraid of this edict and are waiting for someone else to make the decision for fear of falling out of favour with the head of state. All the same, Garri Pogonyailo was right when he said that Yury was the President's personal prisoner, and therefore no law applied. It is he who will decide when and how the prisoner will be freed. That's why they have been dragging their feet for so long".

This morning G. Bandazhevskaya telephoned the governor of the penal settlement, Vassili Koliada, and read article 90 of the penal code to him, according to which, after serving two thirds of his sentence, in other words on 6th January 2005, "the administration of the re-education relegation colony is obliged, within a period of one month, to examine the possibility of applying to the court for parole for Yury Bandazhevsky[199]". Koliada replied: "I understand. Before the end of the month, the commission will meet and present his case to the court; you must understand that I cannot make the decision by myself". G. Bandazhevskaya asked him why Ravkov, who was given exactly the same sentence in the same trial, has been granted his legal rights, without encountering the slightest problem?" "Ravkov is Ravkov. Bandazhevsky is Bandazhevsky" was his response.

---

[199] Reply from the Belarus government to the UN Human Rights Council (Geneva), stating that Bandazhevsky's complaint was admissible.

Yury Bandazhevsky really is "the personal prisoner of the president", as Garri Pogonyailo, lawyer and vice-president of the Helsinki group in Minsk, observed. G. Bandazhevskaya recalls the words spoken by the *Friend* during one of his visits. "If you don't recognise your guilt, you will serve the whole prison term".

11. ORDERS FROM ABOVE

20th January: Galina Bandazhevskaya telephones the governor of the penal settlement again. He seems to be at the end of his tether: "I can't carry on like this. I'm caught in the crossfire. I'm under pressure to keep the prisoner here, but I am obliged by law to refer his case to the court within the time limit, and to present an objective assessment of the detainee. I know my job, and at the required time, I will arrange for the commission to meet, we will make a decision on the basis of my report, it will go to the court, who will give their ruling and I will release the detainee. I have never had a case like this, it's the first time. It's one thing after another! I'm afraid. I'm going to resign".

21st January: G. Bandazhevskaya calls the governor of the penal settlement again. He has slept on it and has had an idea. He says: "On 3rd February, I am going to call a meeting of the commission anyway. Article 90 gives him the right to parole, but there is also article 92, which deals with the detainee's state of health and which would mean he could even be released before the end of his term".

Behind all this could be a remark made by Lukashenko, that he slid into a televised interview, broadcast just today, during the nominations for Vice-Minister of the Interior: "You shouldn't hide behind me—"The boss said so!"—don't cover yourself by using my name. You have the prerogative. You need to take responsibility. You should not involve me in your affairs". They all cover each other, including the "boss". The prison governor knows the system well: Lukashenko only has to raise an eyebrow and the word goes down the line till it reaches the last man, and he will take the rap unless he sorts the matter out in secret. No-one else cares.

Friday 28th January: G. Bandazhevsky calls the governor of the penal settlement. He replies: "We have brought forward the date of the commission to 31st January and WE ARE REFUSING PAROLE. The order came from above. Please understand that this is not my personal whim. The reason that I have to give for the refusal to grant parole is that he does not admit that he is guilty of the crime and has not paid back the money that he owes to the State". G. Bandazhevskaya: "But you are not qualified to make this sort of judgement. You are not a court. You can only base your arguments on the behaviour of the detainee while he has been in your care. He has served his sentence correctly, he deserves parole…" The governor: "Yes, I am violating the law. I have been put in a situation in which I have to violate it. This is the first time this has ever happened to me". When Yury arrived in the settlement, the governor told him he had asked his legal experts to examine his case: they had been dumbfounded at the total lack of consistency in the legal arguments, and the absence of any justification for the verdict. For these people too, it was the first time they had experienced anything like it. Galina allowed herself a little hope. During the weekend she had very high blood pressure and vomited several times.

Today is *Monday 31st;* the commission is meeting at 14:30 hrs local time (13:30 hrs here).

At 16:30 hrs, I telephone G. Bandazhevskaya. "They have refused" The commission met according to all the regulations, the quorum present. The governor read out a positive report about the prisoner and asked him "Why do you not acknowledge your guilt? Why will you not compensate the state financially for the damage you have caused?" Bandazhevsky replied that he was not guilty, and that he would confirm this fact before any court, and if not him, his children and his grandchildren would make sure that the truth prevailed. As for the money he owed the State, it was simply ridiculous: "I have not worked for five years. What money are we talking about?"

14th February 2005. Over the weekend G. Bandazhevskaya wrote to Lukashenko and to the Minister of the Interior, asking them to put an end to the breaches of the law committed by their subordinates; then she wrote her reply to the UN Human Rights Council and contacted the Western press. None of this convinced Yury Bandazhevsky. He believes he will serve another two years and he is preparing himself for it.

In reality, Bandazhevsky's state of health remains worrying and requires appropriate care and specific monitoring.

Given this situation, we decided that the best target for action was the Minister of the Interior, Naumov. We suggested that an avalanche of individual letters be sent to him, based on a text written by those associations who would sign it together[200]. Other organisations could add their signatures.

Everyone was encouraged to write to their own newspapers asking them to use the "Newsletter from relegation" to publicise information about Bandazhevsky, and the action we were taking. We made contact again with as many politicians as possible, asking them to bring the question up again, particularly in Parliament, or in any other public sphere.

A certain number of Academies of Sciences had written to the president of the Republic of Belarus demanding the release of Bandazhevsky.

---

[200] Amnesty International, Enfants de Tchernobyl Belarus, Bandazhevsky Committee, France-Libertés, ACAT, IFHD, GSIEN, Enfants de Tchernobyl, Illzach, the network Sortir du Nucléaire.

On 22nd March 2005 the ambassadors of France and Germany visited him. In a joint communiqué, they underlined *"the growing attention and support across France and Germany for him, expressing the hope that the Belarusian authorities would commute his sentence as soon as possible or at least modify it in such a way that the professor could receive medical care as effective as could be found abroad"*. They insisted that he be permitted to pursue his scientific research.

Finally, on *5th August 2005*, after three and a half years in prison and fifteen months in relegation, Yury Bandazhevsky was given parole with the expectation of being totally free on 6th January 2006.

Chapter IV

CIVIL SOCIETY

HOW THE WAVE GREW

It was a very small group of people, from France and Switzerland, that had direct contact with scientists from the East and were able to pass on information about them to the West. At the beginning they knew very little about the humanitarian, political, medical, and environmental issues, against which Nesterenko was struggling. But after receiving information from him, they set their networks in motion quickly. In 1999, for example, we alerted Amnesty International who recognised Bandazhevsky as a prisoner of conscience. France-Libertés, Danielle Mitterand's foundation, came to our aid too and financed Nesterenko's projects in the Chernobyl villages.

On 2nd July 2000 in Paris, the 14th International Congress of IPPNW (International Physicians for the Prevention of Nuclear War) honoured Professor Bandazhevsky for his work. The President of the Congress, Professor Abraham Béhar, gave Galina Bandazhevskaya the Congress medal, awarded to the Belarusian scientist. The Greens at the European Parliament nominated him for the "Passport for Freedom" which was awarded to him by the European Parliament in June 2001.

The first information about the two "dissident" scientists from Belarus led to a meeting and then collaboration between Belrad and CRIIRAD. The two organisations were born in the aftermath of the Chernobyl disaster, without knowledge of each other, and shared the same aim of exposing the truth and protecting the population from artificial radiation and the government lies. Thanks to CRIIRAD's generosity, the meeting proved to be a lifeline for Nesterenko at a time when he risked having to close his institute, our primary source of information.

In February 2001, CRIIRAD began a support campaign for Bandazhevsky. On 25th May 2002, together with other associations, it organised a demonstration, in front of the Palais des Nations in Geneva. Fifteen organisations including human rights groups, environmental groups and health protection groups, launched an appeal for the scientist to be released from prison. Those participating included Amnesty International, France-Libertes, International Physicians for the Prevention of Nuclear War (IPPNW France and Switzerland), the network Sortir du Nucléaire, Greenpeace, GSIEN, Friends of the Earth, Amandamaji (Finland), of Grandmothers Against Nuclear Power (Finland), Women Against Nuclear Power (Finland), Jose Bové and the Peasant Farmers' Confederation and the Association française des malades de la thyroide (AFMT)[201].

Following the demonstration, on 18th July 2002, a "frank" conversation took place at the director's office at WHO between on the one hand, representatives of IPPNW Switzerland (Professor Michel Fernex and the president, Dr Jean-Luc Riond), and WILPF France (president Solange Fernex), and on the other hand, Dr David Nabarro, executive director of WHO, with Dr R. Helmer and Dr M. Repacholi assisting.

During the interview, the revision of the agreement between WHO and the IAEA was mentioned and Professor Fernex presented an impressive list of WHO's failures at Chernobyl, from the medical point of view. The executive director of WHO was completely taken aback; he had never been informed about the situation that had been described. In the conclusion of the memorandum of the meeting (which is worth looking at on the Chernobyl archives site[202]) Fernex notes: "Dr Nabarro states very clearly that scientists involved in WHO projects should be independent. He demands objectivity and transparency and rejects all outside interference, including from the IAEA. The offer by Dr Nabarro to continue to communicate by Email or by letter is also very constructive". Unfortunately, very soon after this "constructive" meeting, Nabarro was replaced as executive director of WHO. In a similar way, Mme Brundtland was replaced as Director General of WHO. (Following the admission of censorship by Dr Hiroshi Nakajima, she had appointed a commission to verify what had happened at the WHO conference of 1995). It is all very unfortunate. Hardly has a dialogue been initiated than the participants disappear. Perhaps they had been made to disappear?

The only tangible result was the arrival at the Fernex' house, of a heavy and mysterious package, with no indication of who had sent it. It was Christmas and they were on holiday in the mountains when it arrived; the package contained the original files (incomplete) that made up the proceedings from the 1995 WHO

---

[201] Concrete action was taken, in particular: Bandazhevsky's lawyer, Pogonyailo, along with other organisations brought the matter to the attention of the Human Rights Council at the UN as a matter of urgency; a fund was set up to support Bandazhevsky's family and to allow G.Bandazhevskaya to continue her work on the subject of the alterations in children's health at the independent institute Belrad; an appeal was made to French, European and Russian politicians to intervene with the Belarusian authorities; international support for the petition launched by Amnesty International.

[202] "Memo: Meeting at WHO, 18th July 2002" at the director's office at WHO between representatives of IPPNW Switzerland and Dr David Nabarro, executive director of WHO.

conference. Some of them were too old to be published, since they had been piled under Dr Sushkevich's desk for seven years. We had filmed Dr Sushkevich at Kiev. He had explained to Solange Fernex, in a meeting, that WHO did not have the money to publish its documents! If these papers had been published the day after the conference they would have become best sellers overnight and would, no doubt, have had lively repercussions.

As for the Bandazhevsky family, persecuted, thrown out on to the street, and no longer able to remain in Gomel, they found support and protection through Nesterenko. He welcomed Galina Bandazhevskaya to his Belrad Institute and gave her a job, which was commensurate with her professional qualifications, since she had been removed from her chair at Gomel where everyone had avoided her as if she had the plague; she also received material support from our own organisation, Les Enfants de Tchernobyl Belarus and from CRIIRAD, who helped her to find a home in Minsk.

But the network of organisations that had demonstrated in Geneva, although sympathetic by their very nature, did not know very much and had only a superficial understanding of what was really going on in the countries that had been contaminated by Chernobyl. We needed to get our information out to a wider audience in a more systematic way. Preaching to the converted made no sense. Apart from two excellent radio programmes—Ruth Stegassy's "Terre à terre" on France Culture and Arnaud Jouve "Fréquence Terre" on RFI—there had been more or less nothing in the mainstream national French press. Then three dynamic women came to the rescue and filled the gap. It was exactly what was needed.

Maryvonne David-Jougneau (who had been involved for a long time with the problems of dissidents in French institutions) had worked very hard creating the Bandazhevsky Committee, with a group of human rights activists in Grenoble. They, along with France-Libertés and Amnesty International, had kept up the pressure on the Belarus authorities to release Yury Bandazhevsky, and also on the national press, in order to broaden support for Bandazhevsky in the West. Teacher of philosophy and sociology, Maryvonne-David Jougneau had already published an analysis of dissidence within the French administration[203], and another entitled *Antigone ou l'aube de la dissidence*[204]. She was preparing a third book on Socrates, when a friend from Amnesty International gave her a copy of *"News from prison"* that I had begun to distribute. Outraged by the fact that a scientist had been imprisoned for having said things that upset people, and above all, by the lack of any reaction from his peers and from the Western media,

---

[203] *Le Dissident et L'Institution*, L'Harmattan, 1989.
[204] L'Harmattan, 2000.

she put her pen down, asked me to give her more information and set up a support committee.

The Bandazhevsky committee, set up in June 2002, organised its first public demonstration in October in Grenoble, with the help of the Museum of the Resistance and Deportation of Isère, and based around *Mensonges nucléaires* (Nuclear lies), the televised version TSI of the film of the Kiev conference,

"The Right to the Truth" was the theme that united the committee in Grenoble, and the local branch of Amnesty in Isere, and then Amnesty at the national level. Dagmar Daillant, who was responsible for the coordination of Amnesty in Belarus, Carine Hahn, who later replaced him, and Anne Guérin, director of France-Libertés, the Danielle Mitterand Foundation, joined Maryvonne in the appeal to scientists to sign the "Manifesto for the release of Bandazhevsky and for freedom of research". On 25th April 2003, the "Manifesto" appeared in *Le Monde* with 2000 signatures, with this declaration of principles: "The independence of research in the service of humanity is as fundamental a principle as the independence of justice. Bandazhevsky's imprisonment violates both these principles. This is why we, the undersigned, demand Bandazhevsky's release so that he can continue to pursue, without hindrance, his work at his institute". Over the next few months, another 15,000 signatures, including those of eminent scientists, were collected for the manifesto. At the beginning of August 2003, all the supporting organisations came together to present a dossier to the European Parliament to nominate Bandazhevsky for the Sakharov prize. He would be one of seven nominees.

It took two years of hard work before there was any reaction in the press. For weeks and months, we collected written information and films that the group in Grenoble had used in its various initiatives and in its contacts with newspaper editors and local authorities all over France. In June 2003, he was awarded Honorary Citizenship by the city of Paris[205], and it was on this occasion that the first articles appeared in the national press. On 24th June 2003, the wall of silence was finally broken with an excellent article by Hervé Kempf in *Le Monde*[206]. The evening before, the daily newspaper *Liberation* published an interview with Galina Bandazhevskaya by Veronique Soulé on its back page. At the end of the year, Bandazhevsky's work began to be published and gradually, he began to regain confidence and hope. He needed it. The mobilisation of public opinion that he learned about through the thousands of letters he received proved effective.

---

[205] After Clermont-Ferrand, a total of twenty four local authorities followed suit.

[206] "La faute de Youri Bandajevsky" (Yury Bandazhevsky's fault). This article made him very happy. He reread it several times and said enthusiastically. "It is almost as if this man had lived though it with me. He understood everything, he talked about everything that mattered, and did not get any of it wrong".

538

With the creation of an Internet site in French and in English, the Bandazhevsky committee became a centre for information and initiative. It established dynamic links between the varying viewpoints to publicise the affair and broaden the action, asking citizens, organisations, and political and scientific authorities to work together to get Bandazhevsky released from prison in conditions that would allow him to continue his research[207].

For her part, the physicist Bella Belbéoch, secretary of GSIEN, continued to challenge the scientific community and the Belarusian ambassador, Vladimir Senko. On 20th January, she addressed the following scientific and political report to him "[…] on the necessity of freeing Bandazhevsky in the interests of your country and from the point of view of progress in scientific knowledge […] It should not be forgotten that your country has the sad privilege of having experienced a level of contamination unknown up to the present day, and whose harmful effects on the inhabitants will continue for many generations to come. […] We would like to bring to your attention an experimental programme on animals, ENVIRHOM, that has just been launched by the IPSN (Institut de protection et sûreté nucléaire) to study the effects of chronic ingestion of caesium-137, using Bandazhevsky's hypothesis as a basis. Bandazhevsky has played a pioneering role in this type of research, but the exchange of scientific information that is indispensable in a project like this cannot take place when the principal contributor is in prison".

In September 2005, CRIIRAD laid down a challenge to civil society and to our political leaders that Yury Bandazhevsky be provided with his own scientific laboratory in the Republic of Belarus and be made the director. CRIIRAD is appealing to all those who support independent research into the health consequences of the Chernobyl accident: individuals, associations, elected politicians, local authorities.

President Lukashenko seems to have been able to shrug off all this considerable show of support…Or rather, he was fed up of the whole story.

DE FACTO EXILE

In April 2006, three months after his release, within the planned time frame, Yury Ivanovich Bandazhevsky, with the help of the French Ambassador, Chmelewsky, was sent abroad on his own. It was "humanitarian" France, with more nuclear power stations per head of population than anywhere else in the world that offered him a temporary home, in Clermont-Ferrand. His wife, their two daughters and their granddaughter remained in Belarus, as potential hostages.

---

[207] Professor Francois Jacob, President of CODHOS took a stance and wrote to President Lukashenko.

According to the experts in this kind of problem, who were involved in Solzhenitsyn's extradition from the Soviet Union to the West, this was a calculated move to put an end to all the excitement generated by international public opinion, to eliminate the witness, to make it impossible for Bandazhevsky to publicise his scientific discoveries. And as expected, France did not offer him any opportunity to continue his research. But again, like Solzhenitsyn before him, they had underestimated the man with whom they were dealing. His sense of duty to the truth would not allow Bandazhevsky to remain silent.

Partly because of the situation in which they found themselves and partly through choices they made themselves, the physicist V. Nesterenko, the pathologist Y. Bandazhevsky and the cardiologist and paediatrician G. Bandazhevskaya became "dissidents" in a world ruled by force, money and fear. They did not hesitate to defend science against the shameful deceit and silence that emanates from the international scientific community and from governments. All three risked their lives and the well-being of their families. Vassili Nesterenko died prematurely in 2008 as a consequence of the serious contamination to which he was exposed, his body having resisted, extraordinarily, for so long. He devoted himself, until his strength gave way, to saving the children. The Bandazhevsky family, humiliated and dispersed across Europe remain faithful to the Hippocratic Oath. The two daughters have enjoyed brilliant success in their medical examinations.

The lessons of Chernobyl were debated at length at the Kiev conference. I would like to quote from Professor Michel Fernex whose concise but comprehensive assessment of the situation has guided my thinking for the last fifteen years.

The destruction of scientific organisations in Belarus

As long as the World Health Assembly, the governing body of the WHO, does not amend, or even denounce, the Agreement it signed with the IAEA in 1959, that holds it hostage to the nuclear lobby with regard to the medical consequences of radiation, there is no hope of any substantial support for the research groups that need it most.

In Belarus, we are witnessing a dramatic dismantling of those organisations that are researching the consequences of Chernobyl most effectively.

Professor Nesterenko is the physicist who intervened immediately at the nuclear reactor when it was on fire. As a highly qualified physicist, but also on this occasion working as a fireman, he flew in a helicopter in the midst of the radioactive fumes, to investigate the possibility of pouring liquid nitrogen onto the core of the reactor. It is incredible that he survived; of the four other passengers in that helicopter, three have died as a result of the radiation they received and the radioactive contamination. Together with the colleagues at his institute, Nesterenko drew up a map showing the radioactive contamination of the whole territory and formulated proposals to protect the population.

He continued with his work, until his data and his recommendations (including the demand that children, particularly, should be evacuated from within a radius of 100km) were no longer considered acceptable, because they were too alarmist. He lost the institute, his post and his source of income. With the help of Ales Adamovich, Andrei Sakharov, the chess champion Karpov, the Foundation for Peace, Chernobyl Children International of Adi Roche and other donors, Nesterenko, fortunately, was able to set up the independent research institute "Belrad". The Institute helps the victims, and teaches them to protect themselves as much as is possible, given that they are forced to live in a contaminated environment, and attempts, with a medical team, to treat the children The survival of such an organisation depends on the international support it receives, but the nuclear lobby also recruits its detractors.

The Minister of Health, Dr. Dobrishevskaya, who supported the most effective research groups in this field, according to a joint report published in 1996, was also removed from his post.

Professor Okeanov witnessed the same dismantling of the research unit that he directed. It was an irreplaceable organisation revealing the truth about the epidemic of cancers at Chernobyl. In this case, the coincidence of the interventions he made to WHO in 1995, in Minsk in 1996, and above all his refusal to remain silent at the IAEA conference in Vienna the same year, is a clear indication of the identity of those who wanted to see his research unit destroyed.

Professor Bandazhevsky's removal is just the latest in a series. This pioneer of research into the health consequences of the Chernobyl accident has revealed the mechanisms generated by radionuclides incorporated into the organism: after I-131, Cs-137 and Sr-90. With the doctors that he trained at his Institute at Gomel and a number of volunteer researchers, Bandazhevsky has described the illnesses affecting a very large percentage of the population and nearly 90% of the children living in the contaminated territories.

Who will help to bring out into the open this series of destruction of brilliant institutes, these promising careers brought to an end? These systematic repeated attacks that have a negative effect on the country and its people, are no doubt encouraged by those who are put out by work of such quality: there are always people who are jealous or perhaps ambitious. However, in this case it is the richest countries with the most advanced nuclear industry, and the nuclear lobby who have the most to gain, and will benefit most from these attacks.

Given the omnipresence of nuclear power in the world, medicine needs to reclaim its vocation of prevention, care and research. WHO must regain its independence so that it can take action within the terms of its admirable constitution, even in this controversial area. Epidemiological studies need to be undertaken again, and not find themselves interrupted arbitrarily with such destructive effect. Who will monitor the genetic mutations that will start appearing in children born over the next five generations in those countries that have received radioactive fallout? Who will care for the victims and take responsibility for their treatment, or provide better protection for children and pregnant women? The rich countries that have nuclear power stations should be coming to the aid of those populations that are suffering the effects of Chernobyl, in Belarus of course, but also in other countries that have been affected.

The IAEA's current mandate for the promotion of commercial nuclear power should be withdrawn. There are other far more important tasks for this UN agency: the control of plutonium and uranium, and of fissile material from the dismantling of nuclear warheads, and from military and commercial nuclear installations. The IAEA, that should have prevented the proliferation of nuclear weapons in a growing number of countries, has failed to do so. In the future, the IAEA should monitor the management of all the nuclear waste that humanity has already accumulated, within the space of two generations, since the nuclear age began. This monitoring will need to continue for centuries. [208]

This book was published in France in 2006, nine years ago, but the story is still unfolding. The narrative continues as it began, recounting real world events. Opposition to the nuclear lobby has been growing among the public and the scientific community.

---

[208] The Chernobyl catastrophe and health.

Chapter V

POST SCRIPTUM
WHAT CAN BE DONE?

1. THE VIGIL OUTSIDE WHO HEADQUARTERS[209].

Since 26th April 2007, every working day from eight in the morning till six at night, two or three people, take part in a silent vigil outside WHO headquarters in Geneva, to remind this United Nations organisation of its constitutional duty. Every day, for more than eight years now, women and men from different countries in Europe and also from the United States, maintain this Hippocratic vigil as representatives of the group "Independent WHO—Health and Nuclear Power". They will do so until the World Health Organisation recovers its independence from the International Atomic Energy Agency (IAEA). The vigil is held at a crossroads where a great many cars pass by, some with diplomatic corps registration plates. The demonstrators respond politely to the passers-by and distribute information leaflets.

On 20th April 2008, three scientists from Russia and Belarus, A.V. Yablokov, R.I. Goncharova et V.B. Nesterenko stood at this crossroads facing the WHO building and held placards with the following questions and slogans:

- WHO 22 years of SILENCE and LIES about CHERNOBYL
- WHO must honour the terms of its CONSTITUTION
- Cover up and non-assistance = CRIME
- WHO must fulfil its mandate to protect HEALTH
- REVISE the agreement WHA 12–40 between IAEA/WHO
- For the independence of WHO from the IAEA
- Chernobyl—50 deaths or 985 000 deaths. What is the truth?
- CRIME of Chernobyl—WHO accomplice
- 1,000,000 children around Chernobyl, irradiated and ill, ignored by WHO

---

[209] *http://independentwho.org/en/*

2. SCIENTIFIC AND CITIZEN FORUM ON RADIOPROTECTION

In the absence of an adequate response from international organisations and from their own government, Japanese citizens called on independent scientists from other countries for information and advice. It was in order to share knowledge and experience of the Chernobyl and Fukushima disasters that the "Scientific and Citizen Forum on Radioprotection" was organised in Geneva on 12th May 2012 by IndependentWHO. It challenged the issue of "norms" and confronted the official figures with real lived experience, and offered alternative theoretical models that have been put forward by independent scientists. On two occasions now, WHO has rejected, by failing to respond, propositions from IndependentWHO to hold a joint conference between its own experts and independent scientists.

After Remy Pagani, city councillor and mayor of the city of Geneva, which had supported the Forum financially, had welcomed visitors to the Forum, Paul Roullaud (France), co-founder and representative of the collective IndependentWHO described the philosophy of the movement in the following way:

> We have come together today because all over the world, people are suffering the effects of radiation, whether from the fallout from nuclear weapons testing, from the explosion of the nuclear reactors at Chernobyl, at Fukushima and other accidents, from the use of depleted uranium weapons, or from the so called "normal" emissions, in water or air, produced by the nuclear industry. We have chosen to meet here, 200 metres from the World Health Organization headquarters because, in defiance of its own constitution, this international institution denies the victims' suffering, and in this way adds to their tragedy. There is a large body of research documenting the suffering of radiation victims but the WHO, continuing to disdain scientific rigour, chooses to ignore it.
>
> This scandalous attitude has been regularly denounced over the years but in 2006, a group of people from all over Europe decided that not a day should go by in which the criminal consequences of WHO's implacable and intolerable denial of so much suffering, not be denounced as a crime. Many months of preparation went by and then on 26 April 2007, the first Hippocratic Vigil, as it came to be called, was held, 22 years after the start of the Chernobyl health catastrophe.
>
> Since then, more than 300 people have taken part in the Vigil in front of the WHO's headquarters, demanding that this CRIME not be met with indifference one single day more. The Collective IndependentWHO makes sure that this silent vigil is maintained, through rain, wind, snow and ice. For five years, we have denounced this crime, without changing WHO's attitude. From the first day, we knew it would be a very long battle because we are challenging a very powerful international lobby. These five years of the Vigil have at least begun to reveal to the public the relationship between the WHO and the IAEA. WHO's lack of independence from the IAEA, dates from the agreement WHA 12-40, which the two agencies signed on 16 May 1959.
>
> At our twice yearly annual general meetings we unanimously and enthusiastically agree to continue with the Vigil. It would be untrue to say, however, that we never get discouraged or exhausted, and this is mainly because we still have not really got any political support. Yet it is because we still believe in political action that we challenge

WHO about its activities in the area of radioprotection. These can be summed up quickly—there are none—which is, in part, the reason we are holding this Forum.

The proceedings of the "Scientific and Citizen Forum on Radioprotection—from Chernobyl to Fukushima" that brought together 15 independent scientists and doctors from Russia, Belarus, Ukraine, Japan, France, Switzerland, Belgium and the United Kingdom have been published. They can be downloaded from the site http://independentwho.org/en/.

This international forum, organised 27 years after Chernobyl and one year after Fukushima, is the first independent conference on the consequences of a Level 7 nuclear accident. It should be emphasised that this was an independent scientific and citizen forum. It is open to everyone. It keeps the debate alive within international public opinion, because people can read it. The same independent scientists and doctors had spoken at two previous conferences organised by WHO on the subject of Chernobyl. Those two conferences are dead because they took place under the WHO's auspices. WHO is completely subordinate to the IAEA and so these independent voices were stifled: the proceedings of those conferences were never published so no-one could learn from the discussions and the contradictory opinions expressed in the lively debate between high level intellectuals and the ignorant authoritarian representatives of the nuclear lobby.

The Fukushima disaster happened on 11th March 2011. Every year, in the middle of May, all the Ministers of Health come together in Geneva for the World Health Assembly. On 4th May, after four years of blinkered indifference and a few rather pathetic attempts to dislodge the "Hippocratic Vigil" by the Geneva police, the Director-General of WHO asked us to meet with her, and her five other directors in her office *"to listen to [our] concerns and to discuss matters of common interest on the subject of radiation and health"*. After two hours of sterile discussion, fundamentally contradictory but respecting the formalities, she was then able to issue her press release[210], presenting the democratic face of WHO, in which she reaffirmed her independence from the IAEA, *congratulated* us on our *commitment* and our *determination* and promised *to keep open the dialogue* regarding WHO's competence (in the area of radiation and health) and to find out why the *proceedings* had not been published.

On 12th May, in a polite and detailed letter[211], we outlined our dissatisfaction with the legal explanations that she had provided to demonstrate that WHO was fulfilling its statutory obligations in the face of the Chernobyl health catastrophe

---

[210] WHO Director-General meets with advocates for people affected by radiation. See p. 588.
[211] Letter from IndependentWHO to the director general dated 12 May 2011. See p. 589.

and informed her that we would be continuing our action in which we accuse the organisation, of which she is the director, of the crime of non-assistance to populations in danger.

On 4th July, Maria Neira, director of the Department of Public Health and Environment at WHO, wrote to us claiming that the proceedings of the two conferences had been published.

On 7th November 2011, in a letter jointly signed by the 5 members of IndependentWHO who had attended the meeting, we sent Madame Chan and her directors the irrefutable proof that this was a lie[212].

Meanwhile...

3. BANDAZHEVSKY'S PEREGRINATIONS IN THE WEST

After all his torments in prison, Bandazhevsky now faced the torments of exile. As soon as he arrived in France, he began to realise the extent of his powerlessness in the scientific desert in which he now found himself. The Regional Council of the Auvergne (the regional authority in charge of four "departments"), where he was now living, gave him a grant that was enough to buy food, but offered no possibility of continuing his scientific work. The project conceived by CRIIRAD to set up a "CRIIRAD-Bandazhevsky laboratory" in Belarus had come to nothing. The Belarusian authorities refused to register an international independent association—a private organisation which would therefore have been free from government influence, and was to have studied the effect of radioactive substances on humans and animals. Instead, Bandazhevsky had suggested, as a way of continuing the fight to obtain objective scientific data about the effects of radioactivity on the organism, that he and his wife Galina undertake scientific research with children from the Chernobyl area while they were convalescing abroad. In this way they could make use of the experience of different associations in other countries involved in the rehabilitation of children's health, and to strengthen links with them. The aim was to set up a scientific project that was independent of the Republic of Belarus.

The Belarusian authorities interfered with the implementation of the project, by offering their services to CRIIRAD in providing information and helping them to find contacts from within official scientific circles. To Bandazhevsky's surprise CRIIRAD accepted. It was over this issue that the Bandazhevskys and CRIIRAD went their separate ways. On 15th June 2007, in a letter explaining why he was leaving the project, Bandazhevsky wrote:

---

[212] Letter from IWHO concerning non-publication of the Proceedings of the Conferences held in Geneva in 1995 and Kiev in 2001. See p. 592.

Such touching concern for science at Chernobyl is laughable, when the same authorities refused permission for an **independent** scientific laboratory to be set up jointly by myself, Galina and CRIIRAD. And meanwhile the government of Belarus approved the WHO/IAEA report about the medical consequences of the Chernobyl disaster, published on the 20th anniversary of the accident; when it cancelled most of the state subsidies for the liquidators.

By taking on the role of intermediary between scientific circles in Belarus and CRIIRAD, the Belarusian authorities are proposing a compromise, whose aim is to deprive me of CRIIRAD's support in the battle to tell the truth about Chernobyl, and to reduce my influence on public opinion. This compromise excludes any possibility of **independent** scientific research, or of the results of that research being disseminated widely. By collaborating with CRIIRAD, the government of Belarus believes it will be able to exert real influence over me.

All of this, without any concern about the medical consequences of the Chernobyl disaster. The particular project that CRIIRAD is proposing in Belarus, authorised by the government, will be mainly humanitarian, offering whatever aid the government of Belarus permits to be given to the victims, while conducting scientific research, which will be strictly under their control. It will only have an appearance of being scientific. There is no other way of looking at it. I cannot see what interest there can be in conducting scientific research within a project of this kind, when it will be impossible to show objective results, and those results will not be used to improve people's health. The present government in Belarus would not allow that. Collaborating with the authorities in Belarus in this sort of project means consenting to the situation that exists there currently regarding the medical consequences of the Chernobyl disaster, where millions of people are subjected to permanent radioactive contamination because they live in an area where no-one should live.

I cannot close my eyes to the fact that people continue to ingest radioactive elements in the food they eat and that they continue to live in areas contaminated by radionuclides. They are deprived of the necessary preventive and medical help and of rehabilitation. Therefore I will devote my strength and talents to continue the struggle to tell the truth about the medical consequences of the Chernobyl disaster, without making any compromises. That struggle is intimately connected to the struggle for democracy and human rights within the country, and above all with the right to health.

While he languished in inaction and isolation in his flat in Clermont-Ferrand, the local authority that had so generously offered, in the presence of journalists, to pay his rent and give him a mobile telephone during their *humanitarian welcome* in summer 2007, were now asking to be repaid. Yury Bandazhevsky, seeing no opportunities on the horizon in this foreign country, found the strength to drag himself out of the stagnant mire and to leave for Berlin with almost no preparation. He was invited to stay with a kind man who organised convalescent camps for children from Chernobyl. But he soon saw that there was nothing for him in Germany and he left for Strasbourg to stay with a family that he did not really know, who invited him to come and stay with them. In a letter on the 27th October 2007, Bandazhevsky explained:

> I didn't set off to Germany light-heartedly. I was forced to leave Clermont-Ferrand because there seemed to be no possibility of getting a contract for work in France and because of other financial problems. The abortive trip to Germany is simply the continuation of everything that has happened to me over the last 8 years of my life.

In Strasbourg he received a warm welcome and accommodation and it was here, at last, that the scientist was able to mix in well-informed circles. He got to know the Deputies who had supported the idea of the *Passport to Freedom* that had been awarded to him after his arrest. Here, he established relationships, took part in serious discussions, and encountered understanding and support.

On 17th April 2008, Bandazhevsky took part in parliamentary hearings, entitled "Chernobyl today", at the European Parliament in Brussels where he presented his theses and the Deputies adopted a resolution to support a co-ordinating and analytical Centre "Ecology and Health".

The Ukrainian Parliament was very pleased that the European Parliament was showing an interest in the problems of Chernobyl and on 15th July 2009, the Bandazhevsky Centre was officially registered in Kiev as the Scientific Centre for Radiological Medicine and the Centre for Re-adaptation for Victims of Chernobyl of the Academy of Medical Sciences of Ukraine.

The legislative ratification had a specific objective. In September and December 2008, the budgetary Commission of the European Parliament adopted on the first and second reading, the allocation, starting in 2009 of "a grant for the monitoring and analysis of the consequences of the nuclear accident at Chernobyl on public health and the environment in neighbouring countries, in particular Belarus and Ukraine, as well as its socio-economic impacts"

Up until the spring of 2013, it was not clear whether the money would be assigned to Bandazhevsky's Centre, as no specific beneficiary had been designated. On page 229 of this book, I related the story of the million euros destined for the Belrad Institute for a radioprotection project for the most contaminated children in Belarus, as part of the TACIS programme. The dossier was sent in May 2002 to the leader of the delegation of the European Commission in Kiev, disappeared into thin air somewhere in the Ukrainian capital, and never reached its destination in Brussels. I gave that chapter the title "Europe's duplicity". The European Union decided instead to fund the programme CORE, launched on 6th December 2003 in Paris, one of whose aims was to remove the Belrad Institute from its activities in the Belarusian countryside. (see page 277)

4. THE "ECOLOGY AND HEALTH" CENTRE [213]

EXTRACTS FROM BANDAZHEVSKY'S REPORT

Governments in the former USSR, and in the current republics of Belarus, Ukraine and Russia, have not been able to solve the main humanitarian problems resulting from the Chernobyl accident. This has led to the current deterioration in the health of these populations.

The position of concealing the true extent of the consequences of the Chernobyl disaster and the reluctance to implement the necessary measures to protect the health of people is inhumane. **Objective information is urgently needed. Indeed, real action to ensure adequate protection of the health of the population constantly in contact with radiation agents is desperately needed.**

Unilateral acts (in the field of medicine or the socio-economic sphere), taken over the past 20 years in this region, have not yielded positive results in terms of improving people's health. This, in our opinion, is due to **the lack of a comprehensive life-support program** in the areas affected by the Chernobyl nuclear power plant. This requires providing reliable scientific information on health to the international community. It also requires active participation of the international community in promoting the adoption of effective measures to overcome the humanitarian consequences of the Chernobyl disaster.

In this regard, **the main goal of the Centre for "Ecology and Health", established in Ukraine, is to inform the international community about the situation in areas affected by the Chernobyl nuclear power plant.** The Centre also seeks to coordinate the efforts of doctors, ecologists, economists, industrialists, politicians etc. as well as to develop a set of measures aimed at ensuring the safety of people living in the contaminated areas.

The project, entitled "Integrated Model of livelihoods in the area of radioactive contamination", aims to coordinate the efforts of the international community **to devise measures for the population to live safely when exposed to radiation in one of the most highly contaminated areas** of Ukraine. The lessons learned in the region will subsequently be **disseminated to other contaminated areas.**

After analysing the situation created after the Chernobyl disaster, and in consultation with representatives of regional authorities, it was decided that the Ivankov District in the Kiev Region would be chosen to serve as the main pilot area. This district is one of the most heavily contaminated in the region, with very high levels of Cs-137 and Sr-90 radio-nuclides.

---

[213] *www.chernobyl-today.org*

WHY THIS PROJECT IS DIFFERENT FROM OTHERS THAT HAVE ALREADY BEEN UNDERTAKEN OR PROPOSED

1. The project is presented by the co-ordinating and analytical Centre "Ecology and Health", set up by the European Parliament. Members of the European Parliament are on its Monitoring Council.

2. The main aim of the project is to stop the incorporation of Cs-137 radionuclides into the human body, in particular into the bodies of pregnant women and children.

3. The project is based on a coherent solution to the problem posed: the improvement in health of the people who live in areas contaminated by radioactive elements.

4. The project will provide full information at every stage of its development to the international community.

5. The project will provide international cooperation in order to obtain the humanitarian, material and legal aid that will be necessary to complete all stages of its development.

Finally, on 26th April 2013, we learnt in the media of a new development that could have important implications: 27 years after the Chernobyl disaster the European Union was inaugurating cooperative action with Ukraine for the protection of victims, by allocating 4 million euros to Yury Bandazhevsky's Centre "Ecology and Health" in the Ivankov district, situated in the contaminated territories between the power station at Chernobyl and Kiev.

Today, we can rejoice and offer our congratulations to the two Members of the European Parliament, Michèle Rivasi, Vice-President of the Greens/ALE (European Free Alliance) and Corinne Lepage, Group of the Alliance of Liberals and Democrats for Europe, whose insistence in the matter made this possible. But these are crumbs, alms to a beggar in comparison to the aid needed, after 27 years of indifference, to care for the hundreds of thousands of children living there. Given the level of deceit and the lack of care shown by Europe for so long about this health catastrophe, their offer can only be described as paltry, their response almost amnesiac. Europe is spending 500 million euros on a problematic super-sarcophagus and less than 1% of this sum on the lives of the people condemned from childhood to a slow agony, a life cut short and blighted by suffering.

Yet, even this minimal aid programme, arriving so late in the day, would be impossible to implement in Belarus, even though this is the country that received 70% of the radioactive material released onto the European part of the former USSR from the fire at Chernobyl. The EU would use the pretext that it no longer has relations with this country. In reality, everything depends, as always, on political will. In 2003, the German government financed a radioprotection programme for children in Belarus, but the third stage of the programme was cancelled due to pressure from groups who were hostile to the use of pectin (see page 235, in the Chapter "The War against pectin").

5. LONG LIVE THE BELRAD INSTITUTE [214]

The circumstances that allowed the Institute of Radiation Safety "Belrad" to be created will never be reproduced. This means that should "Belrad" disappear through lack of funds, no other organisation will take its place.

The Institute was born during the collapse of the USSR, a particular moment in history and was a response to the enormous damage caused by the radioactive fallout on the inhabitants who were not evacuated from the contaminated areas of Belarus. A similar situation exists in both Ukraine and southern Russia but no comparable organisation came into being in those countries.

On 12th February 2008, six months before he died, Professor Nesterenko, having been invited to a conference organised by the French regional authority Rhone-Alpes, addressed the following appeal to Hélène Blanchard, Vice-President of the Regional Council:

> Twenty two years have passed since the accident at Chernobyl. The health catastrophe for the people who live in the contaminated areas gets worse every year. There is an acute shortage of international projects on radiation protection and rehabilitation of children in Belarus. Over the last few years, I have attended numerous conferences in the rich countries of the West, and I have informed them of the dramatic situation for our children and of the work we do to try to protect them at my institute "Belrad", and although the problem is discussed and promises made, I have never received any financial support from international organisations for the projects I have proposed. The Europeans who participated in the CORE project (Cooperation and Rehabilitation) opposed, under all sorts of pseudo-scientific pretexts, the financing of the harmless pectin-based food additive for children, whose effectiveness in accelerating the elimination of radionuclides from the body has been proven over a number of years. Western scientists collect data, but do not help the children. From the point of view of medical ethics it is inadmissible to measure high levels of caesium-137 in a child's body without providing them with preventive treatment in the form of pectin. We are able to provide a small proportion of the projects that are needed through our association "Enfants de Tchernobyl Belarus" (Children of Chernobyl Belarus), but we do not have nearly enough money. New radioprotection projects for children are needed in the six most contaminated districts. I am putting forward a proposal here for this sort of project and request that the regional authority consider participation and financial support. We need to show how useful a conference like this one is for the children of Chernobyl. After suffering all these years, limiting ourselves to the provision of information only, without being of any practical use to them, would be immoral.

Vassili Nesterenko attended the conference but he did not receive any response to the project that he presented.

In conclusion, I will let our colleagues from this unique institute, that Europe continues to ignore, speak for themselves, to remind us again of its value and why we cannot allow it to be lost.

---

[214] *http://belrad-institute.org/*

The independent institute of radioprotection "Belrad" was created in 1990 to offer practical help to the inhabitants of the contaminated areas. It took some time to set up because various bureaucratic problems needed to be solved and funding, premises and the indispensable qualified personnel found. It took four long years. The Institute was set up by the academic Andrei Sakharov, President of the Soviet Foundation for Peace, the chess champion Anatoli Karpov, the Belarusian writer, Ales Adamovich and the corresponding member of the National Academy of Sciences of Belarus, Vassili Nesterenko. Professor Nesterenko became the director of the Institute, its heart, the generator of ideas and a brilliant organiser.

Unfortunately, Nesterenko, a scientist of international renown, had to deal not only with the scientific side and with leading the Institute, but also with the innumerable battles of attrition with bureaucracy, with misunderstanding, often ignorance, and the refusal of certain government bodies to engage jointly in the problem. These "battles" took a toll on time, money, his strength and his health. A human being's strength and health is not unlimited, and in 2008, Vassili Nesterenko left us forever, leaving behind him a fully functioning Institute, ideas, disciples and many good memories in the minds of thousands of people.

The Belrad Institute has never considered its methods of radiological protection of the population to be the only solution or to be a panacea for all disasters. On the contrary, we are well aware that our work is just a small part of what needs to be done. Positive results can only be achieved by using a whole series of comprehensive protection measures, such as: monitoring of radiation in the environment, food and populations; medical examinations; administrative measures; rehabilitation of contaminated areas; the application of modern methods in agriculture and forestry; education; use of radioprotectors for the quick elimination of radionuclides from the bodies of people and dairy and beef cattle and so on. In this context, the experience and the scope of activities performed by the Belrad Institute could be extremely useful.

However, the Ministry of Health refuses to recognise, with maniacal obstinacy, the existence of the problem and the practical results obtained by the Institute during its twenty years in operation. Such obstinacy over useful work. For example, the fact that by using a pectin-based adsorbent like "Vitapect, the levels of radionuclides in the body of a person consuming locally produced food, can be lowered by between 30% and 35%. Similar results have been obtained in Ukraine, where work was undertaken independently of "Belrad" by specialists from the Academy of Sciences of Ukraine. But, for many years now, the Ministry of Health in Belarus will not even consider the mass production and use of pectin-based products in the territories that have been contaminated by radionuclides, even though they do not suggest any alternative solutions. The Swedish firm "SEMPER" produces food for children containing a pectin additive, but we cannot produce it here because it is not permitted.

In August 2007, a commission from the Ministry of Health came to the Institute. The visit was preceded by a letter to the government from the Ministry of Health. In the letter, the real facts about the Institute's activities were distorted, and amounted to outright falsification and slander. The reaction was unequivocal: inspect and prosecute the Institute's leadership. At this point, the commission came forward not to investigate, make constructive criticisms or suggestions, praise or inquire, but to punish and close the Institute. It did not work. How could it work given that the commission did not even bother to prepare for the inspection and was stumbling about in the dark with even the most elementary questions? All this was reflected in the final document, in what happened as a result of the inspection.

The president of the commission, a civil servant apparently respected somewhere, by someone, unable to understand the difference between a human radiation spectrometer and a radiological diagnosis, kept repeating with a doggedness that might have been admirable in different circumstances, that the Institute was conducting radiological diagnoses, in other words, medical interventions, which were illegal. Either he was never a very good student or he had forgotten all he had ever learnt in his long years sitting in his director's armchair. A woman doctor claimed tearfully, though with a certain indignant tone, that she had heard of a case in which a child had died from taking pectin. In that case, various questions need to be asked: on what grounds had the Ministry of Health authorised the Institute, over many years, to produce and use "Vitapect"? Why is pectin used in the preparation of fruit purees and jam, and why are various pectin-based products on sale in Ukraine and in Russia in pharmacies and in food shops? Where is the logic, where is the concern for children's health, if pectin poses a mortal danger?

It became obvious that the grievances listed in the commission's report were invented and had no basis in law. However the Institute was forbidden from producing "Vitapect" in its laboratory and its network of radiological monitoring centres almost completely dismantled in 2008. In 2007, we were presented with various injunctions according to which we could no longer install our centres in schools because the children would be bringing "dirty" products there. It was proposed that we construct separate buildings or annexes, which was completely impracticable. So food products could not be brought to the school to be monitored for radiation but it was alright to eat them at home? Their logic was stupefying.

We found a way out and began to orient the activities of the Institute in a new direction: we set up mobile laboratories, equipped with all the necessary machines for monitoring children's health, for testing food products, for running courses, seminars, showing training films, photographs and teaching materials. Today the Institute has three mobile units in operation but it is not nearly enough. The problem is finance: it needs to be remembered that almost all the work undertaken by the Institute was financed by our partners abroad: foundations, organisations, associations and social and charitable initiatives from Germany, Norway, Ireland, Spain, Belgium, Japan, Austria, France, Switzerland, the USA, Italy, Great Britain and Lithuania.

To sum up what we have been saying about the Institute, there are three main orientations: looking for finance, fighting bureaucratic obstacles and finally the radioprotection of children living in areas contaminated by radiation.

As time went by, it became harder and harder to find partners from abroad. In 2006, when this book was in the final stages of preparation for publication in France, the situation was critical. The Foundation "France Libertés—Danielle Mitterand" had helped us to stay afloat, but in 2009 the government took away its funding and the Foundation was no longer able to support us. Our association (Les Enfants de Tchernobyl Belarus) remained the only real support (1,500 members without much money) but it was not enough. Alexei Nesterenko, like his father seven years ago, does not know where he will find the money to pay salaries next month.

6. THE NEW YORK ACADEMY OF SCIENCES—THREE MEN BATTLE FOR INDEPENDENT SCIENCE[215]

In December 2009, the New York Academy of Sciences published a book in English by A.V. Yablokov, V.B. Nesterenko and A.V. Nesterenko entitled "Chernobyl: Consequences of the Catastrophe for People and the Environment". It had already been published in 2007 by "Nauka" in St Petersburg, and then, following some revisions and additions, it was published in Kiev in 2011 and in Japan by Iwanami in 2013. By 2007, more than 30,000 scientific papers had been written about the consequences of the disaster, mostly in Slav languages. Nearly 1000 research papers are quoted by the authors of this book, and in total it refers to over 5000 papers, published on paper and on the internet.

The figures obtained from this research are in direct contradiction to the conclusions reached by WHO and the IAEA. The authors estimate that worldwide, between 1986 and 2004, there have been 985,000 deaths that are attributable to the consequences of the Chernobyl accident. This figure needs to be compared with the estimates made in 2005 by the World Health Organisation and the International Atomic Energy Agency: fifty liquidators dead and up to 4000 potential deaths from cancer.

This new book presents an enormous amount of data obtained through independent research in the countries most affected. It represents a serious challenge to the nuclear powers, especially as the strategies that are normally used to discredit independent research on Chernobyl, cannot be used in this case. It will not be easy for the nuclear establishment to dismiss a publication from so prestigious an organisation.

*

A creative, politically active civil society, alert and informed public opinion remain the only hope, the only resource we have to oppose the domination of the industrial and military nuclear lobby, and its uncontrolled corruption at the heart of the UN organisations, whose role is to maintain peace and security in the world... Civil society does not hold the keys to power but it does have financial resources— large fortunes remain unused: there are rich, independent people in possession of

---

[215] *http://stopnuclearpoweruk.net/sites/default/files/Yablokov%20Chernobyl%20book.pdf*

enormous, extravagant wealth who could, if they so wished, finance independent initiatives, that not a single government in the world has so far supported, to defend humanity from the menace that nuclear technology presents to human health. This book began with a message in a bottle. That message remains, an appeal to people's capacity, whether rich or poor, for regeneration.

Just as the Chernobyl disaster is absolutely unique in history, so is the Institute of Radiation Safety, "Belrad". It is the only independent organisation in the world providing scientific proof of the official lies about the health consequences of the catastrophe, while protecting children at the same time. An island within civil society, highly qualified scientifically, surrounded by world wide nuclear totalitarianism. If the work undertaken by "Belrad's" is not safeguarded in a sustainable way, in other words, on a permanent basis, it is condemned to failure. If Belrad were to fail, the academician Vassili Nesterenko's unique challenge would be consigned to oblivion along with it. The particular circumstances that made the creation of "Belrad" possible in 1990—the collapse of the USSR—will never be reproduced. If "Belrad", with its experience and the accumulated data from 15 years of activity, were to disappear through lack of support, no other organisation could ever replace it.

Nine years after its first publication, this book ends, as it were, in mid-sentence. The terrifying experience recounted here is continuing testimony to the reality of the world in which we are living. Five years ago, we heard that an accident had happened at the power station at Fukushima Daiichi in Japan…

April 2015

Documents illustrating WHO's dereliction of duty

Memo: Meeting at the World Health Organisation. July 18, 2002

WHO: Dr. David Nabarro, Executive Director, and two staff of the Dept. of Protection of Human Environment, Dr. Richard Helmer Director and Dr. Michael Repacholi, just back from Belarus.

PSR/IPPNW Switzerland and WILPF: PD Dr. Jean-Luc Riond, President IPPNW (Engaged in the prevention of nuclear war, and of nuclear accidents like Chernobyl), Prof. Michel Fernex, Board member of PSR, the Swiss affiliate of IPPNW (15 years active in WHO Scientific Working Groups (SWG) of Tropical Diseases Research, T.D.R.; former member of the Steering Committee for Malaria, and later for Filariasis), and Ms. Solange Fernex, President of WILPF France (Women International for Peace and Freedom: engaged since half a century in the prevention of nuclear risks).

It became clear that WHO would have welcomed the participation of a representative of Contratom at this meeting. PSR/IPPNW Switzerland and WILPF are supporting the same demands as Contratom, regarding transparency and independence for medical research in the field of ionizing radiation, especially following Chernobyl.

The demand to amend the Agreement between the WHO and the IAEA (Res. WHA. 12.40), as a step to achieve independence in research and publication, is supported by WILPF, IPPNW, Contratom, and many other NGO's. All these NGO's act also to obtain the liberation of Prof. Yury Bandazhevsky, prisoner of conscience in Belarus. In close collaboration in this field, they will continue their fight in favour of Bandazhevsky and other physicians, prisoners of conscience in Belarus.

NGO's consider that after Chernobyl, the WHO teams should have visited the Medical Faculty established in the region with the highest radiocontamination with Caesium, Strontium, Uranium and Plutonium. The Medical State Institute of Gomel was created and directed by Prof. Yury Bandazhevsky. The medical research team of Prof. Bandazhevsky completed over 20 theses on the dysfunction of organs or systems as a consequence of the chronic accumulation of 137Cs in the given organs: pancreas, endocrine glands, thymus, heart or placenta. We have the impression that the WHO delegations had no permission to visit this Medical Faculty, working under the Health Ministry of Belarus.

Dr. Nabarro does not think that the WHO was actually absent in the field in Chernobyl until 1992, and the Legal Department of the organisation does not consider that the Agreement (WHA 12.40) is an obstacle for independence and transparency.

PSR/IPPNW is prepared to discuss this issue. Of course, we are aware of the work undertaken by the WHO Regional Office, essentially through the Helsinki Project Office for Nuclear Emergencies and Public Health, with the remarkable findings of the team of Dr. Keith Baverstock in the early nineties, confirming the epidemic of thyroid cancers in children. Unfortunately, this stochastic effect following Chernobyl was only "accepted" 5 years later by the IAEA and the UNSCEAR, this delay having negative consequences for the therapeutic help to the patients. We also very much appreciate the "Guidelines for Iodine Prophylaxis following Nuclear Accident" (WHO, 1999).

However, the 700 participants at the WHO Conference of November 1995, organized by Dr. Hiroshi Nakajima, received an information sheet (1), indicating that the plans for the IPHECA project **were finalized by the IAEA in May 1991.** The author of this project was not the WHO but the IAEA, possibly due to Article I point 3 of the Agreement with the IAEA, which says: *"Whenever either organisation proposes to initiate a programme or activity in which the other organisation has or may have a substantial interest, the first party shall consult the other with a view to adjusting the matter by mutual agreement".*

IPPNW, WILPF and other NGOs are asking since years to amend and shorten this sentence in the following way: *"...the first party shall inform the other."*

Other modifications of the Agreement were proposed in a letter to the Ministers of Health of all WHO member states present at the WHA 2001 and 2002 (See attachment 2: letter signed by Prof. Abraham Béhar former President of IPPNW France and President of IPPNW Europe, P.D. Dr. Riond, and Prof. Fernex, and the document distributed in 2001). We consider that the wording of the Agreement may explain e.g. why the genetic consequences were not included in the IPHECA research project of May 1991, whereas dental caries in children had to be studied by the WHO.

Dr. Repacholi informed us on his recent trip to Belarus. The demonstration that breast cancer is also linked to radiation in the Chernobyl context is alarming. A lot of basic research is still progressing. The 2 million US$ provided by Japan are spent in Gomel, and should provide further knowledge on thyroid cancer (carcinogenesis).

When trying to discuss the liberation of Prof. Bandazhevsky, Dr. Repacholi was met with a frosty response, as the personalities he met in Minsk were precisely those mentioned in the critical report of Bandazhevsky (Attachment 3). Amnesty International considers that this report on the work performed in 1998, with a grant of 17 billion BY roubles, provided mainly by WHO, IAEA, OCHA, UK and Germany, by the Clinical Research Institute of Radiation Medicine and Endocrinology (including non-independent scientists), was the actual reason for the sudden imprisonnment of Bandazhevsky, soon after he delivered his report, ordered by the Government.

We would like to know whether the WHO agrees or not with the scientific critics of Bandazhevsky on the work done in the above-mentioned Institute, with tax-payers money.

Dr. Nabarro asked us what we would suggest as projects for the most affected part of the population: the children.

A classical approach for the WHO is to convene Scientific Working Groups (SWG) on most relevant pathologies expected in a region. We may therefore make following suggestions :

1. One SWG could be a counterpart of the 1956 meeting on "Effets genetiques des radiations chez l'homme". (Rapport d'un groupe d'etude reuni par l'OMS, published in Geneva, 1957). Instead of H.J. Muller, Nobel prize holder for genetics, A.J. Jeffreys, professor for Genetics could be asked to select the best specialists in this field. In fact, papers by Dubrova could only be published in NATURE, due to the fact that Jeffreys was co-author. The methods discovered by Jeffreys were also used in Chernobyl by the Swedish team of Ellegren et al., and others in Israel for families of liquidators.

Other approaches by the group of Prof. Rose Goncharova in Minsk, should be considered: the continuous increase in the chromosomic alterations in rodents, living in more or less caesium- contaminated regions between Chernobyl and Minsk, after over 20 generations, is striking. In the mean time, the radiological contamination in the environment was progressively decreasing. The reduction of the chemical pollution in Belarus is also marked, since Chernobyl, as the industrial activities diminished, and the use of pesticides in agriculture is reduced or stopped, due to the deteriorating economical situation. Studies in fishes by this group are important, as except for the presence of a little over 1 Ci/km^2 of 137Cs in the mud, there was no chemical nor any other source of pollution in the water of the ponds, where carps are still being studied by this team.

The WHO published in 1996 a Technical Report on "Control of Hereditary Diseases". There, the lack of genetic effects of radiation on the genome after Hiroshima and Nagasaki was again stated, but Chernobyl has to be considered as a fundamentally different problem as the atomic bomb. Facts seem to confirm this difference in all fields of pathology: there is a major difference between the biological effects of an enormous irradiation lasting for a few seconds, and a very chronic low- level irradiation of cells, from inside of the corresponding tissue, e.g.: an endocrine gland or a gonad. This irradiation persists for years.

Geneticists must study actual data, after 10 to 17 year chronic internal irradiation, 10% of the radiation from this isotope being beta rays, more harmful than gamma rays, because they act locally. Scientists must find new ways to calculate the risk per units of dose over years. Calculations based on a homogeneous distribution of Caesium in the orgaism, and only on its activity as a gamma emitter, is wrong. Caesium may concentrate 100 times more in one specific tissue (pancreas, thymus, endocrine glands, heart) than in others (bone, fat tissue, liver).

2. Another SWG could focus on the effect of 137Cs incorporation on the cardiovascular system. There Prof. Bandazhevsky, a pathologist, would be a key figure. Dr. Galina Bandazhevskaya, cardiologist and paediatrician could also make contributions. In areas contaminated by 5 to 15 Ci of 137Cs/km², up to 80% of the children suffer from cardiac symptoms. The disease is due to degenerative changes of the myocardium (The caesium cardiomyopathy was also observed in rats receiving caesium, with the same degenerative changes in the cardiomyocytes of rats, as those found in caesium-contaminated children with sudden death).

As the caesium cardiomyopathy is reversible for a rather long time, it appears urgent that these most common diseases are studied, prevented or treated with the contribution of the WHO. Vascular diseases, especially hypertension in children, also increase significantly with increasing contamination by caesium, complications are cerebral or myocardial infarction.

3. Radiocaesium and reproduction would be an essential subject for a third SWG. The caesium accumulation in the endocrine glands, the gonads and the placenta may explain spontaneous abortion, fragility of newborns, and the increase of congenital malformations, as well as sterility. (The papers of Y. Bandazhevsky and those of G. Lazjuk. unfortunately were not published in Western j ournals).

During the IAEA Conference in Vienna (1996), because he was sitting in the audience wearing a yellow tag (with no speaking rights), instead of the official red tag, Prof. Laziuk was not allowed to protest when the official rapporteur said (if Michel Fernex remembers correctly the wording): "The proof that there is no increase of malformations after Chernobyl, is the absence of a register".

First, this is not a scientific proof, and, on the contrary, Belarus was the only country around Chernobyl with a well functioning national register installed 5 years before Chernobyl. It is worrying that experts from Western countries (financed by the French Commissariat a l'Energie Atomique, CEA) are now "improving" and "correcting" the Belarus Register. The NGO's consider that the sources of financing for "experts" should be examined. If a particular lobby, the tobacco or the nuclear lobby, finances research in the corresponding field, the risks for biased findings are increasing.

4. Chernobyl and the immune system would be a further subject for a SWG. Again, one should concentrate on findings in children. The IAEA specialists claim that stress and vodka are far more important than radiation. This assertion is less convincing for diseases of children. At school in contaminated areas, children receive 2 to 3 meals of "clean" food daily. The follow-up of immunological parameters were started in children immediately after the explosion e.g. by Prof. Titov. in children, and by Prof. Goncharova in wild rodents.

The clinical long-term consequences are allergies, increased incidence of asthma bronchiale, food allergy, and auto-immune diseases. The epidemic of Hashimoto's thyroiditis correlates statistically with the caesium contamination of the soil. Even more dramatic is the increase of the incidence of diabetes mellitus type 1. Diabetes occurs at an earlier age than before: in 1996, it started already in some cases below the age of four; in 2000, it may now occur already in infants of 6 to 10 months of age. This disease, due to an autoimmune ilotitis is not increasing "because doctors are looking for it", as representatives of the lobby usually say: as a rule, children are comatous when entering into the hospital for the first time.

It would be essential to recognise early signs of ilotitis. Bandazhevskaya found a tendency towards hypoglycaemia in highly radio-contaminated children. This could reflect early alterations of beta cells. Prevention of diabetes before the complete destruction of the Langerhans islets would be an important subject to be studied with the WHO.

5. The neuropsychic damages, especially due to the uptake of different uranium or plutonium isotopes by liquidators, may be a subject for a SWG. The Minister of Health of Ukraine stated in Geneva, November 1995, that 10% of the liquidators of his country were already invalid. At the WHO congress of Kiev in June 2001, it appeared in most of the republics of the former Soviet Union, that the proportion of invalids among the liquidators was over 30% !

The deterioration of the neurological status of liquidators is dramatic since 1998, and may lead to the early death of these young and healthy adults, which were aged 33-34 years when mobilized to clean the Chernobyl area. Using different methods, neurologists describe the topography of the lesions, predominantly localized in the left hemisphere, leading to the dramatic deterioration of the mental status of these subjects (they were 600.000, according to data presented in 1995).

Prof. Pierre Flor-Henry from the University of Alberta in Edmonton, finds a destruction of neurones selectively in the left hemisphere of such patients. Children irradiated in utero also seem to have brain damages predominantly in the left hemisphere. Flor-Henry who studied the neuropsychic findings in veterans of the Gulf war, found lesions similar to those of liquidators. E.g.: the so called "chronic fatigue syndrome" is occurring in both groups.

It may be too late to help the victims of the dust contaminated with heavy atoms releasing alpha radiation around Chernobyl. However, for the WHO a study of the effect of particles or heavy atoms chronically producing alpha particles in surrounding tissues, may be relevant: a preventive measure could be the decision to use no more uranium 238 in ammunitions, tanks and planes.

6. Other subjects :

A. The role of a cancer register in case of environmental catastrophes, using Chernobyl as an example, could be taken up by a SWG. After Chernobyl, most of the cancers are still in a latency stage. Studying one or two types of cancers may mask the real problem. Brain tumors in children exposed to radiation were an important subject some years ago, in the Ukraine and in Minsk.

IPPNW regrets that the Cancer Register of Belarus was dismantled, soon after the Vienna Conference of 1996. Researchers, very close to the IAEA, took over the management of the most important part. Many persons think that Okeanov was not dismissed because he spoke too clearly during the Geneva Conference in November 1995, but because he answered questions, after the official presentation at the IAEA meeting, in Vienna 1996, saying the contrary of what the official rapporteur just had stated (absence of any cancer except that of the thyroid, in children exclusively).

Prof. Okeanov lost the direction of his former institute (with accusations similar of those which were used to imprison Prof. Bandazhevsky), and was put aside for some years, the register being dismantled into three pieces.

As he has recently been rehabilitated, some observers think that it would be possible to restore the National Cancer Register. However, Okeanov may still be very fragile. One of his own collaborators, politically much stronger than him, being very close or very supportive of the IAEA policy, is in charge of an important part of the Register since 1996.

B. The consequences of Chernobyl on gastrointestinal functions seem to be important, although we did not collect many documents on this subject. There the stress as a component will be difficult to separate from the radiotoxic effect of caesium. The kidney, the liver and other organs should be studied, as well as wound healing. The difficulty would be to find a correlation with the accumulation of caesium, as most of the hospitals have no anthropogammameters. Where no valid data are available for epidemiologists, experimental work seems to be indicated.

7. All findings by Bandazhevsky's team on radiocaesium incorporation have to be confirmed, e.g. studied again, and his experimental work must be repeated. This would require an international coordination. A SWG could set up such a project.

In hundreds of autopsies of adults, children and foetuses, Bandazhevsky studied the comparative distribution of this radionuclide. The distribution of 137Cs depends on the species (human, animals), but also on age, activity, sex etc.. Under Bandazhevsky's direction, members of the Institute of Medicine of Gomel studied metabolic changes produced by radiocaesium in rodents (and human). Bandazhevsky published findings on different endocrine and haematological disorders induced by radiocaesium, comparing animals and human living in more or less contaminated areas, the similarity between experimetal and clinical findings being impressive.

Congenital defects depending on the 137Cs accumulation in the placenta of Hamster are similar to those observed in foetuses or children of Caesium-contaminated mothers. The 137Cs concentration in the placenta appears to be the most important parameter for the correlation with malformations in human. All experimental work has to be repeated in several species, and in different laboratories. Again the WHO could play a key role in the coordination.

The correlation is often based on the 137Cs contamination of the soil, or on the emission of gamma rays measured in human or animal. Bandazhevsky has shown that these measurements are not sufficient, as 137Cs concentration varies in given organs: The highest values were found in the pancreas, adrenals, thyroid and thymus in infants, the Caesium load being 100 time higher than in bone or fatty tissue, 50 times higher than in the liver. Dosimetry based on speculations, mean values and and mathematics, should be replaced by precise and direct radiametric measurements.

Experimental work could help to design preventive or therapeutic measures for the exposed population. Bandazhevsky already showed that natural apple pectin was able to reduce the uptake of radiocaesium in rats (the same was found with strontium), and to prevent the cardiomyopathy in animal, (and also in children as preliminary findings of Bandazhevskaya tend to show).

Pectin accelerates the elimination of caesium from the organism. The mechanisms still are not well studied. Other absorbents could be tested. Simple pectin extracts from apples, with a purity of 12 to 16% are cheap and capable of mobilizing caesium from the organism of children. it was widely used to protect children in contaminated regions in the Ukraine and recently also in Belarus (see attachment 4: manuscript of Nesterenko et al.).

Dr. Nabarro asked us if we would be in favour of the publication of the proceedings of the WHO Conference of 1995.

The Proceedings, which were to be payed for by the participants, would have been a bestseller in March 1996, the date foreseen and promised by the organizers of the WHO Conference of 1995. In 2002 or 2003. This publication would hold some historical interest, but would no more bring scientific revelations as over 70% of the papers were published elsewhere. Controversies expressed in the discussions would be interesting.

The full presentation of Prof. Okeanov in 1995, may still be important. The presentation of the Minister of Health of Ukraine has not been published, where he indicated that already 10% of the liquidators were invalid (now one third of them are invalid, which is a problem for all former republics of the Soviet Union), and that diabetes mellitus was sharply increasing in his country (not due to excessive food intake, the population becoming rapidly poorer). The speech of Claude Haegy, from the Government of Geneva was also very stimulating.

Other papers would be worthwhile being published, but the number of readers would be 100 time smaller than for the full proceedings published in in March 1996. In the meantime, the knowledge increasedDiscussions would have been interesting. Remarks such as that of Prof. Yarmonenko, saying: "Should in a congress any paper mention effects of low-level radiation, the organizers should immediately exclude the author !" We know now that the genome can be altered by very low-level radiation, and that these changes may persist over generations. This cannot however justify the efforts and cost of publishing the full proceedings of the 1995 WHO Conference, after a delay of seven years.

Dr. Nabarro stated very clearly that he wishes that scientists remain independent in all WHO projects. He demands objectivity and transparency and refuses interference from outside, even from the IAEA. The offer to further allowing us to communicate by telephone or e-mail, was also a very constructive conclusion expressed by Dr. Nabarro.

We are very thankful for the commitment and the concern expressed for the affected population, especially for the children of the Chernobyl area. A message of the Secretary General of the UN, Kofi Annan, states that 9 millions of adults, i.e. two millions of children, suffer from the consequences of Chernobyl, and the tragedy is only beginning. "The legacy of Chernobyl will be with us and our descendants for generations to come".

Michel Fernex—Professeur emerite, Faculte de Medecine de Bale, ex-membre de Comites Directeurs de TDR

(Programme special de Recherche pour les Maladies Tropicales), OMS.

The Chernobyl Catastrophe and Health
By Michel Fernex

Michel Fernex, MD, Professor emeritus, Medical Faculty of Basel, former member of Steering Committees of Scientific working groups on malaria and filariasis, WHO Geneva.

Introduction

After the explosion and the fire in the atomic reactor N. 4 in Chernobyl, radioactive fallout were registered in large parts of the Northern Hemisphere of the planet. The northern part of Ukraine, the South-East of Russia and the territory of Belarus have been contaminated to the greatest extent. The level of the radioactive fallout on the territory of Belarus, a non-nuclear state, was two times as high as that of Ukraine and Russia together.

It would have been essential to thoroughly investigate the consequences of the Chernobyl accident. The Belarusian population, which suffers most, should have been extensively studied from the medical and genetic point of view. Unfortunately, the pronuclear lobby, including the International Agency of Atomic Energy (IAEA) made use of all their influence to reduce the significance or to deny the data coming from the most affected countries. Their goal may have been to avoid paying compensations to the states and to the victims: in Belarus two million people, of 500.000 children, still live in heavily contaminated areas. Moreover, the evacuated population and 800,000 workers in charge of the decontamination of the area close to the exploded reactor, the so-called liquidators, are now scattered in different republics of the former USSR.

It is necessary to understand by which methods the pronuclear lobby and the IAEA achieve their goals and to assess the price of this attitude for Belarus: economic, medical, demographic and social problems of the republic appear to be the consequence of this policy. Twenty five per cent of the national budget is spent for the alleviation of the consequences of Chernobyl. In order to ensure a real protection for the victims, it would be necessary to provide much more help to the population, and in a different way, compared to the actual strategy. Rich countries possessing nuclear technology would have been in a position to take over these expenses. Contrary to other industries, the nuclear industry does not need to contract insurance capable to compensate the consequences of a catastrophe such as Chernobyl. This would cost such an amount, that nuclear electricity would become too expensive. Therefore, it would be fair to assign to the states the payment of the debt of civil responsibility.

It is difficult to understand why the Belarusian authorities seem to follow the demands of the pronuclear lobby. It is much easier to explain why the World Health Organization (WHO) became so inefficient in this field: it is still blocked by an "Agreement" signed in 1959 with the International Agencyfor Atomic Energy (IAEA).

The agreement between the WHO and the IAEA and the Chernobyl disaster.

After the explosion of the reactor, the authorities concealed the information, released it finally much too late and did not consider it necessary to tell the truth [1, 2, 3, 4]. This reaction of the authorities was responsible for *"ignorance and uncertainty"* concerning the radioactive contamination, which followed the explosion of the reactor. Up to the year 2000, the flow of misinformation has not stopped. Thereby it is useful to remember a technical report published by the WHO in 1958 [5]. The report contains a chapter devoted to a "policy in case of an accident" and ends with a wish: *"Nevertheless, from the point of view of mental health, the most favorable solution for future uses of peaceful atomic energy would be the appearance of a new generation, which would learn to adapt to ignorance and uncertainty"*

This apology of ignorance reflects an absence of respect for populations which contradicts the spirit and the letter of the Constitution of the WHO (8). This paragraph was read by Mr. Claude Haegi, representing the government of Geneva at the Conference organized by the WHO on the consequences of the Chernobyl accident, in November 1995 in Geneva. Mr. Haegi also quoted a statement of the Director-General of the IAEA, who, according to the newspaper "Le Monde" of August 28, 1986, four months after the accident, declared that "in view of importance of nuclear energy, the world could bear with one accident of the Chernobyl dimension per year". And Mr. Haegi concluded his speech declaring: *"One Chernobyl is enough. It is necessary to aspire to absolute security"*.

This statement of M. Haegi, as well as so many others presented at the WHO Conference, was to be published in the Proceedings by March 1996. However, these texts have still not been published [6]. Apparently, the papers presented in Geneva could have influenced negatively the IAEA Conference, in Vienna, in April 1996. The only explanation for the non publications appears to be the Agreement between the WHO and the IAEA, signed in 1959.

This Agreement states that the research programs of the WHO should previously be agreed, so that their results would not harm the IAEA main objective, which is: *"To accelerate and enlarge the contribution of atomic energy to peace, health and prosperity throughout the world"*.

This excerpt from the Statute of the IAEA is printed on the first pages of every publication of this Agency, including the Proceedings of the April 1996 conference of the IAEA already published in September 1996 devoted to the Chernobyl accident [7]. The Agreement guarantees that the research will not negatively affect the development of nuclear energy. Article I, §3 of the Agreement, specifies in particular that:

"Whenever either organization proposes to initiate a program or activity on a subject in which the other organization has or may have a substantial interest, the first party shall consult the other with a view to adjusting the matte by mutual agreement."

According to Article III of the mentioned Agreement:

1) The International Atomic Energy Agency and the World Health Organization recognize that they may find it necessary to apply certain limitations for the safeguarding of confidential information furnished to them.

2) Subject to such arrangements as may be necessary for the safeguarding of confidential material, the Secretariat of the IAEA and the Secretariat of the WHO shall keep each other fully informed concerning all projected activities and all programs of work which may be of interest to both parties.

The requirement of Article III, demanding confidentiality, which means silence, is contrary to the Constitution of the WHO. In fact, the purpose of the WHO is specified in chapter I of the Constitution of this Organization: *"The attainment by all peoples of the highest possible level of healt".*

Chapter II, Article 2 specifies how the WHO intends to attain its objective, and defines, in particular, the following functions:

(a) To act as the directing and co-ordination authority on international health work;

(d) To furnish appropriate technical assistance and, in emergencies, necessary aid, upon the request or acceptance of Governments;

(q) To provide information, counsel and assistance in the field of health;

(r) To assist in developing an informed public opinion among all peoples on matters of health;

It is evident, that the provisions of the Agreement prevent open information that is contrary to the Constitution of the WHO. Nevertheless, the Agreement was signed during the 12th World Health Assembly, May 28, 1959. The above quoted clauses can be found in *Basic Documents of the WHO [8].*

A very early publication of the WHO warning against the development of the nuclear industry, has been prepared by a group of outstanding experts in the field of genetics, who met in Geneva in 1956. The winner of the Nobel Prize M. J. M. Muller signed this joint statement. [9]:

"The genome is the most valuable treasure of humankind. It determines the life of our descendants and the harmonious development of the future generations. As experts, we confirm, that the health of future generations is threatened by an increasing development of nuclear industry and the growth of the quantity of radioactive sources

S we also consider the fact of appearance of new mutations observed at people to be fatal for them and for their descendants".

The publication of the proceedings of this conference was not acceptable to the pronuclear lobby. The IAEA decided soon after its creation to put an end to the freedom of expression in this field by concluding an Agreement with different UN organizations, and especially the WHO. This lasts until the beginning of the 21st century.

The attempts of the WHO to disseminate information about Chernobyl in November 1995.

Dr. Hiroshi Nakajima, Director-general of the WHO, organized an international conference *"Consequences of Chernobyl and other radiation accidents and their influence on human health."*, in Geneva, in November 20–23, 1995. Mr. Y. Fujita, governor of the Hiroshima prefecture, was the chairman of the conference. This conference considered the destruction of

Hiroshima and Nagasaki as well as the explosion of the Chernobyl reactor as radioactive accidents, deserving to be compared. Considerable differences were ascertained between these two types of accidents (the above-mentioned three explosions had to be categorized in this context as "Occidents" and not "catastrophes"). As the Proceedings of this Geneva Conference have not been published, it is impossible to refer to the presentations. It is useful to remind its objectives as stated in the program [10]:

- *To present the principal results of the first phase of the international program on health effects of Chernobyl accident (IPHECA).*
- *To compare the obtained results to the results of similar research, related to the health effects of Chernobyl accident.*
- *To improve (and to update) awareness of the type, the total extent and the harm for health of the Chernobyl accident, as known presently and to be foreseen in the future.*
- *To make new results of research concerning consequences of other radioactive accidents, available, in order to give more complete information on their health effects.*
- *To study the effectiveness of the protective measures undertaken in the area of public health during and after the accidents, and to offer recommendations for the future.*
- *To ensure the development and/or to clarify the state of knowledge concerning the consequences of influence of radiation on human health.*
- *To provide information on existing or future research within the framework of the United Nations Scientific Committee on the Effects of Atomic Radiation (UNSCEAR).*

- *To earmark the interesting tendencies and changes, which should become an object of steadfast attention of the researchers.*

This program convinced 700 doctors and experts, many from the most contaminated countries to participate in the work of the congress. The IAEA also mobilized the supporters of the atomic industry. Thus, contrary opinions were expressed, which allowed for hot debates. The representatives of the pronuclear lobby tried to prevent the dialogue and Prof. S. Yarmonenko from the Moscow Oncologic Center, demanded that the organizers would remove from the programs of future congresses those speakers, who intended to speak on the effects of low level radiation on living organisms. This apparently became a rule for all the following international conferences, especially in Vienna, 1996.

The reports, debates and presentations of the posters in Geneva were not published. The large document, which presents on 519 pages the statistical data gathered during the first phase of the WHO pilot project IPHECA [11]: 3Influence of Chernobyl accident on health2, confirms the very slow response of the WHO on the Chernobyl accident. Although the majority of people considered Chernobyl as an extreme incident, demanding urgent measures, the IAEA alone supervised the studies and provided the information on this catastrophe. The IAEA coordinated with the national medical authorities the protective measures for the population, considering as its priority to reduce the expenses.

The WHO was never the "co-ordinating authority" as required by its Constitution. At meetings, where the destiny of the victims was discussed, the WHO was even represented by Prof. Pellerin, a promoter of the development of nuclear industry [1]. Five years after the accident, the WHO, finally started studying the problems in the field, selecting 5 priority subjects, among them dental caries, whereas birth defects and hereditary alterations, which the Committee of experts gathered by the WHO [9], considered as a priority, were carefully overlooked.

As the Proceedings of the WHO-Geneva Conference remain unpublished, it seems to be useful to recall some presentations. M. Martin Griffiths from the **UN Humanitarian Department** in Geneva, stated that people still do not know the truth, and that many are still living in contaminated zones. He requested the WHO to continue its research work and to provide assistance, as he feared that everything would be stopped without adequate financial support. According to M. Martin Griffiths **9 million people are sufferers from Chernobyl**, and the victims of the accident are constantly growing in number.

Dr. Y. Korolenko, Minister of Health of the **Ukraine**, noted that the nuclear fallout contaminated the largest part of his country. Thirty million people drink contaminated water from the Dniepr. Everyone was affected by I-131 and the specialists now perform measures to reconstruct the radiation dose of Cs-137, received by the population. The minister mentioned lesion of the endocrine system and declared that diabetes mellitus had increased by 25 percent (This was not related to diet). Knowing about the social consequences of the insulin-dependent form of diabetes, it is easy to understand the deep concern of the Minister, who recalled the financial situation of his country in that situation, and asked all the states for help.

Prof. E.A. Nechaev from the Ministry of Health and Medical Industry (Moscow) indicated that 2,5 million people were irradiated in the **Russian Federation** following Chernobyl, and that 175.000 people continue to live in contaminated regions. He showed the increased incidence of a very aggressive form of cancer of the thyroid gland in small children, and the increase of birth defects from 220 up to 400 in 100,000 newborns in the contaminated regions. The frequency of similar diseases ranges within 200/100,000 in clean regions of Russia.

Prof. Okeanov from **Belarus** presented the results of his epidemiological research, in particular, data based on the national register of cancer, recognized by the WHO, which has been established in Belarus in 1972. Whereas leukemia increased in Hiroshima within the first years after the bombing with a peak between the sixth and the eighth years in Chelyabinsk the maximum occurred after 15–19 years. Okeanov noticed an increase of leukemia among liquidators only after 9 years, but the peak has not yet been reached. He stated that those liquidators, who worked more than 30 days in contaminated areas, have three times as many leukemia as their colleagues who worked there less than 30 days. The period of **exposure to radiation** seems thus to play an important role. Other forms of cancers are also increasing: cancers of the bladder doubled among the liquidators. The number of cancers of the kidneys, the lungs, and other organs also increased in the Gomel region, an area heavily contaminated by the nuclear fallout with.

The report of this group of Belarusian scientists showed also an increase of cardiovascular diseases among the liquidators, from 1,600 up to 4,000 per 100,000, and up to 3,000 per 100,000 persons living in zones of heavy radioactive contamination. They noticed marked alterations of the immune system, increase of chromosomal aberrations, loss of sight, in particular due to cataract among young subjects. The speakers showed a doubling of mental retardation observed in children as well as mental changes in adults. He insisted on the necessity to study the increase of gastro-intestinal disorders, which he also observed. Among documents received by the WHO, there were unpublished presentations, e.g.: a document of professor Okeanov in Russian of 1994 [12].

All the data submitted in Geneva in November 1995, were not available in March 1996 as officially promised [6]. This delay may well be in relation with the decision of the IAEA, to definitely close the debate about Chernobyl at its own Conference, in Vienna, April 1996 [7].

The publication by the WHO of the Proceedings of its 1995 Conference could have prevented the IAEA from achieving its objective: to put an end to discussions about the health effects of the Chernobyl accident.

The IAEA Conference. April 8–13 1996, in Vienna

The title of this Conference was *"Ten Years after Chernobyl"*. The participants had been selected according to the approval by the ministry of industry and the ministry of international affairs; the ministry of health was not consulted. During the plenary sessions, the speakers expressed contempt and haughtiness towards victims of the disaster. Actions to be taken after the future major accidents, which were considered as unavoidable, were also discussed at the congress. The aim of the discussion on this topic was very clearly formulated: to reduce expenses for the relevant industries, to limit or even avoid evacuation of people from highly contaminated zones, to keep the media under severe control. They believed that "alarmist", "stressful" reports were basically causing practically all the Chernobyl connected health problems.

The speakers for the main reports and especially the chairpersons of the sessions had been instructed to avoid discussions on "difficult" problems related to health, particularly those deriving from the chronic incorporation of Chernobyl radionuclides from the environment in the organism. Those speakers also called for the silence of mass media in case of a catastrophe, since, they believed that "alarmist" reports were basically causing practically all the Chernobyl- connected health problems.

The authors of the main presentations confined themselves to the three types of illnesses (acute irradiation syndrome, mental deficiency in children irradiated in utero, and thyroid cancer, in children exclusively), which had been admitted to be the essential pathological findings due to the increased ionizing radiation caused by Chernobyl. *All the other illnesses put into the large catalogue of psychosomatic diseases associated with unjustified fears, or to some kind of social protest, having nothing to do with radioactivity.*

The **acute radiation syndrome** was one of the rare real "accident's outcome". This syndrome led to discussions to determine if the number of deaths was 31 or 32. These deaths were practically the only ones taken into consideration by the IAEA, as a consequence of the Chernobyl catastrophe.

However, when IPPNW members had rallied in Kazakhstan in order to help the population to stop the Soviet atomic tests, General-in-Chief Ilienko showed memorial shields on the walls of the Officers House in Semipalatinsk. The featured the names of local residents killed during the two world wars and the Afghanistan War. There was a further list of people who died for the nation. The General asked us: "Do you know who are on this list? They are our Chernobyl liquidators !"

The Soviet Union sent 800,000 soldiers, civil experts and foremen, their average age being 33 years, to the site of the disaster to try to decontaminate it, isolate and stabilize the reactor1s ruins. We met the widows of liquidators in Moscow. Several years ago, there were already more than a thousand of them, and they kept gathering new files and photos of other deceased liquidators "Moscovites husbands", who died from new diseases, which they heroically acquired during their service, neither generally acknowledged posthumously nor always glorified by the nation.

As for the liquidators, E. Marchuk, the Ukrainian Prime Minister pointed out at the IAEA Conference [7], that in his country, 3.1 million people were exposed to radiation at the time of the explosion. Many remain in the contaminated area. Among the 360,000 Ukrainian liquidators, 35,000 were already invalid.

Although at the conference IAEA conceded the existence of **neuropsychic diseases in children whose mothers had been exposed to radiation during their pregnancy.** The speakers denied the existence of similar diseases in adults due to the radiation around Chernobyl, although this is a well-known phenomenon. The organizers tried to present victims suffering from neuropsychic diseases (in particular, among the 800,000 liquidators) as malingerers, raising claims for more financial support, or possessed by unjustified fear of radioactivity. IAEA experts first invented the new terminology of **"radiophobia"**. Later, when negative reaction to this concept arose, the term of **"environmental stress"** was created to qualify neurovegetative and subjective disturbances as well as a complex of other illnesses, caused by Chernobyl.

The Permanent Peopled Tribunal (3) judged the behavior of international organizations, especially the IAEA, national commissions for atomic energy as well as governments, which finance them on behalf of the interest of the nuclear industry as follows: **"The absence of concern for these real outcomes of radiation exposure, was in itself one of the ways in which the victims were revictimized after the disaster".**

The IAEA has done its best to allow those responsible for what had happened, as well as the countries possessing nuclear know-how and the Western atomic lobby, to save as much as possible on the expense of the victims of the Chernobyl catastrophe.

Cancer diseases caused by Chernobyl

After many years of obstruction, in particular during the IAEA conference in 1991, the experts of the IAEA had to admit the existence of thyroid pathologies, partially brought about by Iodine 131, discharged into the atmosphere by the blow-up in Chernobyl. According to Bandazhevsky that illness are caused by several radionuclides (e.g. Cs-137, Sr-90 in tissues of different organs. Their toxicity may be synergistic [13]. During discussions on the thyroid cancers, the official speaker of IAEA mentioned that this was "a good cancer". We do not think, that mothers of children, ill with cancer and often having metastases in their lymph nodes and even in their lungs, or that the surgeons who operate those children share this view.

The IAEA tried to show that it would be easy to distribute tablets of stable, non-radioactive iodine among the population in order to prevent thyroid cancer. Doctors were aware of such preventive measures before the catastrophe. However, with the exception of Poland, neither the politicians nor the technical equipment allowed to undertake in time such preventive actions.

During this debate, one of the speakers specified that iodine tablets have to be ingested before the radioactive cloud appears, thus ensuring their maximal efficiency. This seems to be quite problematic, as they call at the same time on the mass media to remain silent in case of a future accident "as to avoid fears". The immediate distribution of iodine pills should be envisaged not only in the radius of 5–30 km, but of 500 km and over.

The IAEA officials considered also long-term effects [7] and made the following conclusion: *"Ten years after the Chernobyl accident, in the three affected countries, there are no serious after-effects caused by the radioactivity as a consequence of this accident, except the dramatic increase in numbers of thyroid cancer diseases in children, exposed to the radioactivity in the most contaminated regions. The death rates do not show any substantial raise due to cancers, which could be related to the accident at the Chernobyl nuclear power plant. In particular, there is no serious increase in the number of blood diseases even among liquidators, i.e. the diseases which were of greatest concern after a radioactivity contamination."*

The wording of this conclusion will be discussed later. However it was contradicted by the co-chair person, Prof. Okeanov, who elaborated on the cancer diseases. His task may well have been on the contrary to remain silent.

The discussion following the "official" report on cancer diseases and the conclusions set up, was strictly restricted to radiometry. The first speaker tried to discuss the problems of cancers, but was forced to leave the floor. When I answered the question as to the topic of my report, I claimed my great interest in radiometry, thus gaining a chance to ask a question to Prof. Okeanov: *"At the WHO Conference [6] in Geneva in 1995 and later at the congress run by an NGO in Minsk in March 1996, you have shown data of significant increase of cancer? Would you care to comment?"*. Okeanov had showed that the global tendency of number of cases of cancers was increasing, liquidators being at the peak of the curve. The incidence of cancers and leukemia was depending on the duration of their exposure to radioactivity.

The increase in numbers of thyroid cancers has been registered in Minsk since 1989. Leukemia in small children, whose mothers had been exposed to radioactivity during their pregnancy, was observed that early, too [15], and followed the mechanism described in the late fifties by Alice Stewart & al (16).

Since 1993–1995, epidemiologists have been observing an increase in the numbers of cancers, mainly among young people, connected with a strong dose of radioactivity after Chernobyl. The liquidators were on the average 33 years old.

The chart of AE Okeanov, which I presented at the IAEA conference, was published in the Proceedings [7].

The number of cases of cancers per 100,000 inhabitants who have been irradiated to some extent compared to 30,000 liquidators exposed more and less than thirty days.

Cancer	Belarus population	Belarusian liquidators (More than 30,000 cases)		
		total	irradiated > of 30 days	irradiated < of 30 days
Colon	12	18,5	20,1	13,4
Urinary bladder	13	31,1	32,1	27,1
Leukemia	10,4	23,3	25,8	16,4
Total	35,4	72,9	77,0	56,9

During the discussion in Vienna Prof. Okeanov confirmed his data and added that the cancer of the thyroid gland in liquidators also increases. He stated that in Gomel, 180 km from Chernobyl, *"an apparent increase in the number of cancer of the colon, the rectum, the lungs, the breast and the urinary system is observed"*.

Okeanov underlined the importance of continuing the epidemiological research. Alas, his Institute, whose quality was recognized by the WHO (France, Germany, Switzerland do not have similar national cancer registers), has been dismembered some time later. This appears to be a deliberate step aimed at the suppression of epidemiological data in connection with Chernobyl.

In Hiroshima, the latency period for cancers of the thyroid gland and leukemia in small children was 4 to 5 years. Petridou & al [15] observed an epidemic of leukemia in small children after the passage of the radioactive cloud in Greece, 1,000 km from Chernobyl. It is therefore very disappointing that the WHO involvement started its studies too late to assess leukemia in infants irradiated in utero.

For the majority of other cancers however, the latency period lasts about 9–30 years and more. This explains the haste with which the pronuclear lobby wanted to stop research in this field. The dismantling of the cancer register, an institution that could demonstrate with accuracy to the world how many tens of thousands of cancers would be caused by Chernobyl serves well the IAEA and the pronuclear lobby.

Let's return to the conclusion of the Proceedings of the IAEA Conference, which contains the following statement: *"No major increase in the incidence or mortality for all cancers has been observed that could be attributed to the accident"*. Not knowing the context, it is possible to assume that the phrase is not a barefaced lie. However, it is preceded by the statement, which clarify the essence: *"... Apart from thyroid cancers, there are no evidence of a major public health impact to date of radiation exposure as a result of the Chernobyl accident in the three most affected countries"*.

Utilizing a misleading technique, **the IAEA selects the false parameter**: the mortality **rate** caused by cancers just 10 years after the catastrophe. At this stage only the **morbidity** rate would be a permissible parameter in this context [17]. Cancer in younger persons is a dramatic event for families, friends and of course for the patient himself. The treatment of such cancers requires long-term hospitalizations, surgical interventions, chemotherapy, and absence from professional activity, which are extremely expensive both for the society and for the families. Meanwhile, the modern methods of treatment allow to cure some forms of cancer and, in many cases, to put off the lethal outcome. Therefore in 1996, the selected parameter should have been the morbidity and not the mortality.

Will the parents, whose children undergo medical treatment for leukemia, even when the child has completely recovered, consider that there were *"no serious health effects"*?

Such outright lies, contained in this conclusion were designed to permit the nuclear lobby to keep on building "reliable" nuclear power plants. Did not the promoters of nuclear reactors tell us over and over again that this industry in completely reliable? Today the nuclear lobby intends to sell nuclear power plants with a new advertising slogan: "they are even more reliable than before". This commercial argument is not related to any scientific reality. We do not want to prove this hypothesis after all those lies we heard so far.

Diseases caused by radionuclides incorporated in the organisms

Chernobyl diseases affect up to 90 % of children in the contaminated zones. These diseases must be categorized, according to the opinion of the IAEA, as not related to radioactive contamination. "If real, these increases (in the frequency of a number of non-specific detrimental health effects) may be attributable to stress and to anxiety resulting from the accident." The stress contributes to symptoms in 80 % of adult population, consulting with practitioners in Western Europe. The IAEA may have hoped to find at least a similar percentage of stress in the inhabitants of the contaminated regions, as well as among the liquidators and the settled out population.

Nevertheless, despite a severe selection of the speakers, and a strict control on the discussions at the Vienna conference in 1996, the chairmen of sessions and **experts invited by IAEA could not achieve unanimity in this area.**

During the first weeks after the accident, the territories of Europe, Scandinavia, the Alps, the Jura, the Balkans and Turkey were contaminated by enormous amounts of iodine 131. Byelorusian doctors revealed quite early the health effects of this contamination.

After 1986 other radionuclides with relatively long-lived half-lives (approximately 30 years for Cs-137 and Sr-90; 240 centuries for plutonium), had already begun to alter the functions of organs (heart, kidneys), nervous and immune systems, and the totality of the genes in cells, especially of those that were closer to incorporated radiation. The map of the nuclear fallout (I-131 and Cs-137), appeared soon after the explosion of the nuclear reactor due to the outstanding work performed by the team of Vasily Nesterenko [18], whose scientifically proved, but alarming reports were the cause of his dismissal.

Facing the dramatic problems of his country, this physicist could fortunately continue his research due to support provided by western charitable funds, and his work to protect the population forced to live in contaminated regions [19]. His information on the contamination was not made public by the authorities. The world was late to find out about it, too late [20].

In the Gomel State Medical Institute under the leadership of its outstanding young rector, Professor Yury Bandazhevsky, the researchers studied the influence of radionuclides incorporated by the human organism, and the pathological changes leading to serious diseases of various organs [21]. Those are the diseases of the majority of the adults and 90 % of children forced to live in zones with a heavy radio-contamination.

Professor Bandazhevsky, a pathologist, showed in experimental models, that the laboratory animals, feeding on contaminated food similar to the one that the inhabitants of the contaminated regions are obliged to eat, present morphological and functional changes similar to the processes observed in the people. His experiments show, in particular, the damage caused by Cs-137, and prove, that for this isotope, whose period of biological half-decay (i.e. the time necessary for a human being to halve concentration of a mentioned radionuclide in organism) is shorter than a year, it is possible to use medical drugs to reduce its toxic and radioactive burden for the organism.

The studies of the Department of Pathology of the Gomel Medical Institute offer also to the researchers a better understanding of the diseases provoked by a chronic intoxication of organs or systems after incorporation of Cs-137. In addition to the whole program of research at his Institute Bandazhevsky supervised 30 candidate dissertations and has published 200 articles and reports, some of them already translated into English [13, 21, 22].

The damage caused by Cs-137 starts already in the prenatal phase. The **placenta** serves as a filter between maternal blood and the blood of the foetus, and protects against foreign molecules such as drugs, but also against this radionuclide during the whole period of pregnancy. Placenta therefore accumulates considerable quantities of Cs-137, more than tissues of the maternal organism [21]. This accumulation of toxic molecules and radioactivity in the placenta, close to the cells responsible for the secretion of hormones necessary for the normal evolution of the pregnancy, appears to be responsible for the abnormal levels of several hormones. Morphological anomalies are more common when high Cs-137 concentrations are found in the placenta. The fetus suffers from anoxia, the risk of abortion increases. Furthermore, incidence of birth defects in children whose mothers live in contaminated zones is twice as high as compared to those, whose mothers live in clean regions.

If the mother lives in zones of radio-contaminated zones, breast-feeding will lead to rapid accumulation of radionuclides in the organism of the child. During childhood, children will continuously incorporate radionuclides, in particular Cs-137 contained in milk, vegetables, fruit etc. This chronic intoxication of different organs leads to frequent diseases, such as abnormal blood pressure, cardiac arrhythmia, but also to allergic diseases, chronic infections, due to immune deficiency.

Bandazhevsky has developed methods to protect those children. This would require from the authorities, on the one hand, to recognize the problem and, on the other, the willingness to help these populations by educational measures, adequate food intake and intermittent treatments. There exist possibilities to remove partially the Cs-137. Professor Bandazhevzky tested several substances: pigments, adsorbents, algae, among others. The best results were achieved with an extract of apple pectin, able to fix Cs-137 and to prevent its absorption. It may also remove it partially from the organism, the mobilization and elimination being mainly with feces.

All these measures reduce the load of toxic radionuclides in the organism; this approach is essential from the medical point of view. The high concentration of Cs-137 in certain organs or tissues lead to irreversible damages, when this radioactive load lasts for years or exceeds certain limits.

From 1996 to 2000, Professor Vasily Nesterenko and his Belarusian Institute of Radiation Safety "Belrad" carried out measures on the internal contamination using spectrometers. More than 50.000 children at schools and kindergarten of contaminated regions of Belarus took part in this project. The Institute of radiation safety found excessive levels of Cs-137 in the organisms of children in contaminated areas. Many accumulated 200–400 Bq/kg Cs-137 in their organism. Children living in Narovlia, Yelsk, Tchetchersk, Vetka, Korma, and Stolin regions had up to 1500–2000 Bq/kg, some children have reached contamination doses of 4000–7000 Bq/kg.

The correlation made by Professor Y. Bandazhevsky, show that if a child's organism has a content of Cs-137 of more than 50 Bq/kg, pathological disorders of the vital organs or systems will occur. That is why the Institute "Belrad" has carried out since 1995 the protection of such children, using a preparation with pectin, vitamins and essential elements, "Yablopect".

The intermittent use of pectin is recommended for children with an internal Cs-137 contamination of more than 20 Bq/kg in their organism. The reduction of radionuclides in a child's organism is 30–40 % after taking 2–3 tablets per day of this preparation for one month, 3 to 4 times in a year with at least 2 months intervals between treatments.

The reduction of the internal dose of Cs-137 in children's organisms, reduces 2–3 times the annual contamination dose.

These tablets diluted in water are well accepted by children (the drink tastes like an apple) and are well tolerated. If this treatment is initiated early enough, the symptoms may be attenuated. The goal would be to prevent the diseases or in advanced cases, to stop their malignant evolution such as cardiac failure, hypertension etc.

This research should raise interest among physician, concerned by the health care of the victims of the Chernobyl accident. Non-governmental organizations (NGOs) of Ireland, Sweden, Belgium have generously contributed to these programs. It is surprising, that this assistance to the victims became an object of spiteful pamphlets and inappropriate irony in some places. When Professor Nesterenko had the possibility to present the results of his research in Western Europe, in order to find new ideas and support, some participants made rather aggressive remarks. There may be a competition between people working the whole year to help local population and scientific tourists from different western countries who collect data during their visits in the contaminated regions.

Electrocardiographic changes were noticed in a large number of first year students of the Gomel Medical Institute, coming from contaminated regions. Unfortunately, those changes tended to worsen during the following 4 years of their studies. The heart muscles (myocardium) concentrate more Cs-137 than other tissues. The circulatory system is affected by the "Cesium cardiomyopathy", described by Bandazhevsky. The preventing of this disease in people living in contaminated areas would be their relocation, or a correct diet, which as a rule is too expensive for the population, and intermittent pectin cures.

The endocrine system is also very sensitive to Cs-137. The thyroid and the placenta have already been mentioned; there are several diseases of the thyroid many are associated with the increase of the antibody titers against the thyroid cells, which leads to hypothyroid function. Such functional disturbances are 100 times more frequent than cancers. They may have a very negative impact on mental and physical development of children.

The immune system is highly sensitive both to internal and external radiation. This protective system relies on the white blood cells, e.g. lymphocytes. Alterations of this cell system may lead to immune deficiencies as in AIDS. Titov et al. have shown, that the production of antibodies is abnormal in contaminated children [23]. The health disturbances of this complex system include **allergic diseases, like**

asthma. Allergy for cow milk and fruits is observed in 50% of the school children and students in Gomel.

Autoimmune diseases occur when the cells, designed to fight intruding organisms such as bacteria, viruses or cancer cells, attack the normal cells of an organ. When Beta cells of the pancreas face a self-attack by lymphocytes, this may cause severe diabetes mellitus.

Many aspects of this pathology were presented during the NGO congress in Minsk (24). As a consequence of Chernobyl the incidence of **diabetes mellitus** had increased by 28%, more or less the same as in the Ukraine; Research conducted by Tatiana Voitovich, endocrinologist in Minsk, shows that after Chernobyl, a new form of diabetes has appeared: **an insulin-dependent, unstable form of diabetes in very young children**. The child is unconscious when he enters into the hospitals. His blood sugar is very difficult to stabilize with insulin injections. This form of diabetes affects the patients condition during all his life. It was very rare before Chernobyl. The number of cases of insulin-dependent diabetes has doubled in the Gomel region.

Ignoring the Problems

At the IAEA conference, the problem of insulin-dependant diabetes has not been quoted among the diseases caused by Chernobyl, although it was described after the bombing in Hiroshima. **The technique for evading this issue during this pronuclear conferenc**e is worth telling. During the discussion, I asked whether there existed any link between diabetes and ionizing radiation. The chairman of the session spoke before the speaker could answer, and said: "You have here experts from all over the world, the best specialists in this field. The fact that none of them has raised his hand to answer your question proves that the *ionizing radiation cannot cause this type of disease*".

In connection with this answer of the chairperson, Prof. Viel (17) exposed methods, used by those who do not want to show a link between ionizing radiation and pathological findings. He quoted similar answers or statements to the one I heard: "The experts were unanimous in the view that... there is no association between radiation and any health". Prof. Viel added that such experts may "conduct inadequate epidemiological researches by integrating epistemological errors". A classical method was to select mortality instead of morbidity, e.g.: to stop the investigations too early when studying cancers, as it was the case after Chernobyl.

The results of such studies show no statistically significant differences. The hypothesis has therefore not been proved. Promoters conclude that it is false, which allows them to pretend that everything is in order and the atomic industry safe.

The dismantlement Gomel Institute

Lesions of the immune system in organisms contribute to the development of cancers in younger subjects. We must keep in mind that cancers are only the visible part of the iceberg, represented by the totality of the diseases caused by Chernobyl. That is why the scientific world was extremely interested by the research of Professor Bandazhevsky [22], which allow to discover or imagine the true dimension of the iceberg. The systematic studies of this research group allowed characterizing new diseases due to cellular damages caused by the accumulation of radionuclides.

Bandazhevsky studied also other radionuclides, e.g. Sr-90, which accumulates in the bones, close to the blood-producing cells, erythrocytes, including mother cells of the immune system. Sr-90 is much more stable in the human organism than Cs-137. Internal contamination of the organism may also be due to particles of plutonium fixed in the lungs, lymphatic or other tissues. Bandazhevsky considers the synergy between the toxicity of the different radionuclides as a complementary problem.

The arrest of Yury Bandazhevsky, Rector of the Gomel Medical on July 13, 1999, shocked those who knew him and his publications. This dynamic teacher and highly motivated researcher devoted himself totally to his work, which he considered to be his debt to his country, and in particular, to the victims of Chernobyl. Bandazhevsky created the Gomel Medical Institute and designed its scientific and research work on the causes of the diseases of the population living in the contaminated areas.

Amnesty International reacted at once, considering **Bandazhevsky as a potential prisoner of conscience** [26]. This opinion was reinforced when the Prosecutor, Oleg Bozhelko, said 9 months after his arrest that he held no proof for his accusation.

Now the Institute has been placed under the direction of a rector "of a follower", who rejected the previous direction of research. This is one of the most brilliant victories of the pronuclear lobby over.

The international solidarity was able to release the prisoner. However Professor Bandazhevsky lost his job, his research instrument, his data, the teaching activity and his income. He is in need of help. His health has been seriously undermined; he has lost 20 kg due to extremely severe conditions in jail. Our solidarity should now allow him to find the way and means to continue his research and to publish his findings. It is also necessary to find money to pay the services of a lawyer

Mutagenic and teratogenic effects

From the ethical point of view, the genetic and hereditary damages are the most disturbing consequence of the radioactive pollution The impact on the genome, i.e. the change in chromosomes or genes, which cause an increase of genetic diseases and birth defects in the coming generations is threatening the workers of the nuclear industry. At all stages of the uranium cycle, from the uranium mining to the management of wastes, including the maintenance of "normally functioning" nuclear facilities, radionuclides are released, in the as gazes or particles, liquids or solids. Radiation lead to an increase of a number of genetic. These were the warnings expressed by the experts invited by the WHO in 1956, when the nuclear industry began to develop [9].

After Chernobyl, changes of the genome were found not only in rodents close to Chernobyl or in Sweden. Children living in contaminated regions, in a radius of 250–300 km from Chernobyl show an increase in mutations; [27]. Dominant mutations may be apparent at birth, or become manifest during life. However most of them are not compatible with survival and can cause abortions. Recessive mutations induce genetic diseases and congenital deformities in the next generations. Thus, it will be necessary to wait up to the third to fifth generations of affected by the Chernobyl fallout to observe the full extent of the damage caused by the Chernobyl catastrophe in the families.

Genetic anomalies in fishes, swallows and rodents

A. Slukvin, a former USSR fishing expert, compared two industrial fish farms for carps. The first was situated 200 km from Chernobyl in a zone with a relatively low level of contamination (about 1 Curie per square km), the second, 400 km away from Chernobyl, in a zone of very low contamination. Since 1988 up to 70% of the fertilized eggs did not produce larvae, and after 6 months, the young fishes in the area where the muddy bottom was contaminated with Cs-137 major deformities were observed in 10 to 20% of the carps, depending on the radiocontamination of the pond [28]. The normal development of carps was still possible 400 km from the exploded nuclear power plant. Prof. Rose Goncharova directed Dr. Slukvin's thesis.

The generations of rodents and birds around Chernobyl follow much quicker than in man. This allows already studying the increase in deformities, caused by recessive genes in animals living in contaminated areas.

A group of Swedish researchers compared a population of swallows, nesting in Chernobyl, with swallows from uncontaminated regions in the Southern Ukraine and a region of Italy. They studied the DNA structure of the minisatellites of adult and young swallows, as did Dubrova in human [27], in chromosomes in the adult swallows and their offspring. The Swedish researchers discovered a statistically significantly higher mutation rate in Chernobyl swallows compared to those living in clean area [29]. Furthermore, they observed an increase of recessive genetic abnormalities in the Chernobyl swallows. Mutants had white spots on their feathers; they had also a much lower chance to survive. Year after year, observations showed a progressive increase in those disorders in the contaminated areas compared to the Southern Ukraine or to the control zone in Italy. The differences were statistically significant.

A number of studies were devoted to rodents living in more or less contaminated areas [30, 31, 32]. The habitat, where these wild rodents (bank voles) live, has a decreasing radioactivity rate, since Caesium-137 is seeping in the soil with rainwater. One could have expected a positive reaction of these animals to these improving radiological conditions. Yet, genetic abnormalities increased from one generation to another [30, 31]. Goncharova and Ryabokon consider this as a kind of reverse adaptation to radioactivity, an increased fragility of the genome.

Baker and his colleagues [32] studied the DNA in one of the genes, transferred to baby bank voles exclusively from their mothers. They observe various mutations from generation to generation, i.e. an alteration of the base of the studied chromosomes, which overpasses 100 times the mutation rate observed until today in any animal species.

For geneticists point of view, human beings and rodents may well be compared. Commenting the publication of Dubrova & al. and that of Baxter & al, Prof. Hillis, from the Texas University, concluded his editorial in *Nature* (April 25, 1996) as follows: *"We know now know that the mutational effects of nuclear accidents can be much greater than suspected and that evolutionary rates in at least parts of n eucariotic genome can be raised well beyond levels previously considered possible"* *(33)*.

The article by Y. Dubrova & al. was published in the same issue Nature [27]. This team working with Prof. A. Jeffreys, Nobel laureate examined children and their parents, living in the contaminated areas 250 -300 km north from Chernobyl. Compared with children in uncontaminated regions. These children of Belarus showed suffered a doubling of mutations in minisatellite loci. The mutation rate decreased with the degree of radioactivity in their parents residence place. A control group was selected in the United Kingdom due to the overall contamination of the Belarusian territory.

Experts think that a low, but chronic dose of radioactivity, is very dangerous thing for the human genome.

In May 1997, the WHO annual report, published on the occasion of the World Health Assembly (WHA) attested that the number of cases of cancers will double within the next ten years. However, this report says that this is due to the growing life expectancy [34]. Such an analysis does not distinguish between the cancers in very old people and those in children and in young adults, which increases most in the Chernobyl regions.

The same publication of the WHO (34) shows an important increase in the number of cases of **diabetes**. In rich countries, Type II concern people with excessive food intake. Without further explanation, this report indicates that the number of insulin-dependent type I diabetes will also increase in young people. Here we should recall the report of Mr. Korolenko, Ukrainian Minister of Health at the WHO conference in 1995, which was not published [6]. He underlined the 25% increase in the number of cases of diabetes after the accident at the Chernobyl nuclear power plant in a population where excessive food intake is rare.

Birth defects in children

At the IAEA conference in Vienna, 1996 [7], the speaker reporting on teratology as a consequence of the Chernobyl accident, made use of the same argument as the lawyers of the chemical industry that produced in the sixties a tranquilizer, Thalidomide, which appeared to be extremely teratogenic, i.e. provoking a number of birth defects in children whose mothers had absorbed it; This drug caused also birth defects in monkeys, birds and insects [35]. The speaker asserted: *"The absence of any register proves that the development of birth defects is not caused by the Chernobyl accident"*.

Of course, the absence of a register is not a proof of the absence of a causal relationship between the increase of birth defects and Chernobyl. But the falsity of this statement is more shocking when Belarus is concerned. Since 1982, i.e. 4 years **before** Chernobyl, Belarus had a national register of birth defects, developed by the Belarusian Institute of Birth Defects and Inherited Diseases, under the leadership of Professor Gennady Laziuk [36]. This Institute records and checks the cases of birth defects, observed in the country. It is compulsory to report ten birth defects, to be detected in children up to 7 days after birth, or in fetuses in case of spontaneous or therapeutic abortions. Following anomalies must reported in any case: anomalies of the development of the central nervous system as major brain damage, dysraphia of the face or spina bifida, polydactylism, absence of limbs or serious defects in their development, rectal stenosis, mongolism and multiple birth defects.

The incidence of birth defects have increased in Belarus in a direct proportion to the contamination by Cs-137 in the regions, where the mother was living during her pregnancy [36]. Rates of birth defects of probable dominant genetic origin, e.g. polydactylism and multiple deformities, compatible with survival, have considerably increased [37]. Deformities probably caused by the teratogenic property of radionuclides are also increasing.

There is practically no region spared from radioactive contamination in Belarus, as 90 % of the contamination is caused by the ingestion of contaminated food. No region of the country can be considered as a control area. That is why the findings registered from 1982 to 1985, constitute the best control data available.

During the WHO conference of November 1995, Dr. Smolnikova from Gomel, in charge of the health of 46 thousand children living in an area contaminated by 40 Curies of Cs-137/km², had already mentioned a high perinatal death-rate and an alarming increase of birth defects in the region [6].

Despite all these reports, the experts from IAEA denied in 1996 any increase of birth defects, related to the Chernobyl catastrophe.

After the epidemic of birth defects caused in Europe by the drug thalidomide (Contergan), and in spite of the fact that thalidomide is not mutagenic, the pharmaceutical industry was forced to exclude, all over the world, substances with mutagenic and teratogenic properties. The fact that similar measures do not apply to the nuclear industry may well be connected with the Agreement signed between the IAEA and other UN organizations, including the WHO. The radionuclides released in the environment by this industry, have mutagenic, teratogenic or cancerogenic properties.

The destruction of scientific structures in Belarus

As long as the World Health Assembly, the governing body of the WHO, does not amend the Agreement, concluded in 1959 with the IAEA, which holds it **hostage to the nuclear lobby** with regard to the radiation induced health effects, there is no hope for independent research groups to receive any substantial support.

The most efficient structures that study in Belarus the health consequences of the Chernobyl accident are being progressively dismantled.

Professor Nesterenko was one of the physicists, who came immediately to the place of the accident. As an expert and, sometimes, he acted as a fireman, flying with a helicopter within the radioactive cloud, pouring containers of liquid nitrogen in the burning reactor. It is incredible that he survived. The three other passengers of that helicopter have died owing to the irradiation. Together with his colleagues, Nesterenko established the map of the radioactive contamination of all the territory and formulated proposals for the protection of the people.

He continued his work, until his data and his recommendations were considered as unsatisfactory, he was considered as an "alarmist" and lost his Institute, his functions and his sources of income. Due to the help of Alex Adamovich, Andrej Sakharov, the chess champion Karpov, the Foundation for Peace, the Northern Ireland Foundation Adi Roche, and others, Nesterenko founded a state-independent research Institute "Belrad", which works to assist victims of Chernobyl, teaching them the best possible methods of self-defense, when they are forced to live in contaminated territories, and tries to rehabilitate children.

The Minister of Health, Dr. Dobrishevskaya, who supported the most efficient research groups in this field, according to a joint report published in 1996 [24], was also not maintained in her function.

Professor Okeanov witnessed the same disorganization of the research structure he was in charge. It was a most valuable instrument, aimed at revealing the true dimensions of the epidemic of cancer diseases caused by the Chernobyl accident. The coincidence with his reports at the conference of the WHO (1995), the NGO conference in Minsk (1996), and his nonobservance of the required silence at the conference of the IAEA in Vienna (1996), shows clearly who wanted get rid or to achieve the destruction of this working instrument.

The removal of Professor Bandazhevsky is the last strike in this destructive series. This pioneer of research of health consequences of the Chernobyl accident has revealed the mechanisms of the action of chronic low dose radiation by incorporated radionuclides in organisms: after iodide-131, Cs-137 and Sr-90. With his group of young researchers from the Gomel Medical Institute and numerous volunteers, Bandazhevsky has described typical diseases occurring in a large proportion of the population and almost all the children living in the highly contaminated regions.

These systematic and repeated strikes, which negatively influence the wellbeing of the country and its population, are, sometimes supported by western scientists perhaps jealous these discoveries. However, those who gain the greatest satisfaction and benefit of such actions, are the richest countries with the most advanced nuclear industry, and the nuclear lobby

It is necessary that the WHO recover its independence, in order to be able to act again in this field according to its Constitution. Epidemiological research should start without delay. Who will study the genetic damages in children in the five coming generations?

Who will devote himself to the rehabilitation of the victims, to their treatment and to the most effective protection of children and pregnant women? Rich nuclear states should come to the aid of victims of Chernobyl in Belarus and in other suffering countries.

It is also necessary to remove the present mandate of the IAEA to promote commercial nuclear industry. This Agency has much more important problems to solve: to keep under surveillance plutonium, uranium and all the fissionable materials, from dismantled nuclear warheads, military and commercial nuclear facilities. The IAEA must also control the problems of the safe storage of the radioactive waste, which humanity managed to produce in only two generations, since the beginning of nuclear age. This surveillance must unfortunately continue for centuries and millenaries.

The bibliography

1) Belbeoch B. and Belbeoch R. : Tchernobyl, une catastrophe. Quelques elements pour un bilan. Edition Allia, 16 rue Charlemagne, Paris IVe , pp 220. 1993.

2) Schtscherbak J. : Protokolle einer Katastrophe (Aus dem Russischen von Barbara Conrad) Athenaum Verlag GmbH. Die kleine weisse Reihe. Frankfurt am Main, 1988.

3) Tribunal Permanent des Peuples. Commission Internationale de Tchernobyl : Consequences sur l'environnement, la sante, et les droits de la personne. Vienne, Autriche, ECODIF- 107 av. Parmentier, 75011 Paris, ISBN 3–00–001533–7, pp 238, 12–15 avril 1996.

4) Yarochinskaya A. : Tchernobyl; Verite interdite (traduit du russe par Michele Kahn). Publie avec l'aide du Groupe des Verts au Parlement Europeen, Artel, Membre du Groupe Erasme, Louvain-la Neuve, Belgique, Ed de l'Aube, pp 143; 1993.

5) OMS. Rapport d'un groupe d'etude : Questions de sante mentale que pose l'utilisation de l'energie atomique a des fins pacifiques. Serie de Rapports Techniques, No 151, pp. 59, OMS, Geneve, 1958.

6) Les consequences de Tchernobyl et d'autres accidents radiologiques sur la sante. Conference Internationale organisee par l'OMS a Geneve, 20–23 novembre 1995. Actes non publies.

7) IAEA. One decade after Chernobyl. Summing up the Consequences of the accident. Proceedings of an International Conference, pp 555, Vienna 8–12 April 1996. Sales and Promotion Unit, International Atomic Energy Agency, Wagramstr. 5 , P.O: Box 100, A-1400, Vienna, Austria.

8) WHO. Documents Fondamentaux de l'Organisation Mondiale de la Sante. 42e edition, pp 182, OMS Geneve, 1999.

9) WHO. Effets genetiques des radiations chez l'homme. Rapport d'un groupe d'etude reuni par l'OMS; pp 183, OMS, Palais des Nations, Geneve, 1957.

10) Programme de la Conference Internationale organisee par l'OMS a Geneve, du 20–23 novembre 1995. Les consequences de Tchernobyl et d'autres accidents radiologiques sur la sante. Le Programme peut etre obtenu a Geneve e/EHG/1995.

11) WHO. Health consequences of the Chernobyl accident. Results of the IPHECA pilot projects and related national programmes. WHO/EHG 95. pp 519. WHO Geneva 1996.

12) Okeanov A.E. et al.: Analysis of results obtained within "Epidemiological Registry" in Belarus. Geneva; the Russian version can be obtained at the WHO (unpublished document WHO/EOS/94.27 and 28) Geneva Switzerland, 1994.

13) Bandazhevsky Yu.I. and Lelevich V.V. : Clinical and experimental aspects of the effect of incorporated radionuclides upon the organism, Gomel, State Medical Institute, Belorussian Engineering Academy. Ministry of Health of the Republic of Belarus, pp 128. 1995.

14) Okeanov. A.E. : Conference a Minsk. Die wichtigsten wissenschaftlichen Referate. International Congress "The World after Chernobyl" Minsk 1996.

15) Petridou E., Trichopoulos D., Dessypris N., Flyzani V.,Haidas S., Kalmanti M., Koiouskas D., Kosmidis H., Piperopoulou F. and Tzortzatou F.: Infant leukemia

after in utero exposure to radiation from Chernobyl. Nature, Vol. 382, 352–353, 1996

16) Stewart A.M.,Webb J., Hewitt D. A Survey of Childhood Malignancies, Brit. med. J. , Vol. i, p. 1495–1508, 28 June 1958.

17) Viel J.-F., Consequences des essais nucleaires sur la sante: quelles enquetes epidemiologiques? Medecine et Guerre Nucleaire, Vol. 11, p 41–44, janv.–mars 1996.

18) Nesterenko V.B. : Chernobyl Accident. Reasons and consequences. The expert Conclusion. Academy of Science of Belarus. pp. 442. Traduit du russe par S. Boos. SENMURV TEST, Minsk 1993.

19) Nesterenko V.B.: Chernobyl accident. Radioprotection of population. Institute of Radiation Safety "Belrad". pp 180, Minsk 1998

20) European Commission, Atlas of Caesium Deposition on Europe after the Chernobyl Accident, Rep. EURO-16733, EC, Luxembourg (1996).

21) Bandazhevsky Y.I. : Structural and functional effects of radioisotopes incorporated by the organism. Ministery of Health Care of the Republic of Belarus. Belarussian Engineering Academy, Gomel State Medical Institute, pp 143, 1997.

22) Bandazhevsky Y.I.: Pathophysiology of incorporated radioactive emissions. Gomel State Medical Institute. pp 57, 1998.

23) Titov L.P., Kharitonic G., Gourmanchuk I.E. & Ignatenko S.I. : Effect of radiation on the production of immunoglobulins in children subsequent to the Chernobyl disaster, Allergy Proc. Vol. 16, No 4, p 185–193, July- August, 1995.

24) Drobyschewskaja I.M., Kryssenko N.A., Shakov I.G., Steshko W.A. & Okeanov A.E. Gesundheitszustand der Bevolkerung, die auf den durch die Tschernobyl-Katastrophe verseuchten Territorium der Republik Belarus lebt. p91–103, dans : Die wichtigsten wissenschaftlichen Referate, International Congress "The World after Chernobyl" Minsk 1996.

25) Vassilevna T., Voitevich T., Mirkulova T., Clinique Universitaire de Pediatrie a Minsk. Communications personnelles, 1996.

26) Amnesty International: BELARUS . Possible Prisoner of Conscience - Professor Yury Bandazhevsky. AI index : EUR 49/27/99, 18 October 1999.

27) Dubrova Yu.E., V.N. Nesterov, N.G. Krouchinsky, V.A. Ostapenko, R. Neumann, D.L. Neil, A.J. Jeffreys (1996). Human minisatellite mutation rate after the Chernobyl accident. Nature, 380:p.683–686, 25 avril 1996.

28) Goncharova R.I. & Slukvin A.M., Study on mutation and modification variability in young fishes of Cyprinus carpio from regions contaminated by the Chernobyl radioactive fallout. 27–28 October 1994, Russia-Norvegian Satellite Symposium on Nuclear Accidents, Radioecology and Health. Abstract Part 1, Moscow, 1994.

29) Ellegren H., Lindgren G. Primmer C.R. & M0ller: Fitness loss and germline mutations in barn swallows breeding in Chernobyl. NATURE, Vol 389, pp. 593–596, 9 October 1997.

30) Goncharova R.I. & Ryabokon N.I.: The levels of cytogenetic injuries in consecutive generations of bank voles, inhabiting radiocontaminated areas. Proceedings of the Belarus-Japan Symposium in Minsk. "Acute and late Consequences of Nuclear catastrophes: Hiroshima- Nagasaki and Chernobyl", pp. 312–321, Oct. 3–5, 1994 31)

31) Goncharova R.I. & Ryabokon N.I., Dynamics of gamma-emitter content level in many generations of wild rodents in contaminated areas of Belarus. 2nd Intern. 25–26 October 1994, Conf. "Radiobiological Consequences of Nuclear Accidents".

32) Baker R.J., Van den Bussche R.A., Wright A.J., Wiggins L.E., Hamilton M.J., Reat E.P., Smith M.H., Lomakin M.D. & Chesser R.K. : High levels of genetic change in rodents of Chernobyl. NATURE , Vol 380, pp. 707–708, 25 April 1996.

33) Hillis D.M., Life in the hot zone around Chernobyl, Nature, Vol. 380, p 665 a 666, 25 avril 1996.

34) The World Health Report 1997 / Conquering suffering, Enriching humanity, pp.162, Distributed at the World Health Assembly (WHA), 1998.

35) Hartlmaier K.M. : Es geht nicht nur um Contergan. I. Mai beginnt der grosse Prozess. Er trifft grundsatzliche Fragen. Zahnarztliche Mitteilungen, Nr. 9, pp. 427–429, 1968.

36) Lazjuk G.I., Satow Y., Nikolaev D.L., Kirillova I.A., Novikova I.V., and Khmel R.D.: Increased frequency of embryonic disorders found in the residents of Belarus after Chernobyl accident. Proceedings of the Belarus-Japan Symposium "Acute and late Consequences of Nuclear Catastrophe: Hiroshima-Nagasaki and Chernobyl"; p. 107–123, Belarus Academy of Sciences, Minsk Oct. 3–5, 1994.

37) Lazjuk G.I. et al. : Genetic consequences of the Chernobyl accident for Belarus Republic (published also in Japanese in Gijutsu-to-Ningen, No 283, p.26–32, Jan./Febr. 1998) Research Activities about the Radiological Consequences of the Chernobyl NPS Accident. p.174–177, Edited by IMANAKA T. Research Reactor Institute, Kyoto University, KURRI-KR-21; March 1998.

May 3, 2000 Michel Fernex,
Address : F-68480 Biederthal, France.

WHO Director meets
IndependentWHO 4May 2011

GENEVA, SWITZERLAND, 4 May 2011—WHO Director-General Dr Margaret Chan met with representatives of the group "Independent WHO" today to listen to their concerns and discuss common interests on radiation and health.

"Independent WHO" is a civil society group advocating for people who have been affected by radiation exposure as a result of the nuclear accident at Chernobyl.

Dr Chan stressed that the mandate of WHO is to protect the health of people and the Organisation works independently to fulfill this role while cooperating with other organisations in the UN system and other partners. She underlined that WHO takes its mandate to protect and advocate for the health of populations all over the world very seriously and does not compromise with the integrity of its functions.

In answer to their concerns, Dr Chan explained that:

- WHO develops guidelines and standards, and although the Organization can advocate for Member States to follow them, WHO cannot take the place of a national government and implement standards in a country or force a Government to do so.
- WHO's responsibilities in these situations are mandated primarily by the International Health Regulations.
- Otherwise, WHO cooperates with the International Atomic Energy Agency (IAEA) on issues of common concern, in a spirit of mutual respect and independence in the light of their respective mandates.
- WHO works closely with the Food and Agriculture Organization (FAO) to ensure food contaminated with radiation does not come to countries and will continue in that role;
- WHO agrees in principle that research should continue on the health effects of radiation and the research should not be influenced by industry.
- WHO will look into why the proceedings of a 2001 meeting on radiation and health were not published.

The Director-General clarified that the legal basis for WHO's cooperation with the IAEA lies in the 1959 agreement between the two agencies, the two international conventions adopted in 1986 after the Chernobyl accident, and the International Health Regulations (2005). Their application allow a balance between cooperation and coordination without interfering with the independent pursuit by WHO of its public health mandate.

Six representatives of the Independent WHO group met with the Director-General for more than two hours. Dr Chan praised the representatives for the group's passion and persistence and promised to keep the lines of dialogue open on the issues within WHO's area of jurisdiction.

Letter to WHO Director-General Margaret Chan May 12, 2011

IndependentWHO
c/o Éric Peytremann
54 rue Ernest Bloch
CH-1207 Genève

Dr Margaret Chan
Director-General of WHO
Avenue Appia
Ch-1211 Genève 27
12 May 2011

Subject: Letter from the collective IndependentWHO after meeting on 4 may 2011 with Dr Margaret Chan, WHO Director-General

Dear Dr. Chan,

We thank you for inviting us and for having received us with great courtesy and in particular for considering us for what we are: "citizen advocates of victims of radioactive contamination."

You explained the institutional position of the World Health Organization headed by you, with its scope and its limitations, defined by both the Agreement with the IAEA and other international agencies and also by two Conventions of 1986 and the International Health Regulations established in 2005.

In this context, we do not doubt your sincerity when you say that our struggle—yours and ours—is common in that you consider it your duty to protect victims of radioactivity and to conduct this task "with entire independence".

After our meeting we none the less decided to continue our vigil in front of WHO. Why?

• Firstly, we did not hear from you concrete proposals that would make it possible, in the short or medium term, to improve the lives of people severely affected by radioactive contamination, particularly that caused by the Chernobyl disaster.

• In your press release dated May 4 you assure that " WHO agrees in principle that research should continue on the health effects of radiation and the research should not be influenced by industry." But that is precisely what we condemn when we ask for a review of your Agreement with the **IAEA** dated 28 May 1959 (WHA

12-40): the **IAEA** , within its mandate, promotes civilian uses of atomic energy and is thus **linked to the nuclear industry** , which prevents it from recognizing objectively both the uncontrolled risks of atomic power and the health hazards of low doses of radionuclides.

• During our meeting on May 4, you acknowledged that " Chernobyl has caused more than fifty dead ." You thereby contradict the joint assessment co-authored by WHO and IAEA (09/05/2005) which refers to "less than fifty deaths and 4,000 potential future deaths." Moreover, a joint statement by WHO and IAEA on Chernobyl, dated April 24, 2009, said the areas affected by the accident are no longer dangerous for the population, which must be "reassured" by "practical advice" and convinced to "return to normal life." Scientists and doctors on the ground, who face the real problems of health of children or those of the liquidators, have a very different vision of reality. The information they provide to us remains alarming.

Why should we not expect the Director-General of the WHO to distance herself from the IAEA by a clear rejection of these estimations of deaths caused by Chernobyl and these evaluations which go against reality?

It is clear to us after our meeting that you lack the means to be independent. Far from being in a position to set up—as we have requested—a team specializing in radiation and health, currently non-existent, you spoke of the "budget deficit of the WHO" which is preparing to lay off staff...

Thus, on the eve of the World Health Assembly opening on 16 May 2011, we reiterate our request for revision of the WHO-IAEA agreement of 1959 which makes you dependent on "experts" of the IAEA and the nuclear industry. We urge you to intervene with the international community to obtain the legal and financial independence to enable you to effectively take charge of health problems linked to radioactive contamination, according to your Constitution.

We have provided you with a copy, signed by the authors, of the book by A. Yablokov, V. and A.Nesterenko on the consequences of the Chernobyl disaster, translated and published by the New York Academy of Sciences[1]. This book is a sum of independent research that provides an assessment of Chernobyl—985,000 deaths—no relation to the one you co-signed in 2005.

At our first meeting with five senior officials from WHO, July 2, 2009, the idea of organizing a forum where conflicting data and analysis on the health consequences of the Chernobyl accident could be discussed. At our meeting on May 4, you have assured you take into account "all sources, both official and unofficial" for your

---

[1] Chernobyl: Consequences of the Catastrophe for People and the Environment, Annals of the New York Academy of Sciences, VOL. 1181, 2009

information. Finally, you referred to the need for transparency in society and "even at WHO!"

Convening such a forum, which becomes increasingly urgent after Fukushima, would meet this need and would demonstrate publicly WHO's will to be independent. It would make public the actual situation in the contaminated territories and determine what steps should be taken in terms of research and health-care.

Please accept, Madam Director-General, the assurance of our deepest respect.
Eric Peytremann
For IndependentWHO

Letter to WHO Director-General Margaret Chan
November 07, 2011

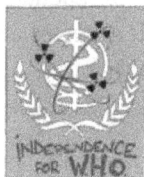

Independent WHO
Correspondent: Wladimir Tchertkoff
Nucleo paese
CH-6945 Origlio
wladimir@vtx.ch
contact@independentwho.org

Dr Margaret Chan
Director-General
World Health Organization
Avenue Appia
1211 Geneva 27
Switzerland

Origlio, 7 November 2011

Subject: non-publication of the Proceedings of the Conferences on the health consequences of Chernobyl, held in Geneva in 1995 and Kiev in 2001

Dear Dr Chan

The World Health Organization's mandate is to promote health in the world and the International Atomic Energy Agency's mandate is to promote nuclear energy. In light of the problems associated with nuclear reactors, many eminent scientists and public health professionals criticize the 1959 Agreement between these two organizations as an obstacle to serious investigation of the consequences of accidents and dissemination of information about the Chernobyl catastrophe which would allow ongoing damage to be documented and future damage to be avoided.

In spite of this major obstacle to the dissemination of knowledge, independent research has continued over the last 25 years and an impressive quantity of data and new knowledge has accumulated. Much of this was presented and discussed at two important conferences organized by the WHO, in Geneva in 1995 and in Kiev in 2001. The Proceedings of these conferences, the importance of which cannot be understated given the exceptional nature of the catastrophe, were never published, as we testified during our meeting with you on 4 May 2011. You promised that an investigation would be undertaken and that we would be informed of its results in due course.

On 4 July 2011, we received a message from Dr Neira addressed personally to Alison Katz, informing her that the Proceedings of the two conferences had indeed been published. On 22 July 2011, we requested you (by email) to inform us whether this message constituted the results of the promised investigation and was the response of the Director-General to IndependentWHO. You did not reply. In order to ensure that this is not a simple misunderstanding of terms *Proceeding (minutes of the discussions of a conference with resolution)* and *abstracts (summaries of presentations)* and that open and serious dialogue is the new style of relationship that you hope to establish with us, we repeat with this letter the arguments and evidence that we presented on 4 May which perhaps, have not been sufficiently clear.

If no response to this letter is received by myself and the Coordination of IndependentWHO (see above email addresses), we will consider that Dr Neira's message constitutes your definitive reply.

Arguments and evidence

1. In Autumn 2007, the Department of Radiation of WHO (RAD) replied (via the DG's Office) to The Guardian newspaper, that the Proceedings of the two conferences had both been published. By way of proof of publication of the Proceedings of the Kiev 2001 conference, RAD cited "the special issue of International Journal of Radiation Medicine (2001, Vol. 3, N1-2, ISSN 15621154)".

This is a reference to the volume of ABSTRACTS which I myself bought on the morning of the 4 June 2001, on the first day of the 3rd International Conference *"Health Consequences of the Chernobyl catastrophe: results of 15 years of research"*, 4–8 June 2001, in Kiev. The conference was filmed in its entirety by a Swiss Television team, under my direction. On Thursday 7 June,

Mrs. Roxana Garnets (UN Chernobyl Programm), who was chairing the Round Table programmed from 16.45 to 18.00, declared *"We have recorded this discussion. We will include the material from this Round Table in the material of the conference and we will publish it[1]"*. The "materials recorded at the conference" constitute the raw material of the promised Proceedings, which have still not seen the light of day. In its reply to the Guardian, unwilling to recognize this shameful fact,

RAD attempted to pass off the *Abstracts* as the *Proceedings*, an absurd substitution of terms, as I think you will concede. In the version of events presented by RAD, the Proceedings were sold before the conference?[2] On 4 July 2011, Dr Neira repeated word for word the terms of the fallacious reply of 2007 in the email message sent

---

[1] Cf. "Le crime de Tchernobyl, le goulag nucleaire" p. 574, W.Tchertkoff, Actes Sud, 2006.

[2] I attach the cover page of this volume which I have in my possession since the morning of the 4 June 2001!

to Mrs. Katz. This fact has scandalized and disappointed us. It confirms that the censorship that we are denouncing continues.

2. In relation to censorship, Dr Hiroshi Nakajima, when questioned by Professor Michel Fernex in Kiev, was in no doubt as to the real nature of the proceedings that were promised to 700 participants at the conference which he had organized in November 1995, nor about the reasons for their nonpublication.

He explained in the clearest terms that the IAEA had not given its agreement because it has a major interest to protect[3]: the conclusions of its own conference to be held in March 1996 in Vienna, which contrary to that of Dr Nakajima, were to demonstrate that no serious health consequences resulted from the Chernobyl accident.

Here is Dr Hiroshi Nakajima's declaration to Swiss television:

M.Fernex.—Why were the proceedings we had ordered not published?

H.Nakajima.—Because it was a conference organized jointly with the IAEA. This was the problem. M.Fernex.—Is this WHO conference here more free than in Geneva?

H.Nakajima.—Here, I am no more Director General, I am a private person.

W.Tchertkoff.—Don't you think that the link between WHO and IAEA impaired the liberty of WHO?

H.Nakajima.—I was Director General, and I was responsible. But it's mainly my legal department... Because the IAEA reports directly to the Security Council of the UN. And we, all specialized agencies, report to the Economic and Social Development Council. The organization which reports of the Security Council—not hierarchically, we are all equal—but for atomic affairs, military use and peaceful or civil use, they have the authority.

The evidence is clear. Dr Nakajima in no way denies the non-publication of the Conference Proceedings; on the contrary, he provides the explanation that we all know: the real hierarchical relationship between the WHO and the IAEA in nuclear matters, codified by the Convention on Assistance of September 1986. This Convention expropriates de facto, from the WHO, its principal functions in health matters in the case of an accident, attributing them to the IAEA. With no competence in health matters, it is the atomic energy agency that acts as directing and coordinating authority "to facilitate prompt assistance in the event of a nuclear accident or radiological emergency to minimize its consequences and to protect life, property and the environment from the effects of radioactive releases." (Convention

---

[3] "Whenever either organization proposes to initiate a programme or activity on a subject in which the other organization has or may have a substantial interest, the first party shall consult the other with a view to adjusting the matter by mutual agreement". Article 1, §3. Of the WHO/IAEA Agreement (Res. WHA 12.40, 28 May 1959).

on Assistance, art.1, 1; art. 3; art. 5). The WHO is not mentioned in any of the Articles of the Convention. If the Proceedings of the Conference had been published as you assert, the ex-Director General would have replied to Professor Fernex five years later: "No no, you are mistaken..."

Dear Dr Chan, we are in the domain of irrefutable fact. Hiroshi Nakajima is responsible for his legal interpretation, as you acknowledged during our meeting on 4 May, and you might have acted differently from him—but you cannot deny that he confirms, in his reply, that the Proceedings were never published.

During the meeting with you, we welcomed your various declarations which corrected WHO's previous position on a number of points. This encouraged us to believe that a serious dialogue could indeed by established between us. But we cannot take seriously Dr Neira's attempt to pass off "a selection of 12 articles" published in the Quarterly Journal of Health Statistics of WHO as the Proceedings of the 1995 Conference—an event of immense wealth and historic significance. We know that the Proceedings of a scientific conference are composed of three parts:
1. Summaries of contributions (abstracts),
2. Summaries of discussions held during the conference
3. Recommendations (conclusions) of the conference.

Professor Michel Fernex was present at the 1995 Conference and he described the essentials to us. This is what made us decide to film the "rejoinder" in Kiev; the arguments and the protagonists were the same but they were confronting a much worsened health situation in the contaminated territories.

Extract from Professor Fernex's report
"In 1995, Dr Hiroshi Nakajima, Director-General of WHO, organized an international conference in Geneva from 20 to 23 November, on *"Consequences of Chernobyl and other radiation accidents and their influence on human health"* At this conference, presided by M. Y. Fujita, Governor of the Prefecture of Hiroshima, the destruction of Hiroshima and Nagasaki and the explosion of a nuclear reactor at Chernobyl were considered as comparable radiological accidents. Important differences were noted between these two kinds of accidents (the three explosions must be termed "accidents" in this *milieu*). The Geneva Conference cannot be referenced[4], as the Proceedings disappeared or were censored, so it is useful to recall its objectives which were clearly set out in the programme[5] :

---

[4] Health consequences of Chernobyl and other radiological accidents. International Conference organized by the WHO in Geneva, 20-23 November 1995. Proceedings not published.

[5] Programme of the International Conference organized by the WHO in Geneva, from 20-23 November 1995. *Consequences of Chernobyl and other radiation accidents and their influence on human health.* The programme is available from WHO, Geneva. WHO/EHG/1995.

- To present the principal results of the first phase of the international program on health effects of Chernobyl accident (IPHECA).
- To compare the obtained results to the results of similar research, related to the health effects of Chernobyl accident.
- To improve (and to update) awareness of the type, the total extent and the harm for health of the Chernobyl accident, as known presently and to be foreseen in the future.
- To make new results of research concerning consequences of other radioactive accidents, available, in order to give more complete information on their health effects.
- To study the effectiveness of the protective measures undertaken in the area of public health during and after the accidents, and to offer recommendations for the future.
- To ensure the development and/or to clarify the state of knowledge concerning the consequences of influence of radiation on human health.
- To provide information on existing or future research within the framework of the United Nations Scientific Committee on the Effects of Atomic Radiation (UNSCEAR).
- To earmark the interesting tendencies and changes, which should become an object of steadfast attention of the researchers.

The conference program convinced the health authorities of the most affected countries and 700 doctors and experts to participate. The IAEA mobilized unconditional supporters of the nuclear industry. Thus, contradictory opinions were expressed which made the debate very lively. Representatives of the nuclear lobby attempted to prevent dialogue and Professor S. Yarmonenko of Moscow's Centre for Oncology with extreme vehemence,- insisted that in the future, organizers of conferences on this subject should exclude any speaker who attempted to discuss the effects of low dose radiation. It appears that this exclusion did indeed become the rule for the international conferences which followed. The presentations, discussions and posters were never published.

The luxurious 519 page document which sets out the facts and figures collected in Phase 1 of the IPHECA pilot project "Health consequences of the Chernobyl accident", confirms that the WHO intervened far too late in Chernobyl, in an accident that a majority of citizens considered an "emergency". For five years, the IAEA appropriated the knowledge, and collaborated with health authorities on the measures to be taken for the population, the concern being to reduce expenditure and discontinue victim compensation.

Not only did the WHO fail to respect its Constitution which stipulates that it intervene in a timely fashion, but it also failed to act as the coordinating and directing authority in all international health matters, as set out in its Constitution. At the meetings where the destiny of populations was to be decided, WHO's expert was

Professor Pellerin, unconditional supporter of nuclear power [6]. Five years after the catastrophe, the WHO was able to start work in "selected" areas, including dental caries in children as one of the priority studies. Hereditary genetic effects, which a WHO expert committee[7] had previously considered a priority, were "forgotten".

As the presentations made at the Geneva Conference remain unpublished, it is useful to recall the words of certain participants, such as Martin Griffiths, of the Department of Humanitarian Affairs of the UN, in Geneva. This speaker noted that the truth had not been told to populations and reminded the conference that people were still living in contaminated areas. He requested that

assistance and studies continue, because without money, everything would cease. He noted that **9 million people were affected and that the negative health consequences continued to increase"** [8]

Dr Chan, you promised us that you would undertake a serious investigation. The most simple, direct and truthful way to do this was to speak to your peer, Dr Hiroshi Nakajima himself. He is the person who was censored. In 2007, I sent him a copy of my book and we invited him to the press conference that we organized in Geneva on 27th June hoping to be able to speak with him in more depth. He sent us a reply via his wife stating that he did not wish to cause you any embarrassment. In the spirit of transparency that you declare you are inaugurating in WHO's relations with the public, perhaps you would consider freeing Dr Nakajima from his sensitivity with regard to yourself.

This is my dedication to Dr Nakajima that I wrote in the copy of the book "The Crime of Chernobyl" that I offered him:

"To Doctor Hiroshi Nakajima, with respect and gratitude for his testimony to the truth; for his research on the impact of an H Bomb on Boston, in which victims are deprived of assistance; for having shown the end of civilization that would result from the nuclear winter and the famine after a limited nuclear war …

I am grateful for your attempt to study the consequences of Chernobyl in 1995, studies which were censored by the IAEA, which manages and minimizes medical problems in this area in the interests of proliferation of commercial nuclear power."

The testimonies assembled in this book demonstrate the suffering of populations irradiated by the Chernobyl accident. This subject has become a source of permanent disinformation which is a shocking and disturbing state of affairs."

---

[6] Belbeoch B. and Belbeoch R. : Tchernobyl, une catastrophe. Quelques elements pour un bilan. Edition Allia, 16 rue Charlemagne, Paris IVe, pp 220. 1993.

[7] WHO. Effect of radiation on human heredity. Report of a Study Group convened by WHO together with Papers presented by various members of the group. WHO, Palais des Nations, Geneva, 1957.

[8] "La catastrophe de Tchemobyl et la sante" in Chroniques sur la Bielorussie contemporaine—L'Harmattan, 2001.

Yours sincerely,
Alison Katz, Bruno Boussagol, Maryvonne David-Jougneau, Paul Roullaud.

Wladimir Tchertkoff For IndependentWHO *contact@independentwho.org*

GLOSSARY

ACTIVITY

Activity is a measure of the amount of radioactivity present in a material (inert or living). It is measured by the number of disintegrations per second. One *becquerel* (Bq) represents one disintegration per second.

The bequerel is far too small a unit to measure the amount of radioactivity in the core of a reactor or in material released following an accident, and therefore the old unit, a curie (Ci) is used, which corresponds to 37 billion becquerels. At Chernobyl several tens of millions of Ci were released. In order to measure the contamination of large areas of land, we use curies per square kilometre (Ci/km^2) **(See page 160,193, 350, 356-359, 363.)** But when we are looking at biological effects, a curie is far too large a unit and we use the becquerel. The radioactivity in food produced on contaminated soil, and in the human body if it has eaten this food (*internal contamination*) is measured in becquerels per kilogram (Bq/kg).

ABSORBED DOSE

RAD AND GRAY
When radiation passes through matter it loses energy. It is on the basis of the amount of *energy* absorbed in the matter that we evaluate the level of radiation, the *absorbed dose*. It is measured in rad or in gray (1 gray = 100 rad)

THE EFFECT ON THE ORGANISM

REM AND SIEVERT

The *effect* of radiation on the organism depends on the absorbed dose and on the type of radiation (α, β, γ). To evaluate the biological impact of radiation, the absorbed dose is assigned a radiation weighting factor that relates to the degree of biological damage produced by that type of radiation. Alpha particles (α) are very effective, and are assigned a radiation weighting factor 20 times that of beta particles (β) or of gamma radiation (γ).

This is how we obtain *equivalent dose*. It is measured in rem or sievert (1 sievert— 100 rem). When there is no possibility of ambiguity, the term "dose" is used instead of the "equivalent dose". To describe the effect on people, the commonly used term "dose" (sievert [Sv], millisievert/year [mSv/yr], or microsievert/hour [μSv/hr]) therefore does not indicate the energy, but the *biological impact* of the energy absorbed in a given tissue, the damage caused to the organism.

Lifetime dose: this is the radiation dose that an individual living in a contaminated area will receive if he lives there for seventy years. (See the controversy surrounding the theory of "35 rem lifetime dose" introduced by Dr Iline, **(page 36-38, 68, 106, 144, 234, 431).**

SOME REFERENCE POINTS

Since 1990 the International Commission for Radiological Protection has recommended an annual dose limit for workers in the nuclear industry of 20 millisieverts (2 rem). Before that it was 50 millisieverts (5 rem).

The annual dose limit recommended for the general public since 1985 is 1 millisievert (0.1 rem). Before it was 0.5 rem (5 millisievert). According to the IAEA norms, no radioprotection measures are necessary below 1 mSv/yr. V. Nesterenko recommends a limit of 0.3 mSv/yr for children (see page 162, 238). According to a law passed in Belarus in 2001, radioprotection measures need to be maintained even if the annual charge drops from 1 mSv/yr to 0.1 mSv/yr **(page 472-473).**

These limits do not define a threshold dividing dangerous from harmless exposures. Any dose of radiation carries a risk and the limits proposed by international experts are based on economic considerations to avoid any obstacles to the development of the nuclear industry (see page 300 "The illusion of norms").

The international norms are based on the principle that the risk to health is proportional to the dose received and that any dose of radiation carries a risk of cancer or to the genes.(ICRP 1990). Exposure to artificial radiation (including nuclear weapons testing) has caused numerous cancers worldwide. According to official United Nations figures it has caused 1.17 million deaths since 1945. The European Committee on Radiation Risk (ECRR) puts the figure at 61.1 million deaths. **(See Yury Bandazhevsky's view on page 189–192.)**

RADIOACTIVE HALF-LIFE

Radioactive half-life is the time required for half the radioactive elements to disappear. At the end of a certain period, only half the initial radioactivity remains; at the end of two periods, a quarter, etc. After ten periods, only a thousandth remains. There is no threshold below which there is no biological effect from radiation.

Half-life varies considerably between one isotope and another, from a fraction of a second to millions or billions of years. Iodine-131 (8.04 days); caesium-134 (2.06 years); caesium-137 (30 years); strontium-90 (29.1 years); plutonium-239 (24,065 years).

Thus, when we consider elements with a short half-life, perhaps a week, these will be giving off considerable radioactivity for several months. Elements whose half-life is a few years to a few decades will remain active for a few decades or centuries. Plutonium, which is not only very long-lived but very toxic, will remain active for hundreds of thousands of years.

(Source: Bella and Roger Belbeoch, *Tchernobyl, une catastrophe*, Ed.Allia, 1993)

INDEX OF NAMES

A. LAZUKO 122
Valery LEGASOV 18, 19, 35, 52, 66, 68, 74, 93, 110, 158
Sylvie LEMASSON 282, 511
Edmund LENGFELDER 232–243, 249, 250, 252, 283
V.I. LENIN 65, 66, 126
M. LENKEVICH 485
Yves LENOIR 63, 64, 153, 300, 417
Georgi LEPIN 144, 204
E. LIGACHEV 70
I. LIKHTAREV 427, 428, 458, 459
Jacques LOCHARD 240, 266, 269, 270, 272, 282, 283, 285, 286
K. LOGANOVSKY 430
M. LOS 514
Natalie LOSEKOOT 477
Alexander G. LUKASHENKO 12, 165, 172, 197, 198, 204–206, 208, 219, 220, 230, 231, 243, 249, 269, 273, 277, 377, 489, 495, 496, 500, 501, 503, 504, 506, 510–513, 517–519, 522, 523, 527, 528, 532, 533, 539
Yulia LUKASHENKO 20, 21
Catherine LUCCIONI 277, 293
Raoul De LUSENBERGER 259, 260
T.D. LYSENKO 36

M

Irina MAKOVETSKAYA 168, 169, 196
Nino MARANESI 302
Maria Kirilovna 316–318
Kiril MARKUS 430
Barry McSWEENEY 256, 257
Marusia 373, 374, 375
M. MASCALEV 189
N. MATUKOVSKY 73
Vassili MAXIMENKO 282
M. MAXIMOV 120
Grigori MEDVEDEV 14, 15, 18, 20
L.S. MELESHKO 220
Dimitri MENDELEEV 431
Louis MICHEL 504
M. MIKHNEVICH 260
Dmitri MIKHNYUK 239

V.F. MINENKO 220
Danielle MITTERAND 169, 244, 535, 538, 553
A. MOISEYEV 67
Karl Z. MORGAN 61, 62
J.M. MULLER 12, 290, 558, 566
Lisa MURAVIOVA 360–362

N

David NABARRO 536, 556–558, 562, 563
Hiroshi NAKAJIMA 414–420, 454, 460, 461, 536, 557, 567, 594, 595, 597
V. NAUMOV 499, 533
M. NEIFACH 194
Alexei (Aliosha) NESTERENKO 80, 83, 112, 540, 590
Ilsa NESTERENKO 80, 89, 92, 112
Vassili B. NESTERENKO 3, 15, 16, 18, 19, 22, 23, 50, 52, 68–96, 109–117, 137, 155–179, 185–187, 189, 192, 195–207, 218–244, 247, 248, 250, 256–276, 279, 281–285, 287, 292–294, 299–310, 313, 322, 324–327, 330–332, 337, 338, 342–350, 357, 361–368, 386, 389, 394, 415, 420, 425, 451, 453, 454, 460, 462, 481, 490, 491, 496, 497, 499, 503, 505, 522, 535, 537, 540, 543, 551, 552, 554, 555, 562, 575–577, 583, 586, 590, 600
Lara NEVMENOVA 168
A. NYAGU 439, 440, 454, 455

O

M. OBUZENKO 344
A.E. OKEANOV 249, 541, 561, 563, 569, 572, 573, 584–586
Henry OLLAGNON 276, 282, 283, 286
V. OREKHOVSKY 224, 225, 243
E. ORLOVA 515–517
V. OSTAPENKO 199, 226, 586

P

André PARIS 454
Vladimir PASHKEVICH 282
Boris PASTERNAK 64

INDEX OF PLACE NAMES

England 18, 37, 96, 152, 244

Europe 2, 8, 12, 16, 17, 19, 26, 31, 39, 49, 50, 59, 79, 128, 166, 169, 187, 191, 246, 256, 281, 288, 293, 302, 447, 454, 504, 505, 506, 511, 520, 521, 522, 540, 543, 544, 550, 551, 574, 575, 577, 583

F

France 2, 7, 12, 14, 38, 67, 96, 128, 159, 167, 170, 179, 183, 184, 218, 220, 240–242, 244, 260, 264–266, 273, 275, 277, 294, 324, 412, 422, 423, 454, 520, 533, 535, 536, 538–540, 542, 544–546, 548, 553, 556, 557, 573

G

Gaishin 360, 362, 363, 365, 376, 379, 384,

Geneva vii, 11, 169 412, 415, 417, 419, 424, 436, 454, 496, 500, 535, 537, 543–546, 560, 561, 563–569, 572, 588, 592, 594, 595, 597

Germany 7, 56, 82, 93, 96, 112, 124, 128, 136, 159, 161, 207, 235, 236, 238–240, 242–244, 249, 250, 252, 277, 283, 284, 294, 324, 338, 366, 416, 446, 466, 520, 534, 547, 548, 553, 558, 573

Glazki 161, 356

Gomel 15, 18, 73, 87, 95, 97, 98, 109, 113–116, 156, 161, 164, 165, 167–

172, 174, 176–179, 181, 183, 195–197, 202, 203, 207, 208, 215–217, 226, 227, 235, 238, 242, 245, 248–251, 259, 272, 274, 288, 289, 292, 293, 308, 314, 329, 344, 345, 357, 387, 397, 404, 406–409, 450, 452, 466, 476, 481, 484, 486, 490, 491, 498, 499, 501, 505, 506, 512, 517, 537, 541, 556, 558, 562, 569, 573, 575, 577–579, 583–586

Gorodnya 161, 270, 272

Great Britain 12, 37, 47, 70, 159, 277, 294, 429, 553

Greece 446, 573

Grenoble 521, 528, 537, 538

Grichinovichi 161

Grodno 169, 171, 177, 212, 251, 484, 508, 517

Grozny 102

Grushevka 161

H

Hanover 82, 338, 366

Hiroshima 17, 52, 53–59, 62, 64, 72, 85, 142, 154, 183, 186, 187, 192, 293, 360, 414, 432, 437, 439, 445, 446, 457, 471, 473, 559, 567, 569, 573, 578, 587, 595

I

Iput' (tributary of Dnieper) 43

Ireland 243, 244, 284, 342, 553, 577, 583

Ispra 256, 259, 260

Italy 7, 128, 159, 167, 256, 277, 284, 332, 553, 580, 581

J

Japan 56, 57, 60, 72, 74, 82, 159, 193, 194, 246, 330, 447, 545, 553–555, 557

K

Kaliningrad 317

Kaluga 434, 465

Kama (river) 319

Kharkov 195

Khatyn 377

Khilchikha 161

Khoiniki 19, 28, 31, 83, 87, 94, 105

Khomenki 161

Khrakovichi 294

Kiev 8, 11, 24, 34, 36, 43, 51, 58, 60, 72, 95, 99, 113, 126, 128–130, 157, 189, 195, 199, 206, 248, 257–261, 289, 324, 335, 415–420, 434, 455, 461–463, 490, 491, 537, 538, 540, 546, 548, 549, 550, 554, 560, 592–595

Kirov 161, 237

Kliapin 360, 384, 389, 390, 391, 395

Kliny 361

Koliudy 399

Kolpita (tributary of Dnieper) 43

Komanov 161

Komarin 294

Konotop 161

Korchevatka 161

Korma 356, 377, 392, 395, 398, 402, 576

Koshara 326

LIST OF ACRONYMS & ORGANISATIONS

ACAT Association chrétienne contre la torture. French.(*Association of Christians Against Torture*), affiliated to the International Federation of Action by Christians for the Abolition of Torture. (FIACAT), 533

ACRO Association pour le contrôle radioactivité dans l'ouest, French. (*Association for Radiation Monitoring in the West*), 240

AEC US Atomic Energy Commission, 60

ALC Admissible Levels of Contamination, 219, 226, 300

AREVA A French industrial conglomerate specialising in energy but in particular nuclear energy. It used to be made up of COGEMA which dealt with the entire fuel cycle from mining, through to reprocessing and Framatome ANP, which built nuclear power stations. In 2001, Cogema was merged with Framatome and CEA Industrie to form the larger group Areva. Areva NP was created by absorbing the nuclear business line of German company Siemens, 58, 266

ARTE Franco-German television network specialising in cultural and arts programmes, 267–269, 273, 275, 418

ASN Autorité de sûreté nucléaire. (*Authority for Nuclear Safety*), 179, 180, 181

ASPEA Association Suisse Pour l'Énergie Atomique. (*Swiss Association for Atomic Energy*), 131

Associations in Europe for the children of Chernobyl

- "Jugends Aktions Netzwerk Umwelt– und Naturchutz e. V. (JANUN) ø Gr. Barlinge 58a–30171 Hannover– Allemagne 244

- "NIKOBELA" Grosse Drakenburger Strasse, 3-31582 Nienburg.– Allemagne 244

- "Association Belgo-Biélorusse pour les Enfants de Tchernobyl–A.S.B.L." 16 rue Marache–5031 Grand-Leez– Belgique *(Belgium-Belarus Association for the Children of Chernobyl)* 244

- "Chernobyl Children's Project" 8 Sidneyville, Bellevue Park, St Luke's, Cork–Ireland 244

- "Enfants de Tchernobyl Bélarus" *(Children of Chernobyl Belarus)*, 20 rue Principale–68480 Biederthal– France 207, 239, 241, 244, 533, 537, 551, 553

- "Les Enfants de Tchernobyl", *(Children of Chernobyl)* 378 rue de Modenheim 68110 Illzach–France 207, 244, 533

- "Solidarité de Biélorussie et de Tchernobyl" *(Solidarity with Belarus and Chernobyl)*, 74 rue de Falaise–14 000 Caen–France 244

BEIR Biological Effects of Ionising Radiation, 62

BelAm Joint 30 year scientific project set up by the Ministry of Energy in USA and the Ministry of Health in Belarus to establish the number of cases of thyroid illness among the inhabitants of contaminated areas in Belarus,   245, 246, 269

BELRAD Independent Institute of Radiological Protection (Belarus),   3, 74, 112, 161, 164, 170, 218, 220–223, 228–235, 239, 240, 241, 243, 244, 248, 256, 258–260, 261, 265, 269, 270,272, 273, 274, 278, 283, 285, 286, 291, 292, 294, 304, 305, 330, 342, 365, 389, 397, 491, 514, 535–537, 541, 548, 551, 552, 555, 576, 577, 583, 586

CEA Commissariat à l'Energie Atomique. French. *(Commisariat for Atomic Energy)*,   58, 165, 179, 180, 190, 264, 266, 267, 270, 348, 422, 423, 451, 560

CEPN Centre d'étude pour l'Évaluation de la Protection dans le domaine Nucléaire *(Centre for the Study of the Evaluation of Nuclear Protection)*,   58, 240, 263, 264, 266, 270, 271, 275, 285, 365

CFDT Confédération française démocratique du travail One of the two main French Trade Unions. *(French Democratic Confederation of Work)*,   180

CIS Community of Independent States. Formed in 1991 by Russia, Belarus and Ukraine as a successor to the USSR,   246

CODHOS Comité de Défense des Hommes de Science de l'Académie des Sciences de Paris *(Committee for the Defence of Human Rights for Scientists at the Academy of Sciences of Paris)*,   539

COGEMA Compagnie générale des matières nucléaires. The French company that worked at all stages of the nuclear fuel cycle from the mining, conversion and enrichment of uranium to the treatment and reprocessing of used nuclear fuel. COGEMA is now called AREVA NC,   58, 180, 266, 270

Comchernobyl Belarus government committee on the consequences of the Chernobyl disaster,   259, 348, 200, 206, 207, 229, 267, 268, 269, 270, 271, 272, 273, 282, 286, 290, 345, 357, 365

CONTRATOM Anti-nuclear organisation based in Geneva, Switzerland,   412, 556

CORE Cooperation for the Rehabilitation of conditions of life in the areas of Belarus contaminated by the Chernobyl accident,   12, 26, 187, 207, 239, 240, 241, 261, 264, 265, 275–298, 364, 548, 551

CRIIRAD Commission de Recherche et d'Information Indépendantes sur la radioactivité *(Commission for Research and Independent Information about Radiation)*,   169, 207, 412, 454, 514, 535, 537, 539, 546, 547

DVTH Deutscher Verbande für Tschernobyl-Hilfe (Fédération allemande d'aide à Tchernobyl),   232, 233

EAD Effective Annual Dose,   224

ECG Electrocardiogram,   173, 174, 248, 298, 387, 388, 389, 391, 392, 395, 396, 397, 398, 399, 401, 402, 403, 473

ECRR European Committee on Radiation Risk. International centre for study and research,   60, 62, 184, 185, 446, 461, 600

EDF Électicité de France,   180, 264, 266, 270

ENS European Nuclear Society,   131

ETHOS Multi-disciplinary French-Belarusian consortium set up by the CEPN and financed by the European Commission for the rehabilitation of areas contaminated by the Chernobyl accident, 256, 257, 264-277, 279-288, 290, 295, 297, 298, 305, 333, 348, 415

EU European Union, 520

EURATOM European Atomic Energy Community, 233, 283

EUROSCIENCE European association for the promotion of Science and technology, 504

FDA Food and Drug Administration, 246

FRANCE LIBERTÉS Foundation set up by Danielle Mitterrand , 22 rue de Milan—75009 Paris, 169, 533, 535, 537, 538, 553

FRG Federal Republic of Germany, 161, 162

GNC Gosoudarstvennyi naoutchnyi centr (Russian Government Scientific Centre), 237

GNP Gross National Product, 290

GSIEN Groupement de Scientifiques pour l'Information sur l'Energie Nucléaire. French. *(Scientist Group for Information on Nuclear Energy)*, 169, 289, 533, 535, 539

HRS Human Radiation Spectrometer sometimes called a Whole Body Counter, 160, 161, 162, 173, 174, 194, 199, 219-232, 243, 250, 257, 303, 304, 367, 472, 473, 474

IAEA International Atomic Energy Agency, 11, 38, 47-49, 51, 58-60, 62-64, 66, 67, 72, 74, 81, 96, 101, 137, 153, 155, 156, 158, 159, 162, 165, 168, 188, 199, 200, 202-204, 206, 218, 224, 230, 233, 234, 238, 246, 265, 283, 287, 297, 298, 300, 358, 386, 412-415, 417-423, 425-428, 430, 439, 440, 451, 453, 455, 460, 461, 463, 485, 536, 540, 541, 543-545, 547, 554, 556-560, 561,

563-568, 570-575, 578, 582-585, 588-590, 594, 596, 597, 600

ICRP International Commission on Radiological Protection, 59-63, 66, 67, 69, 71, 96, 165, 184, 302, 335, 417, 427, 443, 453, 454, 456, 458, 600

IFHD International Federation for Human Rights, 533

IMCC International Medical Commission-Chernobyl [p. 188

IPHECA International Project on Health Effects of the Chernobyl Accident—WHO, 297, 298, 557, 567, 568, 585, 596

IPSN Institut de Protection et de Sûreté Nucléaire French *(Institute of Protection and Nuclear Safety)* [p. 178-181, 183, 184, 539

IRSN Institut de radioprotection et de sûreté nucléaire. French. *(Institute of Radioprotection and Nuclear Safety)*, 179-181, 184-187, 241, 265, 277, 294, 298, 432, 451

JCE Joint Committee of Experts, 234

JCR Joint Research Centre European Commission, 256-259

JÜLICH Research Centre for Radioprotection (Germany), 161, 207, 239, 240, 243, 244, 248, 269, 287, 365, 389, 394, 395, 399, 401

LCRM Local Centre for Radiological Monitoring of food products (Belarus), 114, 206, 207, 233, 264, 266, 267, 270, 272, 273, 275, 279, 283-286, 305, 306, 325, 342, 348

LOS ALAMOS The Los Alamos National Laboratory—LANL is a laboratory of the United States Department of Energy, managed by the University of California, situated at Los Alamos, in New Mexico. It is the biggest institute and the largest employer in New

www.ingramcontent.com/pod-product-compliance
Lightning Source LLC
Chambersburg PA
CBHW070711220326
41598CB00026B/3691